THE STARS

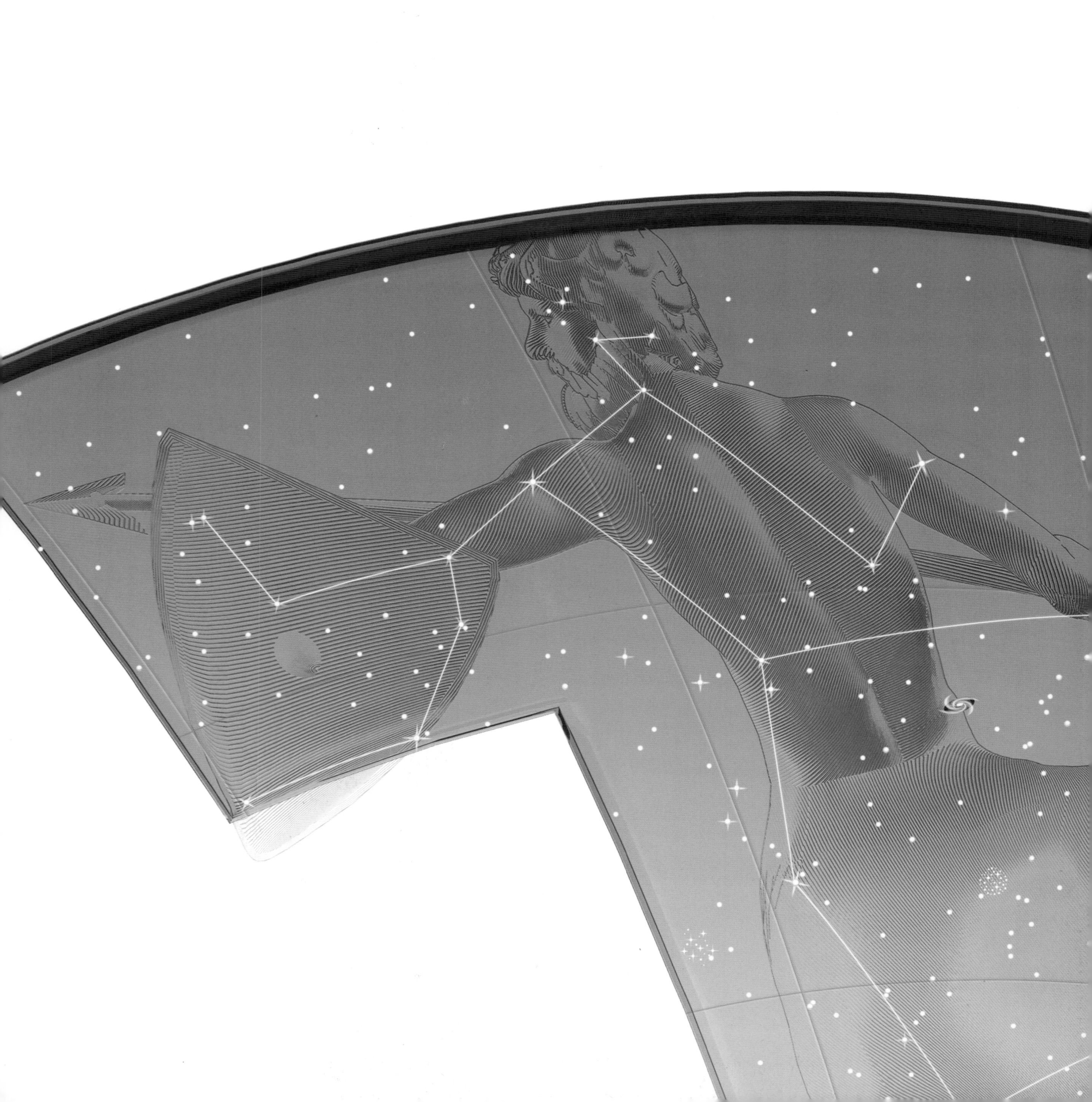

DK

THE STARS

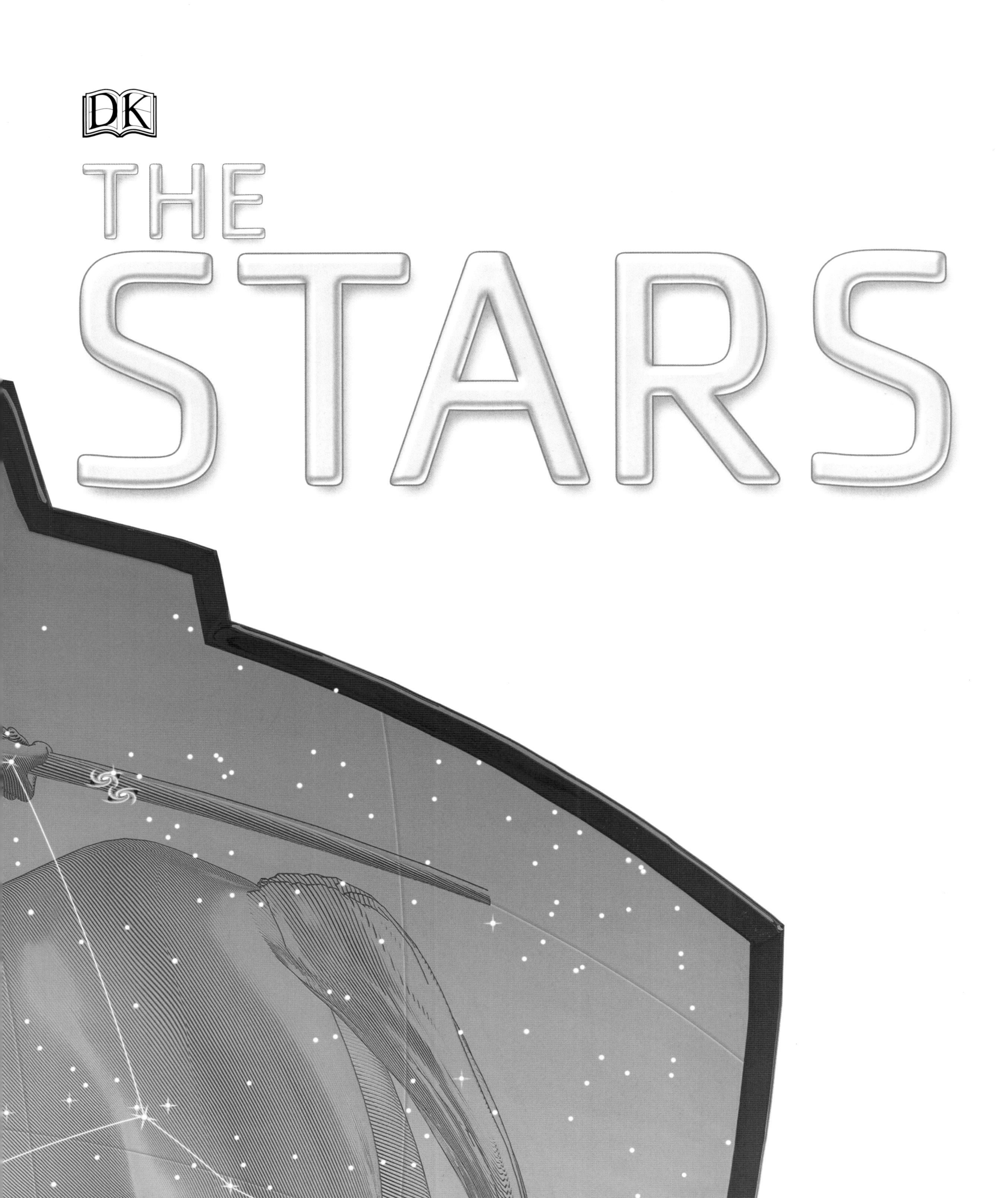

Senior Art Editor Ina Stradins

Project Art Editor Katherine Raj

Project Editor Miezan van Zyl

Editors Martyn Page, Cathy Meeus, Steve Setford, Scarlett O'Hara

Designers Jon Durbin, Alex Lloyd, Steve Woosnam-Savage

Illustrators Peter Bull, Edwood Burns, Mark Garlick, Phil Gamble

Managing Editor Angeles Gavira

Managing Art Editor Michael Duffy

Producers, Pre-Production Luca Frassinetti, Gillian Reid

Senior Producer Mary Slater

Picture Research Liz Moore

Jacket Editor Claire Gell

Jacket Designer Mark Cavanagh

Jacket Design Development Manager Sophia MTT

Art Director Karen Self

Associate Publishing Director Liz Wheeler

Publishing Director Jonathan Metcalf

First published in Great Britain in 2016 by
Dorling Kindersley Limited,
80 Strand, London WC2R 0RL
A Penguin Random House Company

2 4 6 8 10 9 7 5 3 1
001 – 282967 – Sept/2016

A CIP catalogue record for this book is available from the British Library.

ISBN: 978-0-2412-2602-5

Printed and bound in China.

CONTENTS

6 Foreword

8 UNDERSTANDING THE **COSMOS**

10 Out of the Darkness
12 The Cosmos
14 The Big Bang
16 The Origin of the Universe
18 Celestial Objects
20 What is a Star?
22 Star Brightness and Distance
24 Star Sizes
26 Inside a Star
28 The Lives of Stars
30 Starbirth
32 Planetary Nebulae
34 Supernovae
36 Neutron Stars
38 Black Holes
40 Multiple Stars
42 Variable Stars
44 Star Clusters
46 Extrasolar Planetary Systems
48 Multiplanetary Systems
50 Galaxies
52 Galaxy Types
54 The Milky Way
56 The Milky Way in the Spotlight
58 Milky Way from Above
60 Active Galaxies
62 Local Group Collision
64 Colliding Galaxies
66 Clusters and Superclusters
68 Galaxy Clusters
70 The Expanding Universe
72 The Size and Structure of the Univers
74 Dark Matter and Dark Energy
76 Observing the Skies
78 The History of the Telescope
80 Space Telescopes
82 The Search for Life

84 THE **CONSTELLATIONS**

86 Patterns in the Sky
88 Charting the Heavens
90 The Celestial Sphere
92 The Zodiac
94 Mapping the Sky
96 Sky Charts
102 Ursa Minor / Cephus
104 Draco
106 Cassiopeia
108 Lynx / Camelopardalis
110 Ursa Major
112 Canes Venatici
114 The Whirlpool Galaxy
116 Boötes / Corona Borealis
118 Hercules
120 Lyra
122 The Ring Nebula
124 Cygnus
126 Andromeda
128 Triangulum / Lacerta
130 Perseus
132 Leo Minor / Auriga

134 Leo
136 Virgo
138 Coma Berenices / Libra
140 Scorpius
142 Serpens
144 Ophiuchus
146 Aquila / Scutum
148 Vulpecula / Sagitta / Delphinus
150 Pegasus / Equuleus
152 Aquarius
154 Pisces
156 Taurus / Aries
158 Cetus
160 Eridanus
162 Orion
164 The Orion Nebula
166 Gemini
168 Cancer / Canis Minor
170 Monoceros
172 Hydra
174 Sextans / Corvus / Crater
176 Centaurus
178 Crux
180 Lupus / Norma
182 Ara / Corona Australis
184 Sagittarius
186 Capricornus / Piscis Austrinus
188 Grus / Microscopium
190 Sculptor / Caelum
192 Fornax / Lepus
194 Canis Major / Columba
196 Puppis
198 Pyxis / Antlia
200 Vela
202 Carina
204 The Carina Nebula
206 Musca / Circinus / Triangulum Australe / Telescopium
208 Indus / Phoenix
210 Dorado
212 Pictor / Reticulum / Volans
214 Chameleon / Apus / Tucana
216 Pavo / Hydrus
218 Horologium / Mensa / Octans

220 THE **SOLAR SYSTEM**
222 Around the Sun
224 The Solar System
226 The Sun
228 The Inner Planets
230 The Outer Planets
232 The Moon

234 **REFERENCE**
236 Stars and Star Groups
238 Constellations / Milky Way and Other Galaxies
240 Messier Objects
244 Glossary
246 Index
256 Acknowledgments

Foreword
Maggie Aderin-Pocock, MBE, is a space scientist, an honorary research associates at University College London, and co-host of the BBC TV series *The Sky at Night*.

Consultant
Jacqueline Mitton is author, co-author, or editor of about 30 books on space and astronomy and has acted as consultant for many others. She has a degree in Physics from Oxford University and a Cambridge University PhD.

Authors
Robert Dinwiddie specializes in writing educational and illustrated reference books on scientific topics. His particular areas of interest include Earth and ocean science, astronomy, cosmology, and history of science.

David W. Hughes is Emeritus Professor of Astronomy at the University of Sheffield. He has published over 200 research papers on asteroids, comets, meteorites, and meteors, and has worked for the European, British, and Swedish space agencies. Asteroid David Hughes is named in his honour.

Geraint Jones is an astronomer, lecturer, and writer specializing in planetary science. He is a the Head of the Planetary Science Group at the Mullard Space Science Laboratory at UCL London.

Ian Ridpath is author of the Dorling Kindersley *Handbook of Stars and Planets* and is editor of the *Oxford Dictionary of Astronomy*. He is a recipient of the Astronomical Society of the Pacific's award for outstanding contributions to the public understanding and appreciation of astronomy.

Carole Stott is an astronomer and author who has written more than 30 books about astronomy and space. She is a former head of astronomy at the Royal Observatory at Greenwich, London.

Giles Sparrow is an author and editor specializing in astronomy and space science. He is a Fellow of the Royal Astronomical Society.

FOREWORD

Since I was a child, I have always looked up at the night sky and wondered at the outstanding beauty of the stars. As the nursery rhyme states, "twinkle, twinkle little star, how I wonder what you are?".

From that childhood wonder I have forged a career in stargazing, generating instrumentation to help us understand them better. Our understanding of our celestial neighbours is improving all the time and in recent years we have been getting a much better insight into the stars.

Stars are more like us than many of us realize. They have life cycles: they are born in stellar nurseries, they live a full life, and can die in a number of different ways. But the death of stars means life for all of us. Through their lives and deaths, stars create the stuff of life, the chemical elements that make us and all we see.

This book not only takes us on a journey to meet our distant and, seemingly, tranquil neighbours, but also shares their secrets. We may not have the technology to travel to the stars yet but with this book we get a better understanding of their workings.

Maggie Aderin-Pocock

Maggie Aderin-Pocock, MBE, space scientist

◁ Hot blue stars and cooler yellow stars can be seen together in this Hubble Space Telescope image of the globular cluster M15. It is one of the oldest known star clusters, around 12 billion years old. M15 is located 35,000 light-years from Earth, and lies in the constellation Pegasus.

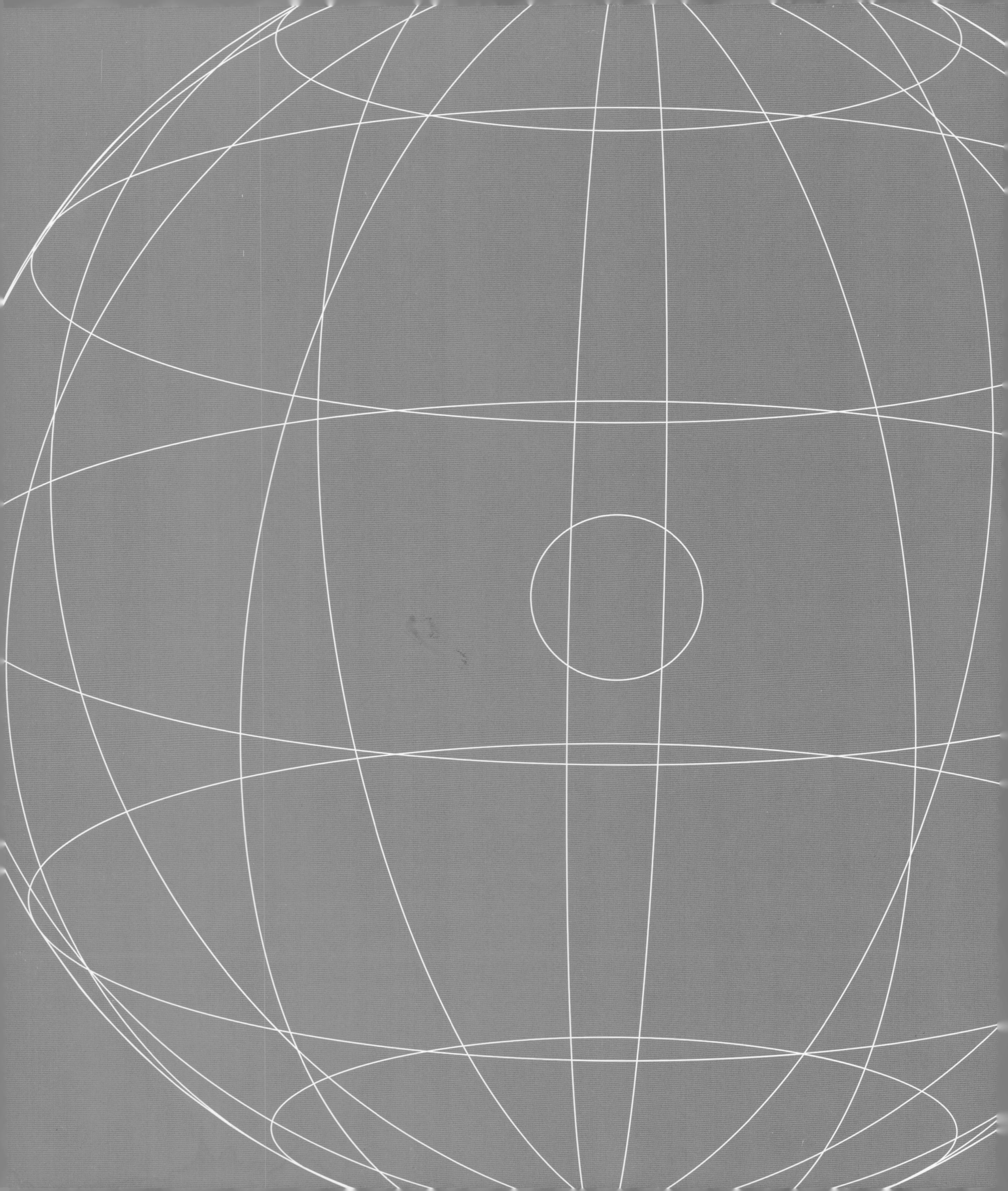

UNDERSTANDING **THE COSMOS**

OUT OF THE **DARKNESS**

After bursting into existence 13.8 billion years ago in the Big Bang, our Universe was for a time entirely dark, since no light-generating objects had yet formed. After a few hundred million years, clumps of matter began to coalesce and heat up, and soon the Universe was bathed in light from the first stars. Today, stars are still the most numerous visible objects in the night sky. We see them as pinpricks of light, seeming to differ only in brightness. However, stars are actually extremely diverse, coming in a vast range of sizes and an array of colours. Some will eventually explode to give rise to such strange phenomena as pulsars and black holes. We now also know that many stars, like our Sun, are orbited by planets, some of which might harbour life. Around the time that the first stars ignited, the first galaxies were also forming. Clusters of stars merged into small galaxies, which in turn merged to make bigger galaxies. All the stars visible to the unaided eye are part of our home galaxy, the Milky Way, a structure so vast that light takes a hundred thousand years to cross it. But this galaxy is just one of untold billions in the cosmos. Gradually, by deploying more and more powerful telescopes and other sensing instruments, astronomers are unlocking the secrets of galaxies too, along with gaining an understanding of the nature of mysterious phenomena, such as dark matter, in which galaxies seem to be embedded.

◁ **Birthplace of stars**
The fiery array of stars near the centre of this Hubble Space Telescope image is a compact young star cluster some 20,000 light-years away. Called Westerlund 2, it contains some of the hottest, brightest, stars known, each with a surface temperature higher than 37,000°C (66,600°F). The cluster lies within a vast, star-forming nebula (cloud of gas and dust) called Gum 29.

THE COSMOS

THE UNIVERSE, OR COSMOS, IS EVERYTHING THAT EXISTS – ALL MATTER, ENERGY, TIME, AND SPACE – AND ITS SCALE IS QUITE MIND-BOGGLING. JUST ABOUT EVERYTHING IN IT IS PART OF SOMETHING BIGGER.

The Universe has a hierarchy of structures. Our planet, Earth, is in the Solar System, which lies in the Milky Way Galaxy, itself just one member of a cluster of galaxies called the Local Group. The Local Group in turn is just a part of a larger structure, the Virgo Supercluster. Astronomers have recently identified a vast region of space they have called Laniakea (meaning "immeasurable heaven" in Hawaiian), which contains the Virgo Supercluster and other superclusters. Intriguingly, all the galaxies in it seem to be flowing towards a region at its centre, called the Great Attractor.

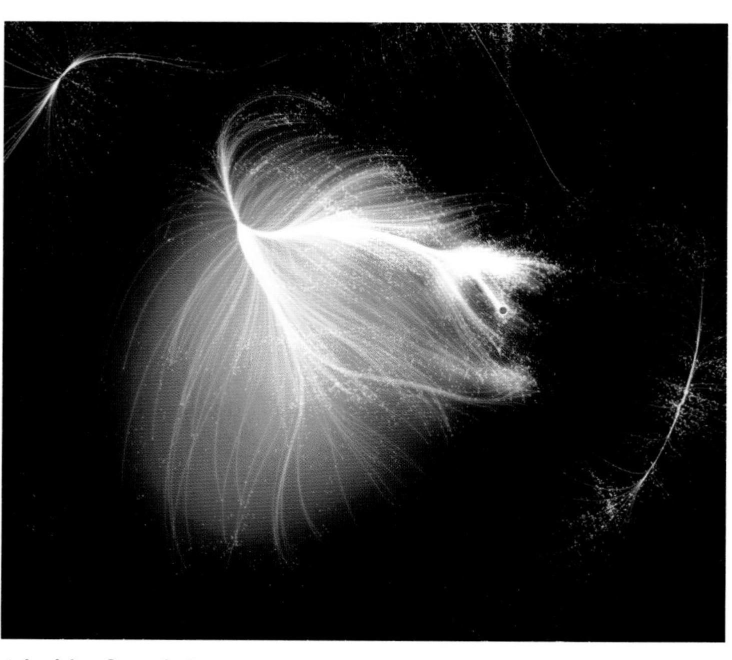

△ **Laniakea Supercluster**
In this depiction of Laniakea (yellow), white lines indicate the flow of galaxies towards a spot near its centre. The approximate position of the Milky Way is shown in red. Laniakea is about 500 million light-years across. It is thought to be surrounded by other similar regions (blue).

Light as a yardstick

Astronomers use light as a yardstick for measuring distances because nothing can cross space faster. Yet even one light-year – the distance light travels in a year, or about 9 trillion km (6 trillion miles) – is dwarfed by the largest structures in the known Universe. Only a fraction of the whole Universe is visible to us: the part from which light has had time to reach Earth since the Big Bang. The true extent of the Cosmos is still unknown and it may even be infinite.

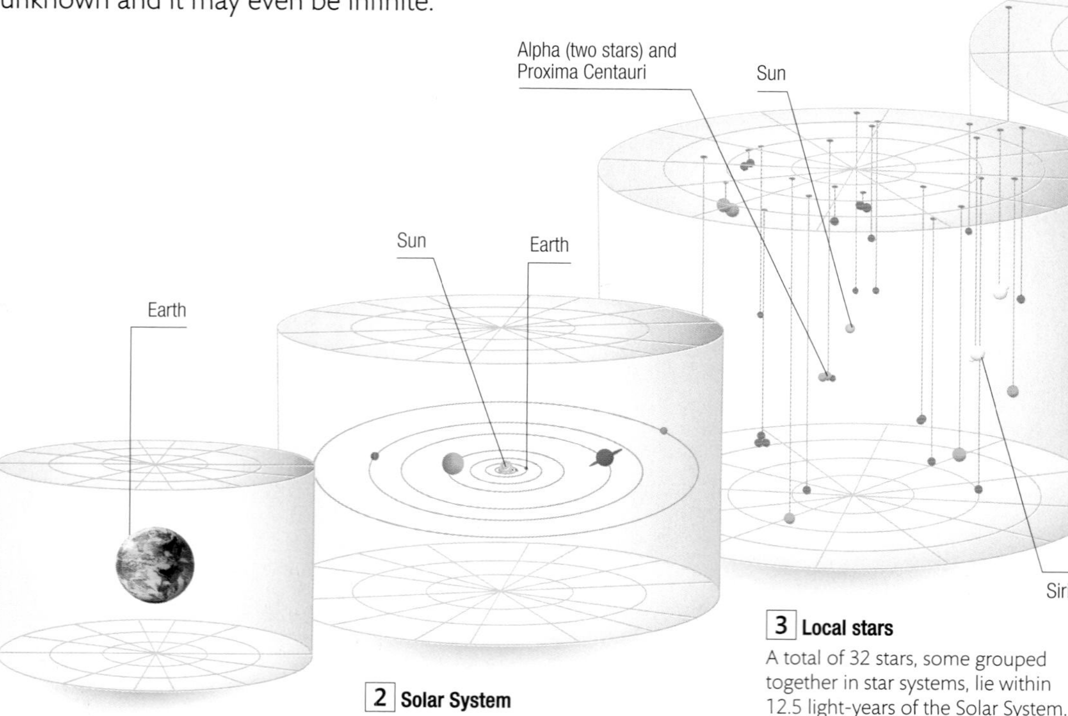

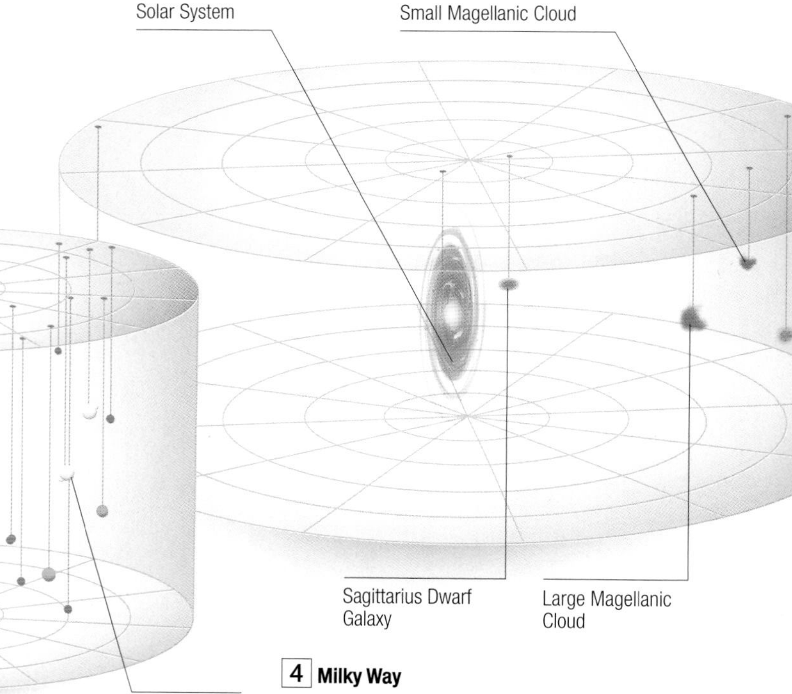

1 Earth
Our home planet is a small rocky sphere floating in the emptiness of space. Earth's closest neighbouring object is the Moon. On average, it is a little more than one second away at the speed of light, so one could say that the Moon is one light-second distant.

2 Solar System
Earth is part of the Solar System, which comprises our local star, the Sun, and all the objects that orbit the Sun. The most distant planet, Neptune, is about 4.5 hours away at light speed, but the Solar System also includes comets that are up to 1.6 light-years distant.

3 Local stars
A total of 32 stars, some grouped together in star systems, lie within 12.5 light-years of the Solar System. They range from dim red dwarfs, invisible to the naked eye, to dazzling, yellow or white, Sun-like stars. A few are suspected to have their own planets.

4 Milky Way
The Solar System and its stellar neighbours occupy just a tiny region of the Milky Way Galaxy, a vast swirling, glittering disk that contains some 200 billion stars, enormous clouds of gas and dust, and a supermassive black hole at its centre. The Milky Way is over 100,000 light-years across. Surrounding it are several smaller, satellite galaxies.

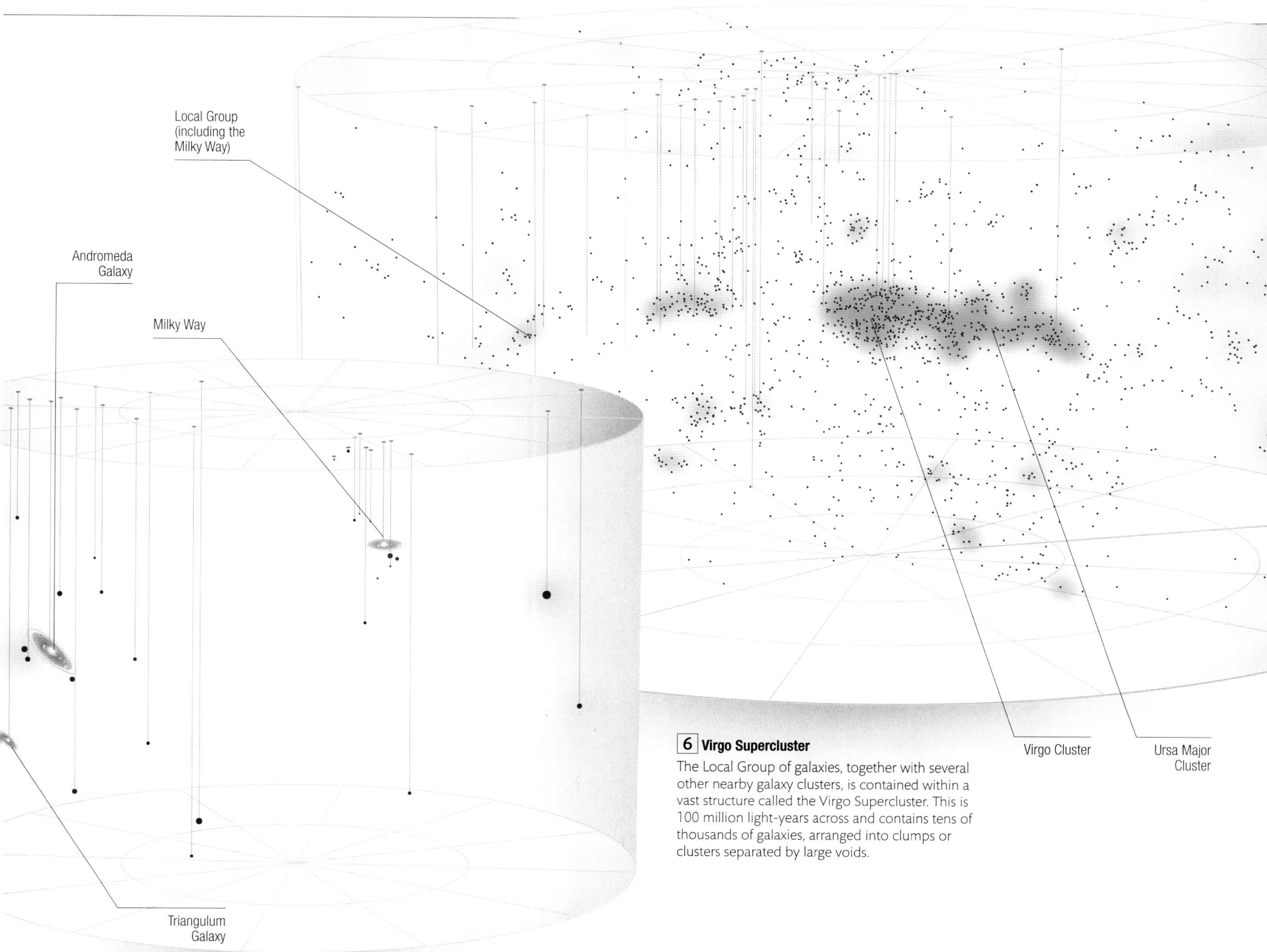

6 Virgo Supercluster
The Local Group of galaxies, together with several other nearby galaxy clusters, is contained within a vast structure called the Virgo Supercluster. This is 100 million light-years across and contains tens of thousands of galaxies, arranged into clumps or clusters separated by large voids.

5 Local group
The Local Group is a cluster of galaxies consisting of the Milky Way, the Andromeda Galaxy (the nearest large spiral galaxy to the Milky Way), another spiral galaxy called Triangulum, and more than 50 other smaller galaxies. All occupy a region of space about 10 million light-years across.

Light from some of the most distant known galaxies has taken over **13 billion years** – most of the **age of the Universe** – to reach us

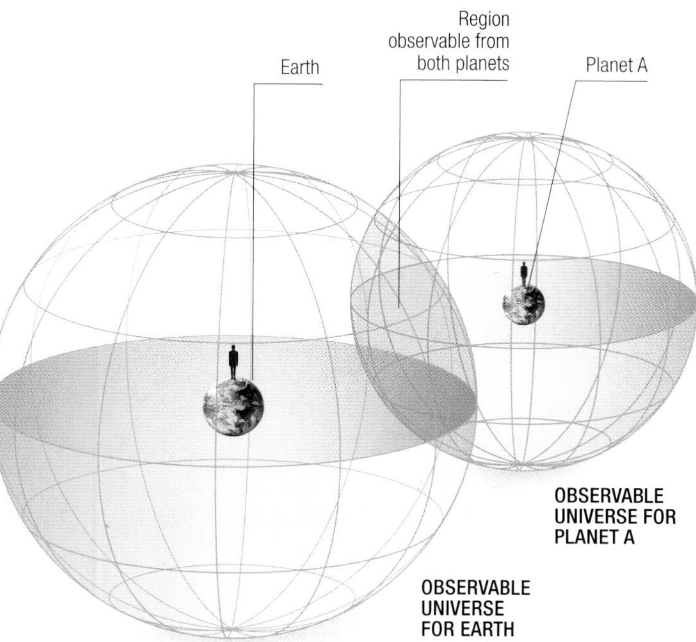

◁ **The observable Universe**
Although the Universe has no edges and may be infinite, the part visible to us is finite. Called the observable Universe, it is the region of space from which light has had time to reach us during the 13.8 billion years since the Big Bang. Physically it is a sphere about 93 billion light-years across, with Earth at the centre. The inhabitants of a planet located outside our observable Universe (Planet A) would have a different observable Universe to us, though the two observable Universes might overlap.

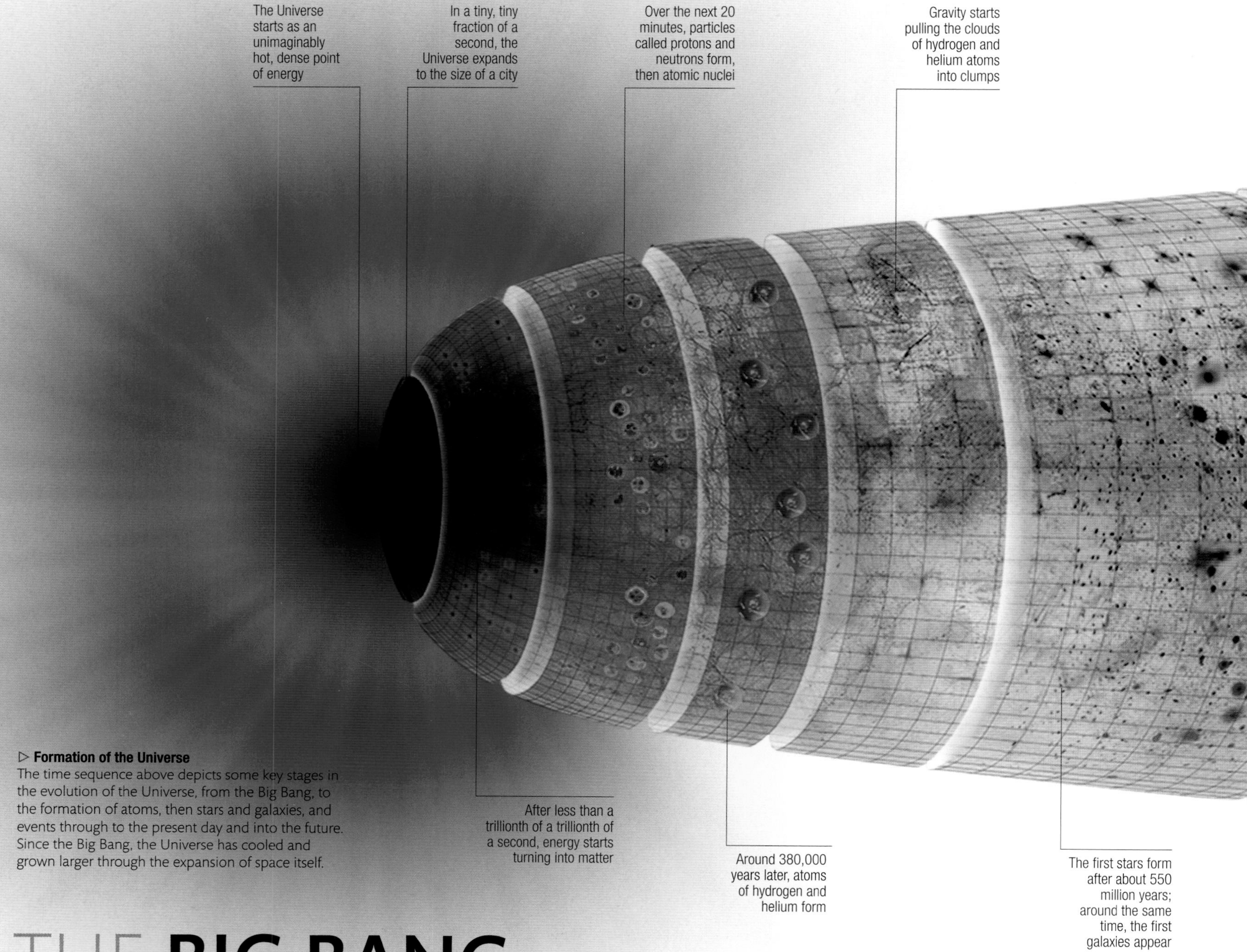
The Universe starts as an unimaginably hot, dense point of energy
In a tiny, tiny fraction of a second, the Universe expands to the size of a city
Over the next 20 minutes, particles called protons and neutrons form, then atomic nuclei
Gravity starts pulling the clouds of hydrogen and helium atoms into clumps
After less than a trillionth of a trillionth of a second, energy starts turning into matter
Around 380,000 years later, atoms of hydrogen and helium form
The first stars form after about 550 million years; around the same time, the first galaxies appear

▷ **Formation of the Universe**
The time sequence above depicts some key stages in the evolution of the Universe, from the Big Bang, to the formation of atoms, then stars and galaxies, and events through to the present day and into the future. Since the Big Bang, the Universe has cooled and grown larger through the expansion of space itself.

THE **BIG BANG**

ABOUT 13.8 BILLION YEARS AGO, AN EXCEEDINGLY DRAMATIC EVENT MARKED THE BEGINNING OF BOTH SPACE AND TIME. FROM NOTHING, THE UNIVERSE SUDDENLY APPEARED AS A TINY POINT OF PURE ENERGY.

Within an instant, in what is known as the Big Bang, the Universe expanded trillions of trillions of times, and then continued to get larger, at the same time cooling from its stupendously hot birth. During the first fractions of a second, a vast "soup" of tiny, interacting particles formed out of the intense energy. Some of these joined to make the nuclei (centres) of atoms – the building blocks of everything we can see in the Universe today. Tens of thousands of years later, actual atoms formed and then, after hundreds of millions of years, the very first stars and galaxies.

◁ **Studying the Big Bang**
Using this complex machine, the Large Hadron Collider, scientists at the European Centre for Nuclear Research (CERN) attempt to re-create the conditions that followed the Big Bang. In the collider, beams of high-energy particles are smashed together and the by-products studied.

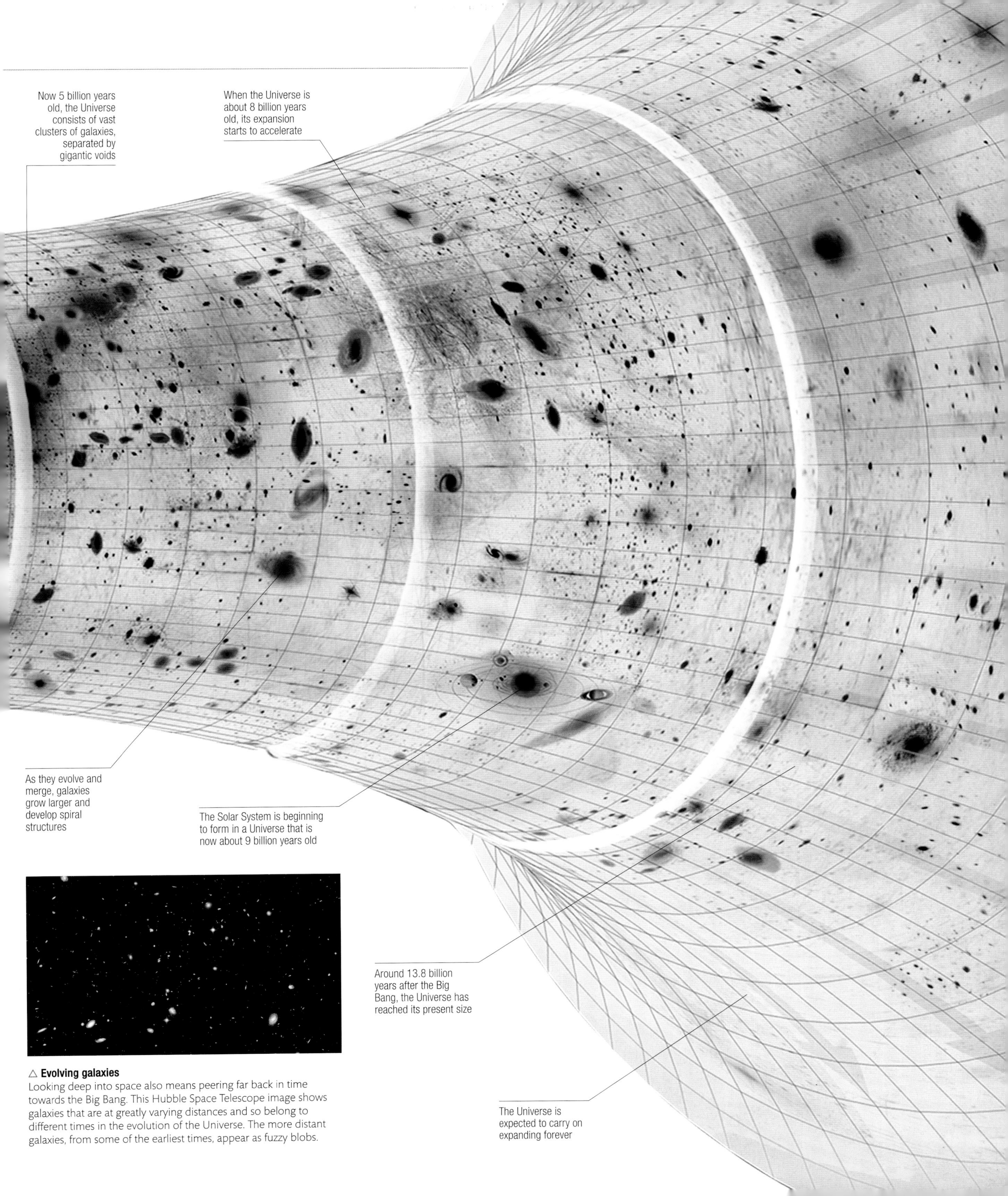

Now 5 billion years old, the Universe consists of vast clusters of galaxies, separated by gigantic voids
When the Universe is about 8 billion years old, its expansion starts to accelerate
As they evolve and merge, galaxies grow larger and develop spiral structures
The Solar System is beginning to form in a Universe that is now about 9 billion years old
Around 13.8 billion years after the Big Bang, the Universe has reached its present size
The Universe is expected to carry on expanding forever

△ **Evolving galaxies**
Looking deep into space also means peering far back in time towards the Big Bang. This Hubble Space Telescope image shows galaxies that are at greatly varying distances and so belong to different times in the evolution of the Universe. The more distant galaxies, from some of the earliest times, appear as fuzzy blobs.

THE NATURE OF THE UNIVERSE

COSMOLOGY – THE STUDY OF THE UNIVERSE AS A WHOLE – IS A FIELD OF ASTRONOMY THAT SEEKS TO ANSWER FUNDAMENTAL QUESTIONS CONCERNING THE SIZE, AGE, AND STRUCTURE OF THE UNIVERSE.

Philosophers and astronomers have been grappling with such questions for thousands of years, with mixed success. The answer to one of the biggest – whether the Universe is finite or infinite in extent – is still not known for certain (though an infinite Universe seems more likely). Other fundamental questions about the nature of the Universe for which answers are now known include how and when the Universe began, whether it has any centre or edges, and whether it encompasses more than just our galaxy.

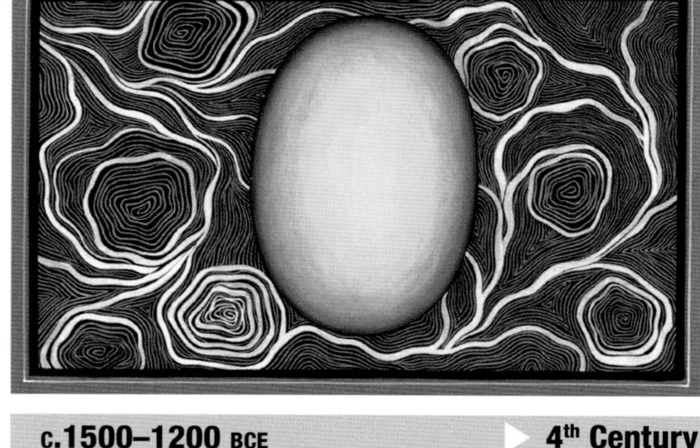

Modern depiction of Hiranyagarbha

c.1500–1200 BCE

Cosmic Egg
Hindu text the *Rigveda* contains a hymn that describes the Universe as originating from a cosmic golden egg or womb known as Hiranyagarbha. This floated in darkness before breaking apart to give rise to Earth, the heavens (space), and underworlds.

4th Century BCE

Aristotle's Earth-centred Universe
The Greek philosopher Aristotle proposes a Universe that is finite in extent, but infinite in time and has a stationary Earth at its centre. Aristotle outlined a complex system containing 55 spheres, the last of which marked out the "edge" of the Universe.

Albert Einstein

1915

General Theory of Relativity
Einstein publishes his General Theory of Relativity, viewed today as the best account of how gravity works on cosmic scales. It proposes that concentrations of mass warp spacetime. He also devises equations that define various possible universes.

1920s

Expanding Universe
American astronomer Edwin Hubble proves that galaxies exist outside our own and observes that distant galaxies are moving away from us at a rate proportional to their distance. Other astronomers conclude that the whole Universe must be expanding.

Georges Lemaitre

1931

Primeval atom
Belgian astronomer and priest Georges Lemaître proposes his "hypothesis of the primeval atom". This suggests that the Universe has expanded from an initial extremely hot, dense state. His model also provides a solution to Olbers' paradox.

1948

The first elements
Russian-American physicist George Gamow and others work out how – starting with just subatomic particles (in this case protons and neutrons) – the nuclei of different light elements could have formed soon after the start of a very hot, dense, but rapidly expanding Universe.

1949

Hoyle coins the term "Big Bang"
British astronomer Fred Hoyle coins the term "Big Bang" for theories that propose the Universe expanded from an exceedingly hot, dense state at a specific moment in the past. The term becomes popular, though Hoyle himself believes in a different theory.

Arno Penzias (left) and Robert Wilson (right)

1965

Cosmic Microwave Background Radiation
Arno Penzias and Robert Wilson, astronomers at Bell Labs in New Jersey, discover the Cosmic Microwave Background Radiation (CMBR) – a faint glow of radiation coming from everywhere in the sky. It comes to be realized that this is a relic of the Big Bang.

Aristarchus of Samos

Giordano Bruno

3rd Century BCE

Sun-centred Universe
The Greek astronomer Aristarchus of Samos puts forward his idea that it is the Sun that sits at the centre of the Universe, with the Earth orbiting it. Aristarchus also suspects that stars are bodies similar to the Sun, but much farther away.

1543

A convincing mathematical model
Polish astronomer Nicolaus Copernicus's book *De revolutionibus orbium coelestium* is published. It contains a detailed and convincing mathematical model of the Universe in which the Sun is at the centre with Earth and other planets orbiting it.

1584

An infinite multitude of stars
Italian philosopher and mathematician Giordano Bruno proposes that the Sun is a relatively insignificant star among an infinite multitude of others. He also argues that because the Universe is infinite, it has no centre or specific object at its centre.

Sketch of Whirlpool Galaxy

1610

Argument against infinite Universe
German astronomer Johannes Kepler argues that any theory of a static, infinite, and eternal Universe is flawed, since in such a Universe, a star would exist in every direction and the night sky would look bright. This argument later comes to be known as Olbers' paradox.

1755

Objects exist outside our galaxy
German philosopher Immanuel Kant suggests that some fuzzy-looking objects in the night sky are galaxies outside the Milky Way Galaxy – implying that the Universe consists of more than just the Milky Way, being considerably bigger.

1905

Spacetime continuum
German physicist Albert Einstein's Special Theory of Relativity proposes that space and time form a combined continuum, spacetime. An inbuilt assumption of his theory is that no location is special – so the Universe has no centre and no edge.

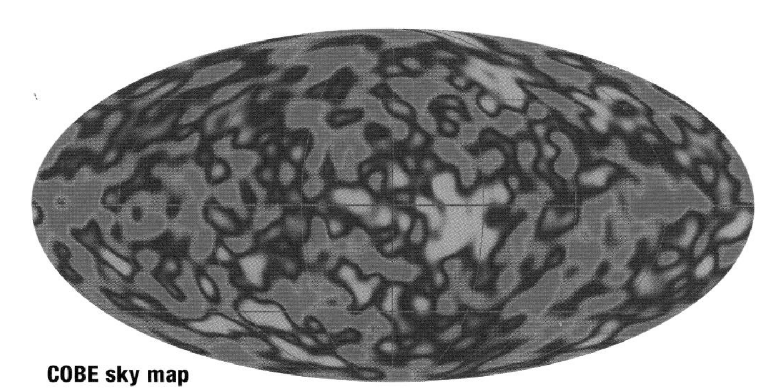

COBE sky map

Computer simulation of gravitational waves

1980

Inflationary Big Bang theory
The American physicist Alan Guth and colleagues suggest that the Universe expanded at a fantastically fast rate during an extremely early phase of its existence after the Big Bang. The theory helps explain the large-scale structure of the cosmos.

1992

Variations in the CMBR
Measurements by the COBE (Cosmic Backgroung Explorer) satellite reveal tiny variations in the CMBR, providing a picture of the seeds of large-scale structure when the Universe was a tiny fraction of its present size and just 380,000 years old.

1999-2001

The existence of dark energy
High-precision measurements of the CMBR and the recessional velocities of galaxies at different distances provide evidence for dark energy – a mysterious phenomenon that seems to be accelerating the Universe's expansion.

2016

Gavitational waves detected
Physicists in the United States announce that they have detected gravitational waves. The existence of these waves supports the Inflationary Big Bang theory and provides further confirmation of Einstein's General Theory of Relativity.

CELESTIAL OBJECTS

SCORES OF DIFFERENT TYPES OF OBJECTS EXIST OUT IN SPACE, RANGING FROM COSMIC RAYS – CHARGED SUBATOMIC PARTICLES WHIZZING AROUND AT EXTREME SPEED – TO VAST, MAJESTIC GALAXY CLUSTERS.

Stars are by far the most numerous objects that can actually be seen, because they emit their own light. Most other observable features of the night sky either consist mainly of stars (galaxies and star clusters) or are visible because they reflect starlight (planets, moons, and comets, for example). In addition, various extremely dim or entirely dark objects, such as brown dwarfs and black holes, are out there, but vary from extremely hard to near-impossible to detect.

△ **Comets**
Comets are chunks of ice and rock that orbit in the far reaches of the Solar System. A few stray close to the Sun – some at regular intervals. Frozen chemicals in the comet then vaporize to produce a glowing coma (head) and long dust and gas tails.

△ **Stars**
A star is an extremely hot ball of gas that generates energy through nuclear fusion of hydrogen (and sometimes other elements). All nearby stars are part of the Milky Way Galaxy, which (as shown above) appears as a band across the night sky.

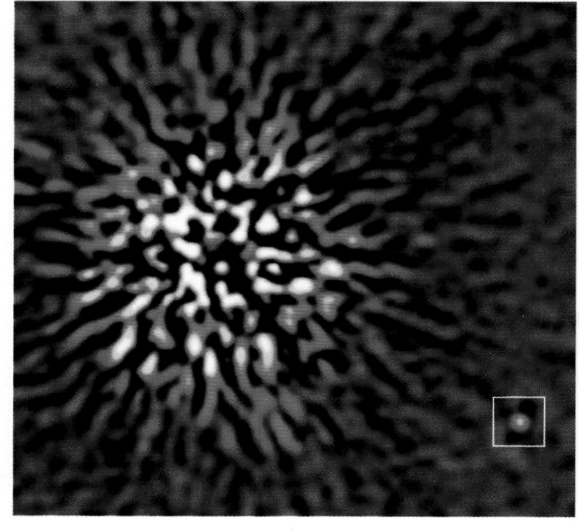

△ **Brown dwarfs**
Brown dwarfs are "nearly-stars". They are more massive than most planets, but not massive enough to sustain the nuclear fusion of ordinary hydrogen, as stars do. This image reveals the dim glow from a brown dwarf (boxed) orbiting a Sun-like star.

△ **Star clusters**
A star cluster is a large collection of stars bound together by gravity. Several thousand have been identified in our galaxy, and they fall into two types: globular clusters (like the one shown above) and open clusters.

△ **Star remnants**
When giant stars die, they leave various types of remnant. This always includes a compact remnant of the original star's core. However, this ghostly-looking object is gas and dust debris ejected from a star when it exploded as a supernova.

Nebulae
Nebulae are clouds of gas and dust in the vast expanses of space between stars. Many contain regions of star formation. In some, light from hot newborn stars excites gas atoms in the nebula, which then begin to emit light in various colours. An example of one of these colourful objects is the Carina Nebula, shown here. It is a prominent naked-eye sky feature in the southern hemisphere.

△ **Planets**
A planet is a near-spherical object that orbits a star. It can be rocky or gaseous but does not generate energy by nuclear fusion. This one is Mars in our own Solar System.

△ **Moons**
A moon is any naturally occurring object orbiting a planet or other body. Hundreds of moons have been identified in the Solar System, including this satellite of Saturn, Mimas.

◁ **Galaxies**
A collection of stars, gas, dust, nebulae, star remnants, planets, and smaller bodies is called a galaxy. Four main types exist – spiral, barred spiral, elliptical, and irregular – the example shown here being a spiral. Called NGC908, it is known to be spawning new stars at a frantic rate.

△ **Galaxy cluster**
Galaxies are grouped into clusters, which are themselves gathered into larger aggregations called superclusters. This galaxy cluster, Abell 2744, contains hundreds of galaxies. The whole cluster is known to be immersed in a vast sea of a mysterious, invisible material called dark matter (see pp.74–75).

Photosphere, the visible surface of a star

Energy-generating core

Interior consisting of extremely hot gas, through which energy gradually moves outward

Prominence, a loop of hot gas emerging from the surface

◁ **Sun-like stars**
Although different-sized stars differ a little in their internal structure, all have the same basic features as the Sun-like star shown here.

WHAT IS A **STAR?**

A STAR IS AN ENORMOUS BALL OF EXTREMELY HOT GAS THAT PRODUCES ENERGY IN ITS CORE AND EMITS THIS ENERGY AT ITS SURFACE.

All the individual stars we can see in the night sky are part of our own galaxy, the Milky Way. Although in cosmic terms these are all "local" stars, they are actually fantastically far away – the closest is nearly 40 trillion km (25 trillion miles) distant, and most are much farther off. Overall in our galaxy there are more than 200 billion stars, of which about 10,000 are visible to the naked eye.

Star appearance and variation

We see all stars in the night sky as just tiny pinpricks of light. Some look brighter than others, but with the unaided eye they don't seem to differ much in colour: all look rather white. In fact, stars are much more varied than might at first appear. They come in a vast range of sizes and temperatures, in an array of colours, and also differ greatly in age and life span. Many of these characteristics of stars are related. For example, a star's surface temperature and colour are closely linked – a star with a relatively low surface temperature glows red, whereas hotter stars appear (with increasing temperature) orange, yellow, white, or blue.

SPECTRAL CLASSIFICATION OF STARS

Class	Apparent colour	Average surface temperature	Example star
O	Blue	over 30,000°C (54,000°F)	Zeta Puppis, also called Naos (Puppis)
B	Deep bluish white	20,000°C (36,000°F)	Rigel (Orion)
A	Pale bluish white	8,500°C (15,000°F)	Sirius A (Canis Major)
F	White	6,500°C (11,700°F)	Procyon A (Canis Minor)
G	Yellow-white	5,300°C (9,500°F)	The Sun
K	Orange	4,000°C (7,150°F)	Aldebaran (Taurus)
M	Red	3,000°C (5,350°F)	Betelgeuse (Orion)

△ **Star spectral classes**
The specrum of light from a star carries a lot of information about the star. By studying its spectrum, scientists can assign any star to a type, called a spectral class, of which the main ones are listed above.

Star classification

Stars can be classified in many ways, but the system preferred by astronomers places the majority into seven main classes (O to M) based on their spectra – the light of various wavelengths received from them. A star's spectrum contains data relating to its colour, temperature, composition, and other properties. In an attempt to see if there is any underlying pattern to the whole range of different stars, in around 1911 and 1913, Danish astronomer Ejnar Hertzsprung and American astronomer Henry Norris Russell independently plotted hundreds of stars on a scatter diagram according to their spectral class on one axis and luminosity (related to brightness) on the other. This revealed something interesting. Most stars fall into, and spend much of their lives in, a part of the diagram called the main sequence. Other parts are filled by giant stars – known to be nearing the end of their life – and by expired giant stars called white dwarfs.

▽ **The Hertzsprung–Russell diagram**
Running diagonally across the diagram is the main sequence – an array of stable stars, ranging from cool red dwarf stars to hotter, bigger, bluish stars. Other parts are occupied by stars that were once on the main sequence but later evolved into luminous giants, and by white dwarfs.

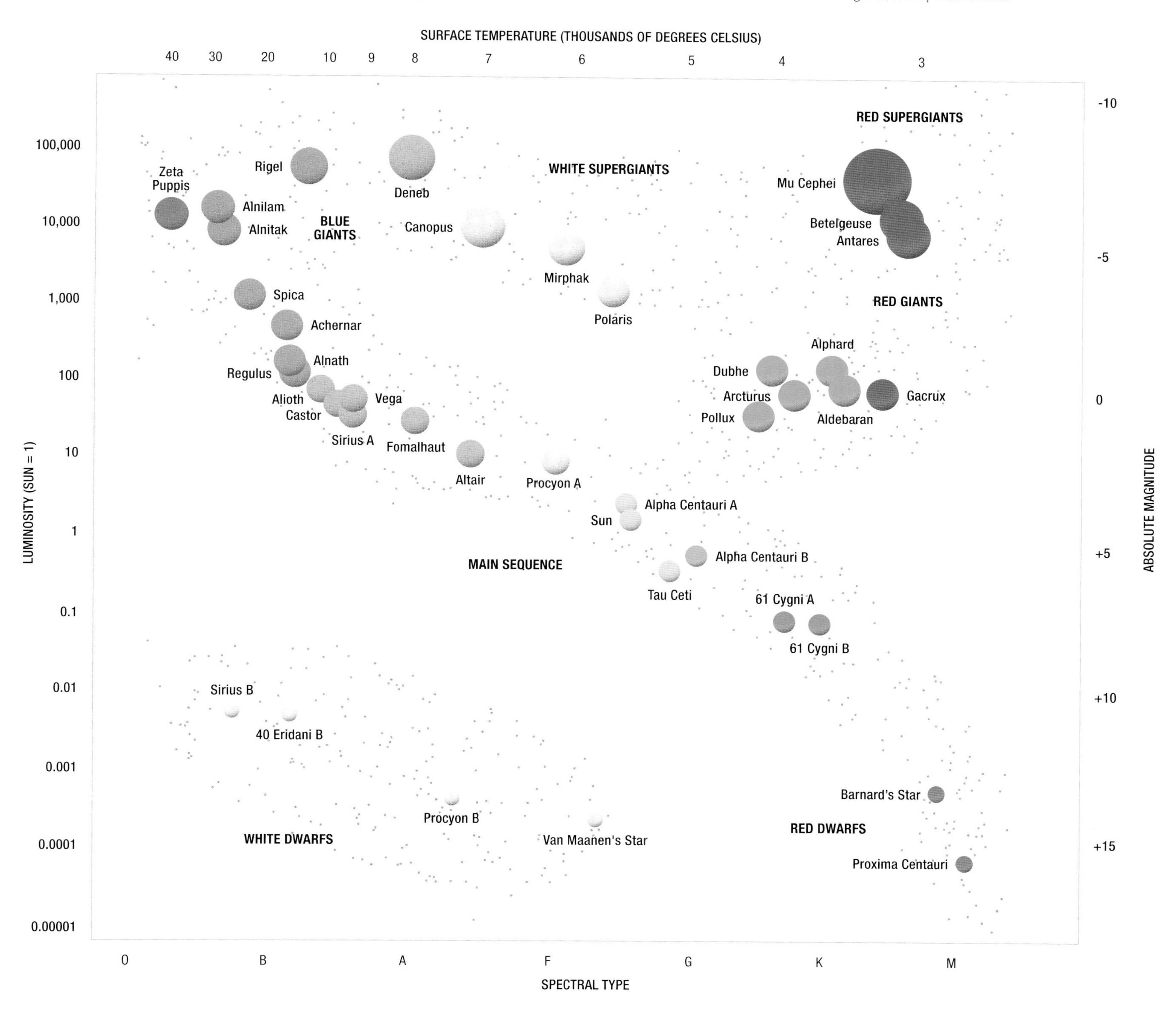

STAR BRIGHTNESS AND DISTANCE

STARS DIFFER HUGELY IN THEIR BRIGHTNESS AND IN THEIR DISTANCE FROM EARTH, EVEN THOUGH ALL, APART FROM THE SUN, ARE EXTREMELY REMOTE. HOW BRIGHT A STAR LOOKS FROM EARTH DEPENDS OF COURSE PARTLY ON HOW FAR AWAY IT IS.

Because stars are so far away, obtaining data about them is tricky. Most of the data about any star comes from studying the light and other radiation coming from it, while the distance to the least remote stars can be worked out by measuring tiny annual variations in their sky positions.

Brightness

There are two different ways of stating a star's brightness: apparent magnitude, which indicates how bright a star looks from Earth, and absolute magnitude, which expresses how bright it would look from a set distance – a better indicator of how brilliant it truly is. On both scales, a change of +1 on the scale means a decrease, and a change of -1 means an increase, in brightness. So, on the apparent magnitude scale, stars just visible to the naked eye score +6 or +5, while very bright stars score about +1 to 0, and the four very brightest have negative scores. The absolute magnitude scale runs from around +20 for some exceptionally dim red dwarfs to around -8 for the brightest supergiant stars. A star's absolute magnitude is related to a measurement called its visual luminosity. This is the amount of light energy that a star emits per unit of time. Luminosity is often stated relative to that of the Sun.

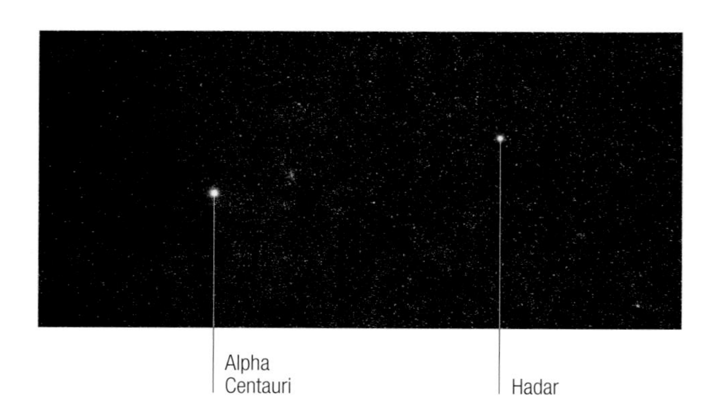

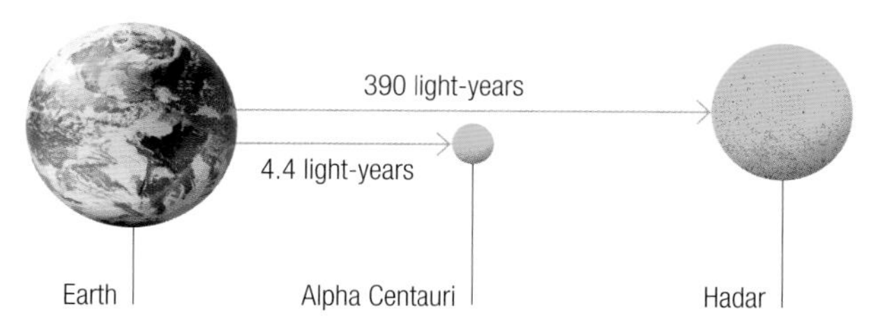

△ **Apparent magnitiude**
The two brightest stars in the photograph at the top – Alpha Centauri (left) and Hadar (right) – appear roughly as bright as each other. In other words, they have a similar apparent magnitude. But intrinsically, Hadar is much brighter because its absolute magnitude is greater. Alpha Centauri looks about as bright as Hadar only because it is about 90 times closer.

▽ **Brightness comparisons**
The apparent and absolute magnitudes, and luminosities, of 11 different stars, including the Sun, are compared in the table below. The stars range from the relatively nearby red dwarf, Proxima Centauri, to distant but fantastically luminous supergiants, such as Rigel.

MAGNITUDE AND LUMINOSITY OF SELECTED STARS

Star (Constellation)	Distance from Earth	Apparent magnitude	Absolute magnitude	Visual luminosity (number of Suns)
The Sun	149,600,000 km (92,960,000 miles)	-26.74	4.83	1
Sirius A (Canis Major)	8.6 light-years	-1.47	1.42	23
Alpha Centauri A (Centaurus)	4.4 light-years	0.01	4.38	1.5
Vega (Lyra)	25 light-years	0.03	0.58	50
Rigel (Orion)	780–940 light-years	0.13	-7.92	125,000
Hadar (Centaurus)	370–410 light-years	0.61	-4.53	5,500
Antares (Scorpius)	550–620 light-years	0.96	-5.28	11,000
Polaris (Ursa Minor)	325–425 light-years	1.98	-3.6	2,400
Megrez (Ursa Major)	58 light-years	3.3	1.33	25
Mu Cephei (Cepheus)	1,200–9,000 light-years	4.08	-7.63	96,000
Proxima Centauri (Centaurus)	4.2 light-years	11.05	15.6	0.00005

△ **Proxima Centauri**
This photograph is of a red dwarf star, Proxima Centauri. At 4.2 light-years away, it is the closest star to Earth other than our Sun. Though brilliant in this Hubble Space Telescope image, relatively speaking, it is a dim star with an absolute magnitude of +15.6 and a luminosity that is only a tiny fraction of that of the Sun.

Distance

Stars other than the Sun are so far away that a special unit is needed to express the distance to them. This unit is the light-year and is the distance light travels through space in a year, which is about 9.5 trillion km (5.9 trillion miles). The 100 brightest stars we can see in the night sky vary from 4.4 to around 2,500 light-years away. The distances to stars can be measured in various ways. For relatively nearby stars, a method called parallax is used (see right). For more remote stars, astronomers have to use more complex indirect methods. Because these methods are less precise, the distances to many stars, even to some of the brightest in the sky, are known only approximately.

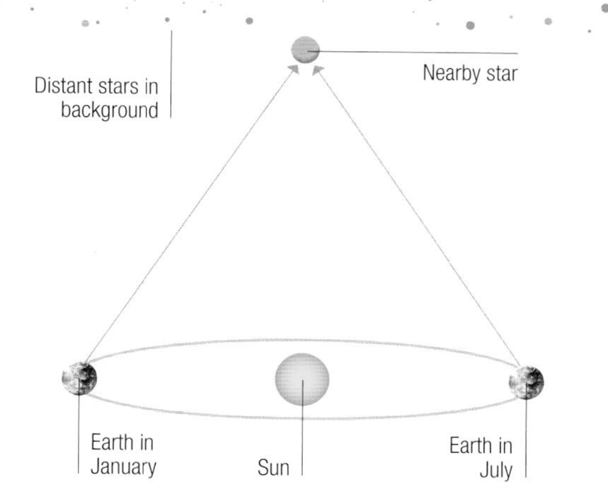

◁ **Parallax method**
If a nearby star is viewed from Earth on two occasions, when Earth is at opposite sides of its orbit around the Sun, the nearby star seems to shift a little against the background of more distant stars. The amount of shift provides the basis for calculating how far away the star is.

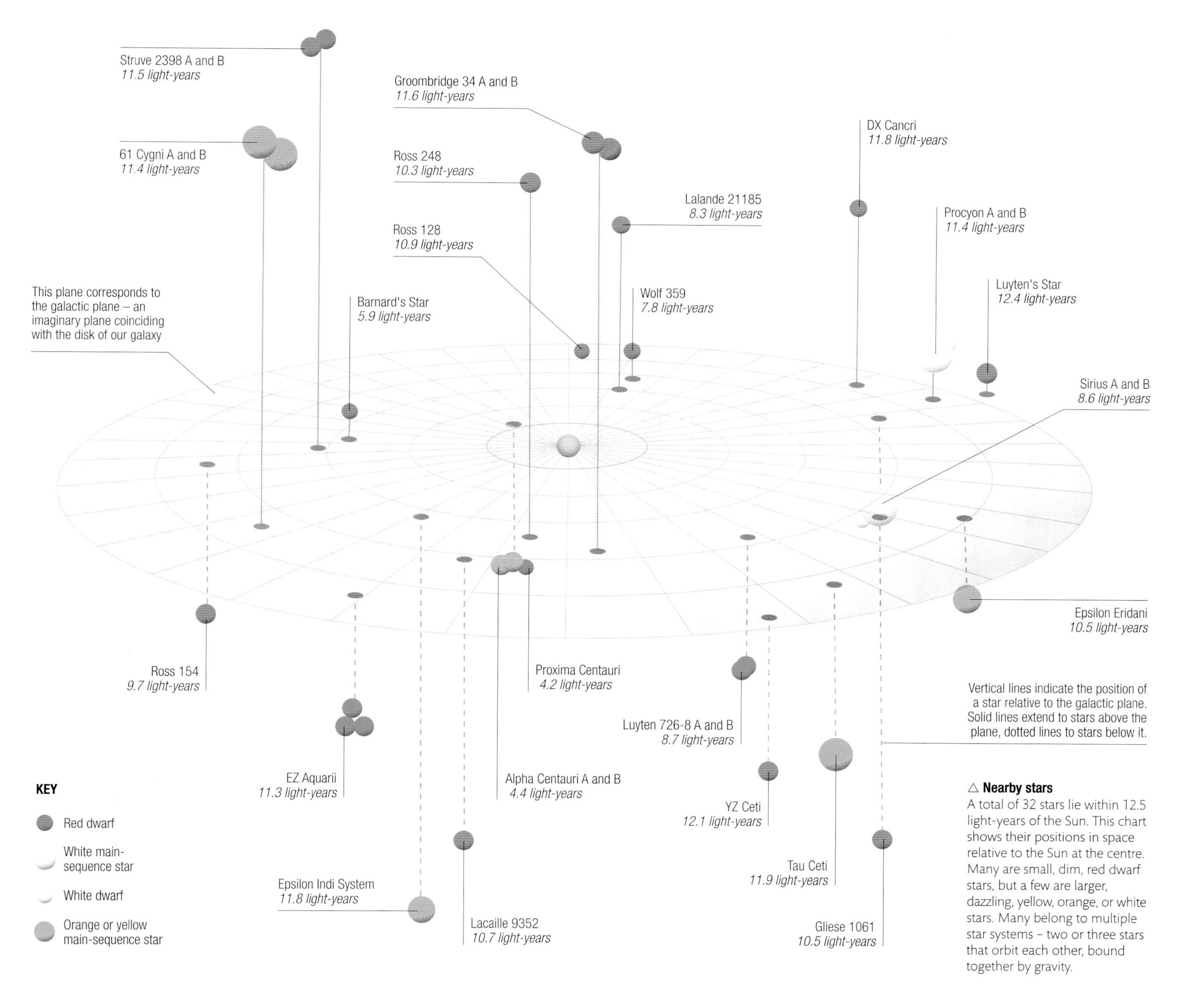

△ **Nearby stars**
A total of 32 stars lie within 12.5 light-years of the Sun. This chart shows their positions in space relative to the Sun at the centre. Many are small, dim, red dwarf stars, but a few are larger, dazzling, yellow, orange, or white stars. Many belong to multiple star systems – two or three stars that orbit each other, bound together by gravity.

A lump of neutron star material roughly the size of a **tennis ball** would weigh as much as **40 times** all the people on Earth

BLUE SUPERGIANT
Rigel A
Having exhausted all the hydrogen in its core, Rigel A – the main component of the Rigel star system – has swollen to 750 times the diameter of the Sun.

RED HYPERGIANT
VY Canis Majoris
This red hypergiant has a radius of around 1,420 times that of the Sun, but it has a much shorter life span.

RED SUPERGIANT
Betelgeuse
Once high-mass stars have used the hydrogen in their cores, they expand into much larger supergiants.

BLUE HYPERGIANT
Pistol star
One of the brightest stars ever discovered, the Pistol Star releases as much energy in six seconds as the Sun does in a year.

△ **Large stars**
Giant, supergiant, and hypergiant stars are all much larger and brighter than main-sequence stars with the same surface temperature. Blue stars tend to be smaller than their red equivalents but are equally bright due to having much higher surface temperatures than the red stars.

0 | 20 million | 40 million | 60 million | km
0 | 20 million | 40 million | miles

STAR SIZES

DESPITE APPEARING AS MERE PINPRICKS IN THE SKY, STARS DIFFER GREATLY IN SIZE, WITH MANY SO BIG THAT THEY DWARF OUR RELATIVELY SMALL SUN. OTHERS ARE SMALLER THAN SOME PLANETS IN OUR SOLAR SYSTEM.

The smallest stars are tiny, super-dense neutron stars that form after a giant star has collapsed. These stars are only 25 km (15 miles) in diameter. Most stars in our galaxy are dwarf stars, some of them with less than one-thousandth of the Sun's volume. The largest stars, the super- and hypergiants, can be as much as 8 billion times greater in volume than the Sun. Stars are grouped into categories based on characteristics such as colour, size, and brightness. A combination of colour and brightness indicates a star's size. For example, a bright blue star is smaller than an equally bright red star, because a blue star is hotter than a red star and needs less surface area for it to be as bright as a cooler red star.

Star B

Star A

Star A's track across face of star B

Dip in light curve

Brightness

Time

△ **Measuring sizes**
By examining the light curve during an eclipse in an eclipsing binary system (see p.43), it is possible to determine how long it takes for one star to pass the other. The time elapsed provides the information needed to work out the diameters of the stars.

ORANGE GIANT
Pollux
The orange coloration of Pollux indicates that it has a lower surface temperature than the Sun.

BLUE GIANT
Bellatrix
Bellatrix is about 20 million years old and has a diameter six times that of the Sun.

YELLOW DWARF
The Sun
Stars in this category are all main-sequence stars and very similar in size to the Sun.

RED GIANT
Aldebaran
Aldebaran is an irregular variable, meaning that its size changes from time to time as the star's forces of gravity and outward pressure tries to balance out.

▷ **Ordinary star**
The Milky Way consists of at least 200 billion stars, of which 90 per cent are in the stable stage (main sequence) of their life cycle. The Sun is a main-sequence star categorized as a yellow dwarf. It has a diameter of 1.39 million km (864,000 miles), but when it runs out of hydrogen it will swell into a red giant before losing its outer layers and finally becoming a white dwarf.

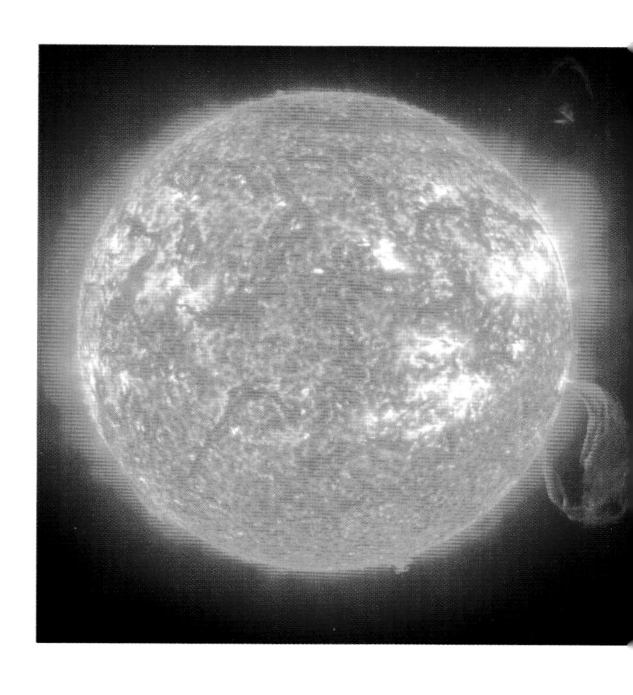

▷ **Dwarf stars**
Most stars are described as dwarf stars. This group of small, dim stars includes stars that are about the size of the Sun and many smaller red dwarfs and white dwarfs – tiny remnants of giant stars that have lost their outer layers. Brown dwarfs are bodies without enough mass to trigger nuclear fusion in their cores and are, in that sense, failed stars.

0 250,000 500,000 750,000 1 million km
0 250,000 500,000 miles

YELLOW DWARF
The Sun

RED DWARF
Proxima Centauri
Red dwarfs are the most numerous star type in our galaxy and will eventually also become white dwarfs.

BROWN DWARF
EROS-MP J0032-4405
Not actually stars, most brown dwarfs are about the same size as the planet Jupiter in our Solar System.

WHITE DWARF
Sirius B
Sirius B is roughly the same size as Earth, but its mass is nearly equal to that of the Sun.

INSIDE A STAR

A STAR IS EFFECTIVELY A MACHINE FOR TRANSFERRING FANTASTIC AMOUNTS OF ENERGY FROM ITS CENTRAL CORE, WHERE THE ENERGY IS PRODUCED, OUT TOWARDS ITS FIERY SURFACE. THIS JOURNEY CAN TAKE 100,000 YEARS OR MORE.

In a star, there is continuous flow of this energy from core to surface, where it escapes into space. The flow creates an outward-acting pressure, without which the star would collapse. The source of energy in the core of a star is the joining together, or fusion, of atomic nuclei (the central parts of atoms) to make larger nuclei.

Energy production and transfer

Nuclear fusion involves a tiny loss of mass, which is converted into energy. In most stars the dominant process is one in which hydrogen nuclei combine to form helium nuclei. From the core of a star, energy moves outwards by radiation and convection. Radiation is the transfer of energy in the form of light, radiant heat, X-rays, and so on, all of which can be thought of as consisting of tiny packets of energy, called photons. Within a typical star, the gaseous material is so tightly packed that photons cannot travel far before they are absorbed and then re-emitted in a different direction. So, energy transferred in this way travels outwards in a slow, zigzag fashion. Convection carries energy towards the surface through circular motions of hot gas outwards and denser cooler gas inwards. Many stars contain layers, with different densities, some transferring energy by radiation, others by convection.

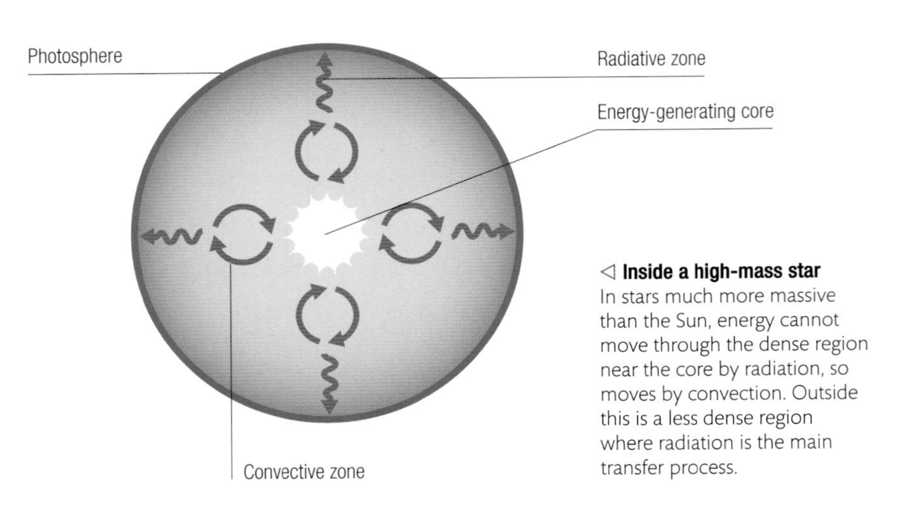

◁ **Inside a high-mass star**
In stars much more massive than the Sun, energy cannot move through the dense region near the core by radiation, so moves by convection. Outside this is a less dense region where radiation is the main transfer process.

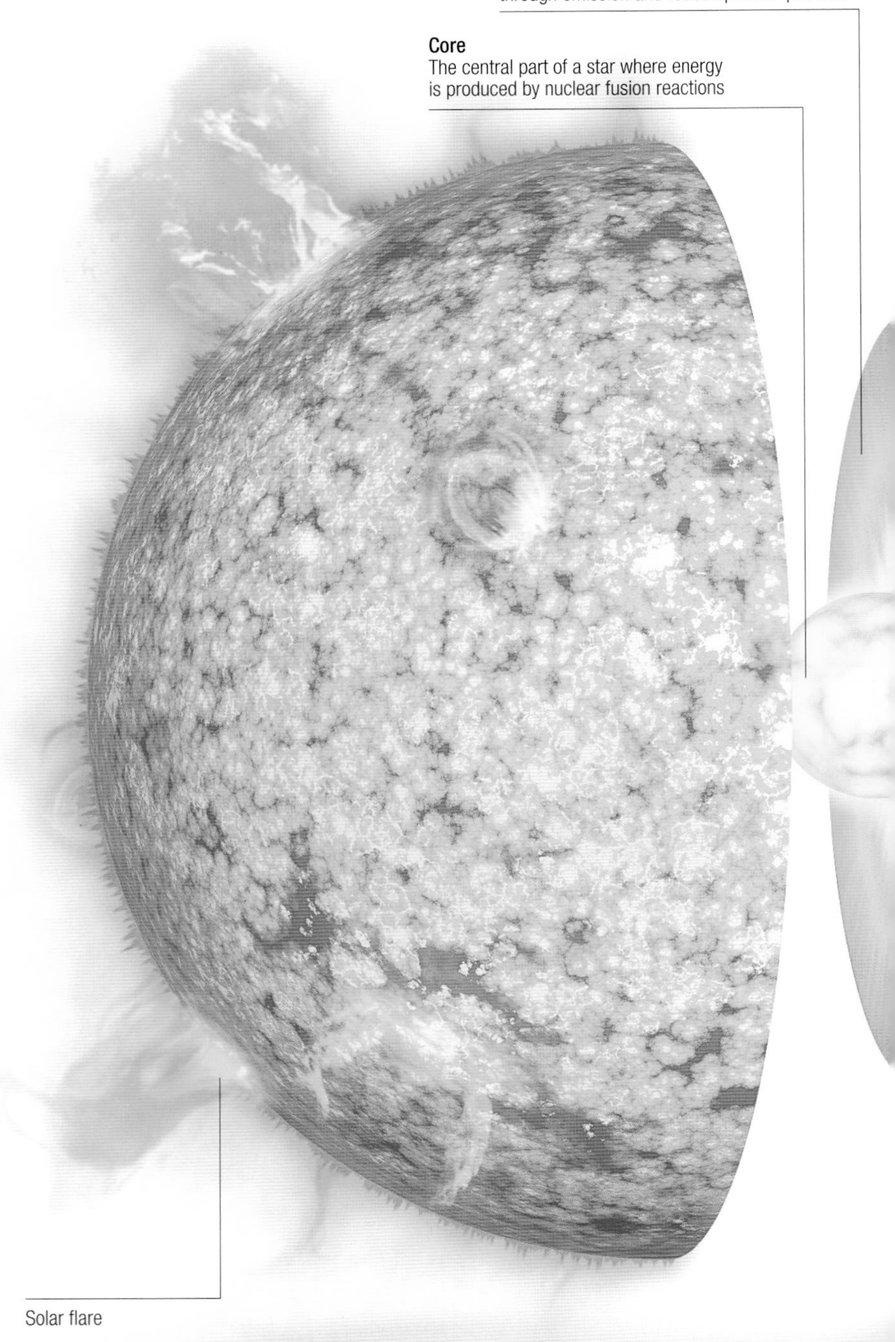

▷ **Inside a Sun-like star**
In a Sun-sized star, the core is surrounded by a radiative zone in which energy gradually zigzags outwards through the emission and reabsorption of photons (packets of radiant energy). On reaching the convective zone, the energy flows to the surface by circular movements of hot gas outwards, and cooler gas inwards. At the star's surface, it escapes as light, heat, and other radiation.

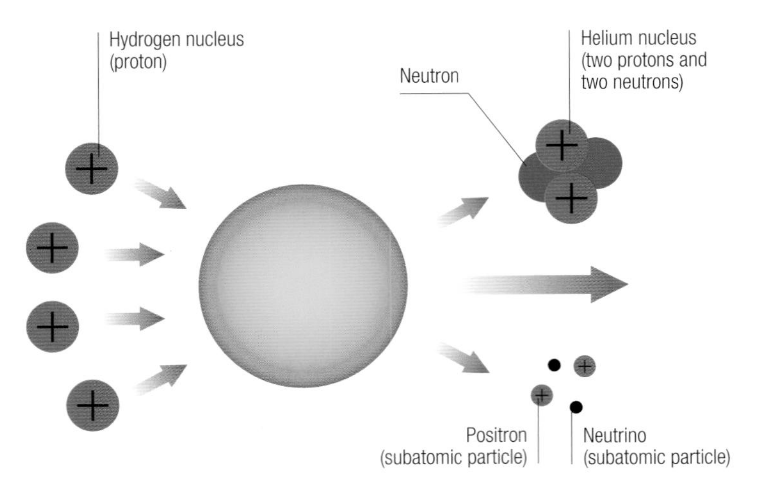

△ **Nuclear fusion in Sun-like stars**
In stars about the size of the Sun or smaller, the main fusion process is called the proton-proton chain reaction. Its overall effect is to convert four protons (hydrogen nuclei) into one helium nucleus, with the release of energy and some tiny subatomic particles.

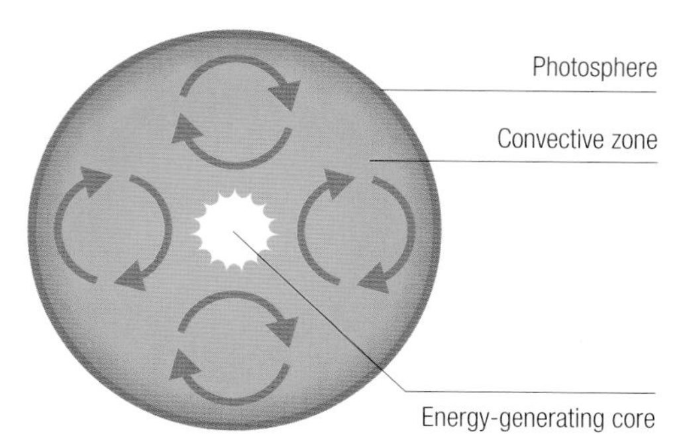
Photosphere
Convective zone
Energy-generating core

△ **Inside a red dwarf**
Inside a low-mass star (a red dwarf), the star's interior is mostly too dense for photons to penetrate far without being reabsorbed. Consequently energy is instead carried all the way to the surface by convection cells.

Forces inside stars

Whatever the mass of a star, two opposing forces keep it in existence. These are gravity, acting inwards, and a pressure force, acting outwards. Normally the opposing forces inside a star are in equilibrium, so it maintains its size over long periods of time. But if something causes the forces to become imbalanced, the star will change size. For example, the cores of most stars heat up towards the ends of their lives: the extra heat boosts the outward pressure so the star swells into a giant or supergiant star.

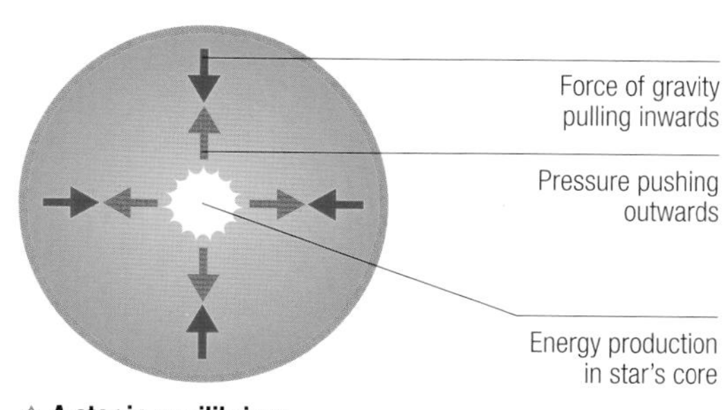
Force of gravity pulling inwards
Pressure pushing outwards
Energy production in star's core

△ **A star in equilibrium**
During most of the life of most stars, the inward-pulling force of gravity is exactly balanced by the outward-acting pressure, and the star maintains its size. If the forces get out of balance, the star is destined to either shrink or swell.

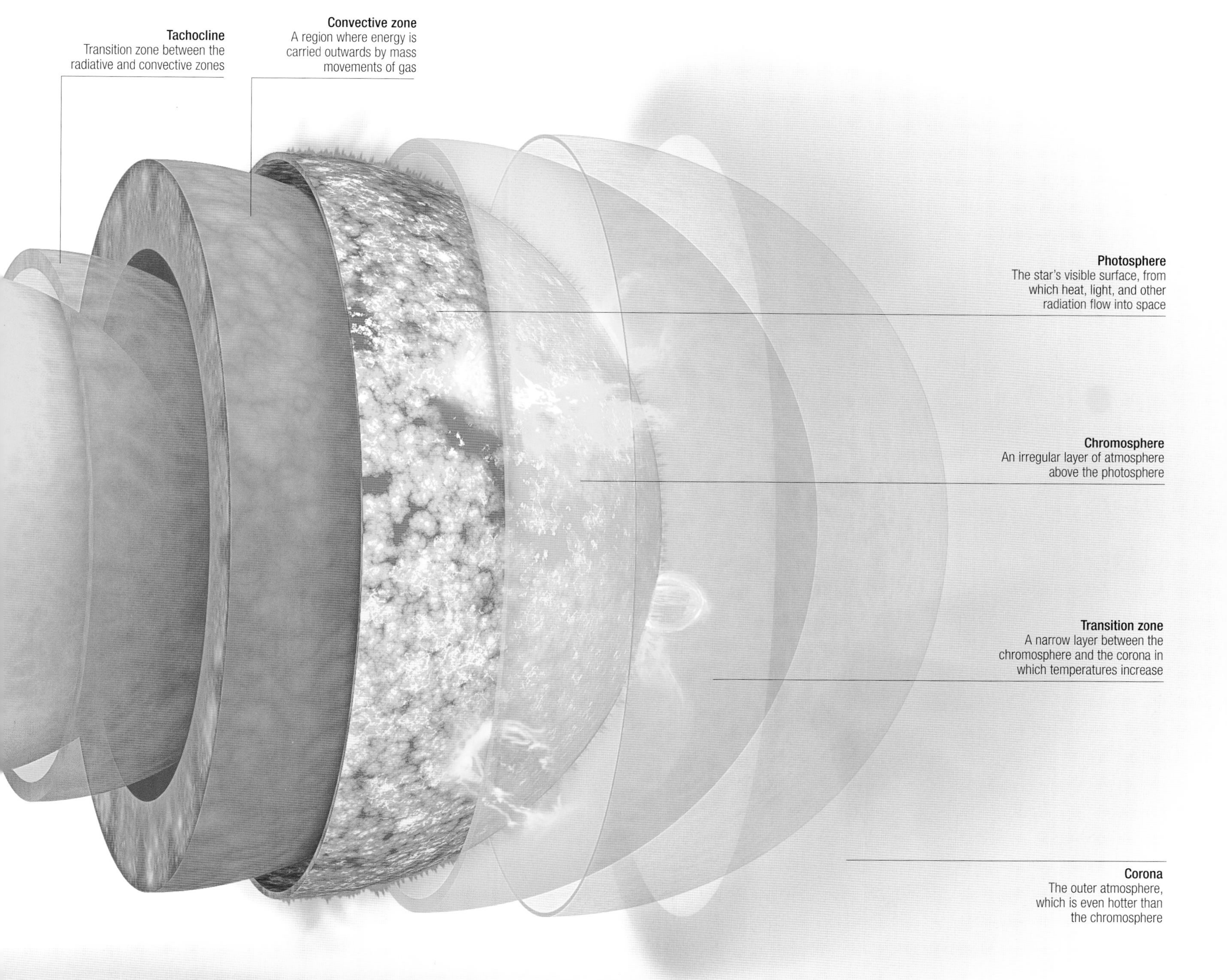
Tachocline
Transition zone between the radiative and convective zones
Convective zone
A region where energy is carried outwards by mass movements of gas
Photosphere
The star's visible surface, from which heat, light, and other radiation flow into space
Chromosphere
An irregular layer of atmosphere above the photosphere
Transition zone
A narrow layer between the chromosphere and the corona in which temperatures increase
Corona
The outer atmosphere, which is even hotter than the chromosphere

THE LIVES OF STARS

ALL STARS START LIFE AS HOT BALLS OF GAS THAT HAVE CONTRACTED DOWN FROM LARGER CLOUDS OF GAS AND DUST UNDER THE INFLUENCE OF GRAVITY. WHAT HAPPENS TO A STAR NEXT DEPENDS ON ITS INITIAL MASS.

Stars that form from the smallest clumps of gas and dust become relatively small, cool objects known as red dwarfs. These are the most common stars in our galaxy and last for tens of billions to trillions of years. As red dwarfs age, it is theorized that their surface temperature and brightness increase until eventually they become objects called blue dwarfs, then white dwarfs. Finally they fade to cold, dead, black dwarfs. However, the Universe is not yet old enough for even a blue dwarf to have formed.

Lives of medium- and high-mass stars

Medium-mass stars (about the size of the Sun) have shorter lives than red dwarfs, lasting for billions to tens of billions of years. They swell into red giants at the end of their lives. A red giant eventually sheds its outer layers to form an object called a planetary nebula, together with a hot, compact, star remnant, known as a white dwarf. The very largest stars have the shortest lives, measured in millions to hundreds of millions of years, because they use up their hydrogen fuel very quickly. In time, they form red supergiants, which disintegrate in stupendous explosions called supernovae. Depending on its mass, the core left by a supernova shrinks to one of two bizarre objects: a neutron star (see pp.36–37) or a stellar black hole (see pp.38–39).

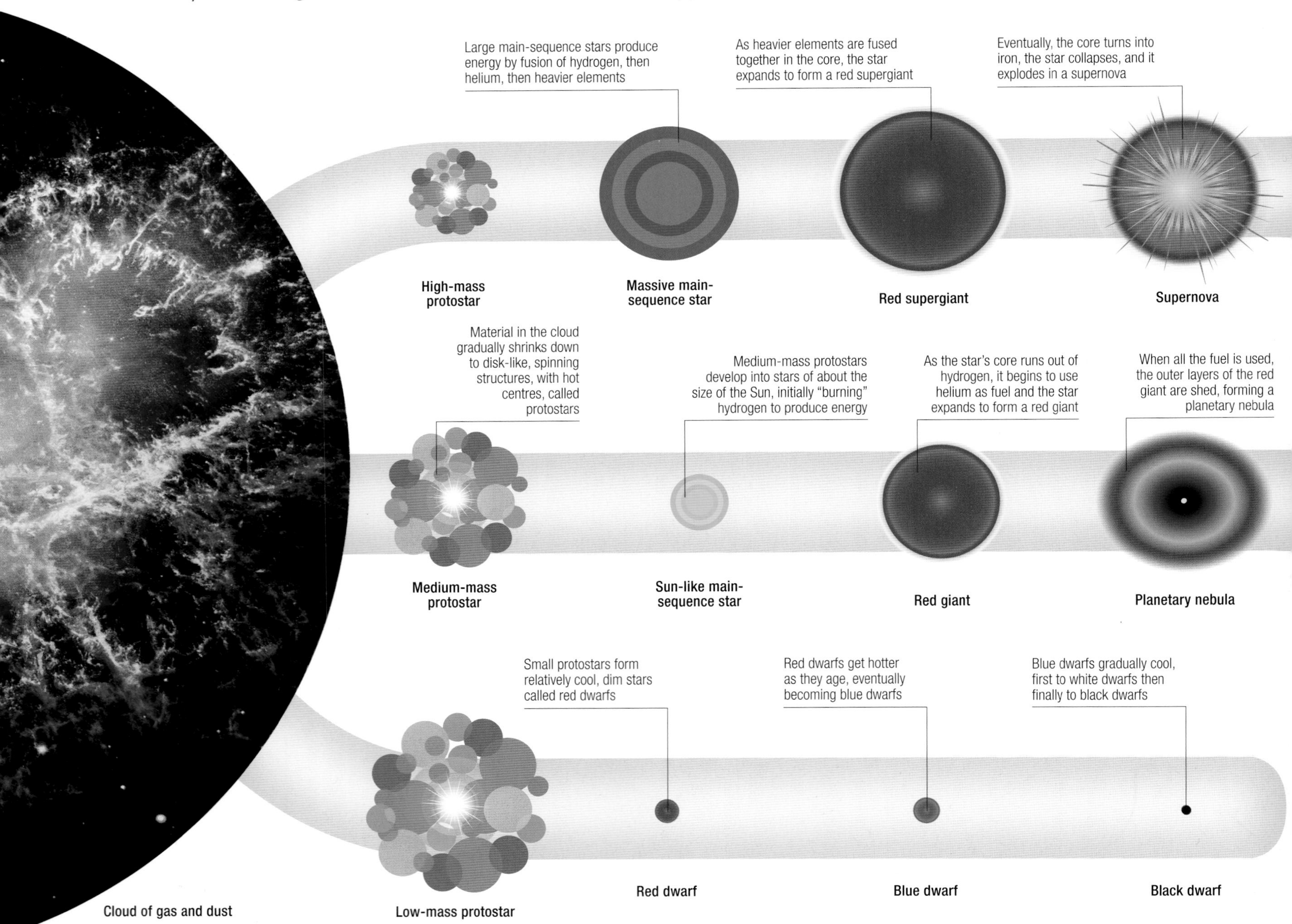
Large main-sequence stars produce energy by fusion of hydrogen, then helium, then heavier elements
As heavier elements are fused together in the core, the star expands to form a red supergiant
Eventually, the core turns into iron, the star collapses, and it explodes in a supernova
High-mass protostar
Massive main-sequence star
Red supergiant
Supernova
Material in the cloud gradually shrinks down to disk-like, spinning structures, with hot centres, called protostars
Medium-mass protostars develop into stars of about the size of the Sun, initially "burning" hydrogen to produce energy
As the star's core runs out of hydrogen, it begins to use helium as fuel and the star expands to form a red giant
When all the fuel is used, the outer layers of the red giant are shed, forming a planetary nebula
Medium-mass protostar
Sun-like main-sequence star
Red giant
Planetary nebula
Small protostars form relatively cool, dim stars called red dwarfs
Red dwarfs get hotter as they age, eventually becoming blue dwarfs
Blue dwarfs gradually cool, first to white dwarfs then finally to black dwarfs
Red dwarf
Blue dwarf
Black dwarf
Cloud of gas and dust
Low-mass protostar

The **smallest red dwarf** stars can live **millions of times** longer than the largest hypergiant stars

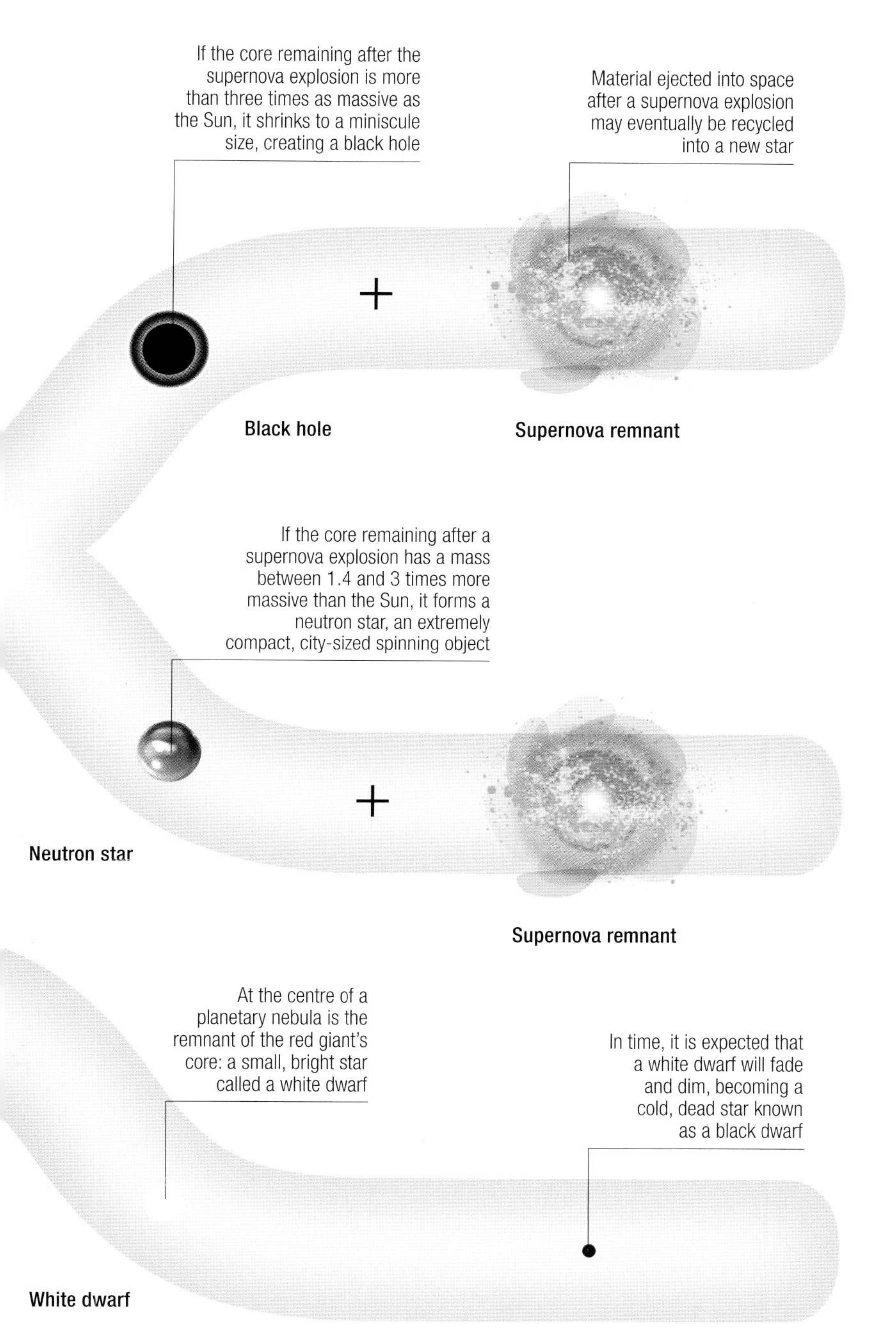
If the core remaining after the supernova explosion is more than three times as massive as the Sun, it shrinks to a miniscule size, creating a black hole
Material ejected into space after a supernova explosion may eventually be recycled into a new star
Black hole
Supernova remnant
If the core remaining after a supernova explosion has a mass between 1.4 and 3 times more massive than the Sun, it forms a neutron star, an extremely compact, city-sized spinning object
Neutron star
Supernova remnant
At the centre of a planetary nebula is the remnant of the red giant's core: a small, bright star called a white dwarf
In time, it is expected that a white dwarf will fade and dim, becoming a cold, dead star known as a black dwarf
White dwarf
Black dwarf

◁ **The lives of stars**
Contrasted here are the life stories of three main categories of stars: (from top) high-mass stars, medium-mass (Sun-sized) stars, and low-mass stars. Stars in each category start off as protostars that have formed in star-forming nebulae, but the course of their lives thereafter can be very different.

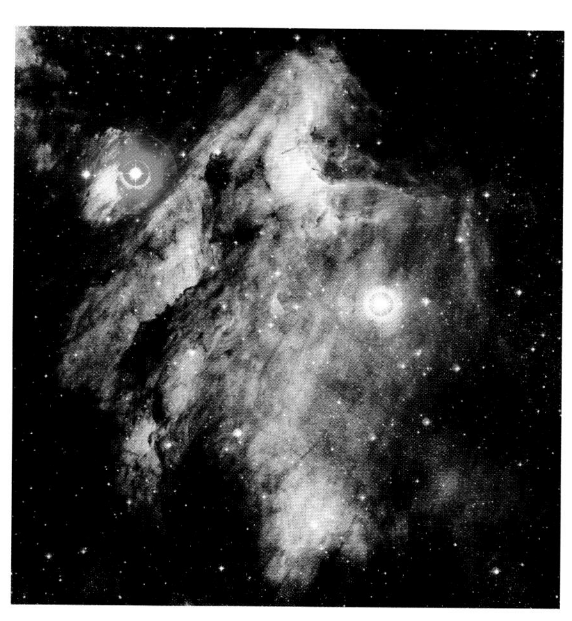

◁ **Star-forming region**
This site of intense star formation is known as the Pelican Nebula because part of it (near the top in this image) resembles the head of a pelican. It lies about 2,000 light-years away. The bright blue objects in the image are stars located between Earth and the nebula.

Stellar recycling

Materials shed from dying stars join the interstellar medium (the name for gas and dust that exists in the space between stars). From there, these materials are recycled into making new stars. Soon after the Big Bang, the Universe contained only the lightest chemical elements: mostly hydrogen and helium. Nearly all other, heavier, elements – such as carbon and oxygen – have been made since then, in stars or in supernova explosions. Through the formation, evolution, and deaths of stars, these heavier elements have gradually become more abundant in the cosmos. Astronomers call the degree to which a star is rich in heavy elements its "metallicity". Young stars tend to have the highest metallicities, as they contain materials that have already been recycled through several star generations.

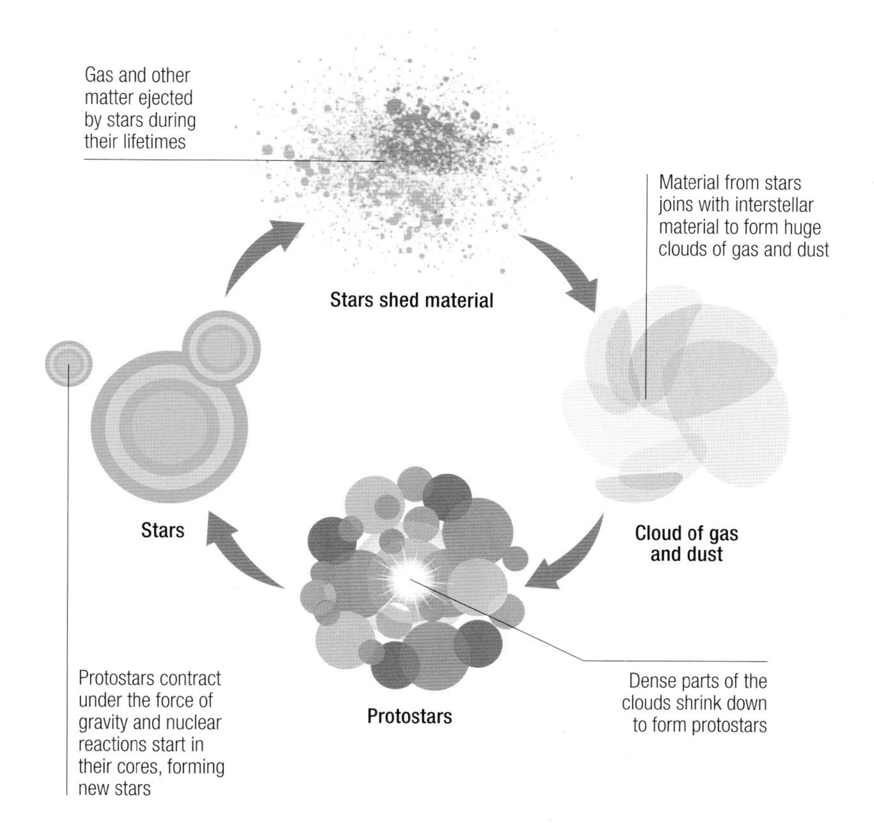
Gas and other matter ejected by stars during their lifetimes
Stars shed material
Material from stars joins with interstellar material to form huge clouds of gas and dust
Cloud of gas and dust
Dense parts of the clouds shrink down to form protostars
Protostars
Protostars contract under the force of gravity and nuclear reactions start in their cores, forming new stars
Stars

▷ **Longer-term cycle**
Stars form partly from materials shed by previous generations of stars. Furthermore, the deaths of massive stars in supernova explosions can trigger changes within the interstellar medium – particularly within star-forming nebulae – that lead to the formation of new stars.

STARBIRTH

STARS FORM OUT OF VAST CLOUDS OF COOL GAS AND DUST, CALLED MOLECULAR CLOUDS, THAT OCCUPY PARTS OF INTERSTELLLAR SPACE. THE PROCESS OF STAR FORMATION WITHIN THESE CLOUDS CAN TAKE MILLIONS OF YEARS.

The molecular clouds where stars are born can be hundreds of light-years across. Most sites of star formation are hidden inside these dense dusty clouds. However, there are places where the radiation from brilliant newly formed stars is clearing the dust away and is lighting up the surrounding gas. We see these star-forming regions as bright nebulae. They include the Eagle Nebula (see opposite) in Serpens, the Orion Nebula (see pp.164–65), and many others. Some specific dark concentrations of dust and gas sometimes seen within molecular clouds are known as Bok globules. These frequently result in the formation of double or multiple star systems (see pp.40–41).

Star formation

For star formation to start within a molecular cloud, a triggering event is needed. This could be a nearby supernova explosion, the passage of the cloud through a more crowded region of space, or an encounter with a passing star. The tidal forces and pressure waves that come into action during these situations push and pull at the cloud, compressing parts until some regions become dense enough for stars to form. Gravity then does the rest of the work of forming each star, pulling more and more material onto the developing knot of matter and concentrating most of it at the centre. As the material grows denser, random motions are transformed into a uniform spin around a single axis. Collisions between particles jostling within the cloud raise its temperature, notably in the centre, and the newly forming star begins to glow with infrared (heat) radiation.

At this stage, the protostar (newly forming star) is quite unstable. It loses mass by expelling gas and dust, directed in two opposing jets from its poles. At its centre, it eventually becomes so hot that nuclear fusion starts, and as the balance between gravity and outward-acting pressure begins to equalize, the protostar settles down to become a main-sequence star.

Astronomers have calculated that on average about **seven new stars per year** are born in the Milky Way Galaxy, most of them somewhat **smaller than the Sun**

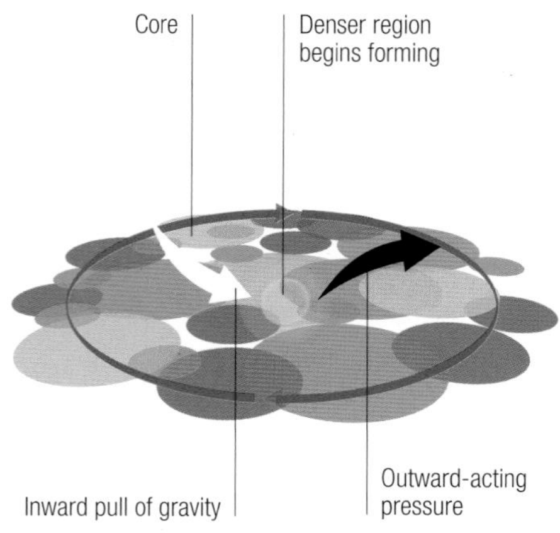

1 Dense region forms in a molecular cloud
Some nearby event, such as a supernova, causes dense clumps to come together inside a molecular cloud under the action of gravity. These clumps will become clusters of stars. They break up further into smaller regions called cores.

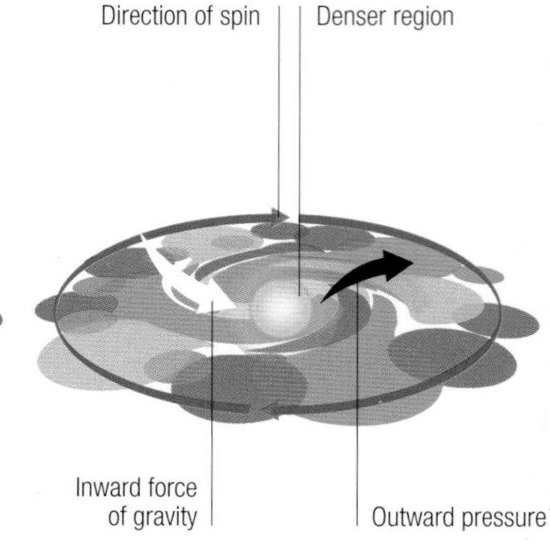

2 Core starts to collapse
Each core then starts to contract under the influence of gravity, and begins to slowly spin. Over tens of thousands of years, this spinning, gradually concentrating mass of gas and dust collapses down to less than a light-year across.

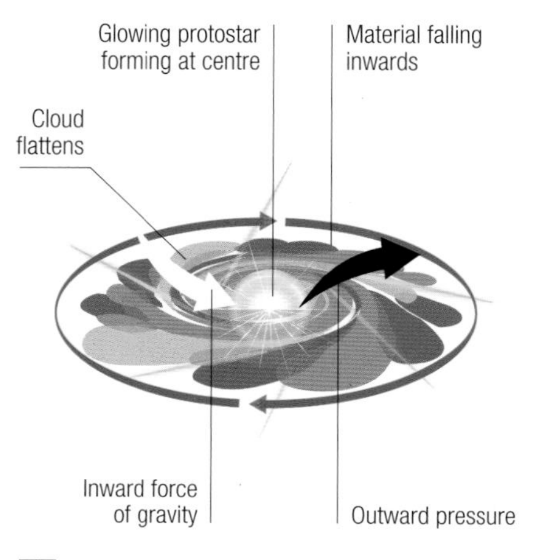

3 Protostar forms
The contracting cloud forms into a flattened, spinning disk, a few light days across, with a hot central bulge, which eventually stabilizes as a rapidly spinning protostar. Material from the cloud falls inwards and feeds onto the star.

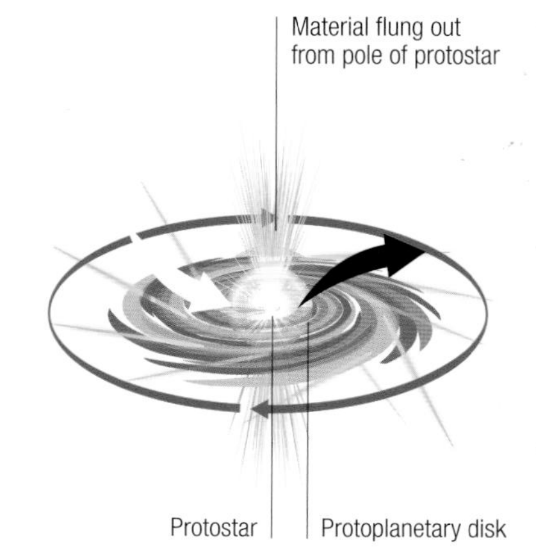

4 Protostar ejects material from its poles
Eventually the protostar spins so rapidly that new material falling onto it is flung back off. This excess material forms two tight jets emerging along the rotation axis. The cloud around the protostar flattens to form a protoplanetary disk.

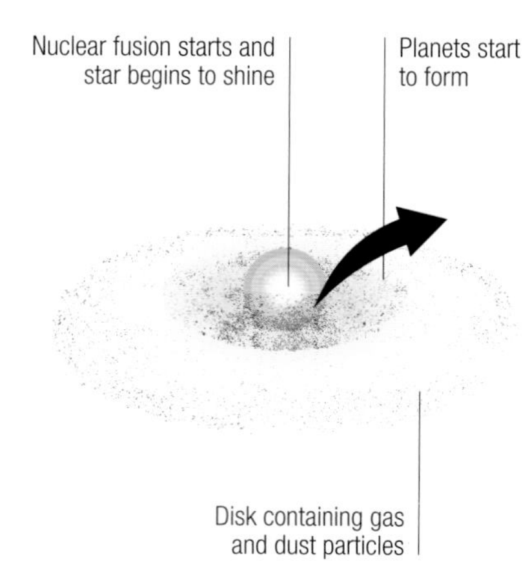

5 Star ignites
When its central core becomes hot enough, nuclear fusion reactions start within the protostar and it begins to shine as a fully-fledged star. Over millions of years, planets may gradually grow from material in the disk of dust and gas.

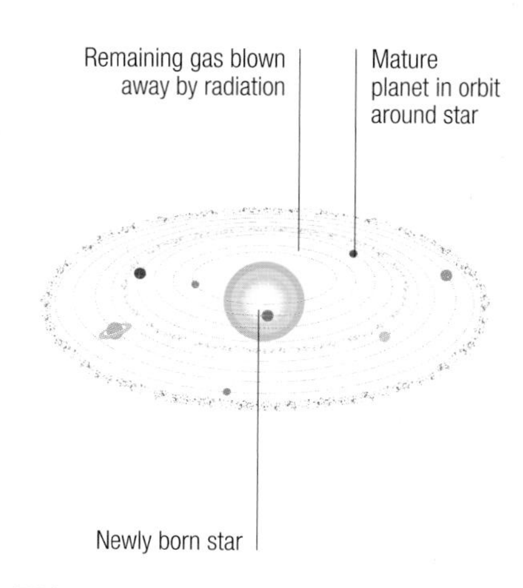

6 Planetary system forms
Radiation pressure from the newborn star blows away the remaining gas (some may accrete onto gas giant planets). Eventually, all that remains is the star, any planets, and possibly some smaller bodies, such as comets and asteroids.

Eagle Nebula
So-called because overall it vaguely resembles the shape of an eagle, the Eagle Nebula (M16) is one of the most spectacular star-forming nebulae in our galaxy. Here, tall pillars and round globules of dust and cold gas mark regions of intense star formation. Already visible are several bright young stars whose light and winds are pushing back the remaining filaments of gas and dust.

PLANETARY NEBULAE

PLANETARY NEBULAE ARE THE HEAVENLY EQUIVALENT OF SMOKE RINGS. RELATIVELY SHORT-LIVED, THEY ARE GRACEFUL CLOUDS OR SHELLS OF GAS PRODUCED DURING THE DYING DAYS OF SUN-SIZED STARS.

Among the finest-looking of celestial objects, planetary nebulae have nothing to do with planets – each is just part of the remains of a disintegrated star. The name planetary nebula comes from the nearly spherical, planet-shaped appearance of some of the first of these objects to be spotted. However, modern telescopes have revealed that they actually come in a wide range of shapes. Some planetary nebulae seem to be genuine rings or spherical shells of gas, but others are butterfly-shaped, hourglass-shaped, or can have any of an apparently infinite variety of other complex structures. What all planetary nebulae have in common is that they result from a red giant star becoming unstable at the end of its life and shedding its outer layers. The instability starts when the star begins to run out of materials to fuse in its core (fusion is the joining together of atomic nuclei to make larger nuclei, with the release of energy).

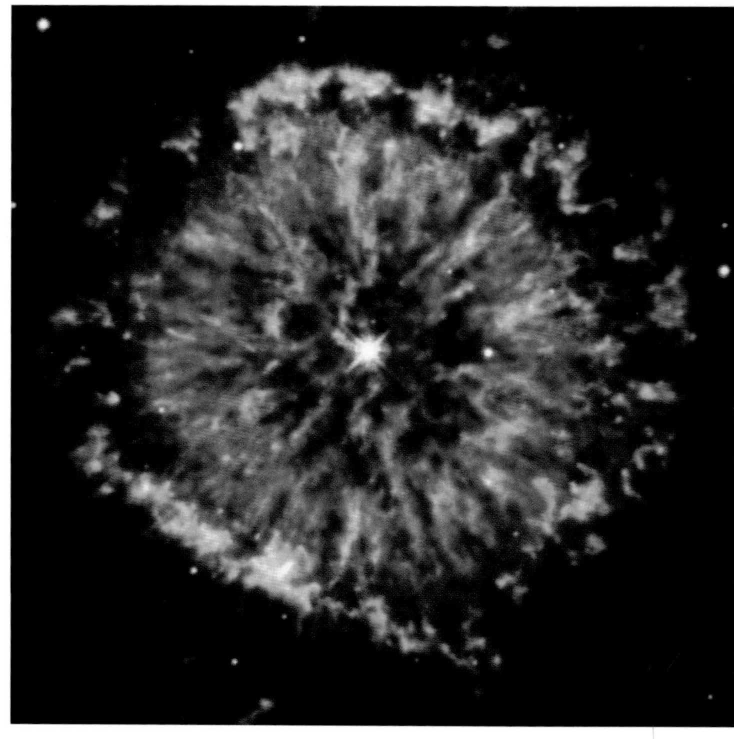

△ **Glowing eye nebula**
The patterning in this planetary nebula (NGC 6751) – including gas streamers moving away from the bright, central white dwarf – make it look like a giant gleaming eye. Blue regions mark the hottest gas, orange regions the coolest. It is around 0.8 light-years across.

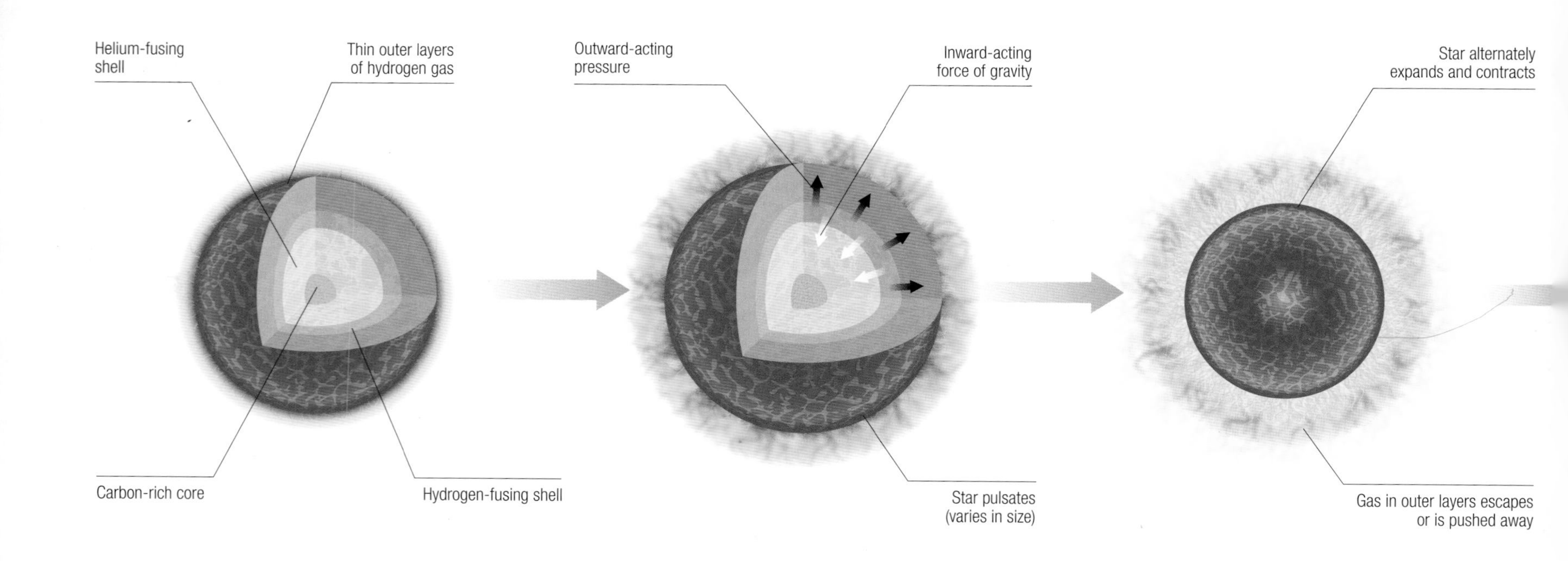

1 Ageing red giant
When a star of about the same mass as the Sun nears the end of its life, its energy production rises and it expands into a red giant as its outer layers puff out. An ageing red giant has a carbon-rich core surrounded by hot, dense shells of gas where helium and hydrogen fusion occur, producing huge amounts of energy.

2 Star becomes unstable
Two forces maintain the size of the star: inward-acting gravity and outward-acting pressure generated by energy output. The energy-producing fusion reactions are sensitive to changes in temperature and pressure, so tiny variations in these can cause instability in the star's size, leading to large-scale pulsations.

3 Star loses material from outer layers
At the extremes of each pulsation, the red giant expands at such a speed that gas in its outer layers can escape the star's gravity altogether, billowing out into space. The gas is also pushed away by the pressure exerted by particles and photons (tiny packets of light) blasted out from the star's hot core.

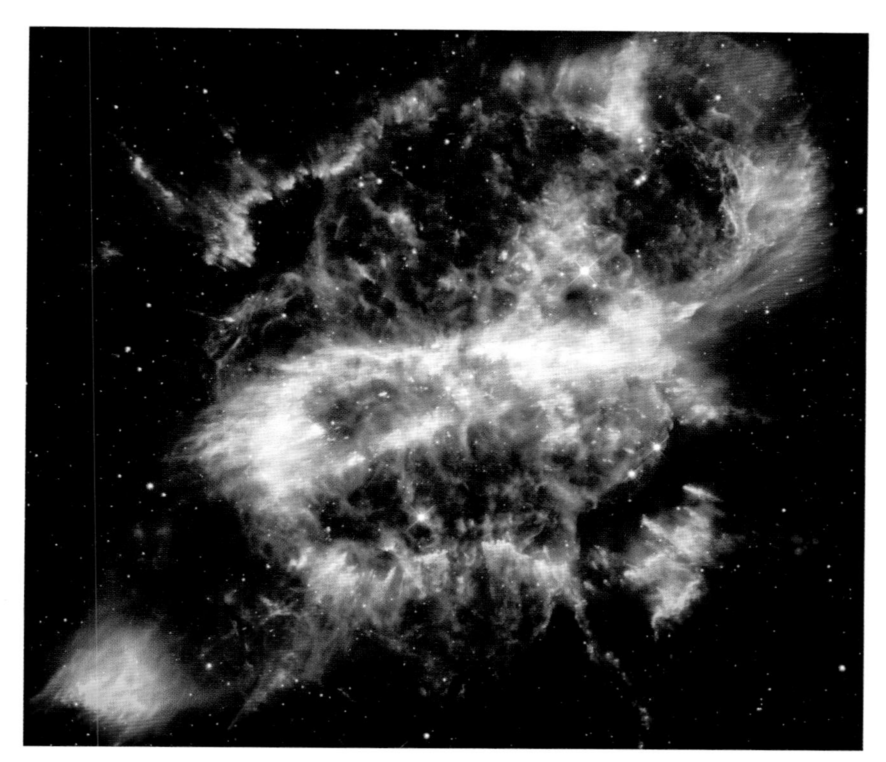

△ **Complex-structured planetary nebula**
This nebula (NGC 5189) has a complex structure, with two separate bodies of gas expanding outwards in different directions. This might be explained by the presence of a second star orbiting the central white dwarf. The nebula lies about 3,000 light-years away.

White dwarfs

When a red giant has shed all its outer layers of gas, forming a planetary nebula, what remains is a hot core consisting, in most cases, of carbon and oxygen. This object is called a white dwarf and is extremely dense – a teaspoon of it would weigh several tonnes. A white dwarf also starts off extremely hot with a surface temperature of anything up to 150,000°C (270,000°F). However, it is not hot enough for internal nuclear fusion reactions to occur. Over extremely long periods of time, a white dwarf gradually cools and fades, eventually (it is envisaged) becoming a cold object called a black dwarf. However, the Universe is not yet old enough for any white dwarf to have cooled to the black dwarf stage.

◁ **Fleming 1**
This planetary nebula is highly unusual in that it contains two white dwarf stars circling close to each other at the nebula's centre. Their orbital motions explain the presence of some remarkably symmetrical jets and other structures that weave into knotty, curved patterns in the surrounding gas.

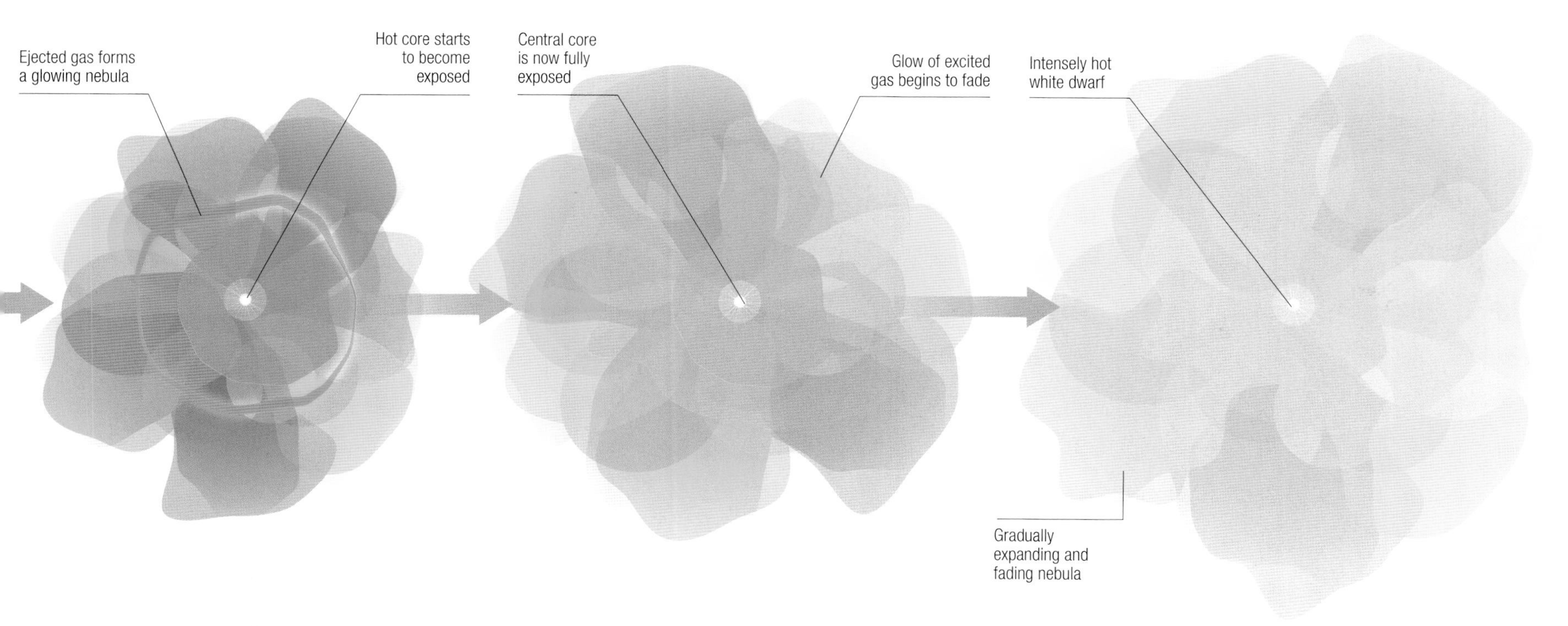

4 Planetary nebula forms
As the star sheds more and more of its gas layers, its core – at this stage usually consisting largely of carbon and oxygen produced by helium fusion – becomes exposed. Intense ultraviolet radiation given off by the core heats the ejected clouds of gas, which begin to glow or fluoresce in a variety of colours due to variations in temperature.

5 Planetary nebula expands
While the nebula expands into space, the excitation from its central star begins to dwindle, and the glow from its gases starts to fade. A planetary nebula typically lasts for a few tens of thousands of years (compared to billions of years for a typical Sun-like star), and during this time it continually evolves.

6 White dwarf remains
Finally, almost all that remains is the exhausted core of the star, known as a white dwarf. Though extremely hot, it looks faint from a distance because of its small size. As the nebula's material drifts away, it becomes part of the interstellar medium – the diffuse matter that fills the space between stars in a galaxy.

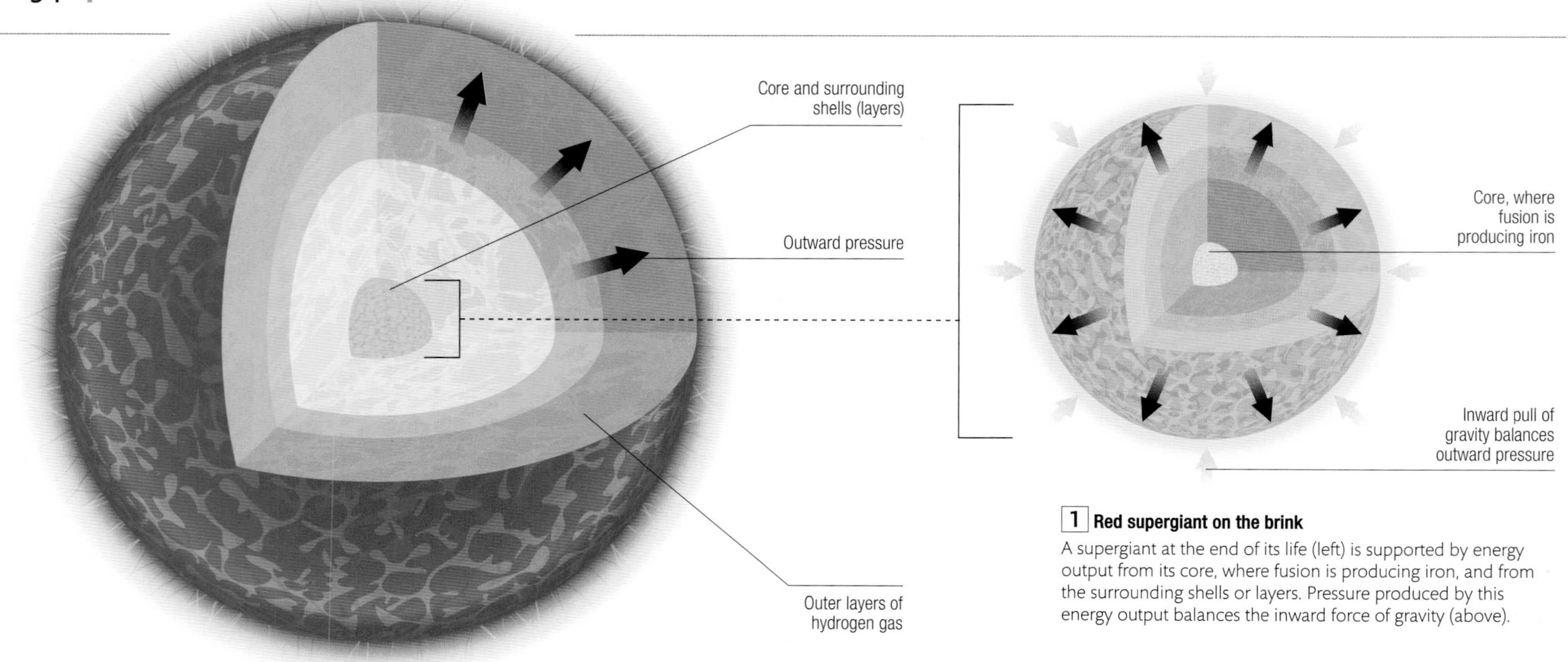

1 Red supergiant on the brink
A supergiant at the end of its life (left) is supported by energy output from its core, where fusion is producing iron, and from the surrounding shells or layers. Pressure produced by this energy output balances the inward force of gravity (above).

SUPERNOVAE

A SUPERNOVA IS THE CATACLYSMIC EXPLOSION OF, IN MOST CASES, A HIGH MASS STAR AT THE END OF ITS LIFE. A SUPERNOVA BLASTS OUT SO MUCH LIGHT AND OTHER ENERGY THAT IT CAN BRIEFLY OUTSHINE A GALAXY.

Supernovae are quite rare astronomical events in individual galaxies. None has been clearly observed in our galaxy since 1604, when a supernova some 20,000 light-years away was visible to the naked eye. However, a growing number of supernovae have been spotted in other galaxies, including one in the Large Magellanic Cloud (a satellite galaxy of the Milky Way) in 1987. A new, bright, supernova might occur in our galaxy at any time.

Types and causes

Supernovae are classified according to their spectra into various types, such as 1a, 1b, and II. Types II and Ib are the main varieties in which very high mass stars explode. As they reach the end of their life, these stars swell into supergiants and obtain their energy from nuclear fusion reactions going on in their cores and in a series of shells or layers surrounding their cores. Eventually they start making iron in their cores, but fuel for this process soon runs out. As iron itself cannot be fused to supply energy, energy output in the core suddenly ceases, and this triggers a massive explosion.

Some chemical elements can be forged only in the **extreme high-energy** conditions of a supernova

Type 1a supernovae

Although most supernovae are caused by the rapid collapse and violent explosions of very high mass stars, one type – known as a Type 1a supernova – has a different mechanism. Supernovae in this category occur in binary star systems (pairs of stars orbiting each other) where at least one star is a white dwarf (see p.20). The transfer of material from a companion star onto a white dwarf, or the collision of two white dwarfs, can both cause Type 1a supernova explosions. These explosions tend to have a uniform light output, which makes observations of them in distant galaxies useful for measuring the distances to those galaxies.

1 Matter transfer between orbiting stars
An ageing star, which has swelled into a red giant, begins to spill some gas from its outer layers onto a white dwarf star it is orbiting. This can lead to bright outbursts, called novas, on the surface of the white dwarf.

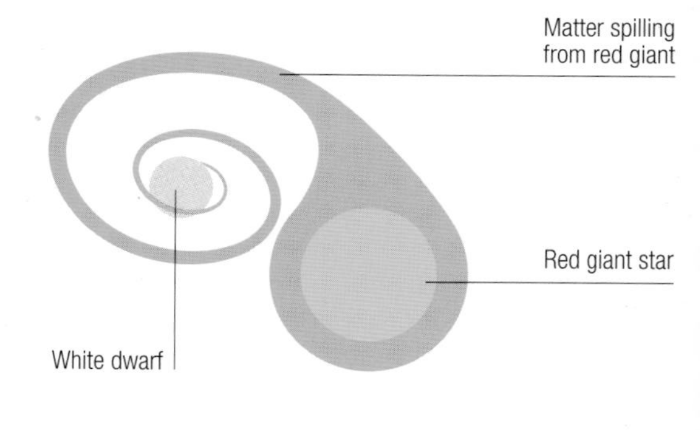

2 White dwarf explodes
The white dwarf's mass gradually increases from the extra gas it is acquiring. Eventually it becomes unstable and explodes as a Type Ia supernova. The explosion may cause the red giant star to be blasted away.

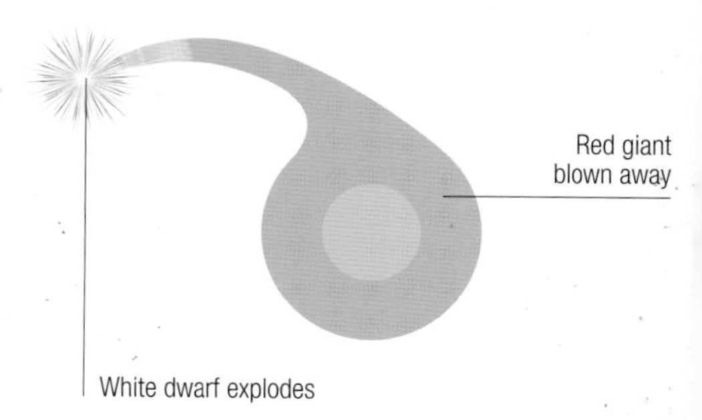

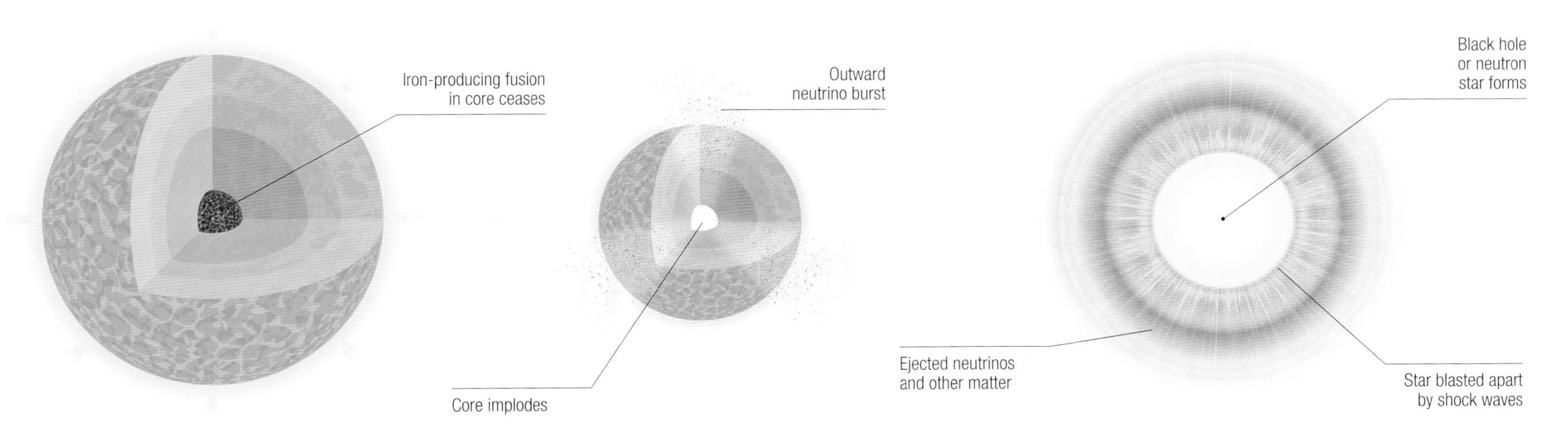

2 Fusion in core stops

Once the iron-producing fusion process slows down, energy output and pressure in the core suddenly drop, since iron itself cannot be fused to produce energy. The whole star becomes vulnerable to collapse.

3 Core collapses, neutrinos released

As the core implodes at almost one quarter of the speed of light, its iron nuclei decompose into neutrons. This event is accompanied by a brief but extremely intense burst of tiny subatomic particles called neutrinos.

4 Star explodes

The collapsing star rebounds from the compressed core with a cataclysmic shock wave that compresses and heats the outer layers. Material is thrown out, while the core becomes either a black hole or neutron star.

Supernova explosion
When a supergiant star explodes, temperatures can reach billions of degrees. In the extreme conditions, atoms of various heavy chemical elements are forged from collisions between subatomic particles. Some elements, such as lead and gold, are naturally made only in supernovae, which are the original source of all atoms of these elements in the Universe.

NEUTRON STARS

A NEUTRON STAR IS AN EXCEEDINGLY DENSE, HOT STAR REMNANT, FORMED FROM THE COLLAPSE OF THE CORE OF A MUCH LARGER STAR – FOUR TO EIGHT TIMES MORE MASSIVE THAN THE SUN – IN A SUPERNOVA EXPLOSION.

Neutron stars are tiny – only about 10–20 km (7–15 miles) across, or about the size of a large city. They are so dense that if a piece the size of a grain of sand was brought to Earth, it would weigh the same as a large passenger aeroplane. Because they are so compact, neutron stars produce extremely strong gravity: an object on a neutron star's surface would weigh 100 billion times more than on Earth. Whereas normal matter is made of atoms – which contain a lot of empty space – neutron stars consist of much more compact matter, mainly the subatomic particles called neutrons.

Axis of rotation
Neutron stars spin rapidly, some as fast as 700 times per second
Surface
A neutron star's gravity is so strong that its solid surface, which is a million times stronger than steel, is pulled into an almost perfectly smooth sphere
Magnetic field
A neutron star has an extremely powerful magnetic field, which rotates at the same speed as the star
Radiation beam
Neutron stars produce beams of electromagnetic radiation from their magnetic poles

△ Features of a neutron star
A neutron star is an extremely dense, spherical, spinning object, with a surface temperature of about 600,000°C (1,080,000°F). The surface is extremely smooth, its highest "mountains" being no more than 5mm (⅕in) tall. Neutron stars produce beams of electromagnetic radiation, which can be light, radio waves, X-rays, or gamma rays.

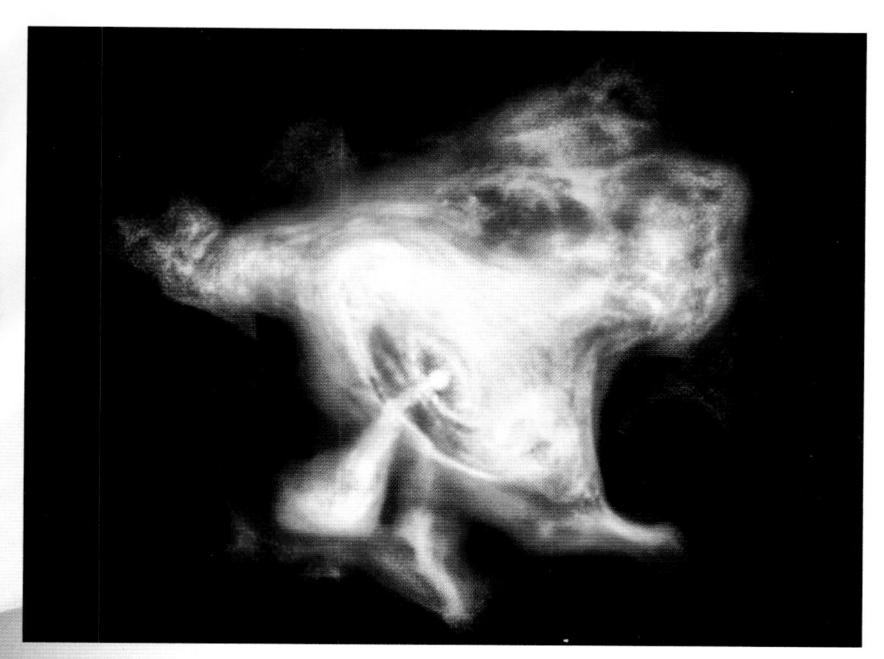

△ **Heart of the Crab Nebula**
In the centre of the Crab Nebula is a neutron star that is spinning 30 times a second and blasting out a blizzard of particles from its surface, as well as radiation beams from its poles. In this image taken by the Chandra X-ray Observatory, the ring-like structures around the pulsar (central blue-white dot) are shock waves produced where the wind of high-speed particles is ploughing into the surrounding nebula.

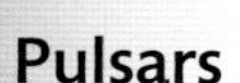

△ **Pulsar 3C58**
This image, taken with a camera that detects X-rays, shows the remains of an ancient supernova explosion. The bright central region, partially obscured by gas that emits X-rays (shown in blue), contains a pulsar. This is producing X-ray beams, which extend for trillions of kilometres to either side and have created loops and swirls (shown in blue and red) in other remnant material from the supernova.

Pulsars

As they spin and sweep their radiation beams through space, neutron stars are like celestial lighthouses. If at least one of the radiation beams points towards Earth at some point in each rotation, then from Earth it will be detectable as a series of radiation pulses. Neutron stars that are detectable in this way are called "pulsars", and the timing of their off/on signals have a precision comparable to that of an atomic clock. The first pulsar was discovered in 1967, but today more than 2,000 are known about in the Milky Way and nearby galaxies.

A neutron star's **gravity** is **so strong** that it **bends light** emitted from its surface. So, if you could look at one you would see part of its far side as well as its near side

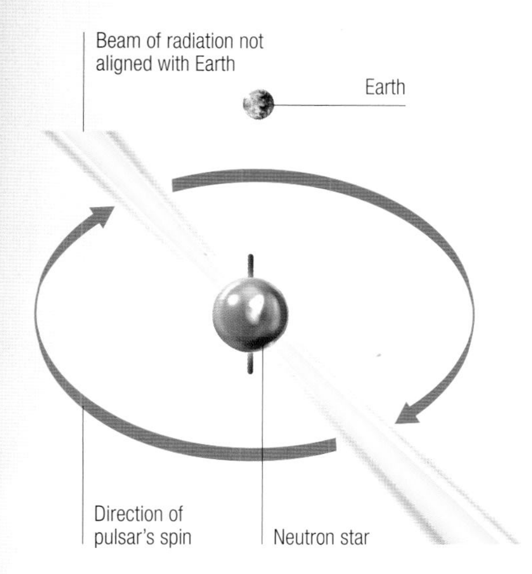

△ **Pulsar off**
As a pulsar rotates, its radiation beams continually sweep through space. At the instant shown here, neither beam points at Earth, so from the perspective of an observer on Earth, the pulsar is "off".

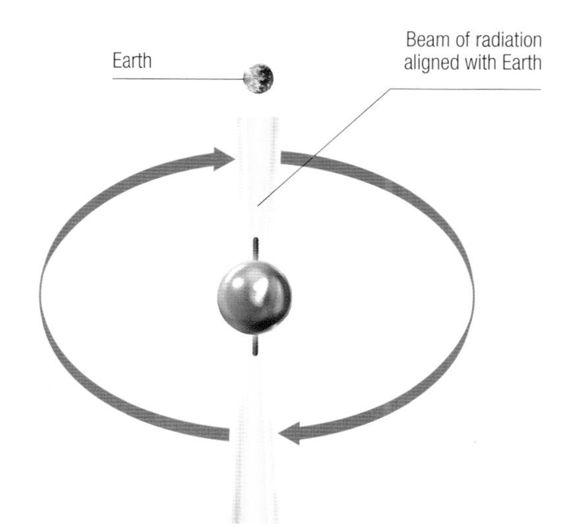

△ **Pulsar on**
A moment later, one of the pulsar's radiation beams is pointing at Earth. With the right equipment, this will be detectable on Earth as a brief signal or pulse of light, radio waves, X-rays, or other radiation.

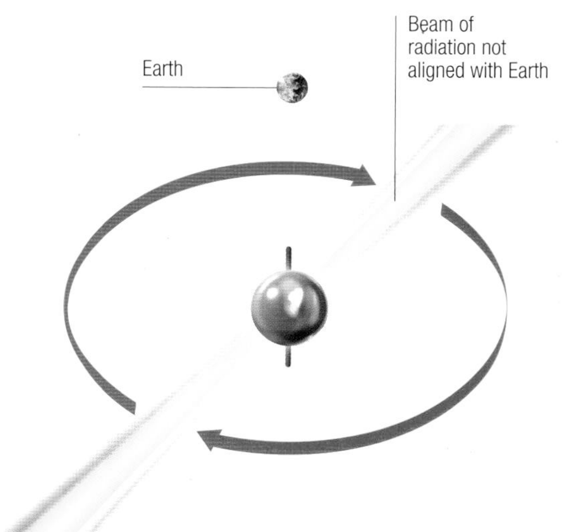

△ **Pulsar off**
Very shortly afterwards, the radiation beam is no longer aligned with Earth, so the pulse or signal switches "off" again. The off/on/off pulses occur at very regular intervals, characteristic of that pulsar.

BLACK HOLES

A BLACK HOLE – ONE OF THE STRANGEST OBJECTS IN THE UNIVERSE – IS A REGION OF SPACE WHERE MATTER HAS BEEN SQUEEZED INTO A MINUSCULE POINT OR RING OF INFINITE DENSITY, CALLED A SINGULARITY.

In a spherical region around the singularity, the gravitational pull towards the centre is so strong that nothing, not even light, can escape. The boundary of the region of no escape is called the event horizon, and anything passing inwards through this boundary can never return. There are two main types of black hole. Stellar black holes form from the collapse of the cores of supergiant stars that have exploded as a supernova at the end of their life. Supermassive black holes are much bigger and are thought to exist at the centre of most galaxies.

Detecting black holes

Because it emits no light, a black hole cannot be observed or imaged directly. However, some black holes can be detected from their strong gravity, which attracts other matter. These black holes may have disk-shaped collections of gas and dust around them that are spiralling into the black hole, at the same time throwing off vast amounts of X-rays or other radiation. The easiest ones to detect are those that produce jets of high-energy particles from their poles.

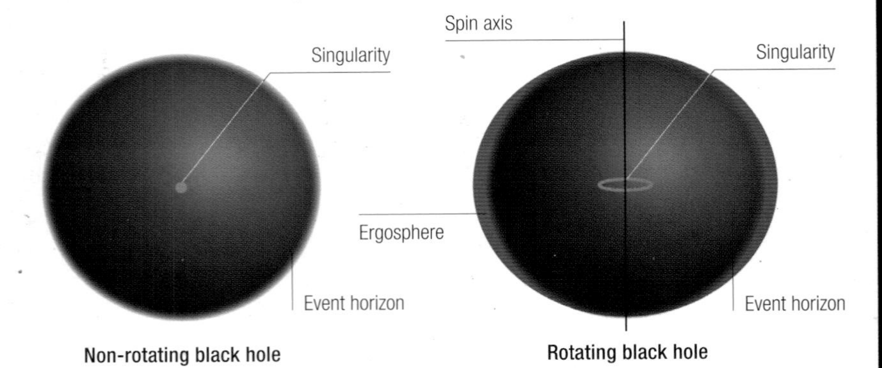

△ **Non-rotating and rotating black holes**
Black holes fall into rotating and non-rotating types – astronomers think that most rotate. In a non-rotating black hole, the singularity is a point of infinite density at the centre of the hole, whereas in the rotating variety, the singularity is ring-shaped. In both types, the event horizon – the boundary of the region of no escape – forms the surface of a sphere. However, around a rotating black hole's event horizon is an additional region, the known as the ergosphere. Anything entering this is dragged around by the black hole's spin.

▷ **Gravitational light bending**
A black hole's gravity is so strong that it warps nearby spacetime (see p.73) and bends the paths of passing light rays. Shown here are the paths of four, originally parallel, light rays travelling near a black hole. The first two have their paths radically altered and the third ray ends up circling the black hole, just outside its event horizon. The fourth ray goes through the event horizon and spirals into the hole.

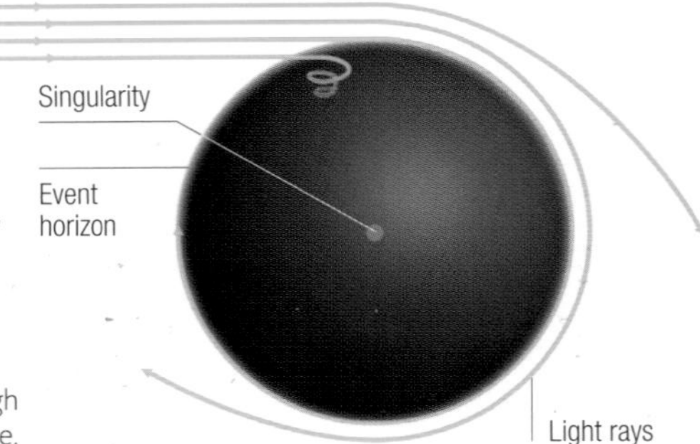

Supermassive black hole
At the centre of galaxy NGC 4258 is a vast black hole into which matter is spiralling, at the same time producing powerful jets of high-energy particles. These jets strike the disk of the galaxy and heat the gas there to thousands of degrees. That is why the centre of the galaxy looks bright, not black. The image combines various types of radiation, including visible light (yellow), infrared (red), and X-rays (blue).

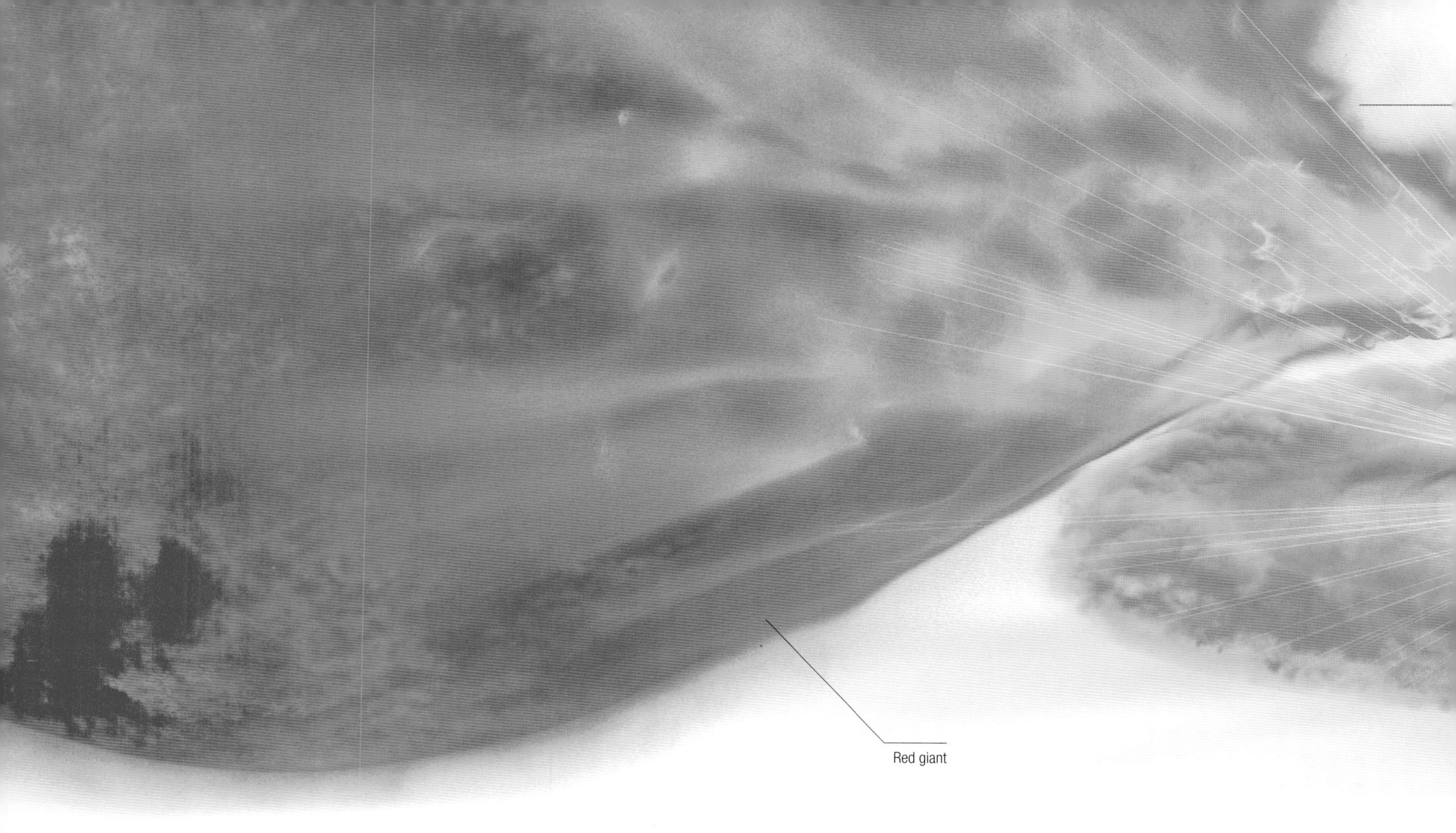

MULTIPLE STARS

OUR SUN IS A LONE STAR WITH NO COMPANIONS. HOWEVER, MOST OF THE STARS WE CAN SEE IN THE SKY BELONG TO MULTIPLE-STAR SYSTEMS. THAT IS, TWO OR MORE STARS ORBITING EACH OTHER, BOUND BY GRAVITY.

The stars in a multiple-star system can orbit one another in various different ways. A pair of stars circling round a common centre of gravity is called a binary system. If the two stars have the same mass, the centre of gravity is halfway between them. More commonly, one star is heavier than the other, and the two stars have orbits of different sizes. In systems of three or more stars, various more complicated orbits are possible. For example, two stars may orbit each other closely, with the third circling the closely orbiting pair at a great distance. Overall, more than half the stars in the Milky Way Galaxy are part of multiple-star systems. These systems are different from star clusters (see pp.44–45), which are large collections of stars only loosely bound by gravity.

True and optical binaries

A star that looks like a single point of light may actually consist of two stars located very close together in the sky. Where these stars are also close together in space and gravitationally bound – they orbit each other – they are known as "true" binaries. An example is Albireo in the constellation Cygnus (see pp.124–25). In contrast, some star pairs that happen to be close in the sky are not close in space, and are not gravitationally bound – they just happen to be in the same direction as seen from Earth. Doubles of this sort are called optical doubles. An example is a star pair called Algedi, or Alpha Capricorni, in the constellation Capricornus (see pp.186-187). Its two components are more than 600 light-years apart.

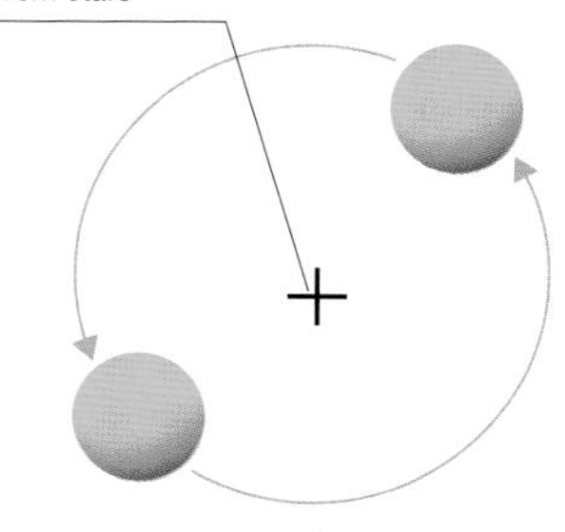

△ **Equal mass**
In binaries that consist of two stars of equal mass, the stars will orbit a common centre of gravity, which lies midway between the two stars.

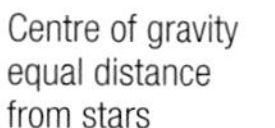
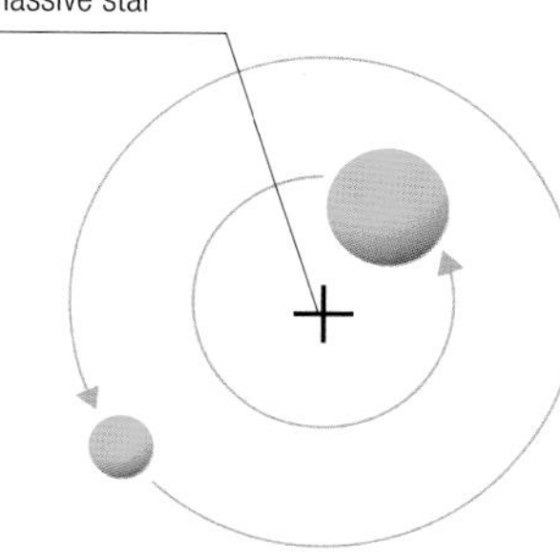

△ **Unequal mass**
If one of the stars in a binary system is more massive than the other, the system's centre of gravity lies closer to the higher-mass star.

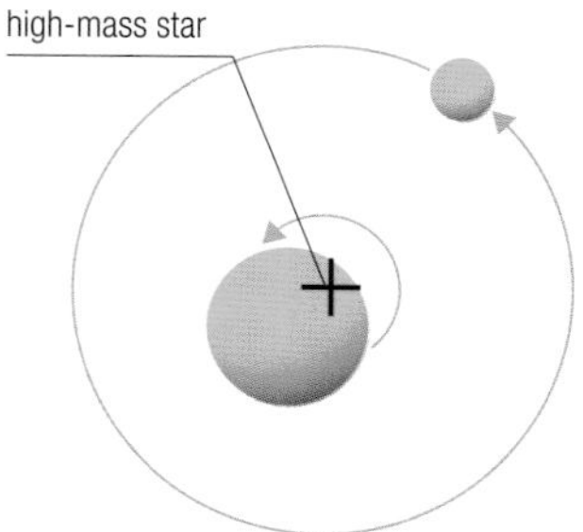

△ **Significant difference in mass**
Sometimes one star is much heavier than the other. In such cases, the centre of gravity may lie at the surface of the more massive star, or even inside it.

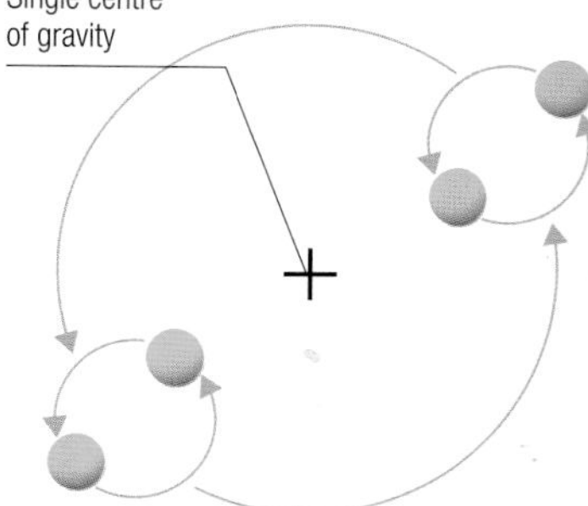

△ **Double binary**
In a double binary or quadruple system, each star typically orbits one companion, and the two pairs orbit a single centre of gravity.

Interacting binaries

Sometimes, two stars are so close that material flows from one to the other. These systems are called interacting binaries. The transferred matter forms a disk, called an accretion disk, as it spirals in towards the receiving object. It may also release X-rays. If one of the stars in the binary is a white dwarf, explosions called novae may occur from time to time on the surface of the white dwarf.

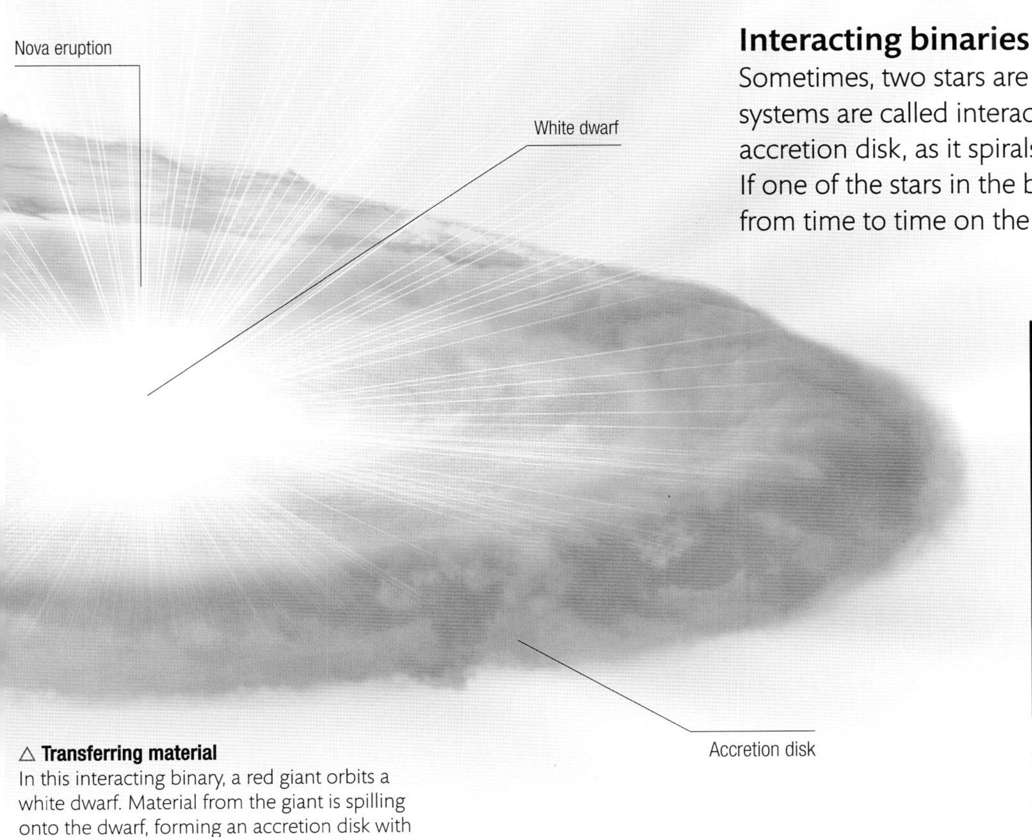

△ **Transferring material**
In this interacting binary, a red giant orbits a white dwarf. Material from the giant is spilling onto the dwarf, forming an accretion disk with occasional nova outbursts.

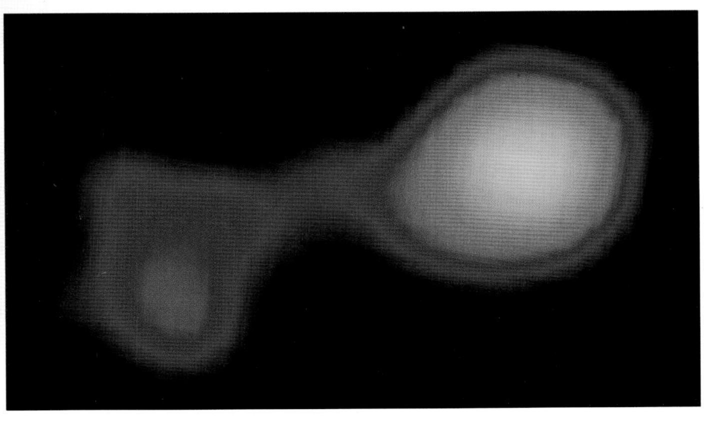

△ **Mira system**
The star system Mira in the constellation Cetus consists of a red giant (which happens to vary in brightness) and a white dwarf, clearly separate in this X-ray image, with some material connecting the two stars.

▷ **HD 98800 system**
This artist's impression of the HD 98800 system shows two pairs of binary stars. All four stars are bound by gravity, but the distance between the two pairs is about 7.5 billion km (4.5 billion miles). A disk of gas and dust, with two distinct belts, surrounds one of the star pairs, and it is suspected that there is a planet orbiting in the gap between the belts.

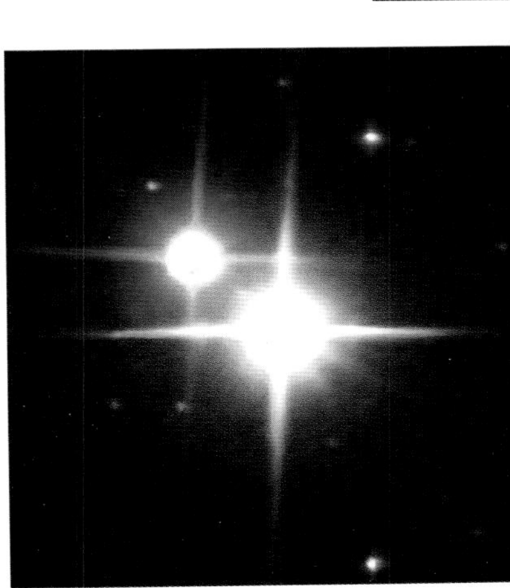

◁ **True binary**
This telescope image clearly shows two bright stars – one gold, the other blue. The two stars are so close in the sky, however, that to the naked eye they look like a single star, which is known as Albireo (Beta Cygni). Astronomers think that Albireo's two components orbit each other, so they constitute a true binary, although each orbit takes about 100,000 years.

▷ **Di Cha system**
This complex star system, some 520 light-years away, contains four stars arranged in two pairs. Only the the two brightest are clearly visible in this Hubble Space Telescope image. However, all four stars are young and surrounded by a wispy wrapping of dust.

VARIABLE **STARS**

MANY STARS DO NOT SHINE WITH A STEADY LIGHT. SOME OCCASIONALLY DIP OR FLARE IN BRIGHTNESS, WHILE OTHERS SLOWLY PULSATE. THESE ARE EXAMPLES OF WHAT ARE CALLED VARIABLE STARS.

Stars varying in brightness, as seen from Earth, fall into two main categories, called intrinsically variable and extrinsically variable. In intrinsically variable stars, the amount of light emitted by a star varies in a regular cycle, or pulsates, or it occasionally flares up. In extrinsic variables, something affects how much of the star's light reaches Earth.

Pulsating variables

These intrinsically variable stars continuously change in diameter, in a regular cycle, because of fluctuations in the forces that affect their size (see pp.26–27). In a class of stars called Cepheid variables, a close relationship exists between the average light output of the star and the length of its pulsation cycle. This relationship allows astronomers to determine distances within our galaxy and to other galaxies.

△ **Cepheid variable**
This star, called RS Puppis, a Cepheid variable, varies in brightness by a factor of five in cycles lasting 41.4 days. As this Hubble Space Telescope image shows, the star is shrouded by thick clouds of dust.

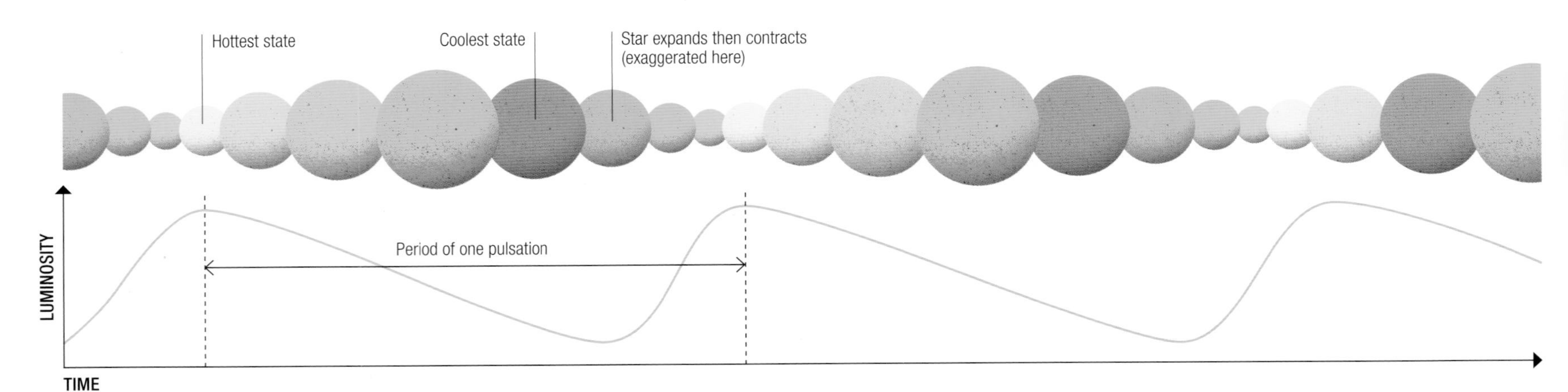

△ **Light curve of a pulsating variable**
The amount of light emitted by a pulsating variable fluctuates in a cycle that, depending on the star, can last for anything from several hours to hundreds of days. The fluctuations are closely related to changes in the star's size.

Flaring or cataclysmic variables

Another type of intrinsically variable star, a nova, or cataclysmic variable, is the sudden brightening of a white dwarf star in a binary system (two stars orbiting each other, see p.41). It is caused by a nuclear explosion on the white dwarf's surface. This occurs because the white dwarf's companion star – usually a giant star – has grown so large that its outer layers of hydrogen gas are no longer gravitationally bound to the star and instead fall onto the white dwarf. Subsequently, fusion reactions start up within the accumulated hydrogen on the surface of the white dwarf, triggering a runaway nuclear explosion. Prior to the outburst, the binary system may have been invisible to the naked eye, and so the outburst brings the system into visibility as a "nova" (which is Latin for "new") star. Some binary systems produce recurrent novae, separated by quiet periods ranging in length from a few years to thousands of years.

△ **GK Persei nova**
GK Persei has produced a nova about every three years since 1980. Surrounding it is an expanding cloud of gas and dust called the Firework Nebula.

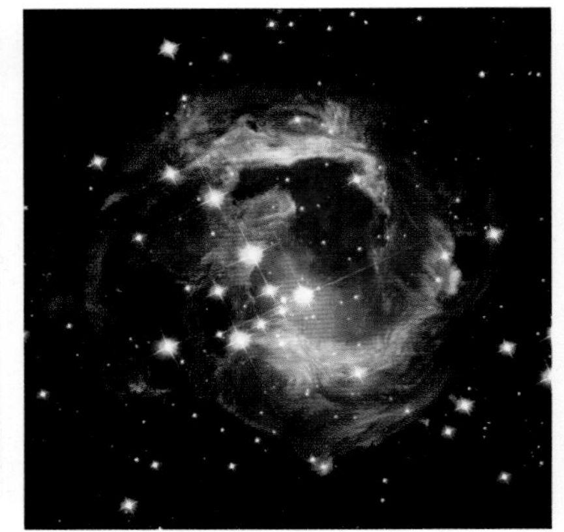

△ **Luminous red nova**
This outburst, from the star V838 Monocerotis, was at first thought to be a regular nova, but it is now suspected be due to two stars colliding.

Binary star systems

Extrinsic variable stars owe their apparent variations in brightness to something other than changes in light output. The most important group of extrinsic variables are called eclipsing binaries. These are binary systems (two stars orbiting each other) with an orbital plane that lines up with Earth. From time to time, one star eclipses (blocks out light from) the other as seen from Earth, causing some dimming. A slight dimming occurs when the brighter star eclipses the fainter star, and a more significant dimming when the fainter star eclipses the brighter one. The first eclipsing binary to be discovered was Algol, in the constellation Perseus. This actually consists of three stars, of which two regularly eclipse each other. Each time the fainter of the two eclipses the brighter, which occurs every 2.86 days, there is a roughly 70 per cent dimming for about 10 hours.

A different and somewhat unusual cause of extrinsic variability occurs where two closely orbiting stars in a binary system have acquired distorted, ellipsoidal shapes. These are called rotating ellipsoidal binaries (see right). An example is the bright star Spica (actually a pair of stars) in the constellation Virgo.

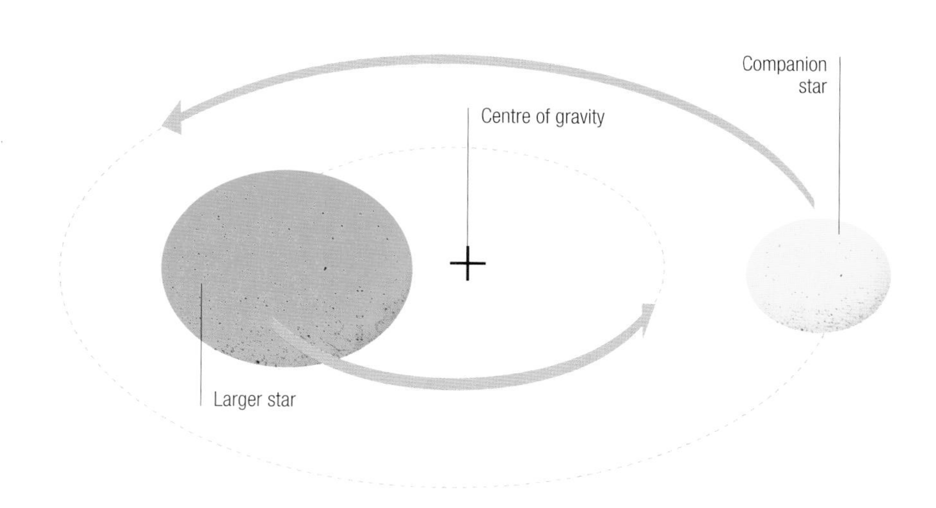

△ **Eclipsing and ellipsoidal variables**
In this type of variable, two stars that are orbiting a common centre of gravity become distorted into ellipsoidal (egg-like) shapes. Sometimes they appear side-on (as here) and at other times end-on (appearing smaller and rounder), which affects how bright they look from Earth.

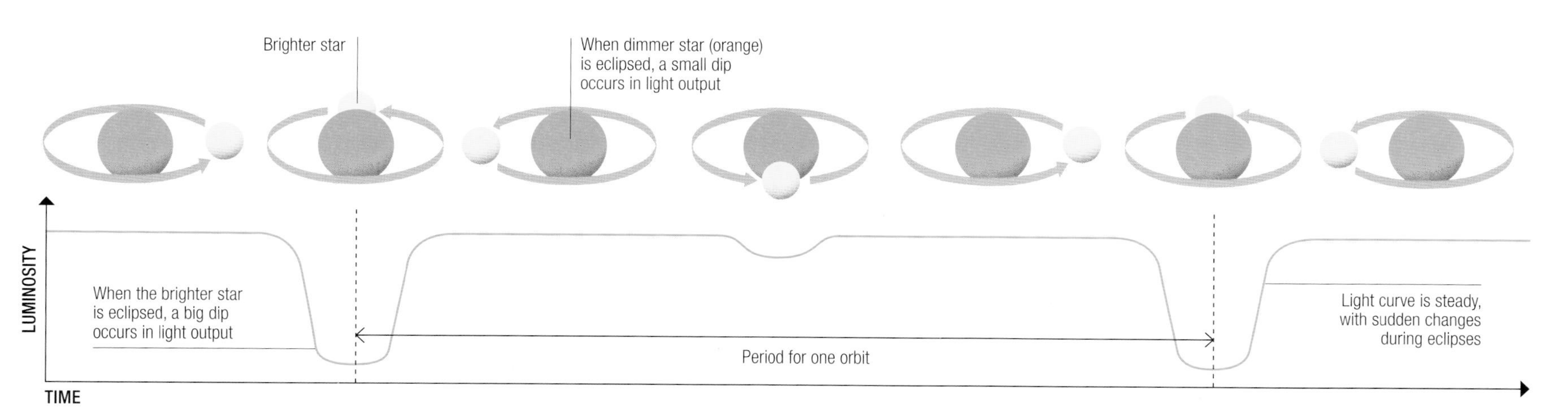

△ **Light curve of an eclipsing binary**
Eclipsing binary stars are detected by regularly occurring apparent dips in a star's brightness. These dips in brilliance occur when one of a pair of stars partially blocks the light coming from the other star, as seen from Earth. The biggest dip occurs when the dimmer of the two stars eclipses the brighter star.

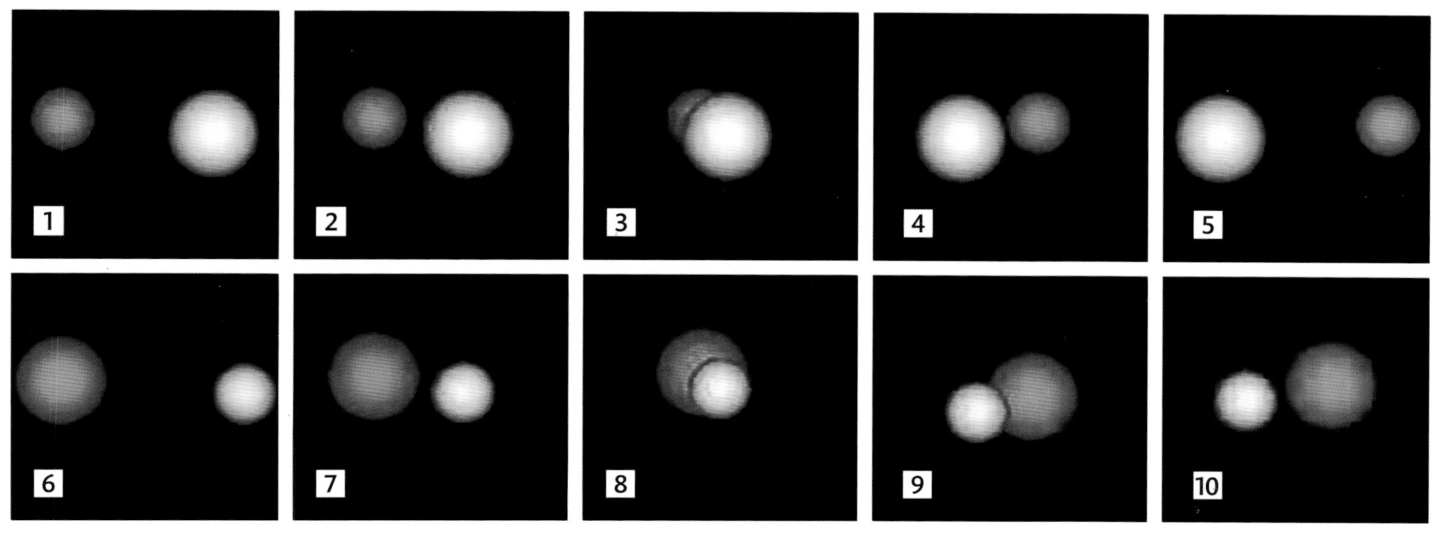

◁ **Binary orbit sequence**
These 10 frames from a movie made with a special infrared-sensitive camera show two young stars orbiting a shared centre of gravity. The images were taken using the ADaptive Optics Near Infrared System (ADONIS) at the European Southern Observatory at La Silla, Chile.

STAR CLUSTERS

A LARGE GROUP OF STARS – ANYTHING FROM A DOZEN TO SEVERAL MILLION STARS – BOUND TOGETHER BY GRAVITY IS CALLED A STAR CLUSTER. THE MILKY WAY GALAXY CONTAINS THOUSANDS OF THESE SPECTACULAR STAR AGGREGATIONS.

Star clusters fall into two types: globular and open. Globular clusters are ancient, dense cities of stars, some containing more stars than a small galaxy. Open clusters, in contrast, are young, contain far fewer stars, and are often the site of new star creation. Many open clusters, and a few globular ones, can be seen in the night sky with the naked eye. Both types can be a magnificent sight when viewed through binoculars or a telescope.

Globular clusters

Globular clusters are groups of between 10,000 and several million mostly very old stars arranged roughly in a sphere. More than 150 clusters like this exist in the Milky Way, and each can last for 10 billion years. The stars in a cluster tend to be concentrated towards its centre, moving in random circular orbits around it.

Many globular clusters consist of a single population of stars that all have the same origin, similar ages, and chemical composition. However, some contain two or more populations that formed at different times – through some of the more massive stars in the initial population dying and materials from them being recycled into a second star generation.

Open clusters

Open clusters are groups of up to a few thousand stars that were formed roughly at the same time from the same cloud of gas and dust, but are more loosely bound by gravity than globular clusters. They survive for a shorter time – from a few hundred million up to a few billion years. Unlike globular clusters, which occur in all types of galaxy, open clusters are found only in spiral and irregular galaxies, where stars are actively being created. Around 1,100 clusters of this type have been identified so far in the Milky Way.

Our galaxy's **largest globular cluster**, called Omega Centauri, contains about **10 million stars**

▽ **Cluster distribution in spiral galaxies**
Star clusters exist in different parts of spiral galaxies like the Milky Way. Globular clusters are found in the halo region, above and below the main disk, and open clusters are seen in the galaxy's disk and spiral arms.

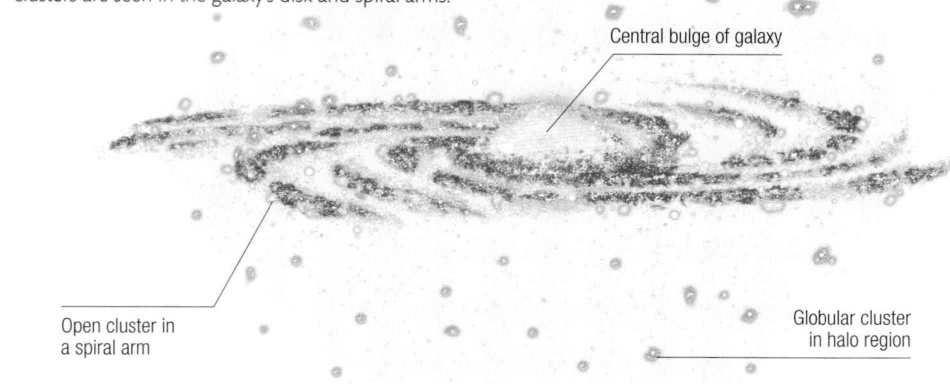

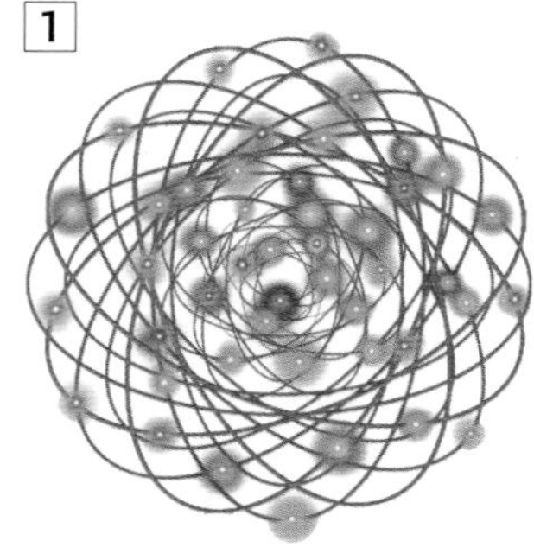

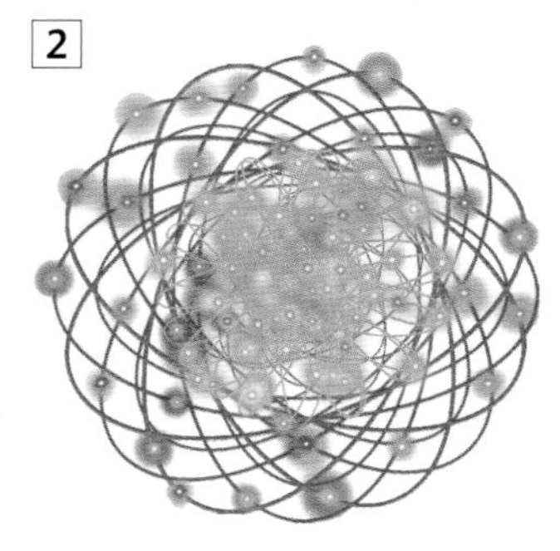

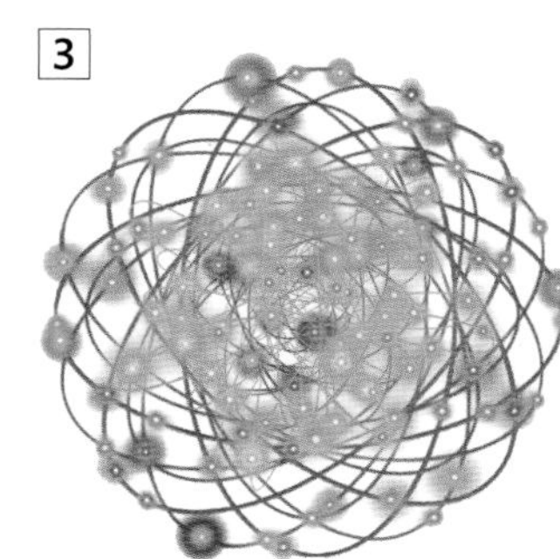

△ **Evolution in a globular cluster**
In this example of cluster evolution, some of the first star generation (red) die. Material from them then forms a second generation (blue), more concentrated at the centre of the cluster. Gradually their orbits change, mixing them with the older red stars.

△ **M7 open cluster**
Also known as the Ptolemy cluster, this array of around 80 stars lies in the constellation of Scorpius. Though 980 light-years away, it is easily seen with the naked eye.

47 Tucanae globular cluster
One of the largest and brightest globular clusters in the night sky, 47 Tucanae is located in the southern hemisphere constellation of Tucana. To the naked eye it looks like a fuzzy patch in the sky, but telescopes reveal it be an immense swarm of several million stars. The cluster's central region is so crowded that many star collisions occur.

EXTRASOLAR PLANETARY SYSTEMS

ANY GROUP OF PLANETS ORBITING A STAR OTHER THAN THE SUN IS CALLED AN EXTRASOLAR PLANETARY SYSTEM. THE INDIVIDUAL PLANETS CIRCLING AROUND IN THESE SYSTEMS ARE CALLED EXOPLANETS.

More than 2,000 exoplanets have been discovered so far, mostly in the last 10 years or so. About half are gas-dominated planets, about the size of Jupiter or Neptune in the Solar System, orbiting close to their host stars. These hellishly hot, star-snuggling gas giants are known as "hot Jupiters" or "hot Neptunes". Many smaller, probably rocky, exoplanets have also been discovered – some about the size of Earth – as well as cold gas giants. The types of stars that exoplanets orbit vary from red dwarfs to Sun-like stars, red giants, and even pulsars.

Perhaps the most remarkable fact about exoplanets is that they can be detected at all. Finding a body many light-years away that emits no light of its own and which orbits a much bigger, brighter body (a star) presents many challenges. So far, relatively few exoplanets have been imaged directly with telescopes, but around a dozen methods have been devised for detecting them indirectly. Three of the most successful of these methods are explained below.

The host star of a hot Jupiter is usually white, yellow, or orange and roughly Sun-sized

On average, **each star** in the Milky Way galaxy has **at least one planet** orbiting it

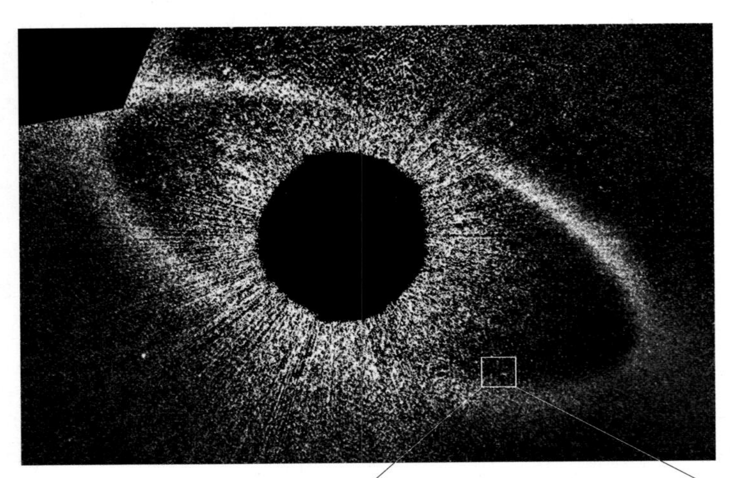

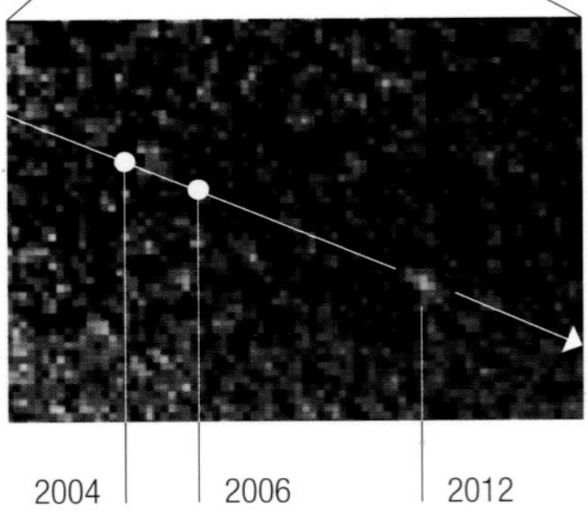

△ **Direct imaging**
The star Fomalhaut has a disk of dust and gas around it, as shown above (the star itself has been blacked out). A planet in the disk has been directly imaged by the Hubble Space Telescope. The planet and its path are shown in the image to the right.

▷ **Transit method**
This approach involves detecting miniscule dips in a star's brightness, caused by transits (movements) of a planet across the face of the star. To do this, an extremely sensitive light-detecting instrument is used.

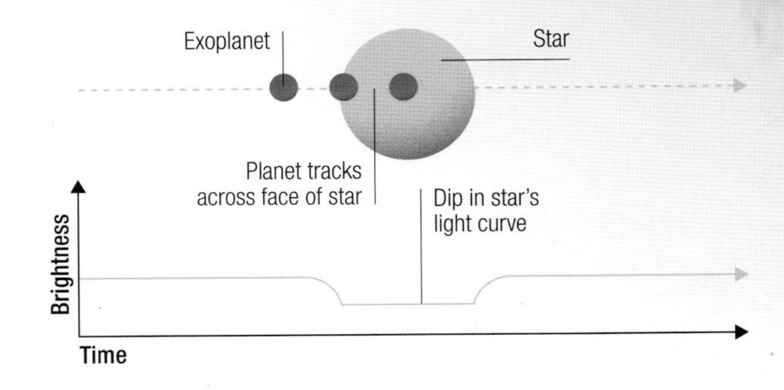

▷ **Gravitational microlensing**
The gravity of a star can bend light coming from a more distant star. This means it can act like a lens and magnify the distant star as it appears from Earth. An exoplanet orbiting the lensing star produces detectable variations in the amount of magnification.

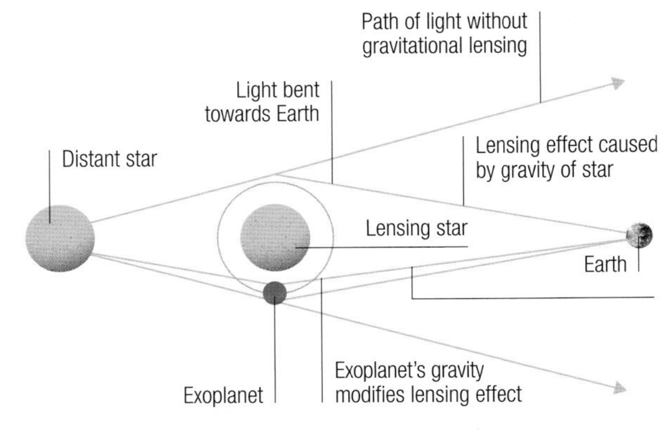

▷ **Doppler spectroscopy**
An exoplanet's orbit causes a "wobble" in the motion of its host star. As a result, light waves coming from the star are alternately slightly lengthened (making them look redder) and shortened (making them look bluer) – a measurable phenomenon.

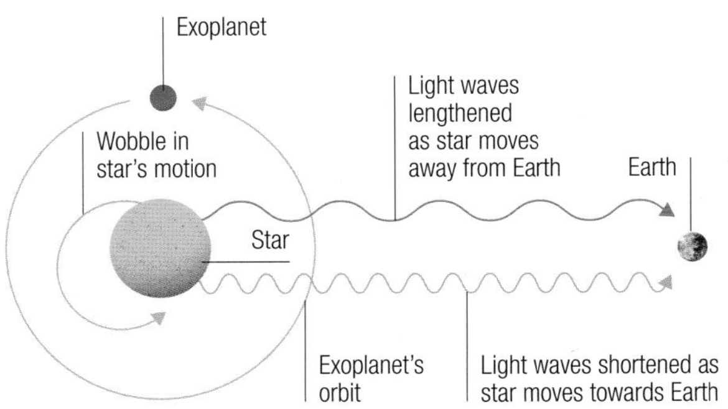

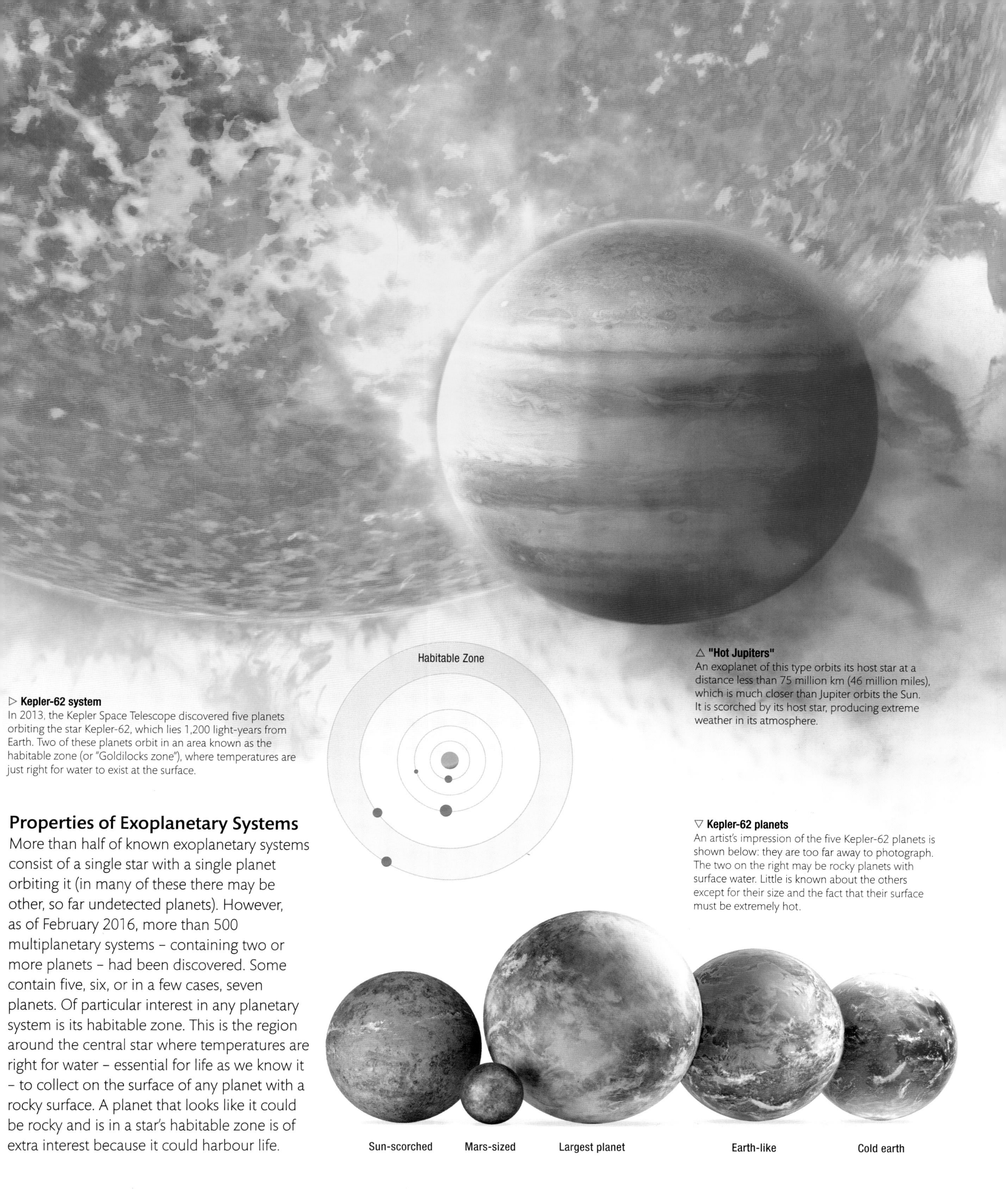

△ **"Hot Jupiters"**
An exoplanet of this type orbits its host star at a distance less than 75 million km (46 million miles), which is much closer than Jupiter orbits the Sun. It is scorched by its host star, producing extreme weather in its atmosphere.

▷ **Kepler-62 system**
In 2013, the Kepler Space Telescope discovered five planets orbiting the star Kepler-62, which lies 1,200 light-years from Earth. Two of these planets orbit in an area known as the habitable zone (or "Goldilocks zone"), where temperatures are just right for water to exist at the surface.

Properties of Exoplanetary Systems

More than half of known exoplanetary systems consist of a single star with a single planet orbiting it (in many of these there may be other, so far undetected planets). However, as of February 2016, more than 500 multiplanetary systems – containing two or more planets – had been discovered. Some contain five, six, or in a few cases, seven planets. Of particular interest in any planetary system is its habitable zone. This is the region around the central star where temperatures are right for water – essential for life as we know it – to collect on the surface of any planet with a rocky surface. A planet that looks like it could be rocky and is in a star's habitable zone is of extra interest because it could harbour life.

▽ **Kepler-62 planets**
An artist's impression of the five Kepler-62 planets is shown below: they are too far away to photograph. The two on the right may be rocky planets with surface water. Little is known about the others except for their size and the fact that their surface must be extremely hot.

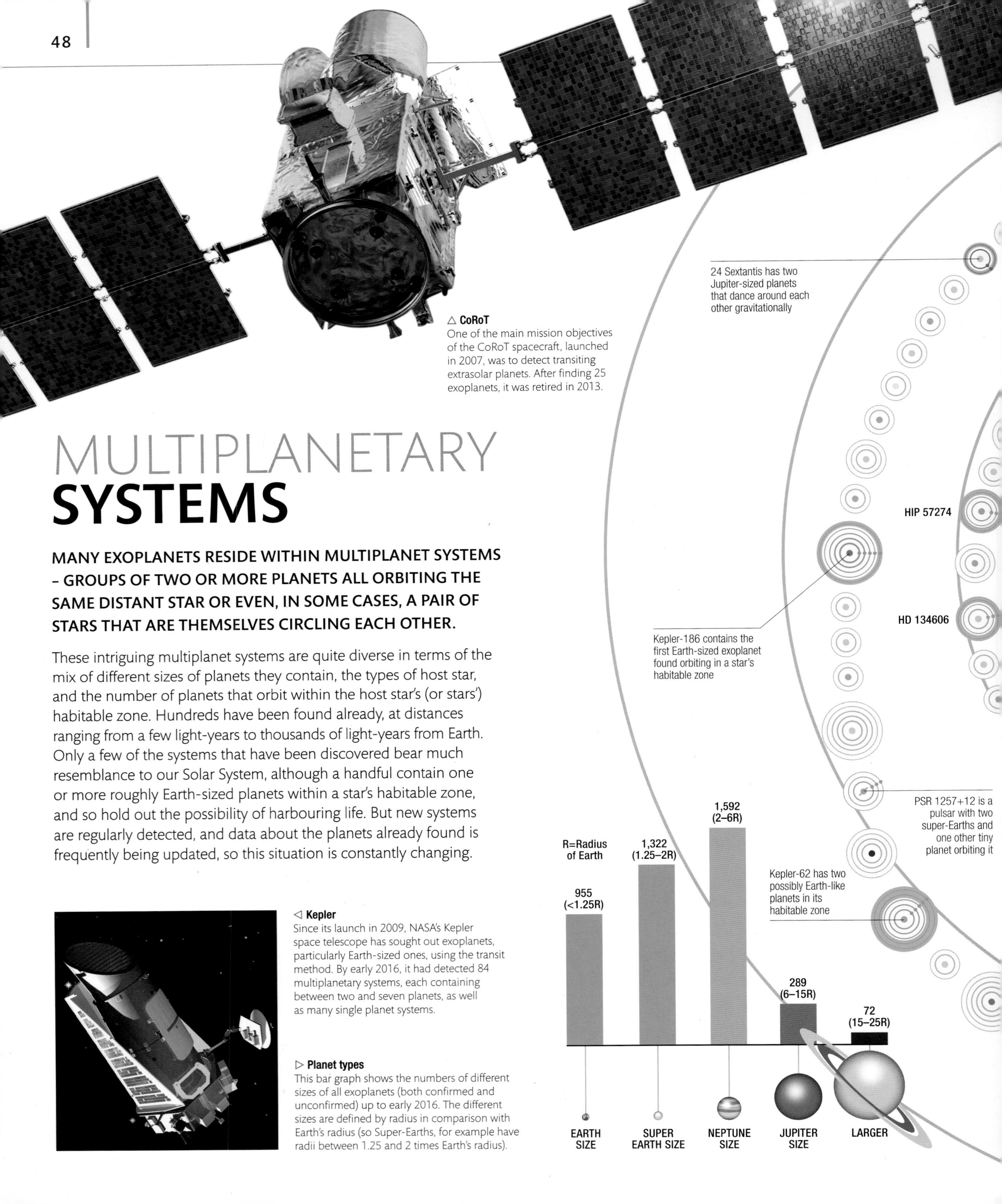
24 Sextantis has two Jupiter-sized planets that dance around each other gravitationally
HIP 57274
HD 134606
Kepler-186 contains the first Earth-sized exoplanet found orbiting in a star's habitable zone
PSR 1257+12 is a pulsar with two super-Earths and one other tiny planet orbiting it
Kepler-62 has two possibly Earth-like planets in its habitable zone
R=Radius of Earth
955 (<1.25R)
1,322 (1.25–2R)
1,592 (2–6R)
289 (6–15R)
72 (15–25R)
EARTH SIZE
SUPER EARTH SIZE
NEPTUNE SIZE
JUPITER SIZE
LARGER

△ **CoRoT**
One of the main mission objectives of the CoRoT spacecraft, launched in 2007, was to detect transiting extrasolar planets. After finding 25 exoplanets, it was retired in 2013.

MULTIPLANETARY **SYSTEMS**

MANY EXOPLANETS RESIDE WITHIN MULTIPLANET SYSTEMS – GROUPS OF TWO OR MORE PLANETS ALL ORBITING THE SAME DISTANT STAR OR EVEN, IN SOME CASES, A PAIR OF STARS THAT ARE THEMSELVES CIRCLING EACH OTHER.

These intriguing multiplanet systems are quite diverse in terms of the mix of different sizes of planets they contain, the types of host star, and the number of planets that orbit within the host star's (or stars') habitable zone. Hundreds have been found already, at distances ranging from a few light-years to thousands of light-years from Earth. Only a few of the systems that have been discovered bear much resemblance to our Solar System, although a handful contain one or more roughly Earth-sized planets within a star's habitable zone, and so hold out the possibility of harbouring life. But new systems are regularly detected, and data about the planets already found is frequently being updated, so this situation is constantly changing.

◁ **Kepler**
Since its launch in 2009, NASA's Kepler space telescope has sought out exoplanets, particularly Earth-sized ones, using the transit method. By early 2016, it had detected 84 multiplanetary systems, each containing between two and seven planets, as well as many single planet systems.

▷ **Planet types**
This bar graph shows the numbers of different sizes of all exoplanets (both confirmed and unconfirmed) up to early 2016. The different sizes are defined by radius in comparison with Earth's radius (so Super-Earths, for example have radii between 1.25 and 2 times Earth's radius).

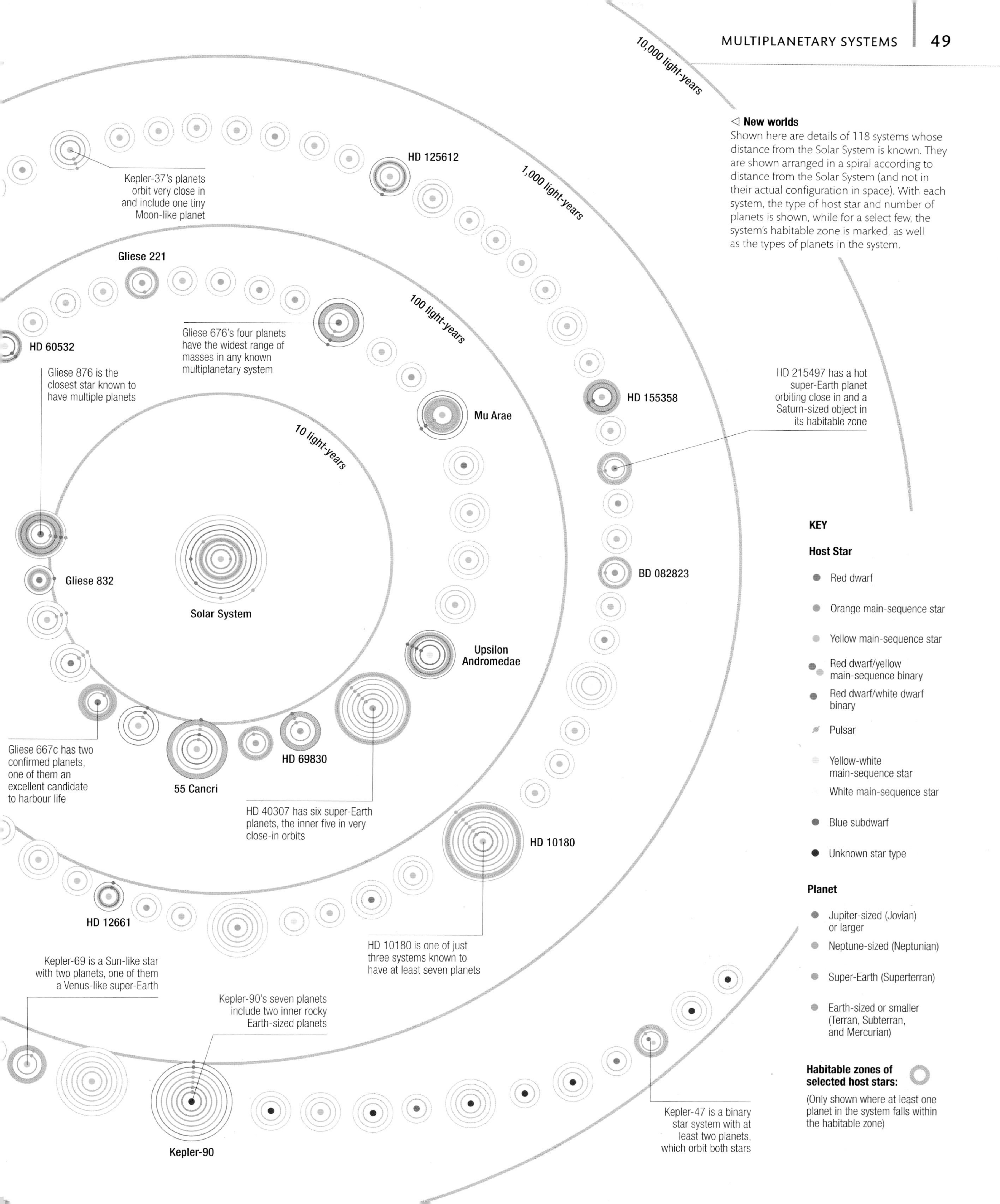
10,000 light-years
1,000 light-years
100 light-years
10 light-years
Kepler-37's planets orbit very close in and include one tiny Moon-like planet
HD 125612
Gliese 221
HD 60532
Gliese 676's four planets have the widest range of masses in any known multiplanetary system
Gliese 876 is the closest star known to have multiple planets
Mu Arae
HD 155358
HD 215497 has a hot super-Earth planet orbiting close in and a Saturn-sized object in its habitable zone
Gliese 832
Solar System
BD 082823
Upsilon Andromedae
Gliese 667c has two confirmed planets, one of them an excellent candidate to harbour life
55 Cancri
HD 69830
HD 40307 has six super-Earth planets, the inner five in very close-in orbits
HD 10180
HD 12661
HD 10180 is one of just three systems known to have at least seven planets
Kepler-69 is a Sun-like star with two planets, one of them a Venus-like super-Earth
Kepler-90's seven planets include two inner rocky Earth-sized planets
Kepler-90
Kepler-47 is a binary star system with at least two planets, which orbit both stars
KEY
Host Star
Red dwarf
Orange main-sequence star
Yellow main-sequence star
Red dwarf/yellow main-sequence binary
Red dwarf/white dwarf binary
Pulsar
Yellow-white main-sequence star
White main-sequence star
Blue subdwarf
Unknown star type
Planet
Jupiter-sized (Jovian) or larger
Neptune-sized (Neptunian)
Super-Earth (Superterran)
Earth-sized or smaller (Terran, Subterran, and Mercurian)
Habitable zones of selected host stars:
(Only shown where at least one planet in the system falls within the habitable zone)

◁ **New worlds**
Shown here are details of 118 systems whose distance from the Solar System is known. They are shown arranged in a spiral according to distance from the Solar System (and not in their actual configuration in space). With each system, the type of host star and number of planets is shown, while for a select few, the system's habitable zone is marked, as well as the types of planets in the system.

GALAXIES

GALAXIES ARE FOUND IN A HUGE VARIETY OF SHAPES AND SIZES, FROM COMPLEX SPIRALS LIKE OUR MILKY WAY TO HUGE BALLS OF ANCIENT RED AND YELLOW STARS, AND SHAPELESS CLOUDS OF GAS, DUST, AND NEW-BORN STARS.

Galaxies are the only places in the Universe where matter is densely packed enough for stars to form, and most stars spend their whole lives within them. Held together by gravity, most galaxies are thought to have a supermassive black hole at their centre.

Types of galaxies

American astornomer Edwin Hubble confirmed the existence of galaxies beyond the Milky Way in the 1920s. He subdivided them into several distinct types distinguished by a code of letters and numbers. Elliptical galaxies (types E0 to E7) are all roughly ball-shaped, but range from rounded spheres to elongated cigars. Today we know that they are dominated by old red and yellow stars. Spirals (types S and SB) are flattened disks with dense areas of star formation in the spiral arms, and older red and yellow stars in the centre. Lenticulars (type S0) have a central hub surrounded by a disk, but no spiral arms, while Irregulars (type Irr I and II) are fairly shapeless clouds rich in star-forming material.

△ **Elliptical galaxies**
Elliptical galaxies, suchs as M60 (shown here with the spiral galaxy NGC 4647), are ball-shaped star systems created by countless stars in overlapping elliptical orbits tilted at a wide range of angles. They have very little of the gas needed to support new star formation, which leaves them dominated by long-lived, low-mass red and yellow stars. They range in size from sparsely populated dwarfs to vast "giant ellipticals", which are the largest galaxies in the Universe.

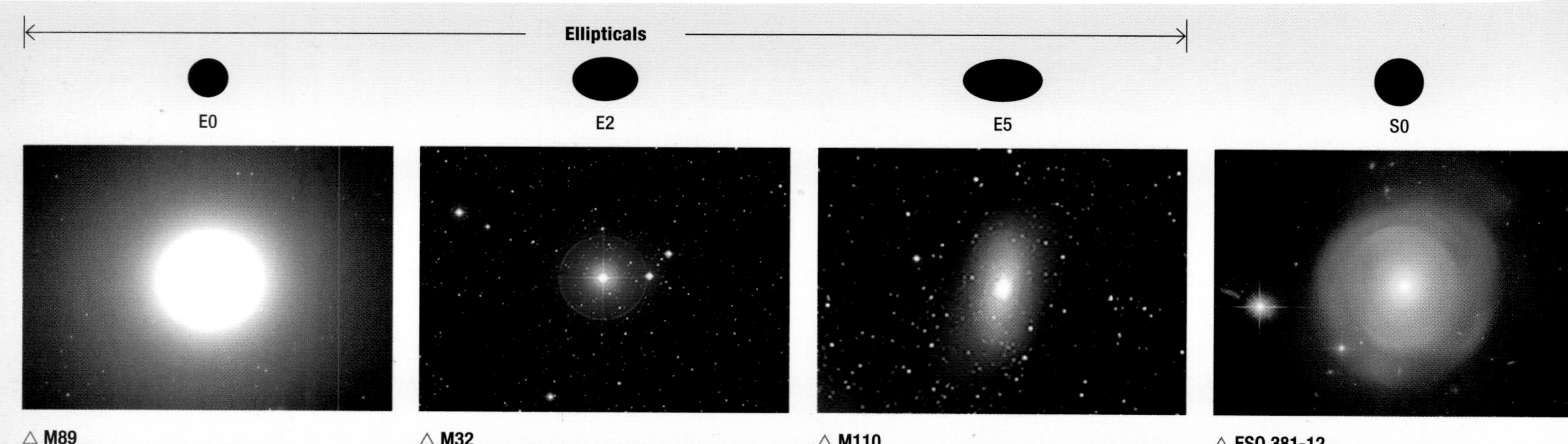

△ **M89**
E0 galaxies, such as M89 in Virgo, are almost perfect spheres of stars. They include the brightest and largest giant elliptical galaxies.

△ **M32**
E2 galaxies, such as the Andromeda Galaxy's satellite M32, have one axis noticeably longer than the other, and tend to be fainter than the brightest E0 galaxies.

△ **M110**
More elongated galaxies, such as M110, also a satellite of the Andromeda Galaxy, are actually somewhat disk-shaped. The orbits of their stars are flattened in one plane due to rotation.

△ **ESO 381-12**
Lenticular (S0) galaxies have a central hub and a flattened disk of stars similar to those in spiral galaxies, but there is little new star formation due to a lack of gas.

△ **Hubble's tuning fork**
Edwin Hubble arranged the various galaxy types in the shape of a musical tuning fork. He thought this illustrated the way that galaxies evolve over time, though the true story is rather more complex (see pp.62–63). Ellipticals are numbered according to their shape, with E0 the most round in shape. There are two distinct types of spiral galaxy – normal spirals (type Sc to Sc or Sd), in which the spiral arms emerge directly from the central hub, and barred spirals (SBa to SBc), in which the arms attach to the ends of a straight bar crossing the hub.

Astronomers think there are as **many galaxies in our Universe** as there are **stars in the Milky Way Galaxy**

Irregular galaxies

Irregular galaxies are relatively shapeless clouds of gas, dust, and stars. The best known examples are the Large and Small Magellanic Clouds, which are the Milky Way's brightest satellite galaxies. Irregulars are rich in the raw materials of star formation and are often undergoing intense bursts of starbirth that make them bright for their size. Larger irregular galaxies show signs of some internal structure, such as central bars or lone, poorly defined arms. Hubble classed these as Irr I galaxies, compared to the truly shapeless Irr II irregulars.

▷ **NGC 1427A**
Dwarf irregular galaxies are thought to play an important role in galaxy evolution. Their stars have relatively few heavy elements, and probably represent raw material left over from the early history of the Universe, which has only recently ignited into star formation. NGC 1427A, for example, is lit up by newborn bright stars whose birth was triggered by its plunge into the Fornax galaxy cluster.

△ **NGC 7217**
Spiral galaxies of type Sa, such as NGC 7217 in Pegasus, have a central hub of older stars surrounded by a disk of stars and gas. Concentrated waves of star formation create tightly wound arms.

△ **M91**
Type Sb spirals have less tightly wound spiral arms emerging directly from the hub. M91 in Coma Berenices has relatively faint spiral arms for a galaxy of this type.

△ **M74**
Sc spirals such as M74 in Pisces have more loosely wound arms but are generally as bright as types Sa and Sb. The loosest Type Sd spirals, however, are usually a lot fainter.

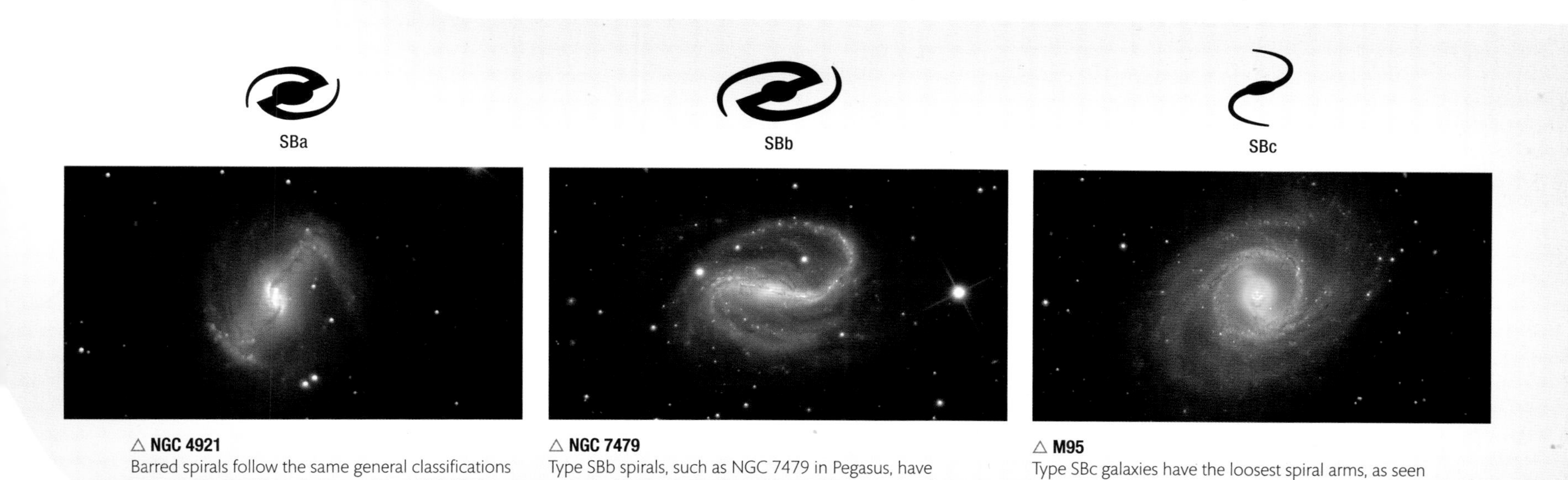

△ **NGC 4921**
Barred spirals follow the same general classifications as bar-less ones. SBa galaxies, such as NGC 4921 in Coma Berenices, have tightly wound spirals.

△ **NGC 7479**
Type SBb spirals, such as NGC 7479 in Pegasus, have a looser spiral structure but retain an obvious central bar emerging from either side of the nucleus.

△ **M95**
Type SBc galaxies have the loosest spiral arms, as seen in the beautiful M95, a barred spiral in some 38 million light-years away in Leo.

GALAXY TYPES

1 Spiral galaxy
At 21 million light-years away, the Pinwheel Galaxy (M101) is relatively close to Earth. It is also roughly 50 per cent bigger than the Milky Way, making it is one of the few galaxies in which individual regions can be studied. The Pinwheel has an extensive system of spiral arms. It appears lopsided, with the core, or nucleus, offset from the true centre, probably as a result interactions with other galaxies in the past.

2 Barred spiral galaxy
This galaxy, NGC 1300, is considered the prototype of the barred spiral galaxy. Instead of spiralling all the way to the central nucleus, the galaxy's two spiral arms are instead connected to each other by a straight bar of stars that includes the nucleus. This detailed Hubble Space Telescope image reveals that the nucleus has its own spiral structure. Gas in the bar is funnelled inwards before spiralling into the nucleus.

3 Elliptical galaxy
Apart from a simple ball shape, elliptical galaxies show little structure. Giant ellipticals such as IC 2006, shown here in an image taken in visible light by the Hubble Space Telescope, initially formed billions of years ago and are thought to have grown larger as they absorbed satellite galaxies. Due to its age, IC 2006 is made up of old, low-mass stars and there is no, or minimal, star formation activity.

4 Lenticular galaxy
This type of galaxy is named after its overall lens-like, or "lenticular", shape. NGC 2787 is one of the closest lenticular galaxies to Earth. This visible-light image shows tightly wound, almost concentric lanes of dust encircling the galaxy's bright nucleus. Several bright blobs of light can be seen on the edge of the galaxy. Each of these is, in fact, a cluster of several hundred thousand stars orbiting NGC 2787.

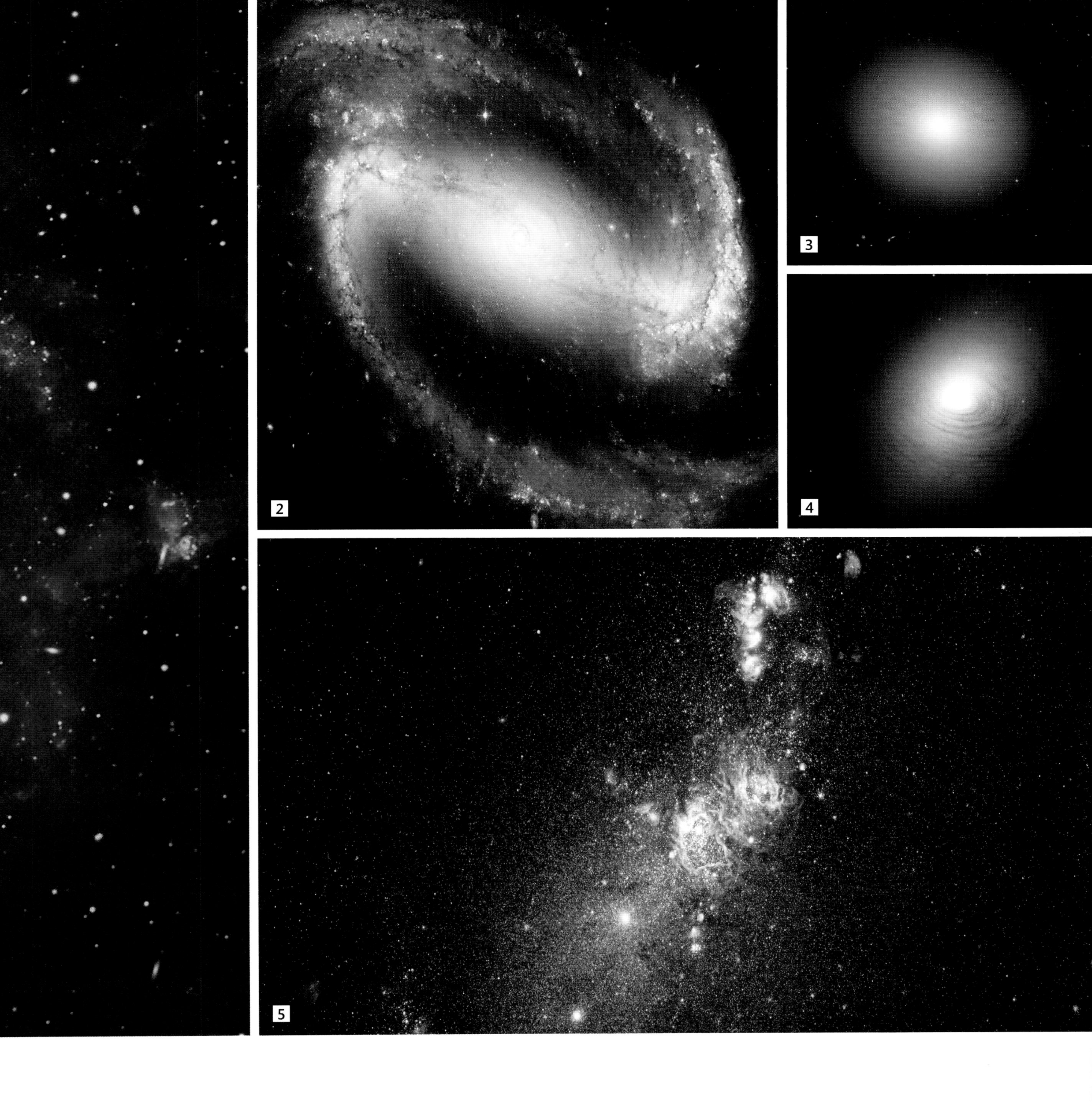

5 Irregular galaxy
Many irregular galaxies are what astronomers call "starburst galaxies", characterized by waves of new star formation. NGC 4214 is one such galaxy. Its abundant supply of hydrogen gas is fuelling the emergence of bright clusters of new stars, while the presence of older red stars provides evidence of earlier episodes of star formation.

The largest of all galaxies, known as the **giant ellipticals,** may each contain many **trillions of stars**

THE MILKY WAY

ALL THE STARS WE CAN SEE IN THE SKY LIE WITHIN THE CONFINES OF OUR HOME GALAXY, THE MILKY WAY. THIS VAST STAR SYSTEM, CONTAINING HUNDREDS OF BILLIONS STARS, IS A BARRED SPIRAL WITH A COMPLEX STRUCTURE AND ABOUT 120,000 LIGHT-YEARS ACROSS.

The Milky Way's visible stars form a disk centred on a bulging hub. Despite its vast diameter, the disk is, on average, only a thousand light-years deeps. From our point of view on Earth, we see many more stars looking across the plane of the disk than we do when we look "up" or "down" from the plane and out into intergalactic space. This is why we see our galaxy as a broad band whose countless faint and distant stars merge together in a milky band of light.

The central bulge of the Milky Way is dominated by low-mass, red and yellow stars with a high metallicity (see p.29), but the surrounding disk is filled with gas, dust, and younger stars. As with all spirals, stars are scattered across the disk, but the brightest are concentrated into the spiral arms. Stars orbit at different rates depending on their distance from the hub, so the arms cannot be permanent structures. Instead, they stand out because they are the active regions of star formation. Here, stars are born, and the most massive and luminous among them pass through their short life cycles before their orbits can carry them out into the wider disk.

Spiral arms

Astronomers have recently confirmed that the Milky Way is a barred spiral galaxy. Its central hub is crossed by a rectangular bar of stars some 27,000 light-years in length. Our galaxy's spiral arms are the result of stars, gas, and dust moving in and out of a spiral-shaped "traffic jam" called a density wave (see opposite).

The latest evidence suggests that the **Milky Way has four spiral arms** – two major and two minor – with distinct differences among their stars

Heart of the Milky Way
The central regions of our galaxy are hidden behind intervening star clouds and dust lanes, but X-rays and infrared can pierce the veil to reveal complex structures, giant star clusters, and an enormous black hole with the mass of several million Suns (embedded in the bright gas cloud to right of the image).

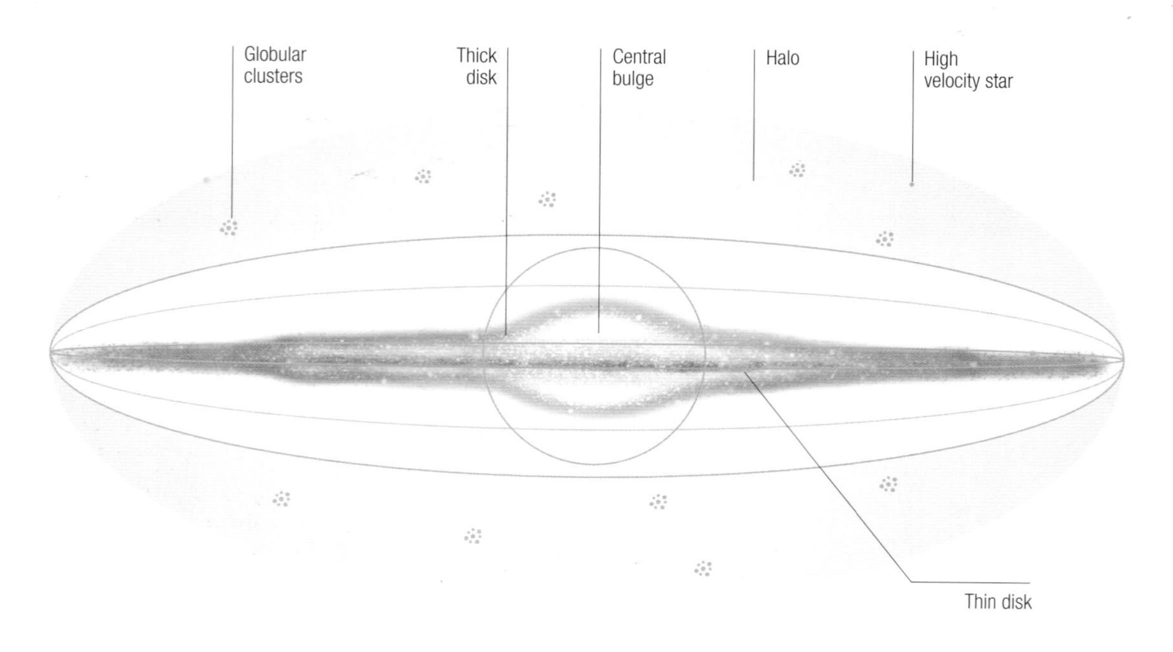

◁ **Cross-section of the Milky Way**
Seen from the side, the Milky Way consists of a disk of stars around a bulging hub some 8,000 light-years across. A broad halo region above and below the galaxy appears largely empty, but is home to globular star clusters, as well as stray high-velocity stars and hot gas ejected from the galactic plane.

FORMATION OF SPIRAL ARMS

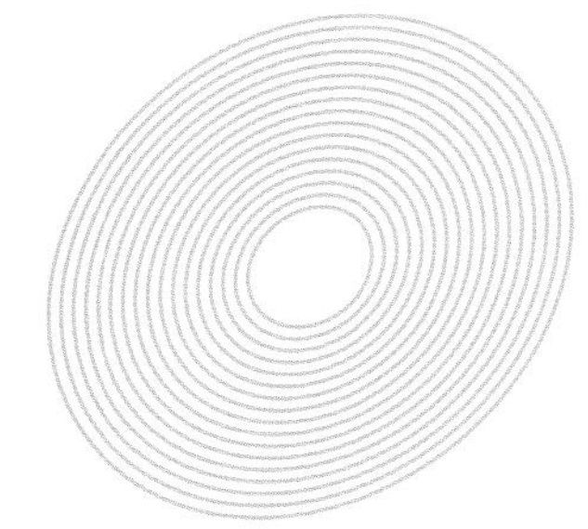

◁ **Perfectly ordered orbits**
In an ideal situation, objects in elliptical orbits around a galaxy's centre would have their longest axes in perfect alignment with each other. Objects naturally move more slowly at the outer edges of their orbit when they are farther from the centre.

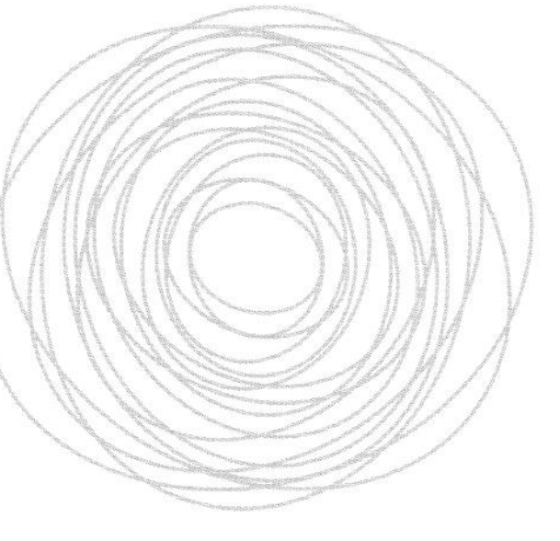

◁ **Chaotic orbits**
In a completely chaotic scenario, the orbits of objects within a galaxy would be aligned in a range of different directions, and no spiral structure would form. The example here shows the same number of orbits as the other two (left and right).

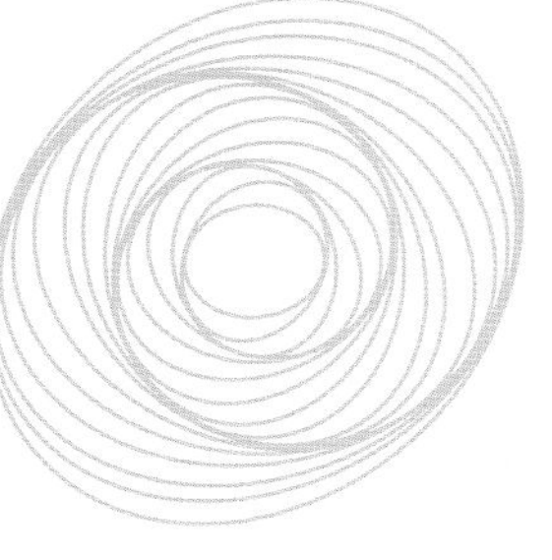

◁ **Density wave**
Spiral structure arises when orbits are pulled into alignments that offset slightly from one another (often due to tidal forces from another galaxy. As a result, orbits slow down and material packs together in spiral-shaped areas of higher density.

MILKY WAY IN THE SPOTLIGHT

The Solar System is part of a barred spiral galaxy called the Milky Way Galaxy. In this wide-angle view, the Milky Way's plane is seen as an arc above the antennae of the Atacama Large Millimeter/submillimeter Array (ALMA) on the Chajnantor plateau in Chile. It glows with the light of a mass of distant stars interspersed with dusty nebulae and patches of glowing gas, where new stars are being born to join the existing billions that make up our galaxy.

ALMA's 66 dishes are either 12m (39ft) or 7m (23ft) in diameter and observe the sky at wavelengths between the infrared and radio parts of the spectrum. ALMA sits at high altitude – 5,000m (16,400ft) above sea level – in a very dry region where the air contains hardly any water vapour to absorb the radiation. The thinness of the atmosphere above it and the low interference from other radio signals also make the plateau ideal for observing at these wavelengths.

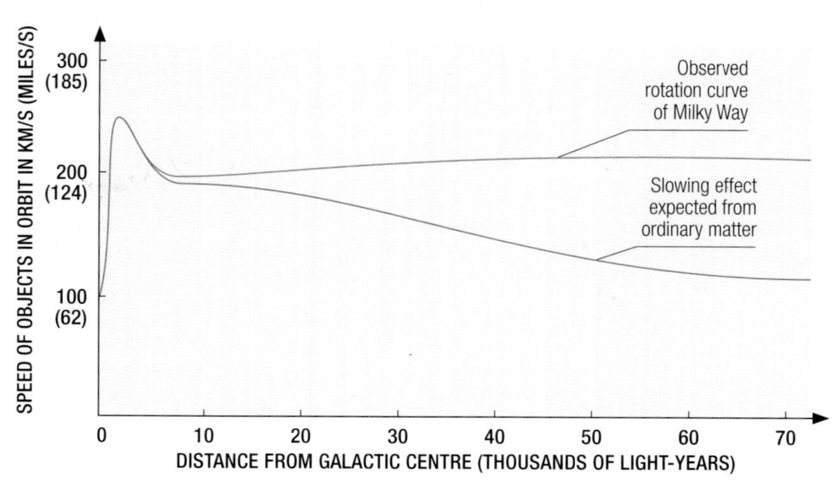

◁ **Rotation curves**
Because the Milky Way is not a solid object, stars and other objects orbit the centre at different speeds. If the distribution of mass matched the concentration of visible objects, then we might expect stellar speeds to fall off with distance like those of planets in the Solar System. In fact, they trail off much more slowly, which is an indication of dark matter lying beyond the visible disk.

THE MILKY WAY FROM ABOVE

OUR GALAXY IS A VAST DISK OF STARS ABOUT 120,000 LIGHT-YEARS ACROSS. THE VAST MAJORITY OF STARS WE SEE IN OUR SKIES, HOWEVER, ARE CONFINED TO A MUCH SMALLER NEIGHBOURHOOD AROUND OUR SOLAR SYSTEM.

The Milky Way contains between 100 and 400 billion stars, most of which are dwarfs with only a fraction of the Sun's mass. Nevertheless, the Galaxy's overall mass is between 1 and 4 trillion solar masses – far more than the combined mass of its stars. While much of the extra mass is accounted for by dust and gas in the galactic disk, all this normal matter is vastly outweighed by so-called dark matter (see pp.74–75).

Studies suggest that there are at least 100 billion planets orbiting the stars, and perhaps many more. Most of the visible material is in a central bulge and a flattened disk just 1,000 light-years deep, with stray stars, globular clusters, and large amounts of dark matter in the halo region.

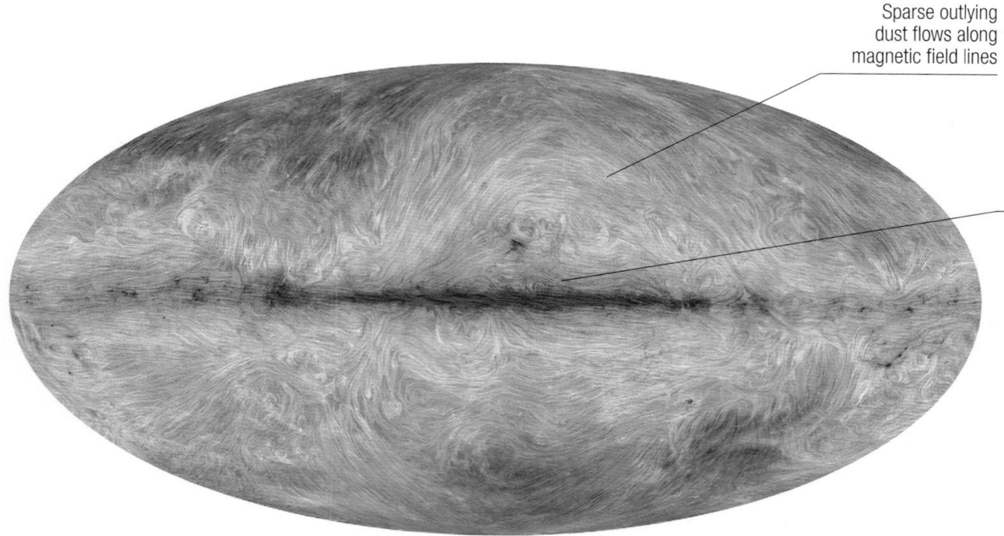

◁ **Dust and magnetism**
Although dust makes up a relatively small proportion of our galaxy's composition, it plays an important role in the formation of stars and planets. What's more, its particles tend to align in relation to local magnetic fields. Microwave emission from dust particles can therefore be used as a way of revealing the Milky Way's overall magnetic field, as shown on this map from ESA's Planck satellite.

THOUSAND LIGHT-YEARS
40
30
20
OUTER ARM
PERSEUS ARM

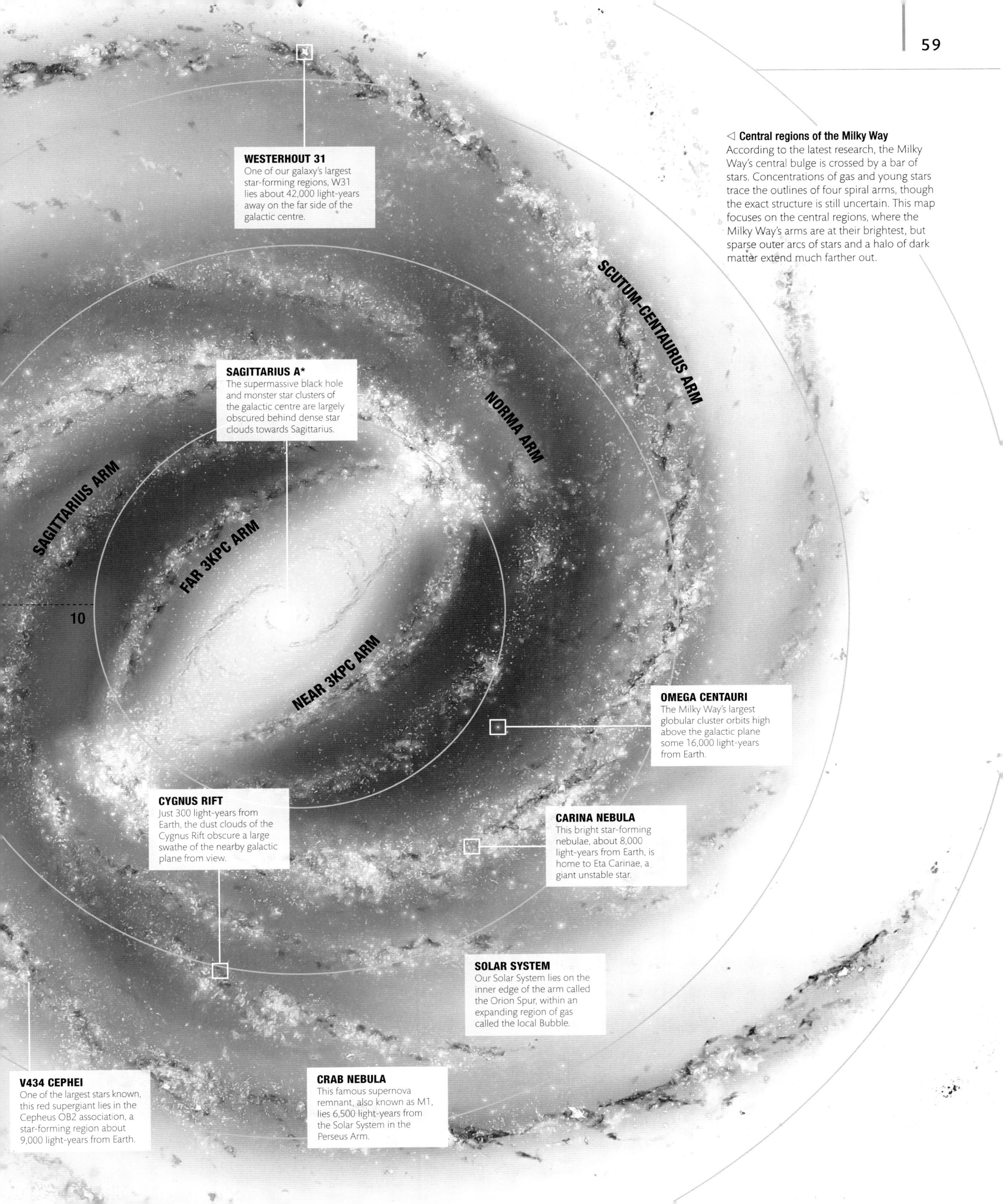

◁ **Central regions of the Milky Way**
According to the latest research, the Milky Way's central bulge is crossed by a bar of stars. Concentrations of gas and young stars trace the outlines of four spiral arms, though the exact structure is still uncertain. This map focuses on the central regions, where the Milky Way's arms are at their brightest, but sparse outer arcs of stars and a halo of dark matter extend much farther out.

ACTIVE **GALAXIES**

MANY GALAXIES SHOW SIGNS OF ENERGY OUTPUT FROM THEIR CENTRAL REGIONS THAT CANNOT BE EXPLAINED BY STARS ALONE. THESE ACTIVE GALAXIES APPEAR VARIED, BUT CAN ALL BE EXPLAINED BY THE SAME MECHANISM.

Active galaxies generate vast amounts of energy – not only visible light but also as radio waves, X-rays, ultraviolet radiation, and gamma rays – from regions near their centre. These galaxies are divided into four major types: Seyfert galaxies, radio galaxies, quasars, and blazars. Seyfert galaxies are otherwise normal-looking spiral galaxies with a bright and concentrated source of radiation embedded in their centre, and unusual overall energy output that cannot be explained through starlight alone. Radio galaxies, in contrast, are distinguished by two large clouds of radio-emitting gas on either side of a central galaxy (in some cases, narrow jets can be seen linking them to the heart of the galaxy). Quasars are very distant galaxies with an intense star-like source of light, far brighter than a Seyfert galaxy, at their centre. They are also sources of radio waves, and vary in brightness over hours or days. Finally, blazars are broadly similar to quasars, but have distinctive differences in their radiation that mark them out.

Astronomers think that all these different types of activity are actually caused by the same kind of object – an engine called an active galactic nucleus (AGN). The speed at which AGNs change their energy output means that they can be no larger than our Solar System, and the only object capable of releasing such vast amounts of energy in such a small region of space is a superheated "accretion disk" of matter falling into a supermassive black hole. The amount of material falling into the black hole, and the AGN's alignment as seen from our point of view on Earth, determine which type of active galaxy we see.

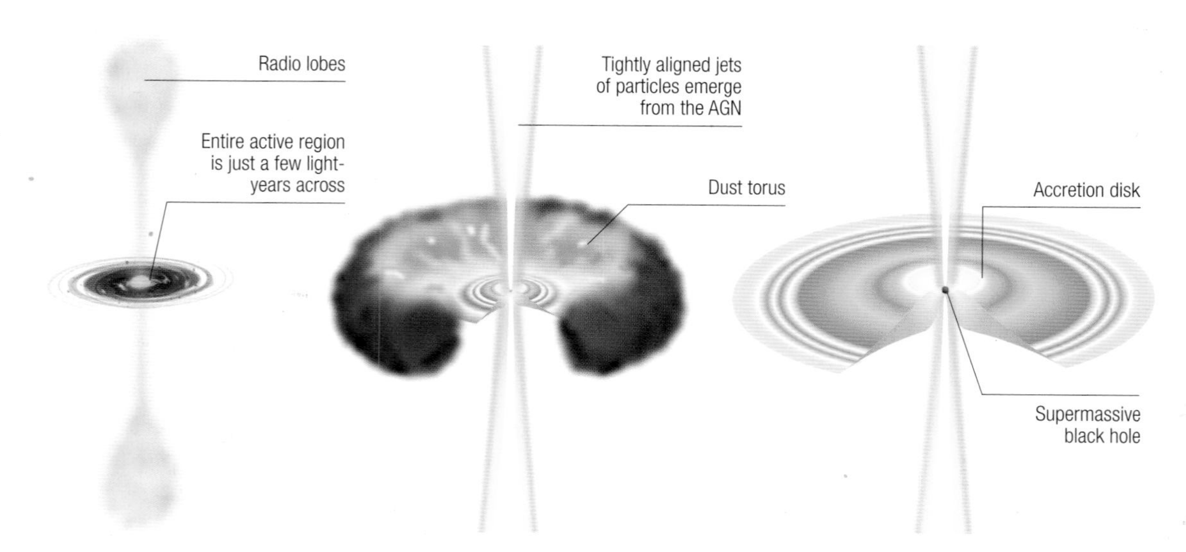

△ **Whole galaxy**
From a distance, an active galaxy may be surrounded by two huge lobes of radio emission, created as jets of particles ejected from the central disk encounter gas in intergalactic space. When the central AGN is visible, its light output can dwarf that of the surrounding galaxy.

△ **Dust torus**
The active galaxy's central regions are surrounded by a thick doughnut-shaped ring or torus of light-obscuring gas and dust. If we see this ring edge-on, then the surrounding radio lobes are the only visible sign of unusual activity, manifesting as a radio galaxy.

△ **Nucleus**
At the heart of the active galactic nucleus is a supermassive black hole. With the gas of a million or more Suns, it pulls material to its doom, heating it to millions of degrees where it emits intense radiation. If the particle jet points directly towards Earth, the AGN creates a blazar.

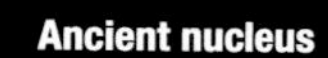

Ancient nucleus
Light from quasars is redshifted by huge amounts, revealing, due to the expansion of the Universe (see pp.70–71), that they are billions of light-years away and incredibly bright. As a result we are seeing them during a much earlier stage of cosmic evolution. Astronomers suspect that most galaxies go through a quasar phase early in their history.

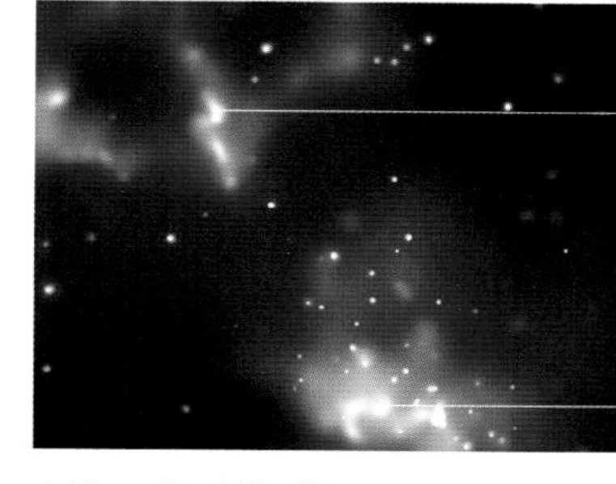

△ The active Milky Way
Our own galaxy's supermassive black hole has long ago swept up the material from its immediate surroundings and become dormant, but objects such as stray asteroids still occasionally wander into its grasp. When this happens, matter is violently torn apart, resulting in intense bursts of radiation.

LOCAL GROUP COLLISION

The Milky Way is a member of a small galaxy cluster called the Local Group, alongside the spiral Andromeda Galaxy (M32), the smaller Triangulum spiral (M33), the Large and Small Magellanic Clouds, and several dozen dwarf galaxies of various types. Andromeda and the Milky Way are by far the heaviest galaxies in the group, and are being pulled together by gravity with ever increasing speed. Approaching at 110 km per second (68 miles per second), the two galaxies are doomed to a head-on collision in about 4 billion years, when the night skies of a future Earth may bear witness to the astonishing scene shown in this artist's impression. Collisions between stars will be rare, but colliding gas clouds will trigger waves of new star formation, and as the mass of the two galaxies becomes concentrated at the centre, the galaxies will form a single giant system, sometimes nicknamed "Milkomeda".

COLLIDING **GALAXIES**

THE RELATIVELY SHORT DISTANCES BETWEEN SOME GALAXIES COMPARED TO THEIR SIZE MAKES COLLISIONS FAIRLY COMMON. THESE SPECTACULAR EVENTS TRIGGER HUGE WAVES OF STAR FORMATION AND PLAY A KEY ROLE IN THE STORY OF GALAXY EVOLUTION.

When galaxies collide, the stars within them are so widely spaced that they rarely hit one another. However, huge clouds of star-forming gas smash together in head-on collisions that compress and trigger vast new waves of star formation. The powerful gravity of the coalescing gas pulls the stars back towards the centre, causing the colliding galaxies to merge over hundreds of millions of years.

Galaxy evolution

There is evidence that galaxies change from one type to another over time – irregular galaxies were far more common in the early Universe, spirals dominate today, and elliptical galaxies are most common in the heart of galaxy clusters (see pp.66–67). As a result, astronomers think that collisions play a key role in galaxy evolution: gas-rich spirals initially formed from smaller irregular galaxies, while ellipticals are created by collisions between spirals, which send their stars into chaotic orbits, trigger vast waves of new star formation, and ultimately drive gas away into intergalactic space, where it can no longer form new stars.

Depending on the energy of the collision and the nearby environment, the merged galaxy's gravity may be able to pull back sufficient material from its surroundings to generate a new disk, and eventually to restart star formation, creating a new spiral that will eventually become part of another merger.

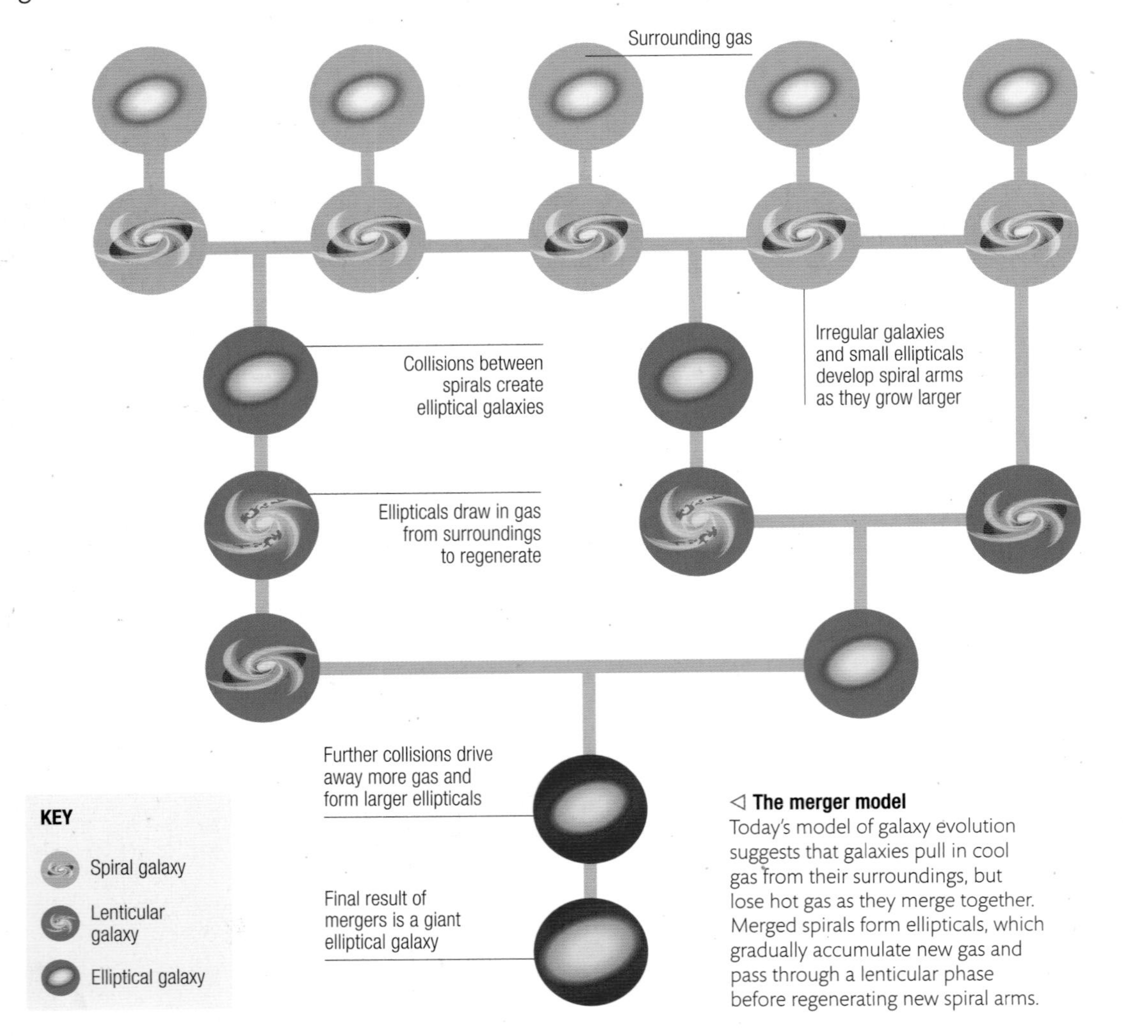
Surrounding gas
Collisions between spirals create elliptical galaxies
Irregular galaxies and small ellipticals develop spiral arms as they grow larger
Ellipticals draw in gas from surroundings to regenerate
Further collisions drive away more gas and form larger ellipticals
Final result of mergers is a giant elliptical galaxy
KEY
Spiral galaxy
Lenticular galaxy
Elliptical galaxy

◁ **The merger model**
Today's model of galaxy evolution suggests that galaxies pull in cool gas from their surroundings, but lose hot gas as they merge together. Merged spirals form ellipticals, which gradually accumulate new gas and pass through a lenticular phase before regenerating new spiral arms.

△ **Close encounter**
Near misses between galaxies are even more common than direct collisions. During these events, tidal forces are created that strengthen features such as spiral arms.

△ **Unwinding arms**
As galaxies come together, the orbits of stars within them are disrupted. Spiral arms unwind and their individual stars are scattered into intergalactic space.

△ **Starburst**
Most stars end up in chaotic orbits while gas is rammed together in huge star-forming clouds. Anchored by supermassive black holes, the galactic cores merge together.

△ **Elliptical ending**
The heating effects of the collision drive gas away from the galaxy, choking off the burst of star formation and leaving an elliptical system dominated by fainter, longer-lived stars.

Colliding galaxies
This pair of interacting galaxies, collectively named Arp 273, is around 300 million light-years from Earth in Andromeda. Arp 273 reveals the early stages of a galactic merger. One of the larger galaxy's spiral arms is already unwinding into space, while the smaller galaxy is undergoing an intense burst of star formation, creating "super star clusters" that will evolve over time into globular clusters.

GALAXY CLUSTERS AND SUPERCLUSTERS

MOST GALAXIES ARE FOUND IN CLUSTERS OF ANYTHING FROM A FEW DOZEN TO A THOUSAND OR MORE GALAXIES GROUPED TOGETHER BY GRAVITY. CLUSTERS BLEND AT THE EDGES TO FORM SUPERCLUSTERS, CREATING WEB-LIKE FILAMENTS AROUND APPARENTLY EMPTY SPACE.

Galaxy clusters are the largest structures in the Universe that are created by the force of gravity, pulling galaxies together over millions of light-years of space to form huge concentrations of mass and matter. Because of this, they tend to fill a fairly similar volume of space (10–20 million light-years across) regardless of how many galaxies they contain. Superclusters and even bigger structures reflect the large-scale distribution of matter caused by the Big Bang itself (see pp.70–71).

Types of clusters

The Milky Way Galaxy belongs to a low-density cluster called the Local Group – it is one of three large spirals surrounded by 50 or so smaller galaxies, most of which are tiny and faint dwarf systems. Most small clusters seem to follow this pattern and are dominated by spirals and irregular galaxies. However, clusters containing large numbers of galaxies are very different, tending to be dominated by elliptical galaxies full of red and yellow stars. This is an important clue that ellipticals are created by the collision and merger of other types of galaxies within dense clusters.

Local Group (Milky Way)
Fornax Cluster
Eridanus Cluster

◁ **Coma cluster**
Roughly 320 million light-years from Earth, the Coma Cluster is a group of about 1,000 galaxies, most of which are elliptical or lenticular. This infrared image reveals large numbers of dwarf galaxies, too faint to detect in visible light.

◁ **Lensing effect**
The distorted images caused by lensing allow astronomers to use gravity as a natural telescope for spotting faint and distant objects. They can also be used to work out how mass is distributed within the lensing cluster.

Actual location of distant galaxy
Light passing near galaxy cluster is bent back towards Earth
Path of light without lensing effect
Apparent direction and distorted shape of galaxy as seen from Earth
Light spreads out from galaxy in all directions
Light reaches Earth from different directions

△ **Gravitational lensing**
The huge concentration of mass in galaxy clusters gives rise to an effect known as gravitational lensing. Large masses alter the shape of space near to them (see p.73). Light passing through a massive galaxy cluster changes direction, so galaxies beyond the cluster look distorted and sometime magnified.

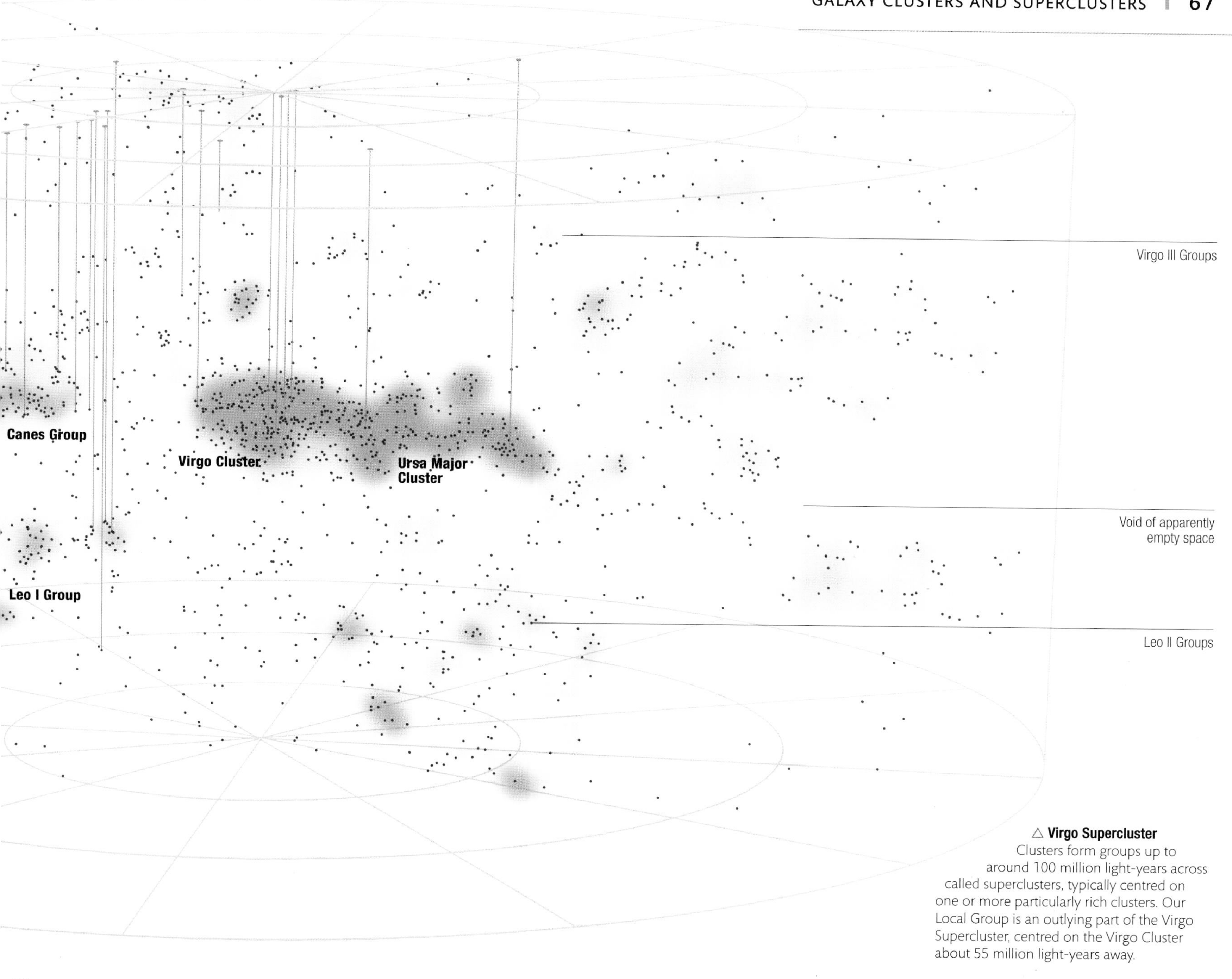

△ **Virgo Supercluster**
Clusters form groups up to around 100 million light-years across called superclusters, typically centred on one or more particularly rich clusters. Our Local Group is an outlying part of the Virgo Supercluster, centred on the Virgo Cluster about 55 million light-years away.

Cluster gas

Images of dense galaxy clusters taken by orbiting X-ray telescopes reveal that much of the apparently empty space between galaxies in most clusters is filled with superheated gas at temperatures of 10 million degrees or more. This hot gas is thought to originate in the cluster's individual galaxies, and to escape when it is heated up during collisions between cluster members. Hotter gas particles move faster and find it easier to escape from the gravity of their original galaxies, though not from the cluster as a whole. X-ray gas tends to accumulate at the centre of a cluster over time, and is richest in the densest galaxy clusters, where it may weigh twenty times more than all the visible galaxies put together.

▷ **Evolving cluster**
This image of galaxy cluster IDCS J1426 combines visible, infrared, and X-ray views (in green, red, and blue respectively). It shows how the X-ray gas has largely disconnected from the visible galaxies, except where it concentrates around a recent galaxy collision.

GALAXY CLUSTERS

1 Virgo cluster
The closest major galaxy cluster to Earth, the Virgo Cluster contains perhaps 2,000 galaxies scattered over the constellations Virgo and Coma Berenices. Its larger galaxies are a mix of spirals and ellipticals (with the latter formed by collisions between the former). Its largest member is the giant elliptical galaxy M87, which lies at the centre of the cluster, about 53 million light-years from Earth.

2 Abell 383
The dense cluster Abell 383 lies about 2.5 billion light-years away and is a powerful source of X-rays, thanks to a vast cloud of superheated gas stripped away from its individual galaxies. The cluster's enormous mass allows it to act as a gravitational lens, bending the space around it and deflecting the path of light rays from more distant galaxies to produce distorted arcs of light.

3 Stephan's Quintet
This group of five galaxies in Pegasus is deceptive. While four of its galaxies form a tight physical group some 290 million light-years from Earth, the blue spiral at upper left is actually a much nearer foreground object. The other four members – three spirals and an elliptical – will almost certainly merge into a single giant elliptical galaxy in the next billion years or so.

4 MOO J1142+1527
At a distance of 8.5 billion light-years from Earth, this monster cluster is so distant that its light has been redshifted (see p.72) almost to invisibility. As a result it was only discovered in 2015 when observations from two separate infrared space telescopes were combined. With a similar mass to El Gordo (see right), it may be one of just a handful of giant clusters that formed in the first few billion years of cosmic history.

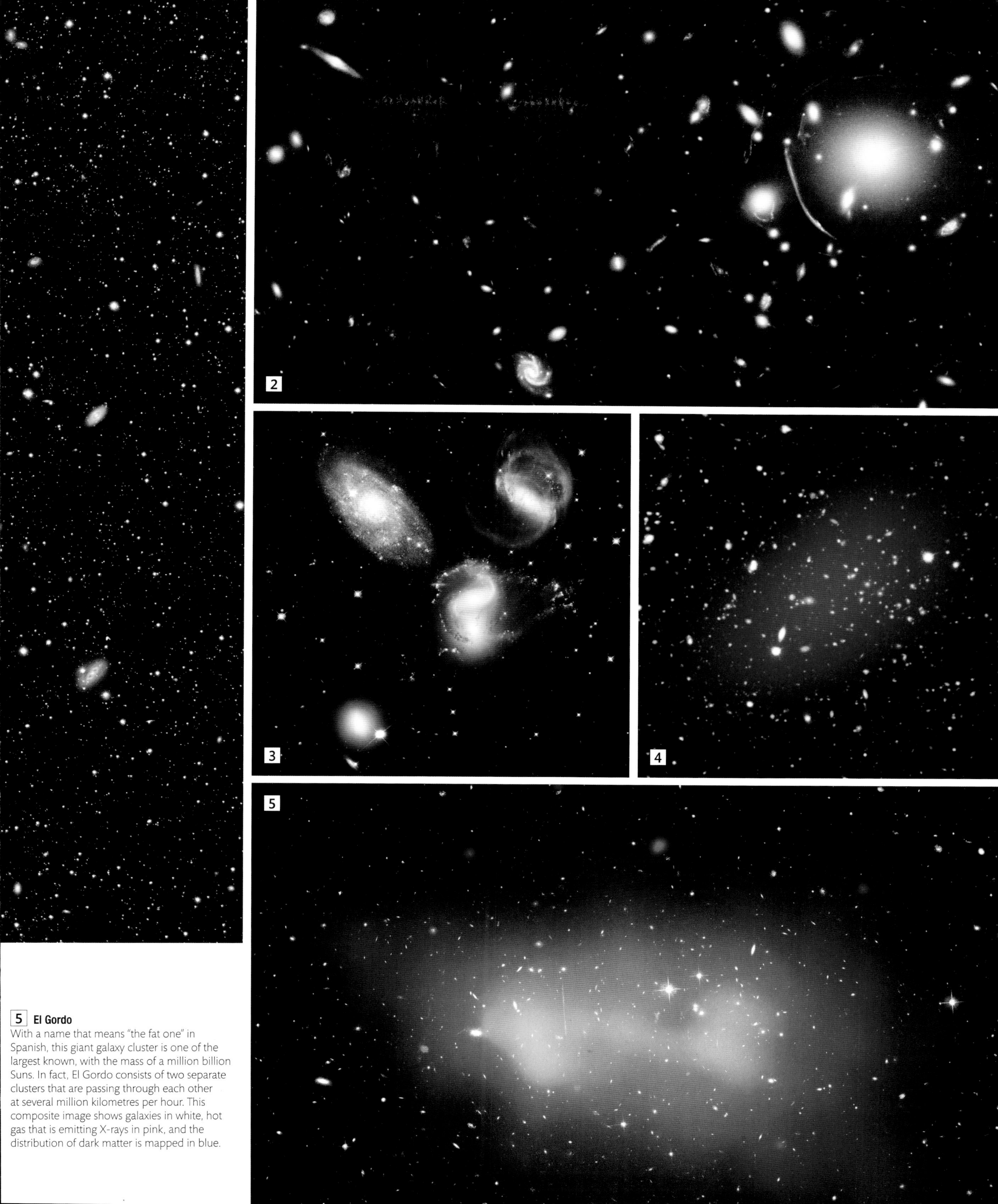

5 **El Gordo**
With a name that means "the fat one" in Spanish, this giant galaxy cluster is one of the largest known, with the mass of a million billion Suns. In fact, El Gordo consists of two separate clusters that are passing through each other at several million kilometres per hour. This composite image shows galaxies in white, hot gas that is emitting X-rays in pink, and the distribution of dark matter is mapped in blue.

THE EXPANDING UNIVERSE

AS A GENERAL RULE, THE FARTHER A GALAXY IS FROM EARTH, THE FASTER IT IS MOVING AWAY FROM US. THIS IS VITAL EVIDENCE THAT THE UNIVERSE AS A WHOLE IS STILL EXPANDING. AND WHAT'S MORE, THAT EXPANSION IS ACCELERATING.

The key evidence for the expansion of the cosmos comes from the Doppler effect (a way of measuring the speed at which any light-emitting object is moving towards or away from Earth). This reveals that more distant galaxies are moving away from Earth more rapidly – a discovery best explained by the idea that space as a whole is expanding and carrying galaxies away from one another – rather like currants in a cake moving apart as the batter rises in the oven.

The Doppler effect and redshift

The Doppler effect is a shift in the wavelength and frequency of waves, such as sound or light, moving past an observer. In everyday life, we experience Doppler shift when an emergency vehicle's siren speeds past: waves move past us more rapidly and the pitch is higher as it moves towards us, but they reach us more slowly as it retreats and so the pitch drops.

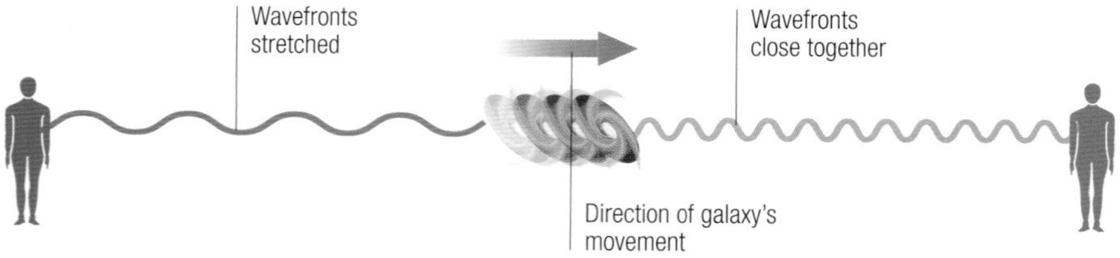

The Universe is expanding at about **20km (12 miles) per second** every million light-years

Wavefronts stretched
Wavefronts close together
Direction of galaxy's movement

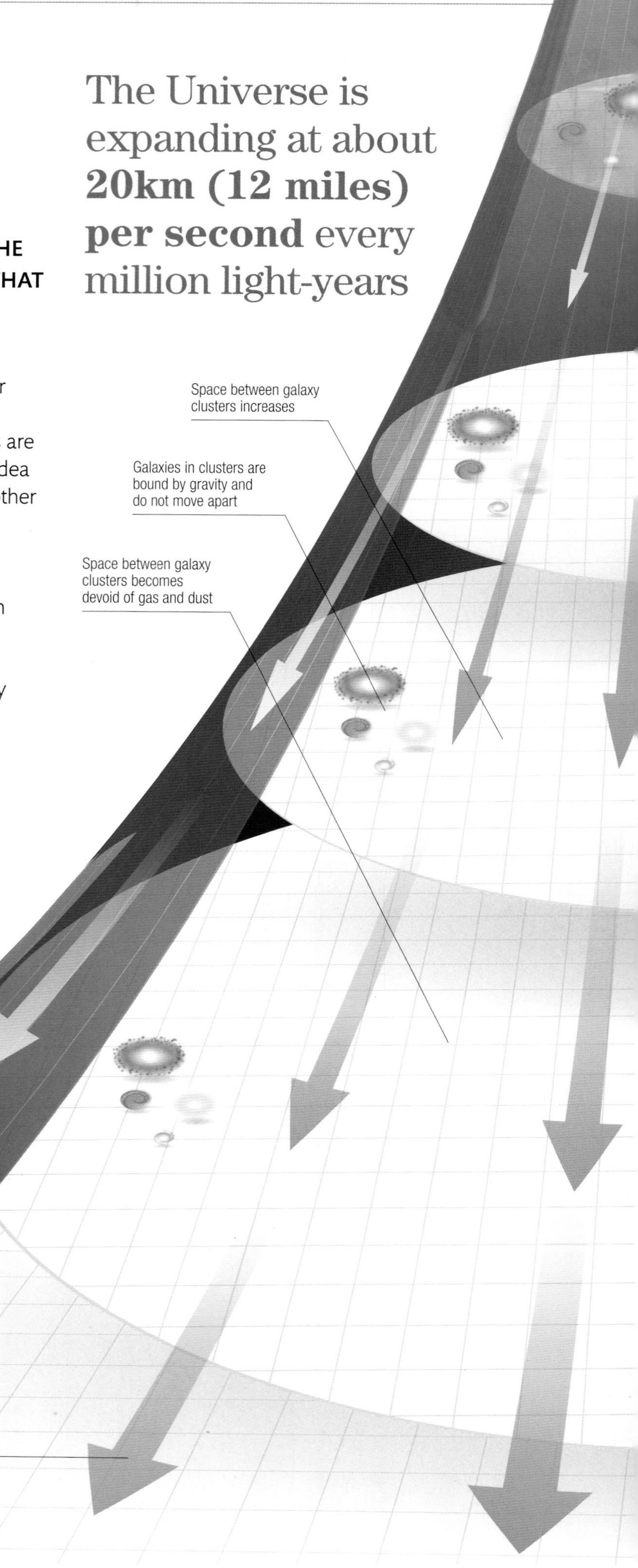

△ **Redshift and blueshift**
Light from galaxies moving away from us has its wavelength stretched and therefore appears redder. When nearby galaxies move towards us their wavelengths are compressed and they appear bluer. These shifts can be precisely measured from the way they affect spectral lines in a galaxy's light.

Space between galaxy clusters increases
Galaxies in clusters are bound by gravity and do not move apart
Space between galaxy clusters becomes devoid of gas and dust
Universe continues to expand

▷ **Cosmic expansion**
The expansion of the Universe is not simply a case of galaxies moving apart in space – most of it is caused by space itself expanding. This is a result of the Big Bang explosion that created not only matter but space and time themselves.

Lookback time

Although light is the fastest thing in the Universe, its speed is still limited, so that it crosses about 9.5 million million kilometres (5.9 trillion miles) in a year. Coupled with cosmic expansion, this transforms our telescopes into time machines. The farther we look across space, the longer light has taken to reach us, and the farther back in time we are looking. Hence the most distant galaxies appear the most ancient and primitive.

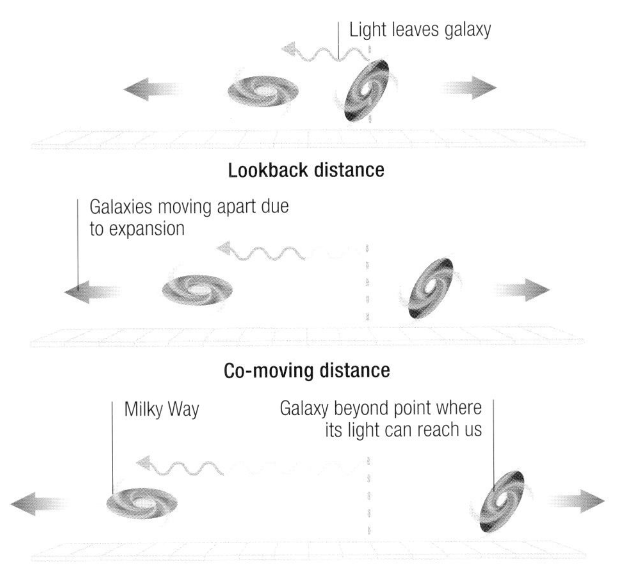

◁ **Stretching space**
On cosmic scales, most of the redshift in distant galaxies arises not from pure Doppler shift, but also from the way that light has stretched in its passage across expanding space.

◁ **Receding galaxy**
Although light may take a certain time to pass between distant galaxies, by the time it arrives the two galaxies may be much farther apart. The true separation of galaxies is called their co-moving distance.

◁ **Shifted to invisibility**
The most distant objects of all – the very first stars and galaxies formed in the early days of the Universe – have undergone extreme redshift, which renders them invisible to even the most advanced telescopes.

Mapping from redshift

Because more distant galaxies show larger redshifts (an effect called Hubble's Law), redshift itself can be used as a way of estimating distance for large numbers of galaxies that are too distant to measure in other ways. Mapping the redshifts of galaxies in different parts of the sky shows how galaxy superclusters form a web of elongated chains and sheets, known as filaments, around vast and apparently empty regions, called voids.

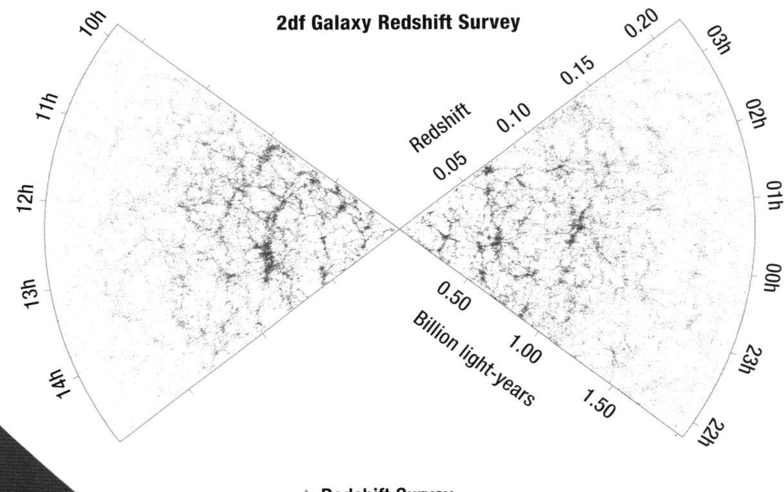

△ **Redshift Survey**
This map from the 2dF Galaxy Redshift Survey plots the redshifts of more than 200,000 galaxies across two broad swathes of sky, revealing structures that are hundreds of millions of light-years across. These structures are far too large to have formed by gravitational effects since the Big Bang. Instead they are thought to have been ingrained in the initial distribution of matter from which everything formed.

SIZE AND STRUCTURE OF THE UNIVERSE

THE EXTENT OF THE UNIVERSE WE CAN SEE AROUND US IS LIMITED BY THE SPEED OF LIGHT AND THE RATE OF COSMIC EXPANSION, BUT THE UNIVERSE AS A WHOLE GOES FAR BEYOND THESE LIMITS.

The Big Bang explosion that created the Universe some 13.8 billion years ago produced not just matter, but also space and time. As a result, there is a limit to how far we can theoretically see across space because we can only see regions whose light has had time to reach us. This places us at the centre of a spherical bubble called our observable Universe, but space itself extends far beyond this boundary. In fact, every location in the cosmos is at the centre of an observable Universe of its own.

Cosmic microwave background radiation

Immediately after the Big Bang, the Universe was an opaque, expanding fireball from which no light could escape. As a result, the most distant region of the cosmos that we can actually see corresponds to the period about 380,000 years later, when the "fog" of the early Universe cleared. Light from the edge of the fireball can still be seen if we look far enough in any direction, but during its 13.8-billion-year journey to reach us it has become redshifted (see p.70) so that the fireball's light now makes the whole sky glow at microwave wavelengths.

Observable Universe

Our observations of the Universe are limited to those objects whose light has had time to reach us over the past 13.8 billion years. However, thanks to cosmic expansion (see pp.70–71), the farthest regions of the Universe are moving away from us at the speed of light itself. As a result, light from regions beyond the observable Universe will never be able to reach Earth, no matter how much time passes.

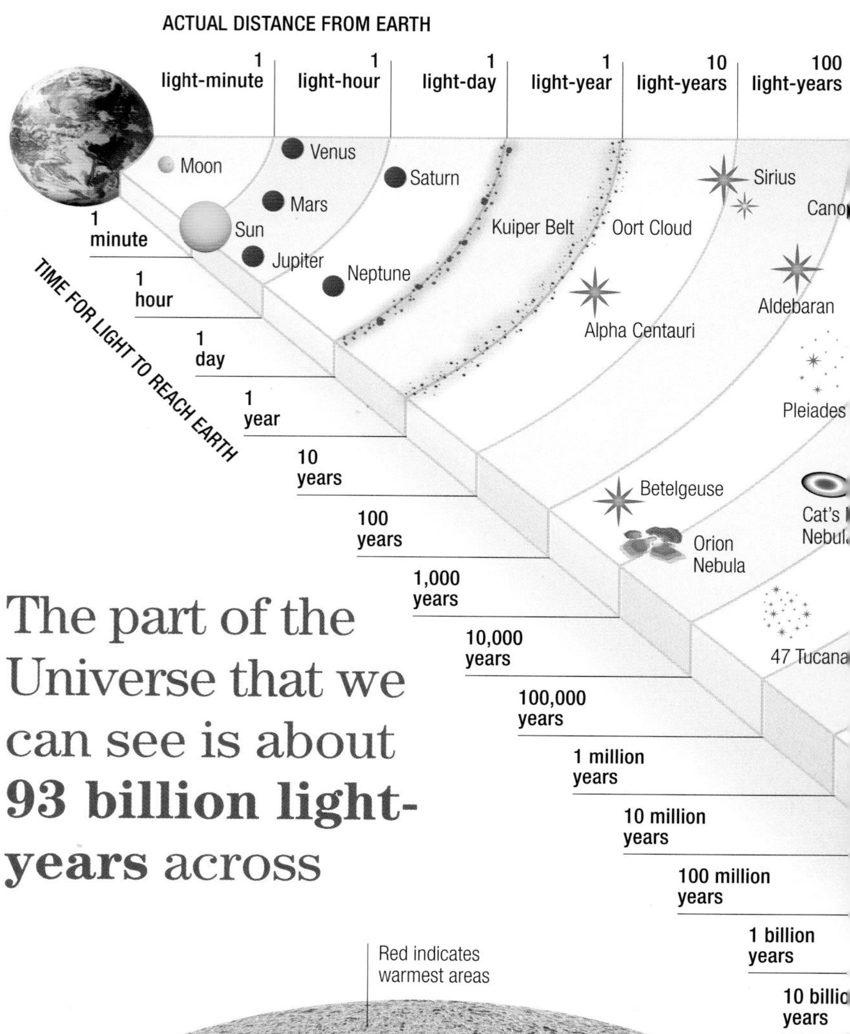

The part of the Universe that we can see is about 93 billion light-years across

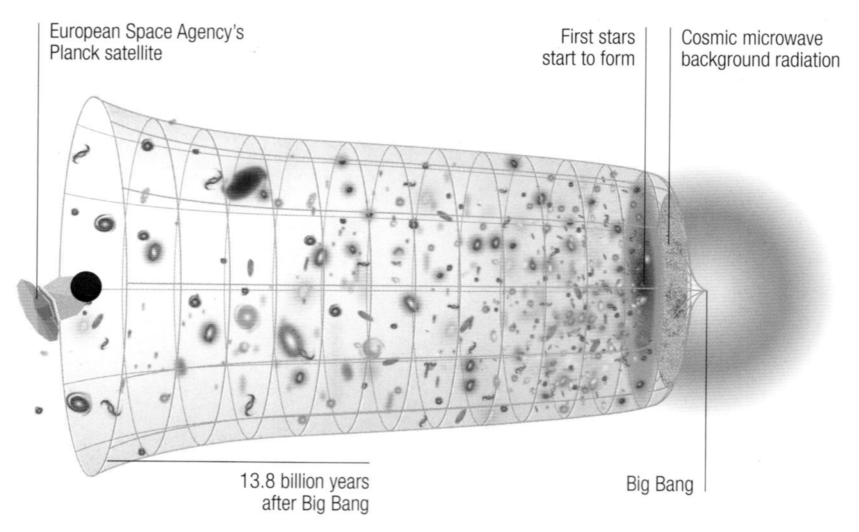

△ **Looking back in time**
Looking farther away in space is one way of studying features of the young Universe that do not survive today. At great distances, we can see young galaxies busy forming, and soon we may even see radiation from the very first stars. The cosmic microwave background marks the earliest and most distant light radiation we can detect.

Red indicates warmest areas

Dark blue regions are the coldest

△ **Radiation map**
The European Space Agency's Planck satellite mapped variations in the microwave background radiation corresponding to infinitesimal temperature differences. These reveal tiny variations in the temperature and density of the earlier Universe, showing where structures were starting to form even at this early time.

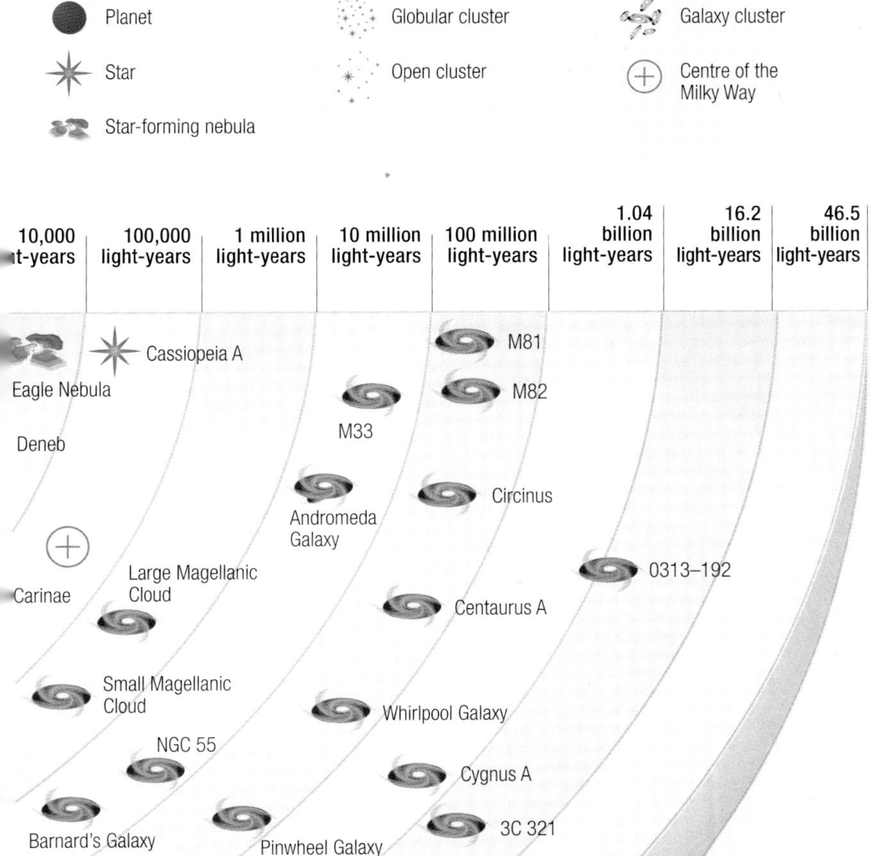

△ **The visible edge**
During the time their light has been travelling toward us, the expansion of space has carried distant objects even farther away. As a result, the most distant objects we can see at present in any direction are now about 46.5 billion light-years away.

Spacetime

Albert Einstein's theories of special and general relativity explain the fabric of the Universe as a four-dimensional "manifold", in which three familiar dimensions of space (length, breadth, and height) can be traded off with one another and with the fourth dimension of time itself. This is only apparent in extreme situations, such as when objects move at close to the speed of light, but also when large amounts of mass are present. According to Einstein, the force of gravity around massive objects is a result of the way they distort spacetime.

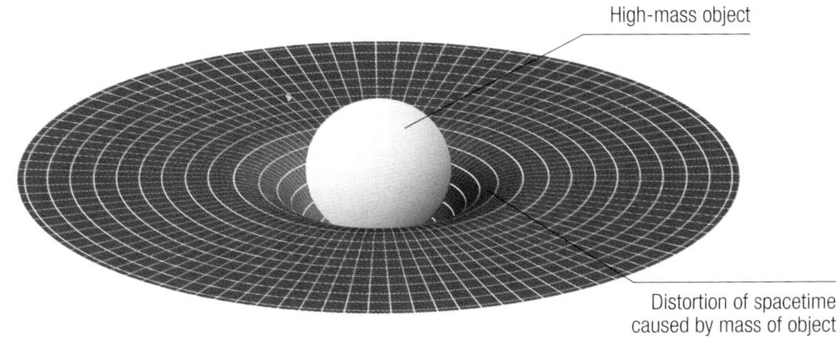

△ **Warped space**
The idea of spacetime is hard to visualize in four dimensions, but it gets easier if you visualize space as a flat rubber sheet. Massive objects create dents in the sheet (gravitational fields), and these deflect the paths of other moving nearby, or even rays of light (see p.64).

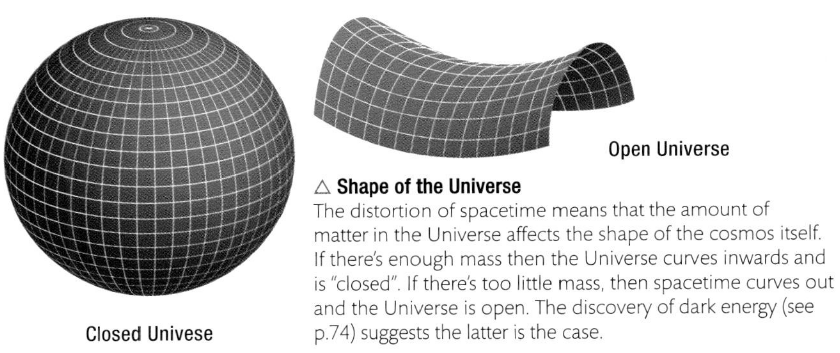

△ **Shape of the Universe**
The distortion of spacetime means that the amount of matter in the Universe affects the shape of the cosmos itself. If there's enough mass then the Universe curves inwards and is "closed". If there's too little mass, then spacetime curves out and the Universe is open. The discovery of dark energy (see p.74) suggests the latter is the case.

More than one universe

Our familiar cosmos certainly stretches beyond the limits of what we can see, but is this Universe the only one? Or, are we part of a wider multiverse? One theory, known as eternal inflation, suggests that our Universe is just one of many. Bursts of inflation energy, such as the one at the beginning of our Universe, are continuously producing new "bubble Universes".

▷ **Eternal inflation?**
If our Universe is one of many created from the same raw material, then it's possible that the walls of separate bubble Universes would occasionally collide and interact.

Mapping dark matter
Dark matter cannot be imaged directly, therefore much of our understanding of it comes from the effects of so-called gravitational lensing. This is the way in which concentrations of mass distort the fabric of spacetime and deflect the light rays from more distant objects. The distribution of dark matter around a galaxy cluster in Pisces called Cl 0024+17 is shown here in lighter blue.

DARK MATTER AND DARK ENERGY

THE LUMINOUS MATERIAL OF THE UNIVERSE IS A TINY PART OF ITS OVERALL COMPOSITION – IT ALSO CONTAINS LARGE AMOUNTS OF INVISIBLE MASS KNOWN AS DARK MATTER, AND ANOTHER MYSTERIOUS SUBSTANCE CALLED DARK ENERGY.

Since the 1930s astronomers have suspected the existence of dark matter. Such material is not just dark, but completely immune to interactions with light, and only makes its influence felt through the force of its gravity. More recently, cosmologists have found that cosmic expansion (see pp.70–71) is accelerated by an effect called dark energy, which seems to counteract gravity.

The nature of dark matter

The first evidence for dark matter came from two sources: the way that galaxies orbit inside galaxy clusters and the speed at which stars orbit in the Milky Way. Both suggest the Universe contains about five times more dark than luminous matter. Some of this may be accounted for by compact, faint objects such as dead stars and stray planets, but most of it probably consists of unknown subatomic particles – tiny objects that interact with normal matter only through gravity.

What is dark energy?

In the 1990s, cosmologists discovered that cosmic expansion, which began in the Big Bang (see pp.14–15) has sped up over time, rather than slowing down as we might expect due to the gravity of all the dark and luminous matter within the Universe. The substance that drives this expansion is known as dark energy, but it's still poorly understood – it may be a "cosmological constant" (a uniform property of spacetime itself), or a "quintessence" (a localized force that can vary from place to place). Which theory is correct could have an important effect on the fate of the Universe.

△ **Bullet cluster**
This collision between two distant galaxy clusters reveals the motion of dark matter. Most of the cluster's luminous matter takes the form of clouds of gas that emit X-rays (pink), but the distribution of dark matter (blue) broadly matches that of the visible galaxies (white).

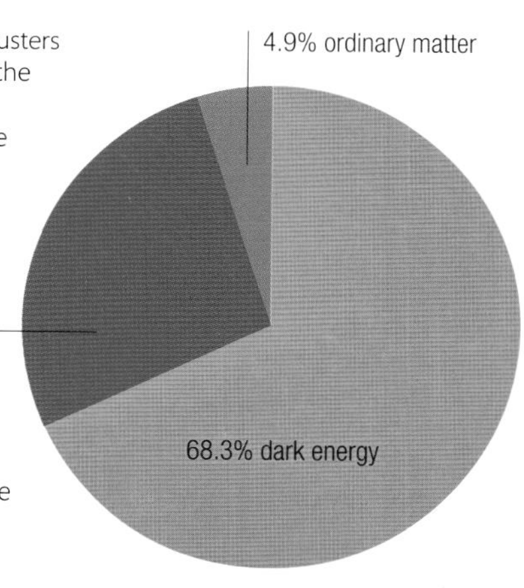
4.9% ordinary matter
26.8% dark matter
68.3% dark energy

▷ **Composition of the Universe**
Using Einstein's famous equation $E=mc^2$, cosmologists can estimate the overall balance between dark energy and the energy locked up in dark and luminous matter.

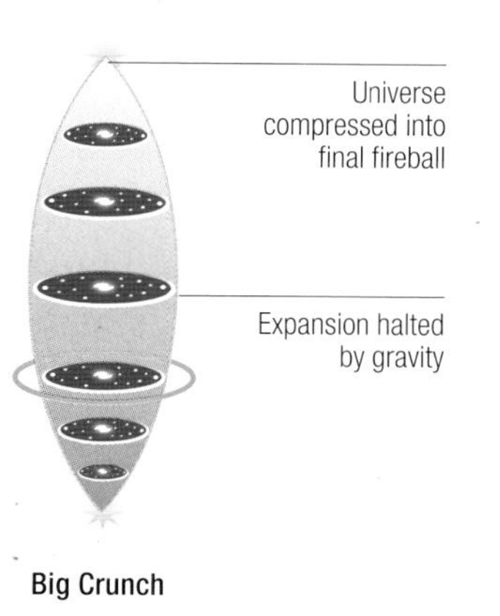
Universe compressed into final fireball
Expansion halted by gravity
Big Crunch

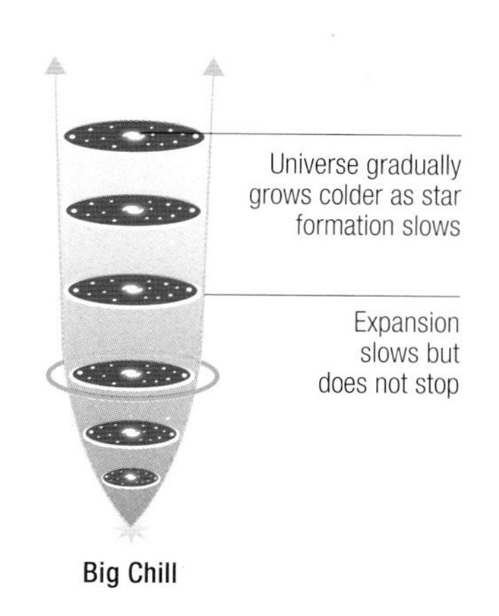
Universe gradually grows colder as star formation slows
Expansion slows but does not stop
Big Chill

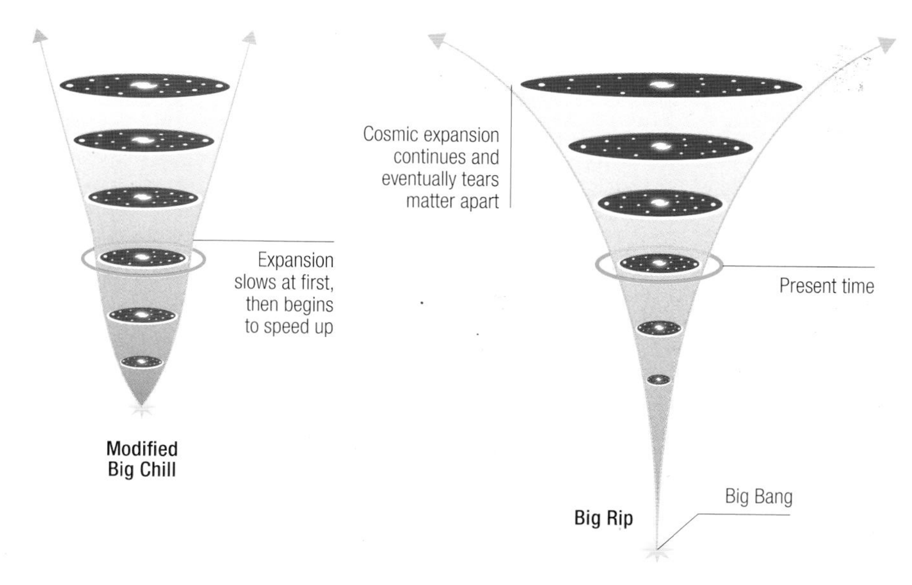
Expansion slows at first, then begins to speed up
Modified Big Chill
Cosmic expansion continues and eventually tears matter apart
Present time
Big Bang
Big Rip

△ **Fates of the Universe**
The precise balance between combined dark and luminous matter (whose gravity slows down the expansion of the Universe) and dark energy (which tends to speed up the expansion) will ultimately determine the way our Universe comes to an end.

OBSERVING THE **SKIES**

LIGHT IS OUR MAIN SOURCE OF INFORMATION ABOUT DISTANT OBJECTS IN THE UNIVERSE, AND GROUND-BASED TELESCOPES REMAIN IMPORTANT TOOLS FOR CAPTURING LIGHT AND STUDYING FAR-AWAY STARS AND GALAXIES.

Very few objects from space ever reach Earth, and most of those only come from our immediate planetary neighbourhood. Studying light that reaches our planet from distant space is therefore one of the best ways of learning about objects in the wider Universe. Many other forms of radiation are absorbed by Earth's atmosphere and astronomers use space-based observatories (see pp.80–81) to study objects in these other wavelengths.

By gathering light across a large surface and concentrating it into a much smaller image, telescopes allow us to see objects that are much fainter than those we can see with the unaided eye. Magnifying these small images then enables us to distinguish much finer detail. However, modern scientific telescopes are very different machines from those used by backyard stargazers. Most are reflector designs that use a series of mirrors to bring light onto converging paths and create a focused image on a detector instrument. The telescope is supported by a mount or cradle that swings back and forth and allows it to keep pace with the path of objects across the sky. A technique called interferometry allows astronomers to links two or more telescopes together and detect even finer details.

▷ **Refractor and reflector telescopes**
The first telescopes, invented by Dutch spectacle-makers in the Netherlands around 1609, used a lens-based refractor design. One lens (the objective) collects light and bends it to a focus, while another (the eyepiece) magnifies image. The simplest reflector designs, invented by Isaac Newton around 1668, use a curved mirror to collect light and direct it to a secondary mirror, which then directs it to a lens-based eyepiece.

Refractor telescope

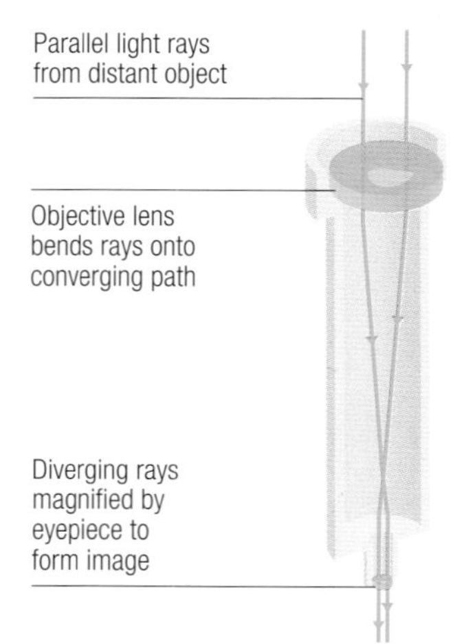
Parallel light rays from distant object
Objective lens bends rays onto converging path
Diverging rays magnified by eyepiece to form image

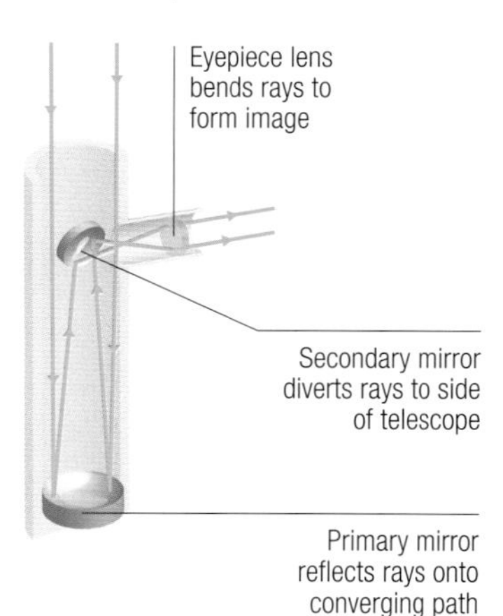
Reflector telescope
Eyepiece lens bends rays to form image
Secondary mirror diverts rays to side of telescope
Primary mirror reflects rays onto converging path

◁ **Learning from light**
Professional astronomers rarely use an eyepiece to look directly through a telescope. Instead, they use the instrument to channel light to various detectors. These can include digital cameras, photometers (which precisely measure the brightness of light from individual objects), and spectrometers that analyse colours of light, enabling scientists to learn about the chemistry of stars.

Yerkes Observatory
102cm, Wisconsin, 1893
Largest successful refractor telescope

Hale Reflector
508cm, California, 1948
Landmark single-mirror reflector telescope

Multi Mirror Telescope
4.5-metre equivalent, Arizona, 1979
Pioneering multi-mirror telescope, converted to single mirror in 2000

Hubble Space Telescope
2.4 metres, Low-Earth Orbit, 1990
First large telescope in space

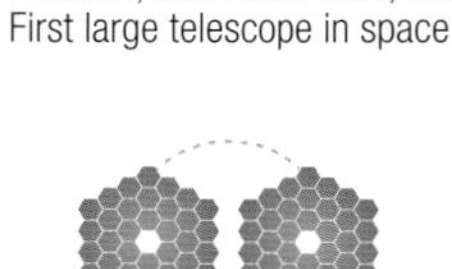

James Webb Space Telescope
6.5 metres, Low Earth Orbit, 2018
Planned successor to Hubble

Keck Telescope
2x10 metres, Hawaii, 1993/1996
First large telescope with interfermometer

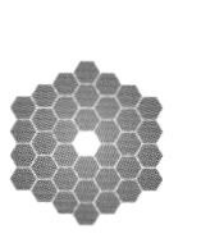

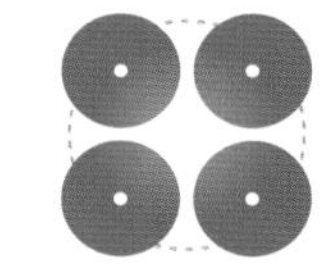

Gran Telescopio Canarias
10.4 metres, La Palma, 2008
Largest single-aperture telescope

Very Large Telescope
4x8.2 metres, Chile, 1998–2000
Largest overall collecting area

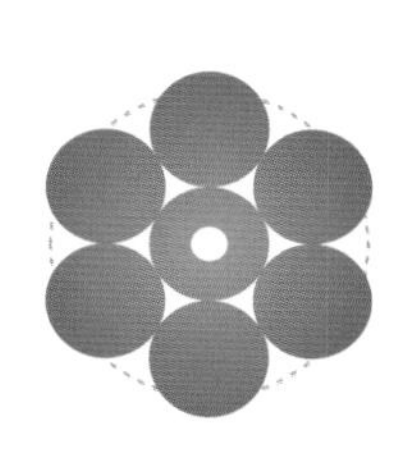

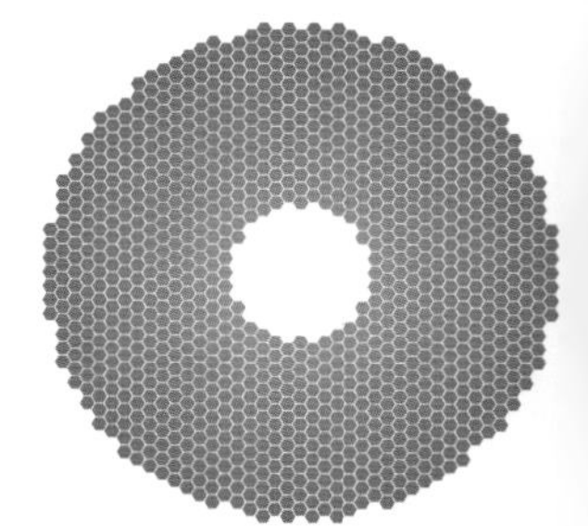

Giant Magellan Telescope
24.5-metre equivalent
To be built in Chile, 2025

European Extremely Large Telecope
39.3 metres
Under construction in Chile, 2024

△ **Collecting areas**
Telescope collecting areas were limited by technology for much of the 20th century, but instruments have grown rapidly in the past few decades.

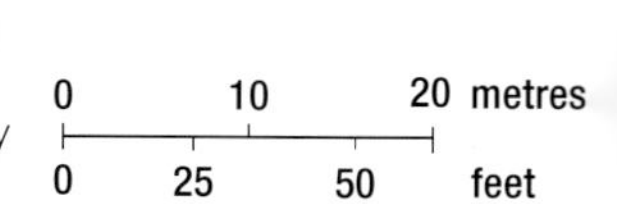
0
10
20 metres
0
25
50
feet

A typical 20-cm (8-in) reflecting telescope gathers **830 times more light** than the human eye alone

Research telescope
Most modern research telescopes, such as the European Extremely Large Telescope illustrated here, are located at mountaintop sites that put them above the bulk of Earth's turbulent, light-absorbing atmosphere. Using segmented mirrors increases a telescope's collecting area and the faintness of objects it can image.

◁ **Radio telescopes**
Radio signals from space were discovered in the 1930s and are measured today using enormous bowl-shaped antennae. The much longer wavelengths of radio waves mean they need a much bigger collecting surface to resolve fine details, but radio antennae can be linked together in arrays, like this one in New Mexico.

THE HISTORY OF THE TELESCOPE

TELESCOPES ARE THE VITAL TOOLS OF AN ASTRONOMER'S TRADE, IMPROVING ON HUMAN EYESIGHT AND ALLOWING IMAGES AND DATA TO BE PROCESSED IN DIFFERENT WAYS, AND RECORDED FOR POSTERITY.

Dutch lensmaker Hans Lippershey is usually credited with inventing the telescope around 1608, but it was Italian physicist Galileo Galilei who first turned it towards the sky. Since then, telescope technology has gone through huge advances – the introduction of the mirrored reflector design, mounts that can keep a telescope in sync with the movement of the stars, spectroscopy to analyse the chemical fingerprints of starlight, and photography to keep a permanent record of observations. More recently, computer control and space-based observatories have helped push the limits of telescope technology.

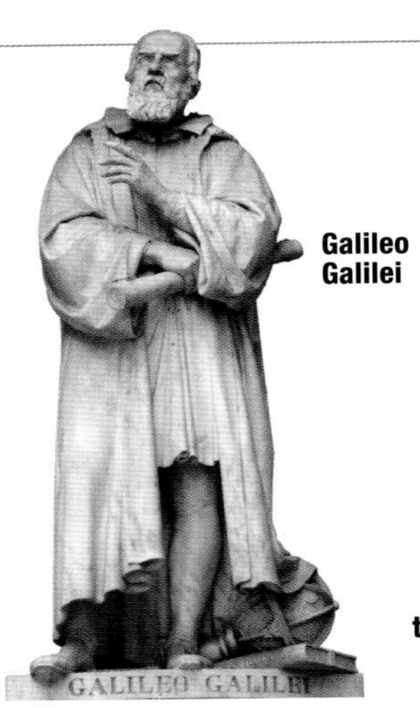

Galileo Galilei

Aerial telescope built by Johannes Hevelius

1609

Galileo's telescope
Galileo's first lens-based telescope produced an image multiplied by a factor of three, but later designs improved rapidly. He used his instruments to make discoveries including mountains on the Moon, spots on the Sun, and countless stars invisible to the naked eye.

1673

Aerial telescopes
One way of improving the magnification of telescopes was to use larger lenses separated by a greater distance. In the mid-17th century this led to enormous aerial telescopes with lenses suspended on open frames up to 31m (150ft) long.

High-altitude twin Keck telescopes on Mauna Kea, Hawaii

1949

Mountaintop telescopes
Astronomers had long recognized that observing from high altitudes helped reduce the atmospheric turbulence affecting starlight, but it was only in the second half of the 20th century that observatories on remote mountaintops became practical.

1970

Orbiting observatories
The launch of the first X-ray astronomy satellite, Uhuru, heralded a new age of space-based astronomy, studying the Universe at wavelengths that are blocked by Earth's atmosphere. These included not only X-rays, but also ultraviolet and infrared radiation.

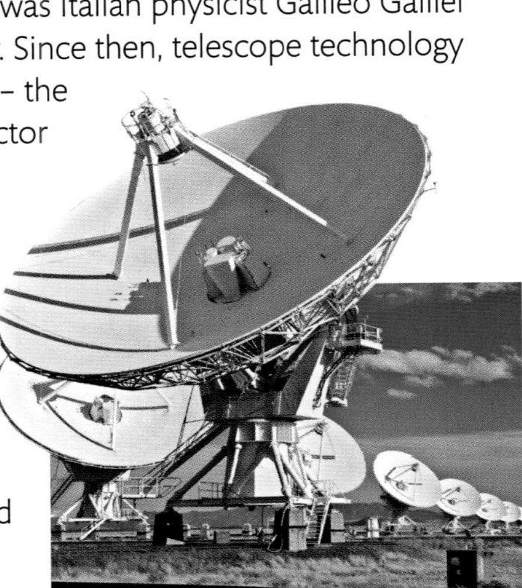

Very Large Array, New Mexico

1980

Telescope arrays
A technique called interferometry allows for the combination of signals from several telescopes to mimic the resolving power of a single, impossibly large instrument. The technique was pioneered for use with the long-wavelength radio waves at the Very Large Array.

1980s

Segmented-mirror telescopes
The weight of traditional mirrors brought the growth of telescopes to a halt in the mid-20th century, but from the 1980s onwards, engineering breakthroughs allowed telescopes to reach even bigger sizes. Key to this was the ability to align honeycomb-like mirror segments to mimic a single reflecting surface.

Hubble Space Telescope

1990

Hubble Space Telescope
The idea of placing a large optical telescope above Earth's atmosphere, where it would experience perfect observing conditions, was suggested as early as 1946. The ability to repair and upgrade the Hubble Space Telescope in orbit has kept it functional for more than a quarter century. While modest in size compared to today's ground-based giants, Hubble's location allows it to deliver both stunning images and revolutionary scientific discoveries.

Replica of Newton's reflector telescope

1668

Newtonian reflector

British physicst and mathematician Isaac Newton designed the first telescope that used a curved mirror rather than a lens to collect light. This permitted a much more compact telescope design called the Newtonian reflector.

1781

William Herschel

From the late 18th century, British astronomer William Herschel developed new metals for his reflecting telescope mirrors. These allowed him to produce the finest telescopes so far, and to make new discoveries, including the planet Uranus.

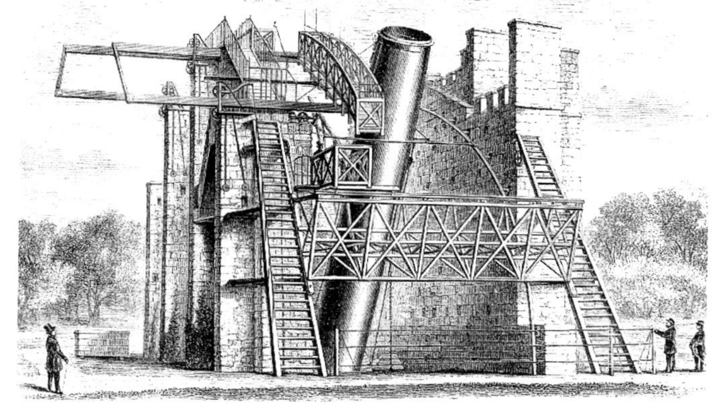

William Parsons' telescope

1845

The Leviathan of Parsonstown

Irish astronomer William Parsons built this enormous reflecting telescope with a 1.8m (6ft) mirror on his estate at Birr Castle in Ireland, but it could only point in a limited range of directions because of the walls needed to support it. It remained the world's largest telescope for more than 70 years.

The Hooker Telescope

1917

Hooker Telescope

The 2.5m (8ft) Hooker reflector at Mount Wilson Observatory was the first telescope to combine giant size with manoeuvrability. It remained the world's largest telescope until 1949, and was key to the discovery of the expansion of the Universe.

1933

Radio astronomy

American physicist Karl Jansky's discovery of radio signals from the sky, associated with the rising and setting of the Milky Way, marked the beginning of radio astronomy. The long wavelength of radio waves means that very large collecting areas are needed.

V-2 rocket launch

1949

Space-based astronomy

In the late 1940s, US astronomers used captured German V-2 war rockets to carry radiation detectors on short trips above Earth's atmosphere. These confirmed that radiation from space, such as X-rays, is blocked by the atmosphere.

1998

The Very Large Telescope

The European Southern Observatory's VLT in Chile is made up of four separate 8.2m (27ft) reflecting telescopes. It marked a breakthrough in the manufacture of large single-element mirrors and can also combine light from multiple telescopes.

Artist's impression of the European Extremely Large Telescope

2014

Future giants

Currently under construction in Chile's Atacama Desert, the European Extremely Large Telescope will have a primary mirror diameter of more than 39.3m (129ft) – constructed out of 798 separate cells. It will only become operational in 2024.

Artist's impression of the James Webb Space Telescope

2018

James Webb Space Telescope

NASA's successor to Hubble, the infrared James Webb Space Telescope, should allow us to see farther into the depths of the Universe than ever before. Its huge sunshield can protect it from not only solar heat, but also radiation from the Earth.

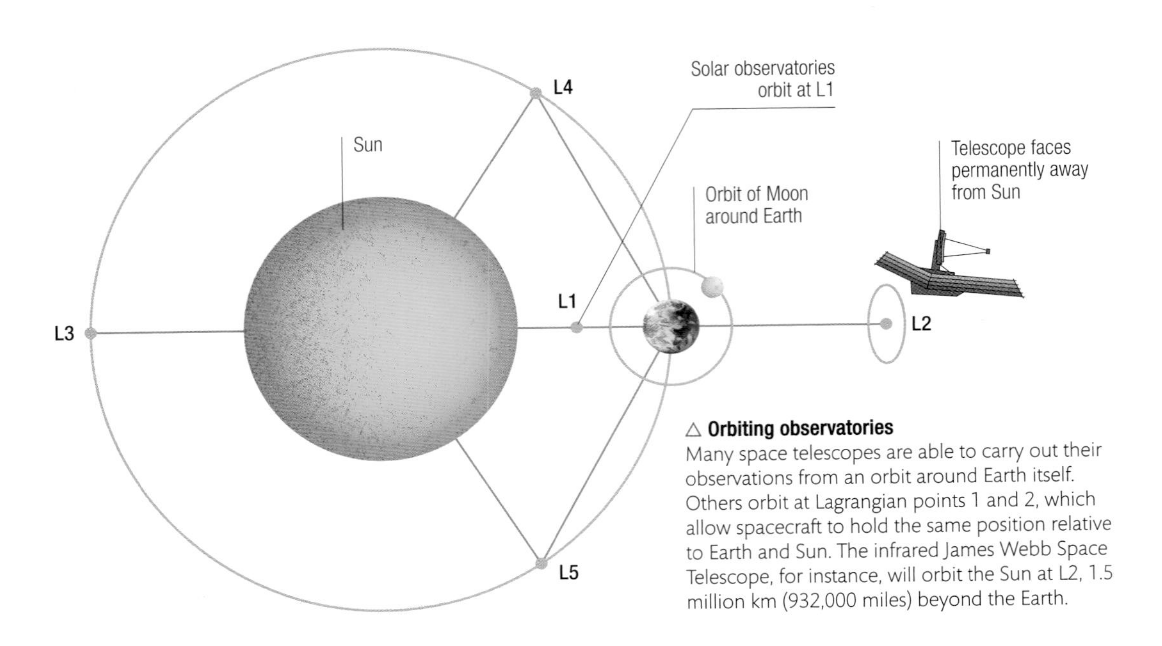

△ **Orbiting observatories**
Many space telescopes are able to carry out their observations from an orbit around Earth itself. Others orbit at Lagrangian points 1 and 2, which allow spacecraft to hold the same position relative to Earth and Sun. The infrared James Webb Space Telescope, for instance, will orbit the Sun at L2, 1.5 million km (932,000 miles) beyond the Earth.

SPACE **TELESCOPES**

TELESCOPES ORBITING ABOVE EARTH OPEN UP NEW VISTAS ON THE UNIVERSE. THEY NOT ONLY REVEAL OBJECTS WHOSE INVISIBLE RADIATIONS ARE BLOCKED BY OUR PLANET'S ATMOSPHERE, BUT ALSO ALLOW ASTRONOMERS TO CARRY OUT MORE PRECISE STUDIES IN VISIBLE LIGHT.

The visible light with which we view the Universe is simply a type of electromagnetic wave. These are packets of energy that move through space in the form of self-reinforcing electric and magnetic fields. The properties of light we perceive as colours depend on the wavelength of these waves, but the broader electromagnetic spectrum goes far beyond them to encompass much longer and shorter waves. Robotic space telescopes allow astronomers to study these elusive wavelengths, but they are often very different from Earth-bound instruments. For example, infrared telescopes need extreme cooling so their own heat does not swamp the weak signals, while X-rays and gamma rays will pass straight through most traditional mirrored telescope designs.

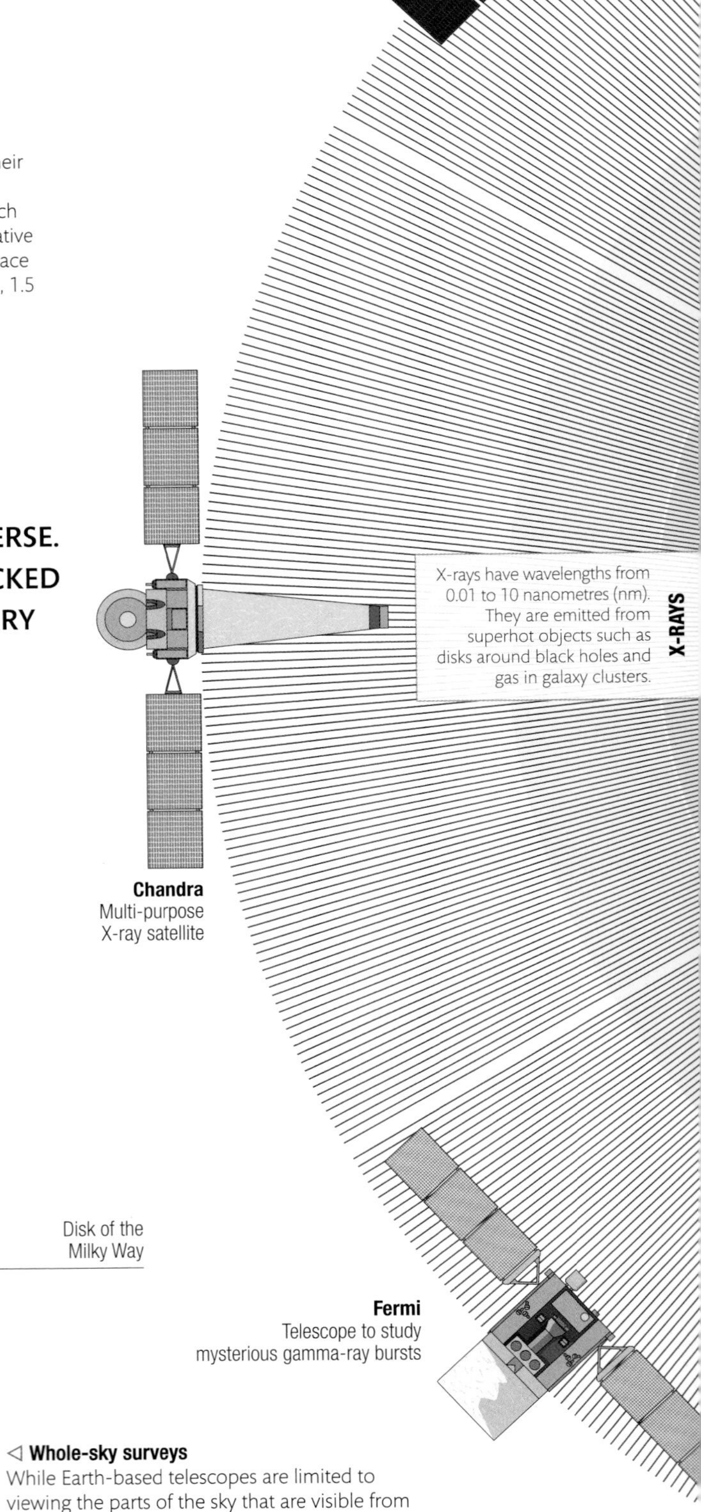

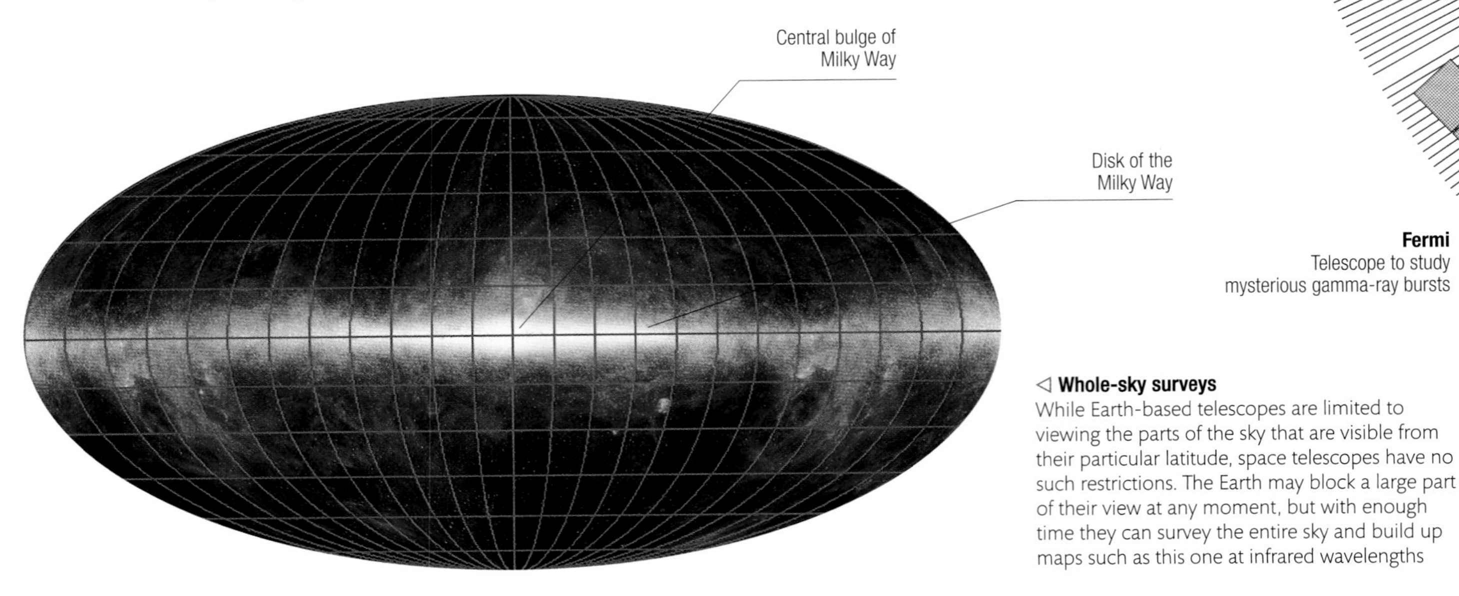

◁ **Whole-sky surveys**
While Earth-based telescopes are limited to viewing the parts of the sky that are visible from their particular latitude, space telescopes have no such restrictions. The Earth may block a large part of their view at any moment, but with enough time they can survey the entire sky and build up maps such as this one at infrared wavelengths

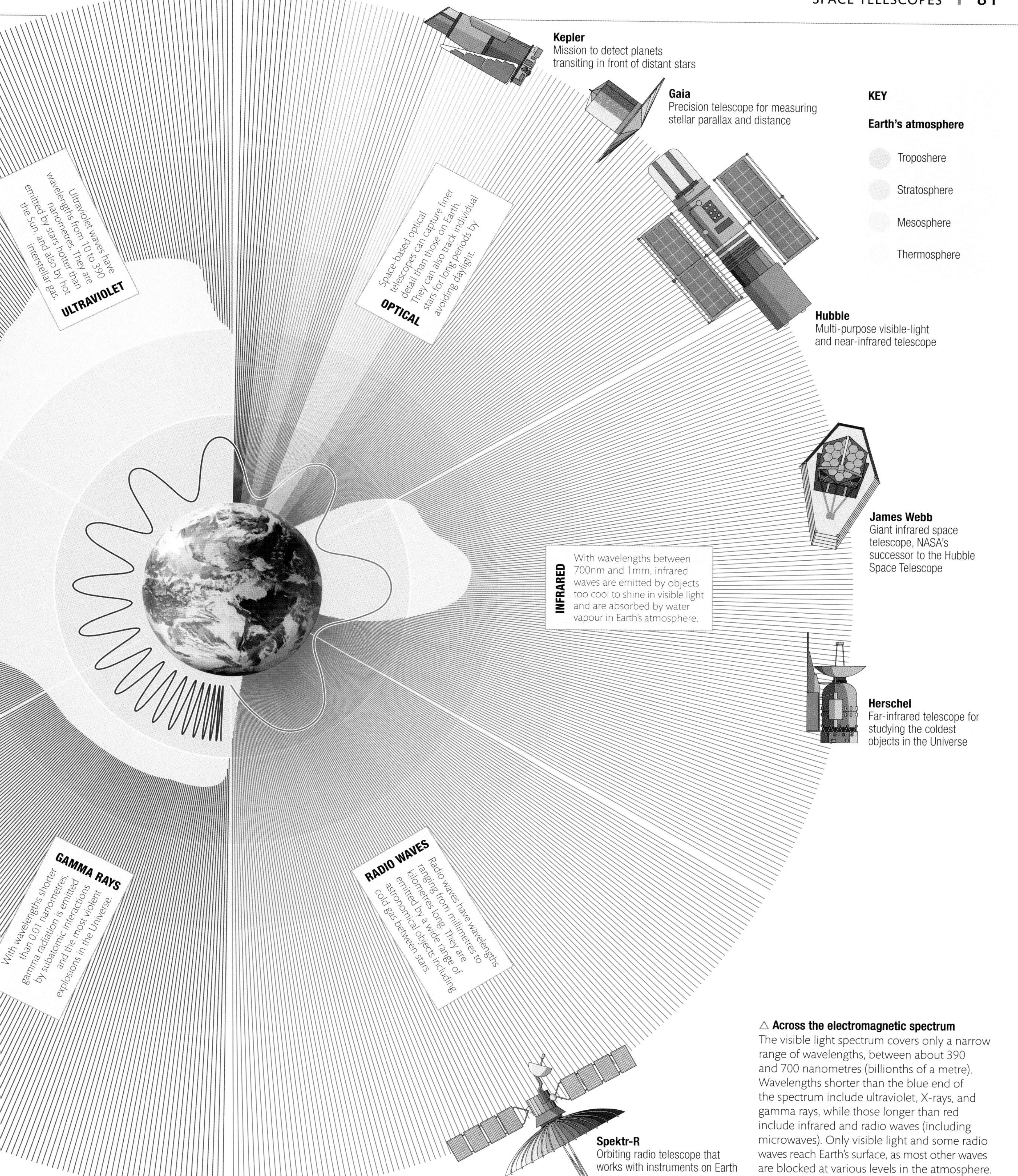

Kepler
Mission to detect planets transiting in front of distant stars
Gaia
Precision telescope for measuring stellar parallax and distance
KEY
Earth's atmosphere
Troposphere
Stratosphere
Mesosphere
Thermosphere
Hubble
Multi-purpose visible-light and near-infrared telescope
James Webb
Giant infrared space telescope, NASA's successor to the Hubble Space Telescope
Herschel
Far-infrared telescope for studying the coldest objects in the Universe
Spektr-R
Orbiting radio telescope that works with instruments on Earth
ULTRAVIOLET
Ultraviolet waves have wavelengths from 10 to 390 nanometres. They are emitted by stars hotter than the Sun, and also by hot interstellar gas.
OPTICAL
Space-based optical telescopes can capture finer detail than those on Earth. They can also track individual stars for long periods by avoiding daylight.
INFRARED
With wavelengths between 700nm and 1mm, infrared waves are emitted by objects too cool to shine in visible light and are absorbed by water vapour in Earth's atmosphere.
GAMMA RAYS
With wavelengths shorter than 0.01 nanometres, gamma radiation is emitted by subatomic interactions and the most violent explosions in the Universe.
RADIO WAVES
Radio waves have wavelengths ranging from millimetres to kilometres long. They are emitted by a wide range of astronomical objects including cold gas between stars.

△ **Across the electromagnetic spectrum**
The visible light spectrum covers only a narrow range of wavelengths, between about 390 and 700 nanometres (billionths of a metre). Wavelengths shorter than the blue end of the spectrum include ultraviolet, X-rays, and gamma rays, while those longer than red include infrared and radio waves (including microwaves). Only visible light and some radio waves reach Earth's surface, as most other waves are blocked at various levels in the atmosphere.

THE SEARCH FOR **LIFE**

PEOPLE HAVE ALWAYS BEEN FASCINATED BY THE POSSIBILITY OF LIFE BEYOND EARTH, BUT THE ODDS OF ITS EXISTENCE, AND THE PROSPECTS FOR ITS DETECTION, HAVE BEEN GIVEN SEVERAL HUGE BOOSTS IN RECENT YEARS.

The search for life in the Universe has been transformed by the discovery of volcanically active ocean moons in our own Solar System, and countless exoplanets orbiting other stars, offering potential homes for different forms of life.

Requirements for life

Traditional theories assumed that life got started in the so-called "primordial soup" of a shallow, warm, chemical-rich seas on the surface of the early Earth. This appears to be the ideal condition to provide the three necessities of life: carbon-based chemicals, water, and an energy source in the form of sunlight. Today, the carbon and water requirements still seem reasonable, since they allow the development of complex chemistry. But the discovery of "extremophiles" – organisms that feed on chemical energy in the pitch darkness of deep-sea volcanic vents, or even deep inside hot subterranean rocks – have changed ideas about what life is, and the conditions it needs to survive.

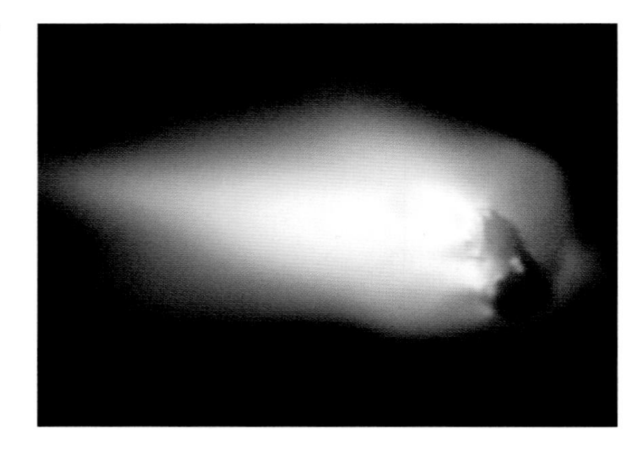

◁ **Transporting life**
Some astronomers have speculated that life on Earth might not have needed to have evolved from scratch. Instead, either life itself, or at least complex chemicals that would help it get started, might be transferred between planets inside meteorites or comets.

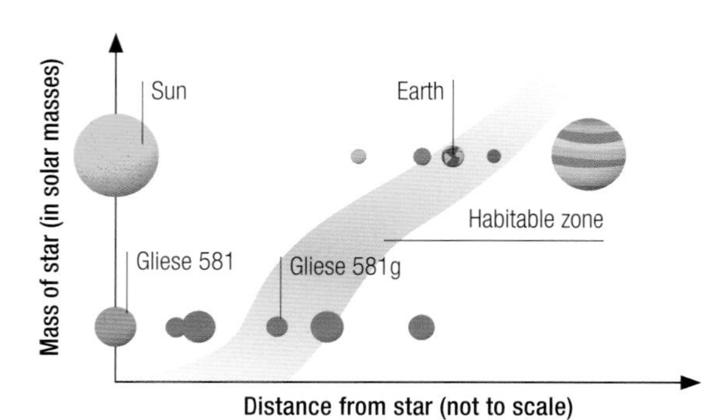

◁ **Goldilocks zone**
The most hospitable worlds around any star are likely to be Earth-like planets orbiting within a habitable "Goldilocks zone" around their stars, where temperatures are just right for liquid water to survive on the surface. A few such worlds have been identified, and they are probably abundant in the Milky Way.

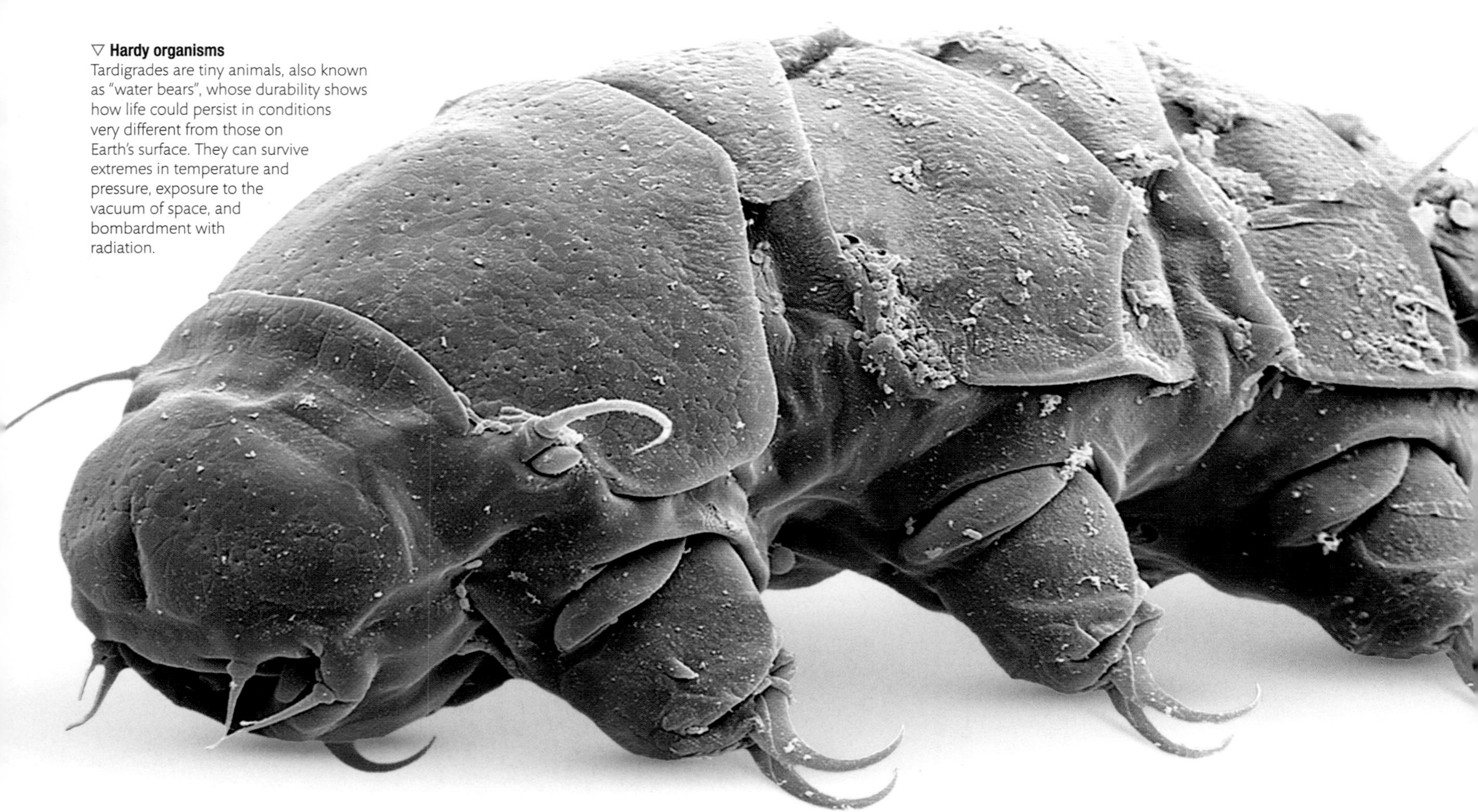

▽ **Hardy organisms**
Tardigrades are tiny animals, also known as "water bears", whose durability shows how life could persist in conditions very different from those on Earth's surface. They can survive extremes in temperature and pressure, exposure to the vacuum of space, and bombardment with radiation.

Signature of life

Any form of life must sustain itself through a series of chemical reactions known as the organism's metabolism and, over time, this inevitably transforms the environment of the planet around it. For example, oxygen is a naturally reactive chemical that tends to get locked away as mineral compounds within rocks, so Earth would have no oxygen in its atmosphere if not for the evolution of life and the metabolic reactions of photosynthetic plants and algae over billions of years. Atmospheric oxygen is therefore a potential chemical biosignature for life on other worlds. Astronomers have already measured the atmospheres of a few exoplanets, but future telescopes should make this far more commonplace.

△ **Seas of Enceladus**
In 2005, NASA's Cassini space probe discovered huge plumes of water ice erupting from Saturn's small moon Enceladus, indicating of a hidden ocean kept liquid by heating from powerful tidal forces. These make Enceladus one of the Solar System's most likely habitats for life.

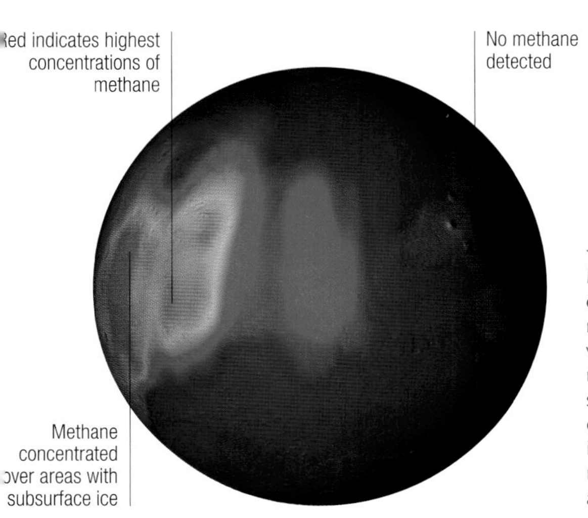

◁ **Methane on Mars**
Methane is a gas that can only be produced by living microorganisms or active volcanism. It breaks down rapidly on exposure to sunlight, so the recent discovery of methane patches in the atmosphere of Mars raises intriguing questions about the Red Planet.

Estimates of the number of communicating civilizations in our galaxy range **from many millions, to just one**

Intelligent life

The Search for Extraterrestrial Intelligence (SETI) uses a variety of methods in the hope of tracking down evidence of intelligent aliens in the Universe. The most common is to scour the sky in search of artificial radio signals, but such signals are only likely to be found if aliens are deliberately beaming them towards us. Alternative approaches include looking for technosignatures (signs of technology), such as pollution in planetary atmospheres or even changes in the light output of stars created by alien engineering.

▷ **Messages in space**
Two space probes launched in the 1970s, Pioneer 10 (1972) and Pioneer 11 (1973), carried plaques with a pictogram message. This message was meant for any intelligent life that might intercept or recover one of the probes at some point in the future.

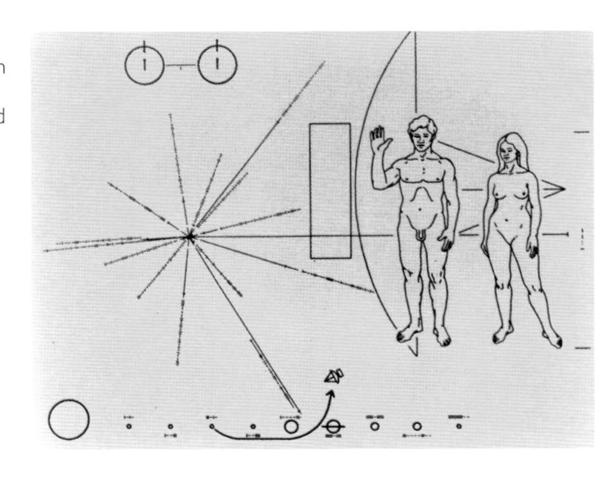

▷ **Drake's equation**
In 1961, SETI pioneer Frank Drake devised the Drake Equation, a way of assessing the number of civilizations that might be communicating by radio signals in the Milky Way at any one time.

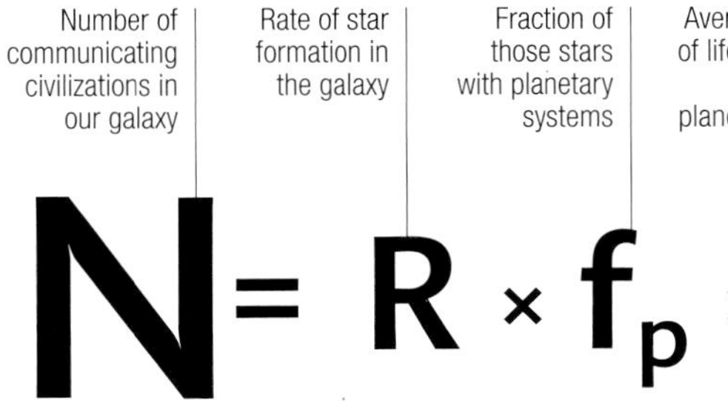

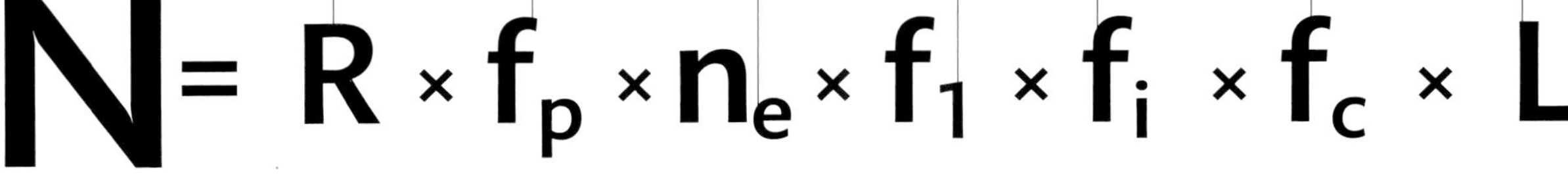

$$N = R \times f_p \times n_e \times f_1 \times f_i \times f_c \times L$$

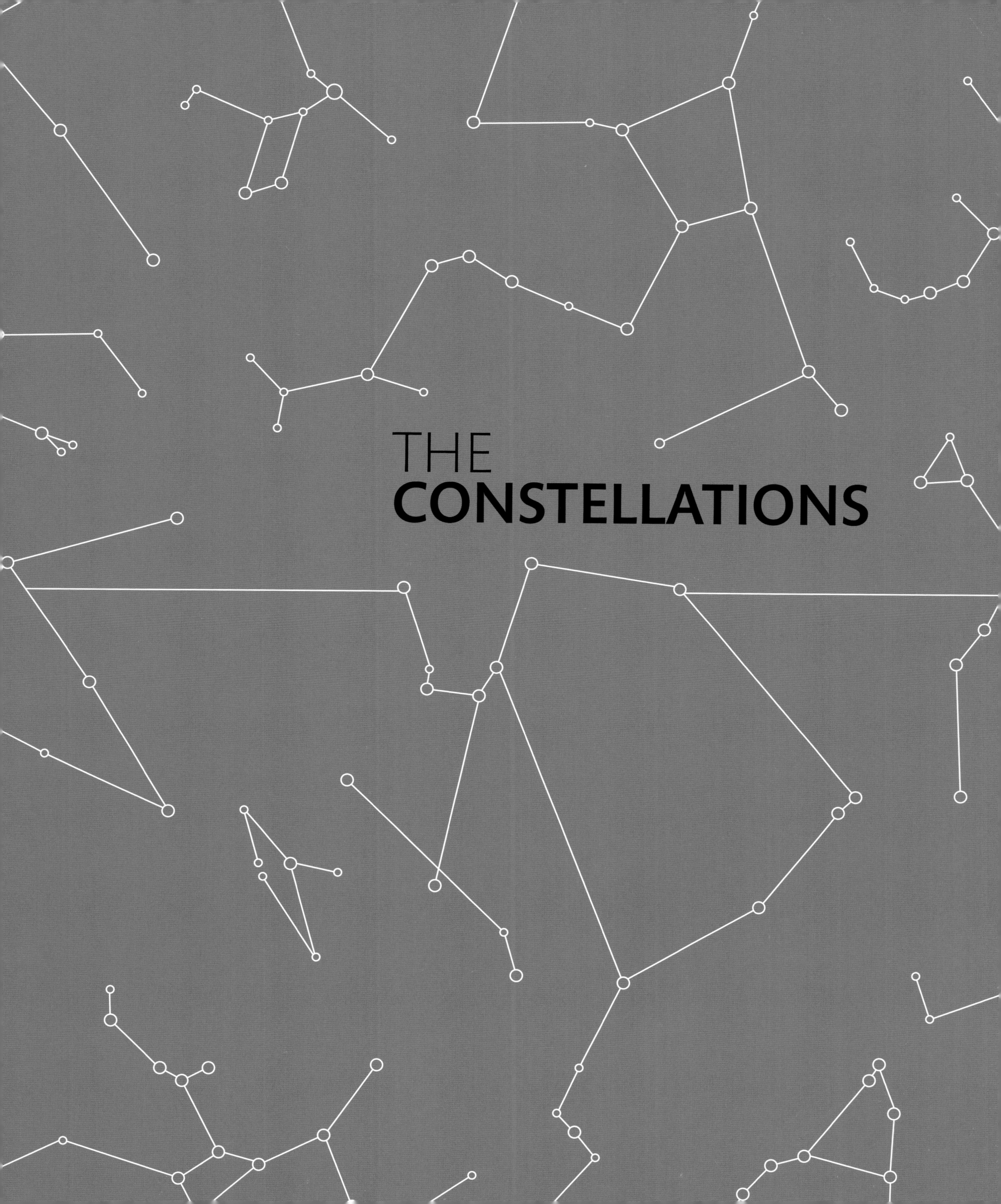

THE CONSTELLATIONS

The first constellations were simple patterns of stars, picked out by imaginative humans thousands of years ago. A good knowledge of the sky had practical uses. Bright stars and constellations were navigation aids for travellers at night, their risings and settings provided a simple clock, and their annual progression around the sky acted as a calendar. The sky also became a picture book in

PATTERNS IN **THE SKY**

which storytellers could imagine the starry outlines of gods, heroes, and mythical beasts. All civilizations had their own constellations, based on their own culture. Those we use today stem from a group of 48 known to the Ancient Greeks around 2,000 years ago. These were supplemented by others invented by astronomers in the 16th to 18th centuries, particularly in the far southern sky, which the Greeks could not see. In the 1920s, the International Astronomical Union, astronomy's governing body, officially recognized a total of 88 constellations that fill the sky from pole to pole with no gaps between them. Although constellations have outgrown their original purpose in this age of computer-controlled telescopes on Earth and in space, they still serve as a useful way of identifying the general area of sky in which a celestial object lies. They also provide a connection with the original stargazers who first looked at the sky and tried to understand the Universe around them.

◁ **Star trails**
A whirling pattern of lights is seen in the sky above the Atacama Large Millimetre/submillimeter Array (ALMA) in Chile. The streaks are star trails captured by a long-exposure photograph. Though the stars appear to circle the southern celestial pole, the movement is really that of the Earth rotating on its axis.

CHARTING THE HEAVENS

PEOPLE HAVE OBSERVED THE HEAVENS FOR THOUSANDS OF YEARS, AND MANY CULTURES HAVE LINKED THE PATTERNS THEY DISCERNED AMONG THE THOUSANDS OF VISIBLE STARS TO THEIR OWN MYTHOLOGY.

Today, the International Astronomical Union (IAU) recognizes 88 constellations and together the constellations form a complete sphere (see pp.94–95) around the Earth. Our modern system of constellations is based on the 48 figures described by the ancient Greek astronomer Ptolemy. Other civilizations also visualized patterns in the sky and linked those to their myths and legends, but only the Greek system is recognized today. It was not until the 16th century, when sailors started to navigate and explore the southern hemisphere, that whole new areas of the celestial sphere were mapped and new constellations created.

Babylonian clay tablet

Globe with early Greek constellations

3000–1000 BCE

Dawn of astronomy
Sumerian and Babylonian astronomers watch the yearly motions of the Sun and stars and create the first constellations, such as GUD.AN.NA, the modern Taurus. Their observations are recorded in cuneiform script on clay tablets like this one.

400–250 BCE

First Greek constellation system
Eudoxus, a Greek astronomer, introduces Babylonian constellations to the West in amended form in a book entitled *Phaenomena*. His original text is long lost, but it was turned into an instructional poem by another Greek, Aratus, and later translated into Latin.

Hercules by Bayer

1592–1612

New constellations
Petrus Plancius, a Dutch cartographer and astronomer, introduces 15 new constellations. Twelve of them lie in the far southern sky that is invisible from Europe and include stars plotted by Dutch navigators Pieter Dirkszoon Keyser and Frederick de Houtman.

1603

First all-sky star atlas
Johann Bayer, a German lawyer and amateur astronomer, publishes the first celestial atlas to cover the whole sky, *Uranometria*. He assigns a full page to each of the 48 Ptolemaic constellations, with an additional page for the 12 new southern constellations.

Edmond Halley

1679

Halley's southern survey
English astronomer Edmond Halley makes the first accurate survey of the southern sky from the island of St Helena. His catalogue contains 341 stars, and he introduces a new constellation, Robur Carolinum (Charles's Oak), but it is not accepted by other astronomers.

Hevelius's Leo Minor

1690

Hevelius's new constellations
Johannes Hevelius, a Polish astronomer, publishes a catalogue of over 1,500 stars, larger and more accurate than that of Tycho Brahe, along with a new star atlas. Hevelius introduces ten new constellations, seven of which are still accepted by astronomers today.

1725

Flamsteed's atlas and catalogue
John Flamsteed, England's first Astronomer Royal, produces the first major catalogue of stars observed with the aid of a telescope. It is published posthumously along with *Atlas Coelestis*, and they become the standard references for the next century.

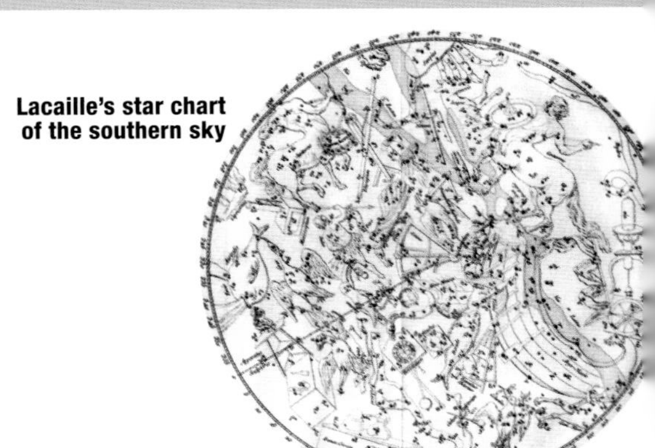

Lacaille's star chart of the southern sky

1751–52

More southern constellations
Nicolas Louis de Lacaille, a French astronomer, surveys the southern sky from the Cape of Good Hope, publishing a catalogue of nearly 2,000 stars along with a star chart. He introduces 14 new southern constellations, all still recognized by astronomers.

Hipparchus observes the night sky

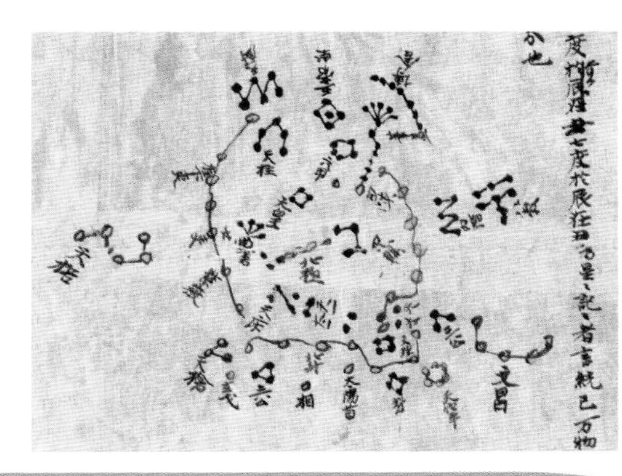
Ancient Chinese constellations

c.150 BCE

First great star catalogue
Hipparchus, a Greek astronomer, compiles the first great star catalogue of antiquity, grouping 850 stars into over 40 constellations. Hipparchus also divides the stars into six levels of brightness, the origin of the system of stellar magnitudes.

c.150

The Almagest
Greek astronomer Ptolemy produces a summary of Greek astronomy called the *Almagest*, which includes a revised version of Hipparchus's star catalogue with 48 constellations. It is the standard work on Western astronomy for nearly 1,500 years.

c.650

Oldest star chart
The oldest surviving star chart was drawn in 7th-century China on a paper scroll. Chinese constellations were smaller and more numerous than those in the West, with over 250 against Ptolemy's 48. Chinese astronomers also recorded hundreds more stars than the Greeks.

Taurus as depicted by al-Sufi

The northern sky by Dürer

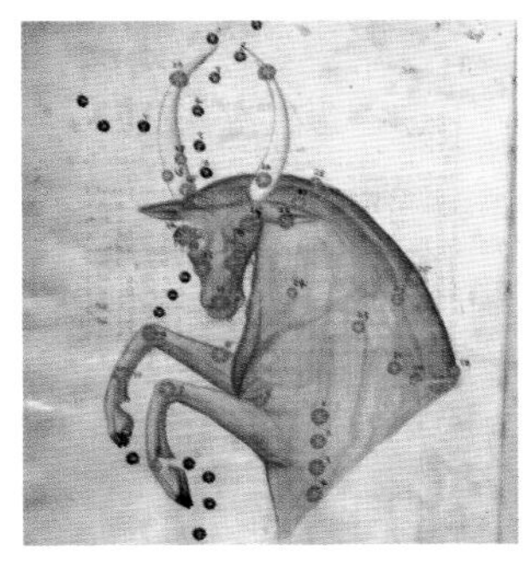

Tycho's observatory Uraniborg

964

Arabic star charts
Al-Sufi, a Persian astronomer also known in the West as Azophi, produces an updated version of the Greek *Almagest*, entitled *The Book of the Fixed Stars*. This includes illustrations of each constellation, something the *Almagest* lacked, drawn in Arabic style.

1515

Dürer's star chart
Albrecht Dürer draws the first European printed star chart, based on the catalogue in Ptolemy's *Almagest*. One half depicts the zodiac and northern constellations, the other shows the southern sky. Constellations are shown reversed, as on a celestial globe.

1598

Tycho Brahe
Tycho Brahe, a Danish astronomer, produces a new and improved catalogue of over 1,000 stars, ten times more accurate than the one in Ptolemy's *Almagest*. He still uses naked-eye sighting instruments, as the telescope has not yet been invented.

The constellation Pegasus in *Uranographia*

The Gaia spacecraft

1801

Greatest star atlas
The greatest of the old-style pictorial star atlases is published in 1801 by Johann Elert Bode, director of Berlin Observatory. Called *Uranographia*, it contains 17,000 stars divided into over 100 constellations, five of them invented by Bode himself.

1922–30

The final list
The newly formed International Astronomical Union (IAU) fixes the number of recognized constellations at 88, covering the entire celestial sphere, and draws up official boundaries for them. From now on, no more constellations can be added.

1989–93

Star cataloguing from space
A European Space Agency satellite called Hipparcos, named in memory of Hipparchus, compiles a catalogue from orbit of the positions, motions, and brightnesses of over 100,000 stars with unprecedented accuracy.

2013

The Galaxy in 3D
The European observatory Gaia is launched. It will spend five years measuring the distances and motions of over a billion stars to build up a three-dimensional map of our Galaxy.

THE CELESTIAL SPHERE

ALTHOUGH STARS LIE AT DIFFERENT DISTANCES FROM EARTH, FOR RECORDING THEIR POSITIONS IN THE SKY IT IS HELPFUL TO PRETEND THAT THEY ARE ALL STUCK TO THE INSIDE OF A VAST SPHERE THAT SURROUNDS EARTH.

This enormous imaginary globe is known as the celestial sphere. Every star in the sky other than the Sun, as well as other very remote objects such as galaxies, has a position on the surface of this sphere that remains more or less "fixed" – that is, it hardly changes except over extremely long periods of time. Other, closer objects, such as the Sun and other Solar System bodies, do appear to continuously "wander", at varying speeds, against the background of stars on the celestial sphere, but they do so in a predictable way.

The sky as a sphere

Like the real sphere of the Earth, the celestial sphere has north and south poles, an equator, and the equivalent of latitude and longitude lines. It is like a celestial version of the globe. The positions of stars and galaxies can be recorded on it as well, just as cities on Earth have their positions of latitude and longitude on a globe. The idea of the sphere also helps astronomers, or indeed anyone, better understand how their location on Earth, the time of night, and the time of year affect what can be viewed in the night sky.

Celestial sphere
The huge sphere on the surface of which stars are imagined to be "fixed"

Ecliptic plane
An imaginary plane on which Earth moves as it orbits the Sun

Orbits of Solar System planets
Most of the other planets orbit the Sun very close to the ecliptic plane

The Sun

The ecliptic
The circle where the ecliptic plane meets the celestial sphere

△ **The ecliptic**
One of the major circles on the celestial sphere is called the ecliptic. It marks where the ecliptic plane (the plane on which Earth orbits the Sun) meets the sphere's surface. In their motion – as seen from Earth – against the backdrop of stars on the sphere, the Sun always remains on, and planets stay close to, the ecliptic.

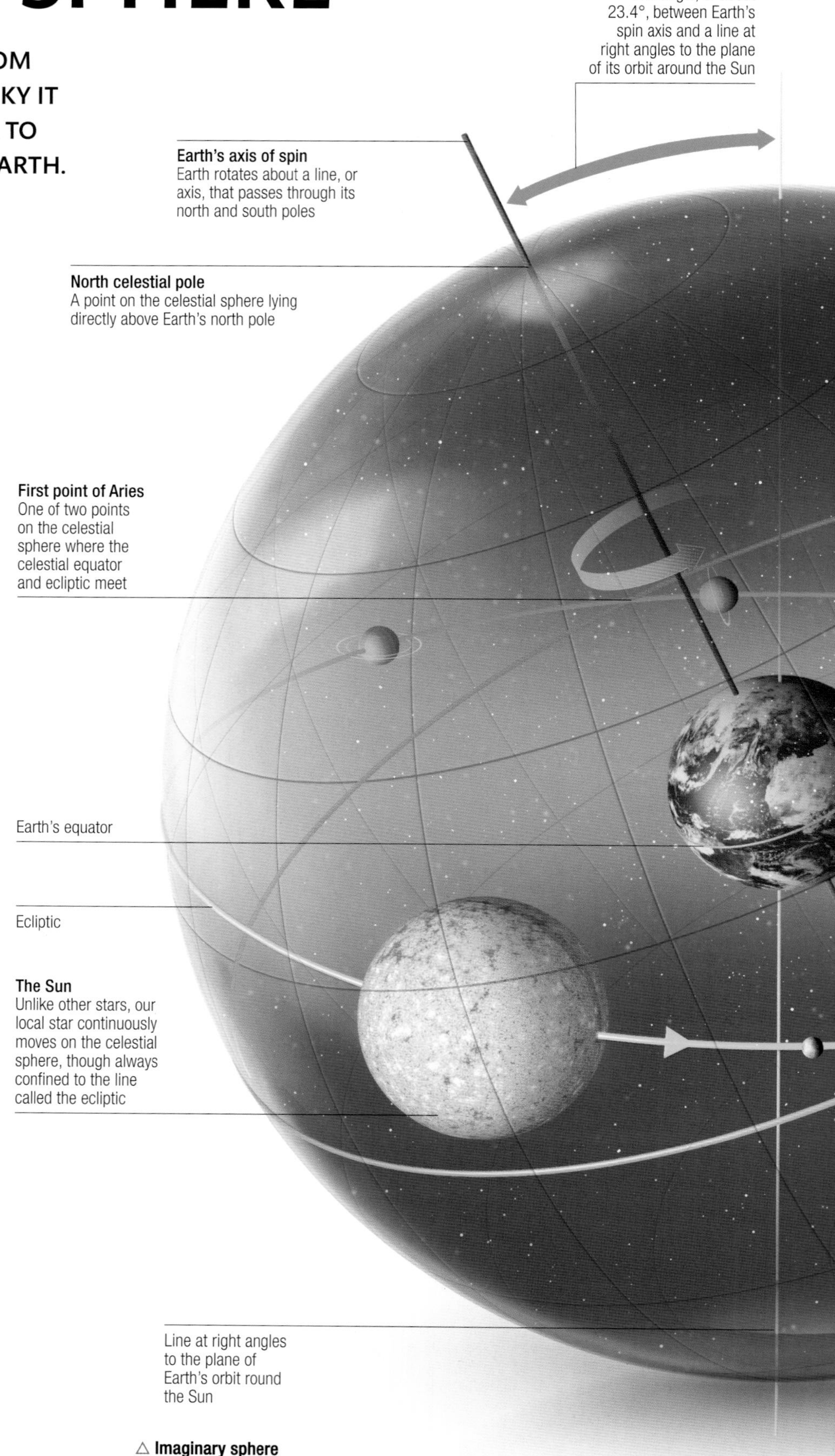

Earth's angle of tilt
The angle, of about 23.4°, between Earth's spin axis and a line at right angles to the plane of its orbit around the Sun
Earth's axis of spin
Earth rotates about a line, or axis, that passes through its north and south poles
North celestial pole
A point on the celestial sphere lying directly above Earth's north pole
First point of Aries
One of two points on the celestial sphere where the celestial equator and ecliptic meet
Earth's equator
Ecliptic
The Sun
Unlike other stars, our local star continuously moves on the celestial sphere, though always confined to the line called the ecliptic
Line at right angles to the plane of Earth's orbit round the Sun

△ **Imaginary sphere**
The celestial sphere is a purely imaginary concept, with a specific shape but no particular size. Astronomers use exactly defined points and curves on its surface as references for describing or determining the positions of stars and various other types of celestial object.

Celestial equator
A great circle on the celestial sphere that lies directly above Earth's equator

Surface of celestial sphere

First point of Libra
One of two points where the celestial equator and ecliptic meet

South celestial pole
A point lying directly below Earth's south pole

Apparent star movement

A person standing still and looking up at the night sky sees a slow, curving movement of stars and other objects across the sky. This apparent motion occurs because Earth is spinning within the celestial sphere. The pattern of motion seen varies according to the observer's location. Movements appear similar in both hemispheres, except that whereas in the northern hemisphere stars appear to circle anticlockwise around the north celestial pole, in the southern hemisphere they circle clockwise around the south celestial pole.

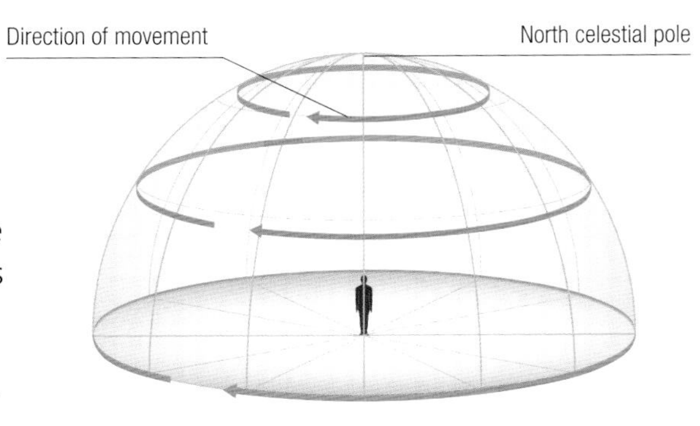

△ **Apparent motion at the north pole**
From the observer's viewpoint, the stars seem to circle anti-clockwise around a point directly overhead – the north celestial pole. Stars near the horizon move around the horizon.

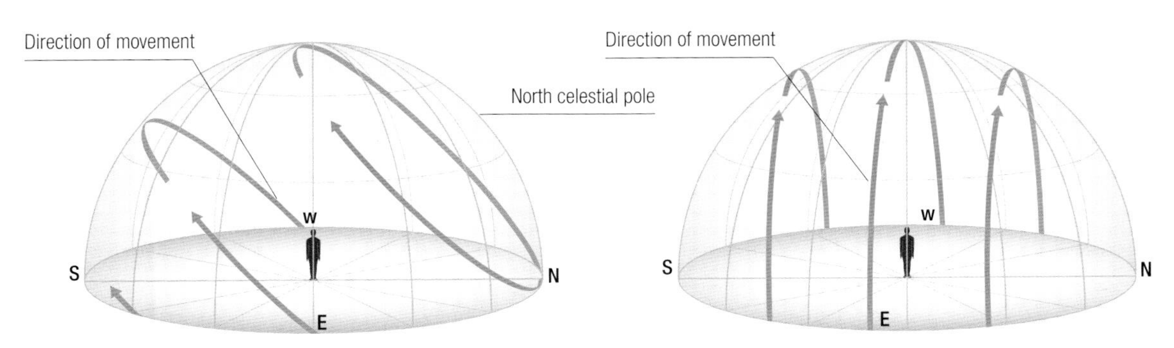

△ **Apparent motion at northern Hemisphere mid-latitudes**
For this observer, most stars rise in the east, cross the southern sky, and set in the west. But stars in the northern part of the sky circle anti-clockwise around the north celestial pole.

△ **Apparent motion at the equator**
For an observer standing on or close to the equator, the stars appear to rise vertically in the east, swing overhead, and then drop vertically down again and set in the west.

Celestial coordinates

Astronomers can record the position of any object on the celestial sphere using a system of coordinates similar to that of latitude and longitude. The coordinates used by astronomers are called declination and right ascension. Declination is measured in degrees north or south of the celestial equator. Right ascension is measured in degrees east of the celestial meridian – a line that passes through both celestial poles and a point on the celestial equator called the First point of Aries.

▷ **Pinpointing a star's position**
The measurement of declination on the celestial sphere is very similar to measuring latitude on Earth's surface, while the measurement of right ascension is quite similar to the expression of longitude. The star shown here has a declination (Dec) of +45°, and a right ascension (RA) of 1 hour, or 15°.

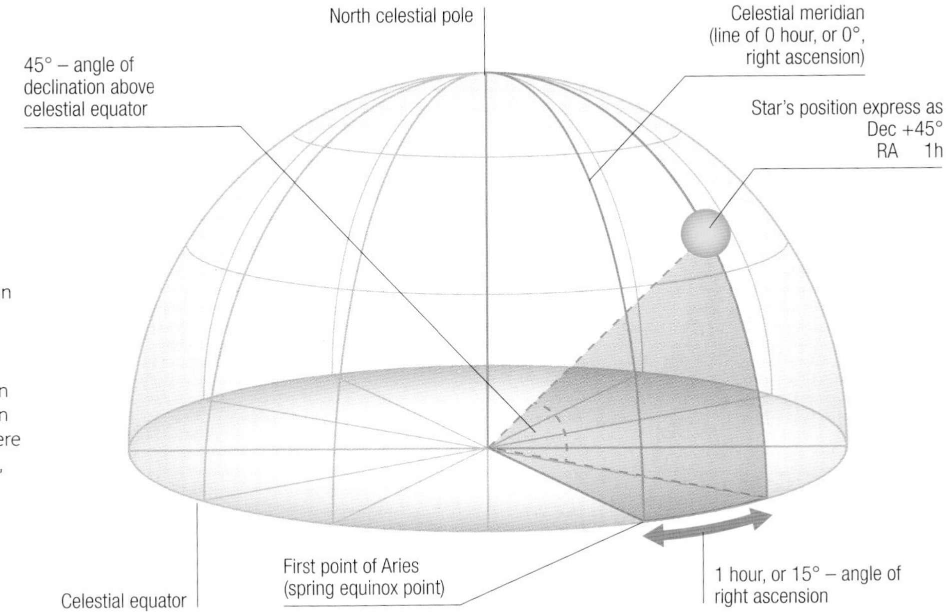

THE ZODIAC

ALTHOUGH IT IS NOT OBVIOUS BECAUSE OF THE SUN'S GLARE, AS EARTH ORBITS THE SUN, THE SUN SEEMS TO MOVE AGAINST THE BACKDROP OF STARS, ALWAYS STAYING WITHIN A BAND OF THE CELESTIAL SPHERE CALLED THE ZODIAC.

During the course of this annual journey around the celestial sphere, the Sun moves along a circle called the ecliptic (see p.90). An imaginary band around the celestial sphere that extends for about 8-9° on either side of the ecliptic is called the zodiac. The ecliptic passes through 13 constellations that lie, at least in part, in the zodiac and these are known as zodiacal constellations. The astrological zodiac is divided into 12 equal segments, called "signs", and excludes the constellation Ophiuchus.

The Sun spends a period of time in each zodiacal constellation but the dates it does so do not correspond with those ascribed to the astrological signs. This is due to the effects of precession and because the constellations are not all the same size.

Whenever the Sun is moving through a particular area of the zodiac, the stars in that part of the celestial sphere cannot be seen because of the glare. Rather, the most easily observed parts of the celestial sphere are always those on the opposite side to Earth from the Sun. These are the parts visible in the middle of the night. Over the course of a year, as Earth orbits the Sun, the portions of the celestial sphere - including the different parts of the zodiac - that can be viewed from Earth at night quite dramatically alter.

SUN'S PROGRESS

Constellation	Dates in each constellation	Constellation	Dates in each constellation
Aries	19 April – 13 May	Scorpio	23 – 29 November
Taurus	14 May – 19 June	Ophiuchus	30 November – 17 December
Gemini	20 June – 20 July	Sagittarius	18 December – 18 January
Cancer	21 July – 9 August	Capricorn	19 January – 15 February
Leo	10 August – 15 September	Aquarius	16 February – 11 March
Virgo	16 September – 30 October	Pisces	12 March – 18 April
Libra	31 October – 22 November		

△ **Days of the zodiac**
The dates the Sun passes through the 13 zodiacal constellations are completely different from the dates associated with the astrological signs of the zodiac.

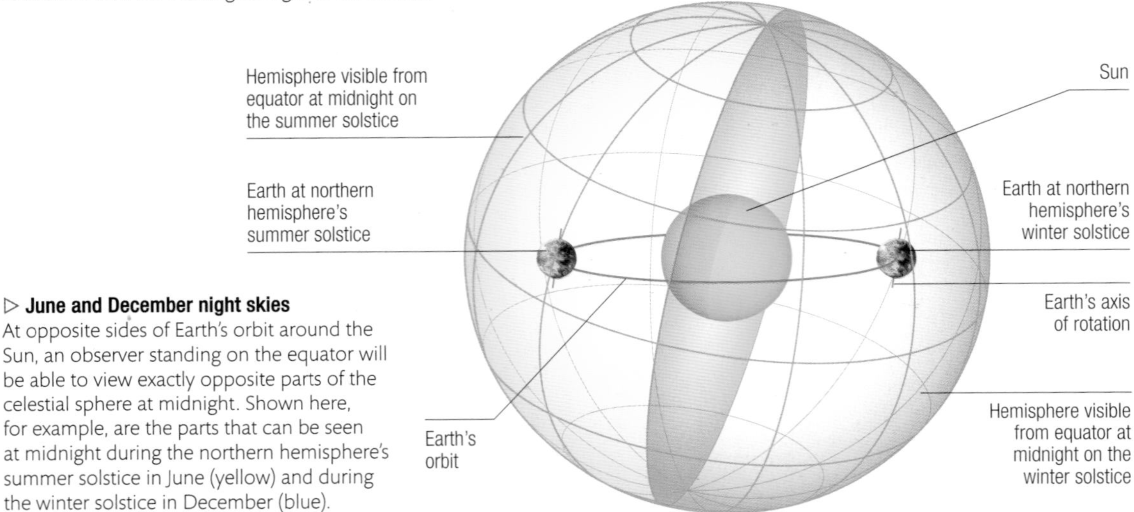

▷ **June and December night skies**
At opposite sides of Earth's orbit around the Sun, an observer standing on the equator will be able to view exactly opposite parts of the celestial sphere at midnight. Shown here, for example, are the parts that can be seen at midnight during the northern hemisphere's summer solstice in June (yellow) and during the winter solstice in December (blue).

Ophiuchus, the 13th constellation of the zodiac

Winter solstice in northern hemisphere, the point where the Sun is farthest below the celestial equator

Libra

Ophiuchus

Scorpius

Sagittarius

Capricornus

▷ **Band of the zodiac**
The zodiac constitutes about one-sixth of the surface area of the celestial sphere (its depth is exaggerated here). The ecliptic runs through its centre. As well as the Sun, the celestial paths of the Moon and the planets of the Solar System are also restricted to the zodiac.

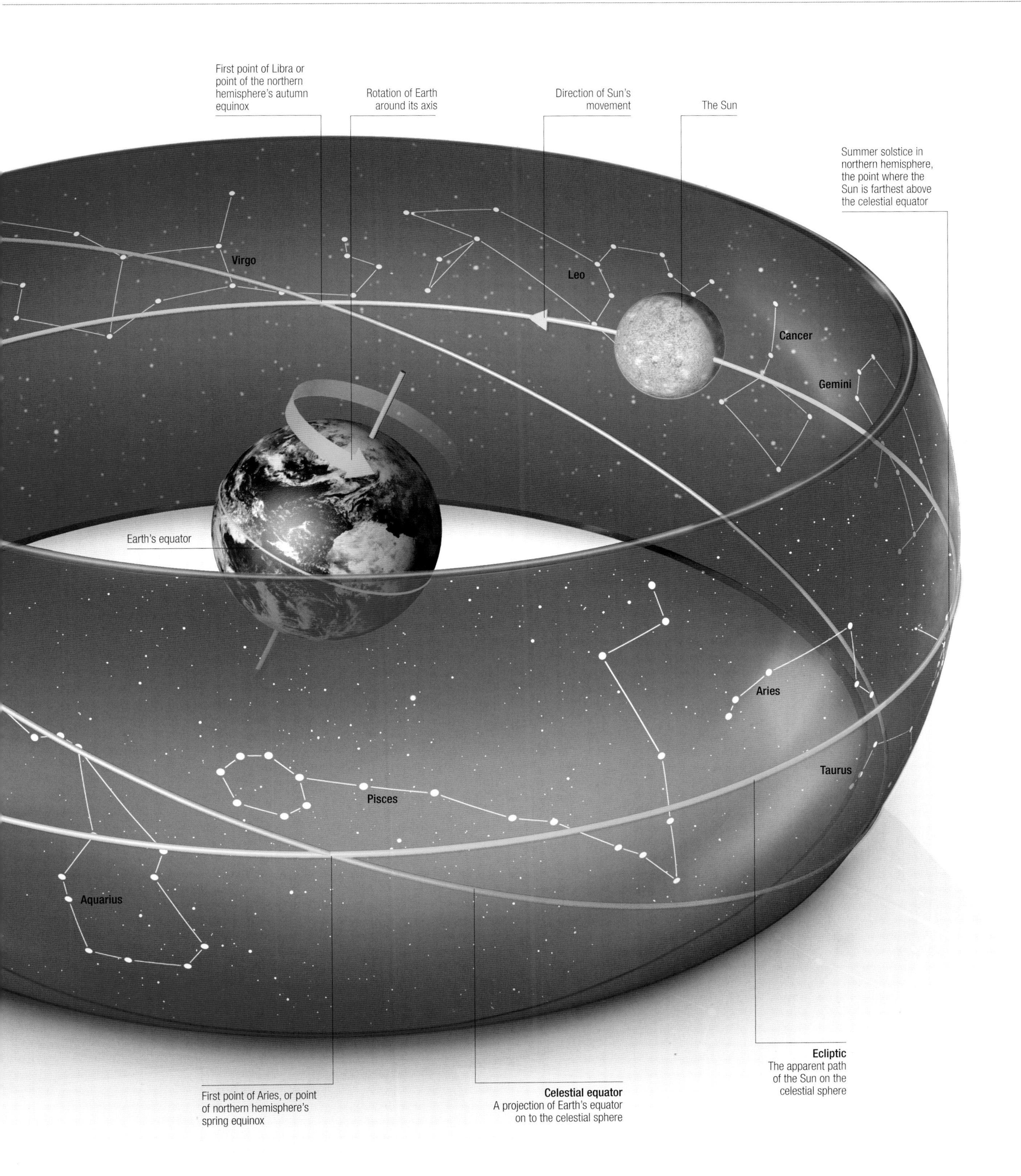
First point of Libra or point of the northern hemisphere's autumn equinox
Rotation of Earth around its axis
Direction of Sun's movement
The Sun
Summer solstice in northern hemisphere, the point where the Sun is farthest above the celestial equator
Virgo
Leo
Cancer
Gemini
Earth's equator
Aries
Taurus
Pisces
Aquarius
First point of Aries, or point of northern hemisphere's spring equinox
Celestial equator
A projection of Earth's equator on to the celestial sphere
Ecliptic
The apparent path of the Sun on the celestial sphere

MAPPING THE SKY

TO FIND OBJECTS IN SPACE AND TO MAKE MAPS OF THE SKY, ASTRONOMERS USE A FRAME OF REFERENCE CALLED THE CELESTIAL SPHERE. THIS SPHERE IS AN IMAGINARY SHELL CENTRED ON THE EARTH UPON WHICH ANY OBJECT IN THE SKY CAN BE LOCATED.

We know that objects in space can lie at any distance from Earth, but in order to position them on a map we can think of them as all being stuck to the inside of the celestial sphere. Just like the Earth itself, the sphere can be divided up with lines of longitude and latitude, including an equator. Similarly, just as the land area of the Earth is separated into countries, the celestial sphere is divided into areas called constellations.

The celestial sphere is an imaginary sphere surrounding Earth

Constellation boundaries are straight and either horizontal or vertical

▷ **The constellations**
For millennia, humans have joined stars with imaginary lines to make recognizable patterns, or constellations. These patterns include the outlines of animals and mythical beasts and heroes. In the early 20th century, the International Astronomical Union gave formal definition to 88 constellations, giving them official names and setting the positions of their boundaries. In this modern system, a constellation is an area of sky rather than a pattern of lines between stars.

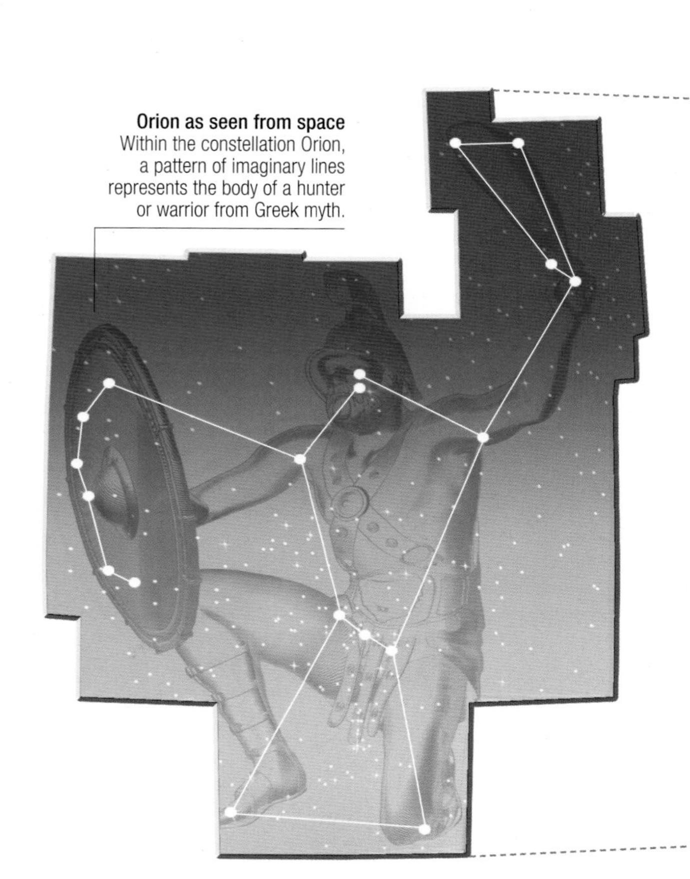

Orion as seen from space
Within the constellation Orion, a pattern of imaginary lines represents the body of a hunter or warrior from Greek myth.

▽ **Observer's location**
From a particular place on Earth, up to half of the celestial sphere can be seen at any one time, with the rest hidden by the Earth itself. Whether or not a particular constellation is visible also depends on an observer's location. For example, the whole of the constellation Canis Major can be seen between latitudes 56 degrees north and the south pole. From a belt to the north of this, only part of the constellation can be seen, while in the region around the north pole, none of the constellation is visible.

Constellation not visible

Part of constellation visible

Whole constellation visible

CANIS MAJOR VISIBILITY FROM EARTH

Canis Major

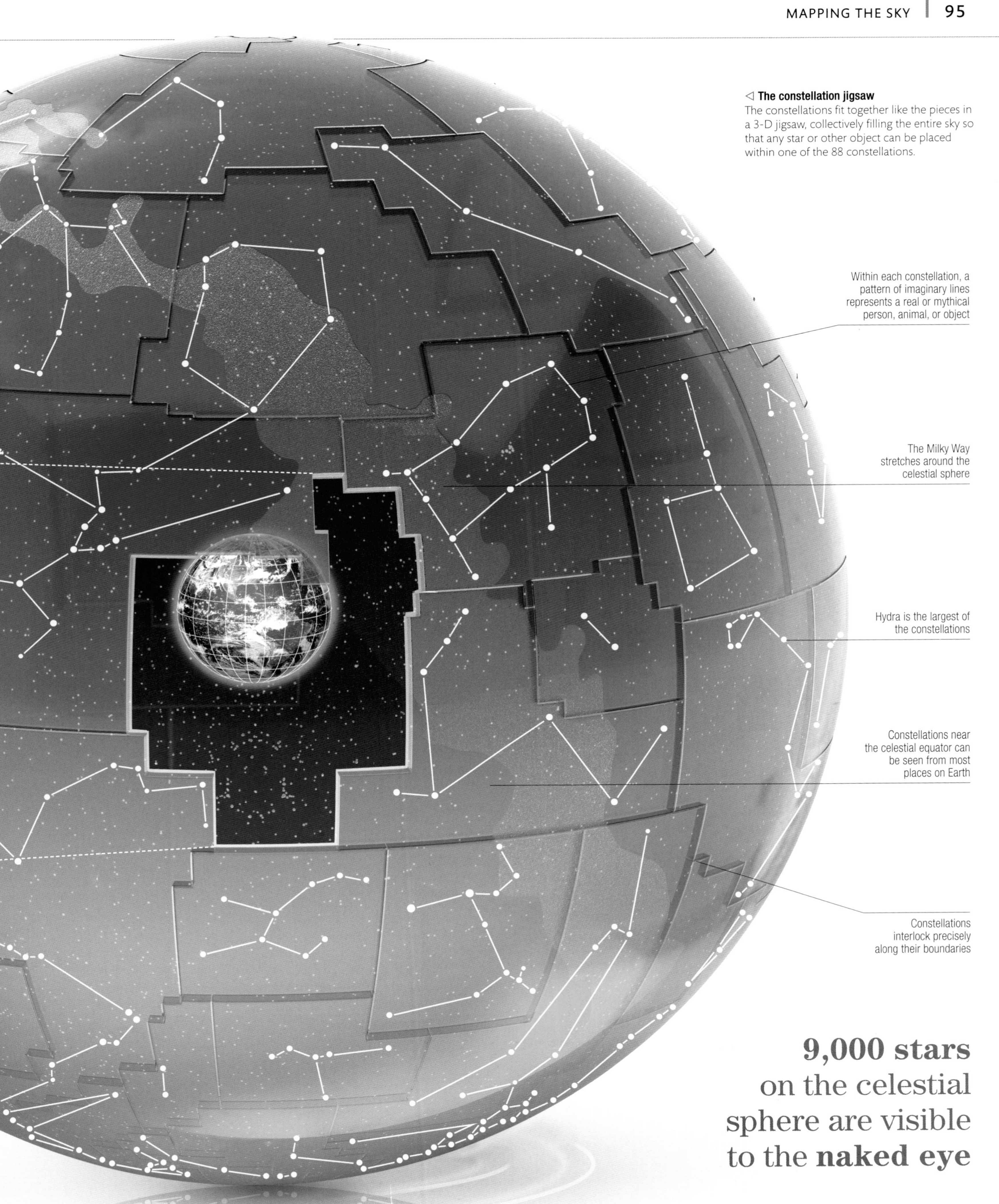
Within each constellation, a pattern of imaginary lines represents a real or mythical person, animal, or object
The Milky Way stretches around the celestial sphere
Hydra is the largest of the constellations
Constellations near the celestial equator can be seen from most places on Earth
Constellations interlock precisely along their boundaries

◁ **The constellation jigsaw**
The constellations fit together like the pieces in a 3-D jigsaw, collectively filling the entire sky so that any star or other object can be placed within one of the 88 constellations.

9,000 stars on the celestial sphere are visible to the **naked eye**

SKY CHARTS

THE SIX CHARTS ON THE FOLLOWING PAGES COVER THE WHOLE CELESTIAL SPHERE; ONE FOR EACH OF THE NORTH AND SOUTH POLAR REGIONS, AND FOUR FOR THE BELT OF SKY BETWEEN.

Visibility, magnitude, and distance
Each constellation has a data panel, which gives key information about the constellation, including the latitudes from where it is fully visible, and the months when it is highest in the sky. Each of the main stars has a brightness symbol together with its apparent magnitude, and a distance symbol with its distance from Earth in light-years.

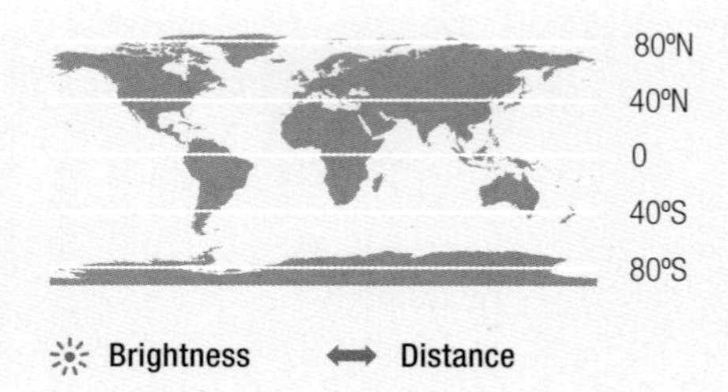

Constellation chart key
The individual constellation charts show the main stars of the constellation, including the stars that make up its pattern and other notable stars. The apparent magnitude (brightness) of the stars is indicated by the key shown right. The charts also include key deep-sky objects, such as galaxies, nebulae, and star clusters, the symbols for which are also shown in the key on the right.

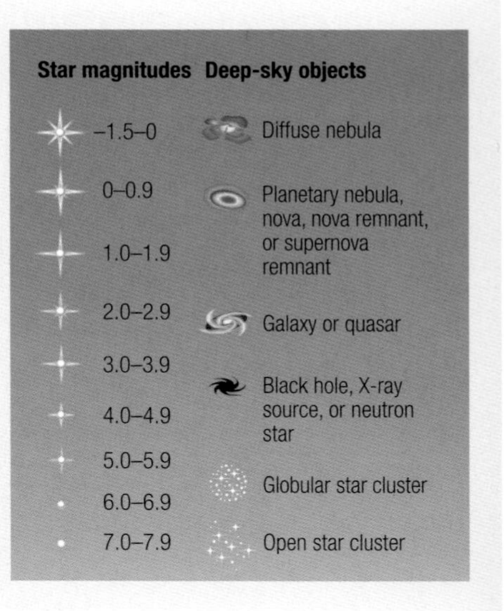

Star magnitudes
Deep-sky objects
–1.5–0
0–0.9
1.0–1.9
2.0–2.9
3.0–3.9
4.0–4.9
5.0–5.9
6.0–6.9
7.0–7.9
Diffuse nebula
Planetary nebula, nova, nova remnant, or supernova remnant
Galaxy or quasar
Black hole, X-ray source, or neutron star
Globular star cluster
Open star cluster

Together, the six charts show the entire sky surrounding Earth and the location of all 88 constellations. The two circular maps shown here are each centred on a celestial pole. The other four maps on the following pages cover the equatorial regions; each is centred on a quarter of the celestial equator. Individual constellations are profiled in the pages following the maps.

CHART 1

NORTH POLAR SKY

Centred on the north celestial pole, this chart shows the constellations of the north polar sky. It covers the area from declination 90° at the pole, southwards to declination 50°. The star Polaris, in Ursa Minor, is less than 1° from the pole and almost in the centre of the chart. Polaris and the other stars around it are circumpolar; they never set below the horizon for observers in the northern hemisphere. How much of the sky is circumpolar depends on the observer's latitude; the amount increases the farther north you are.

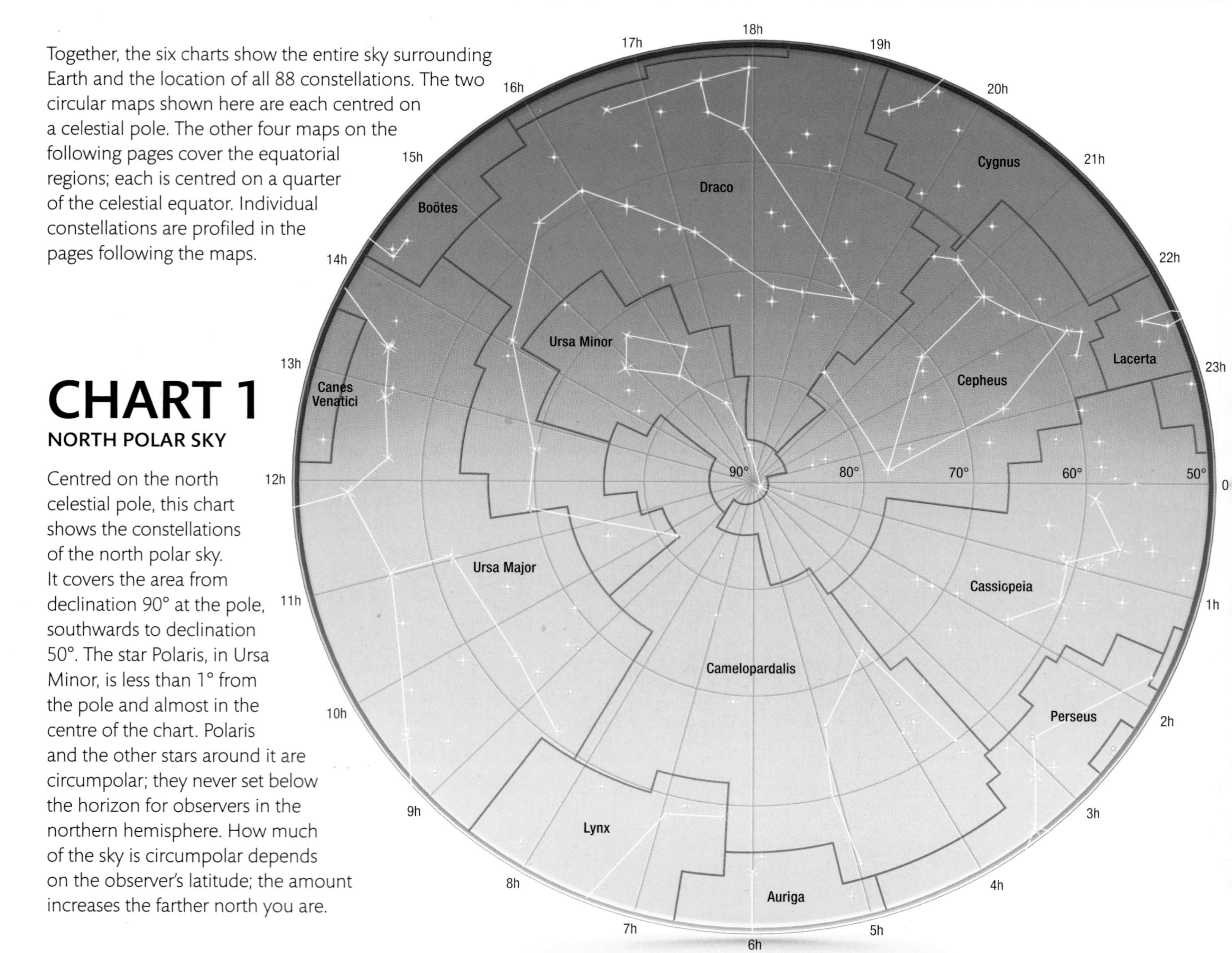

Luminosity scale
Major constellations include a scale that shows the luminosity (the total energy emitted, in multiple's of our Sun's energy) of key stars, including the least and most luminous of the constellation's pattern stars.

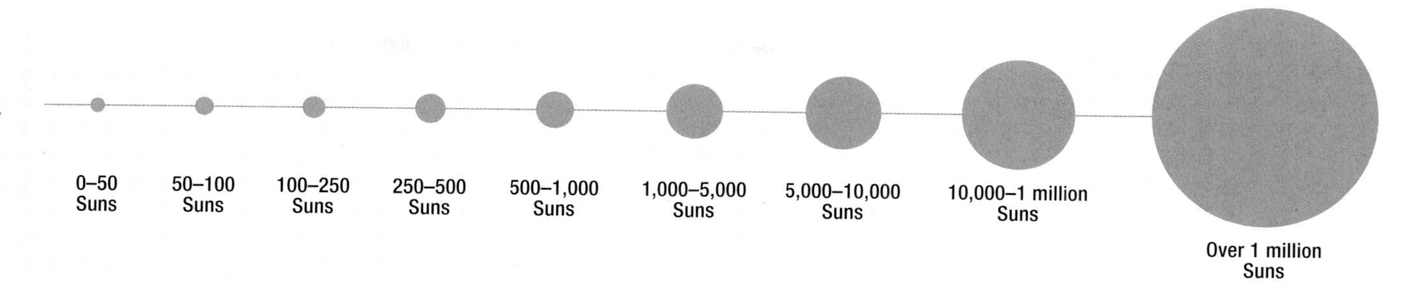

Constellation locator charts
Each constellation includes a locator chart in the data panel that shows where the constellation lies on the celestial sphere. The locator charts are numbered to correspond to the large charts on these introductory pages.

Greek alphabet
The constellation charts use letters of the Greek alphabet to identify bright stars, according to the commonly used system originally invented by the German astronomer Johann Bayer.

Alpha	α	Eta	η	Nu	ν	Tau	τ
Beta	β	Theta	θ	Xi	ξ	Upsilon	υ
Gamma	γ	Iota	ι	Omicron	ο	Phi	φ
Delta	δ	Kappa	κ	Pi	π	Chi	χ
Epsilon	ε	Lambda	λ	Rho	ρ	Psi	ψ
Zeta	ζ	Mu	μ	Sigma	σ	Omega	ω

CHART 2
SOUTH POLAR SKY

Centred on the south celestial pole, this chart shows the constellations of the south polar sky. It covers the area from declination -90° at the pole, northwards to declination -50°. The sky around the pole is lacking in bright stars, and no one star is close enough to identify the pole's position. Stars in the area around the pole are circumpolar for observers in the southern hemisphere; they remain visible in the night sky, never setting below the horizon. The farther south the observer is located, the greater the amount of sky that is circumpolar.

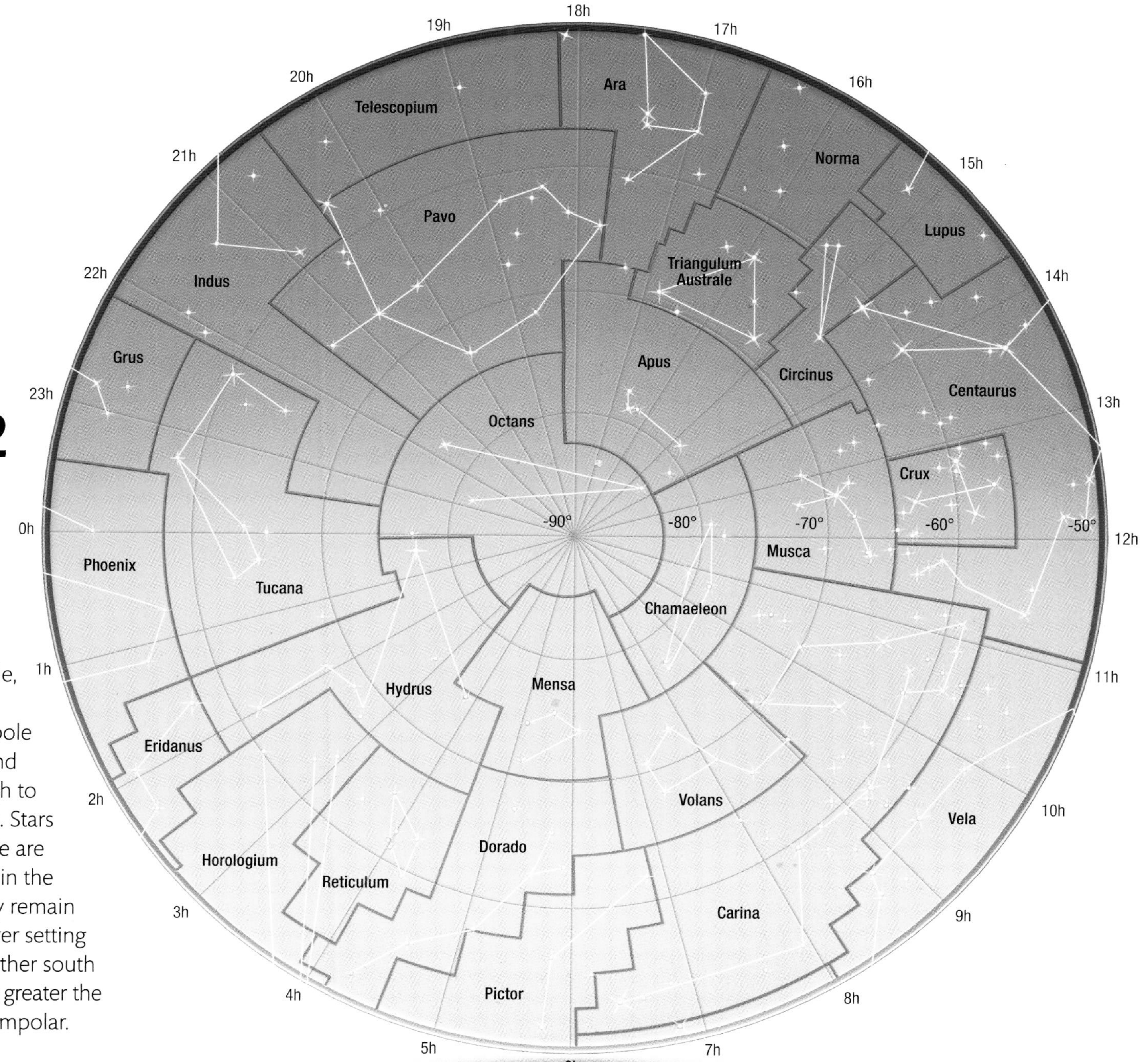

CHART 3

EQUATORIAL SKY

The region of sky in this chart is best placed for observation on evenings in September, October, and November. The map is centred on a part of the celestial equator that is crossed by the ecliptic, the Sun's path. The crossing point is where the Sun moves from the southern to the northern sky in late March each year. It is the point where lines of right ascension are measured from, and is the celestial equivalent of 0° longitude on Earth.

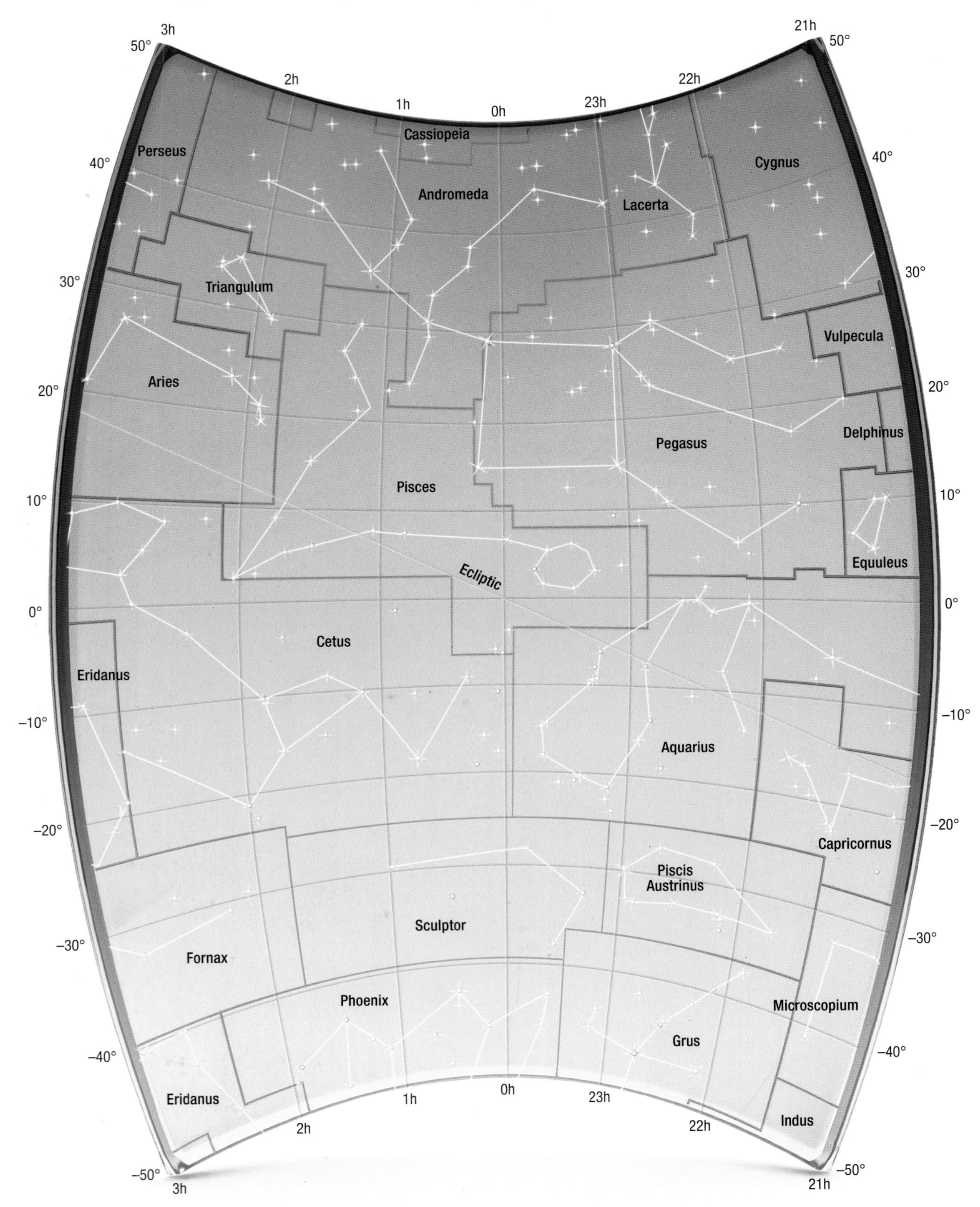

CHART 4
EQUATORIAL SKY

The region of sky in this chart is best placed for observation on evenings in June, July, and August. The Sun's path is always south of the celestial equator in this part of the sky. Each year it reaches its most southerly declination in Sagittarius, around 21 December, when it is the longest day in the southern hemisphere and shortest in the northern. Rich Milky Way star fields cross this region from Cygnus in the north to Scorpius in the south.

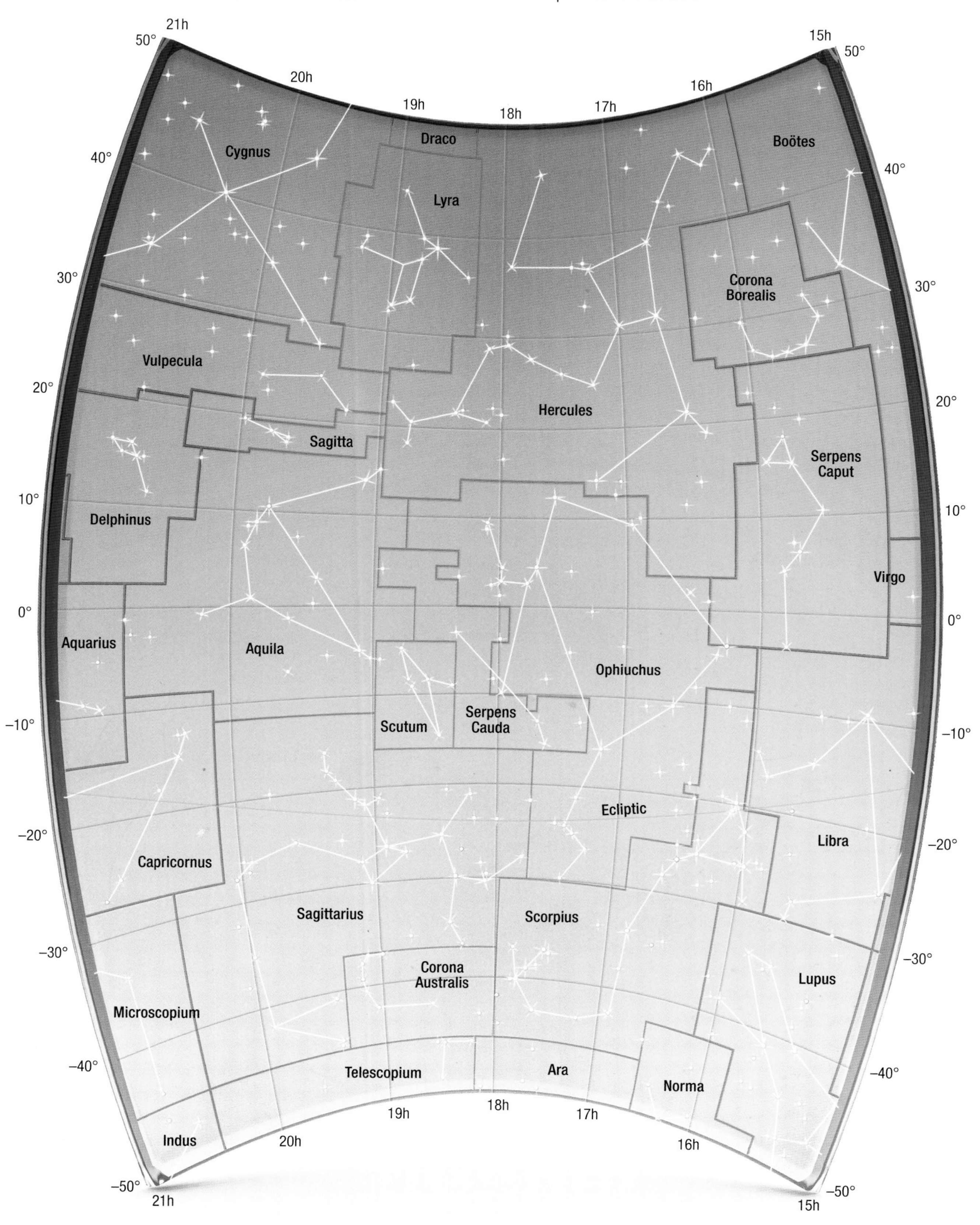

CHART 5

EQUATORIAL SKY

The region of sky in this chart is best placed for observation on evenings in March, April, and May. The map is centred on a part of the celestial equator crossed by the ecliptic, the Sun's path. The crossing point, within Virgo, is where the Sun moves from the northern to the southern sky in September. Day and night are then of equal length across the planet. The appearance of Arcturus, Boötes' bright star, marks the arrival of northern spring.

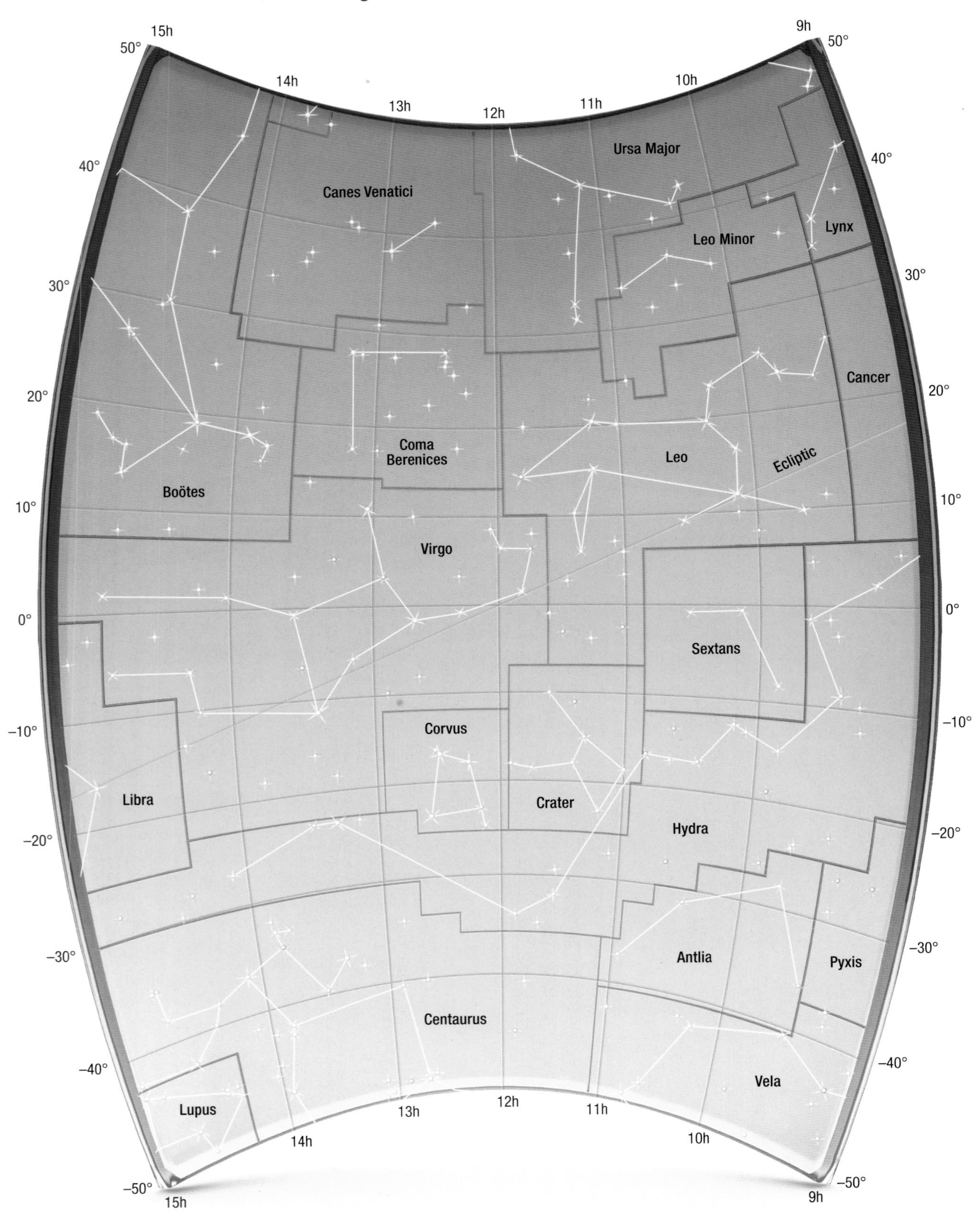

CHART 6

EQUATORIAL SKY

The region of sky in this chart is best placed for observation on evenings in December, January, and February. The Sun's path is always north of the celestial equator in this part of the sky. Each year it reaches its most northerly declination on the Taurus–Gemini border. This occurs around 21 June, which is the longest day in the northern hemisphere and the shortest day in the southern hemisphere.

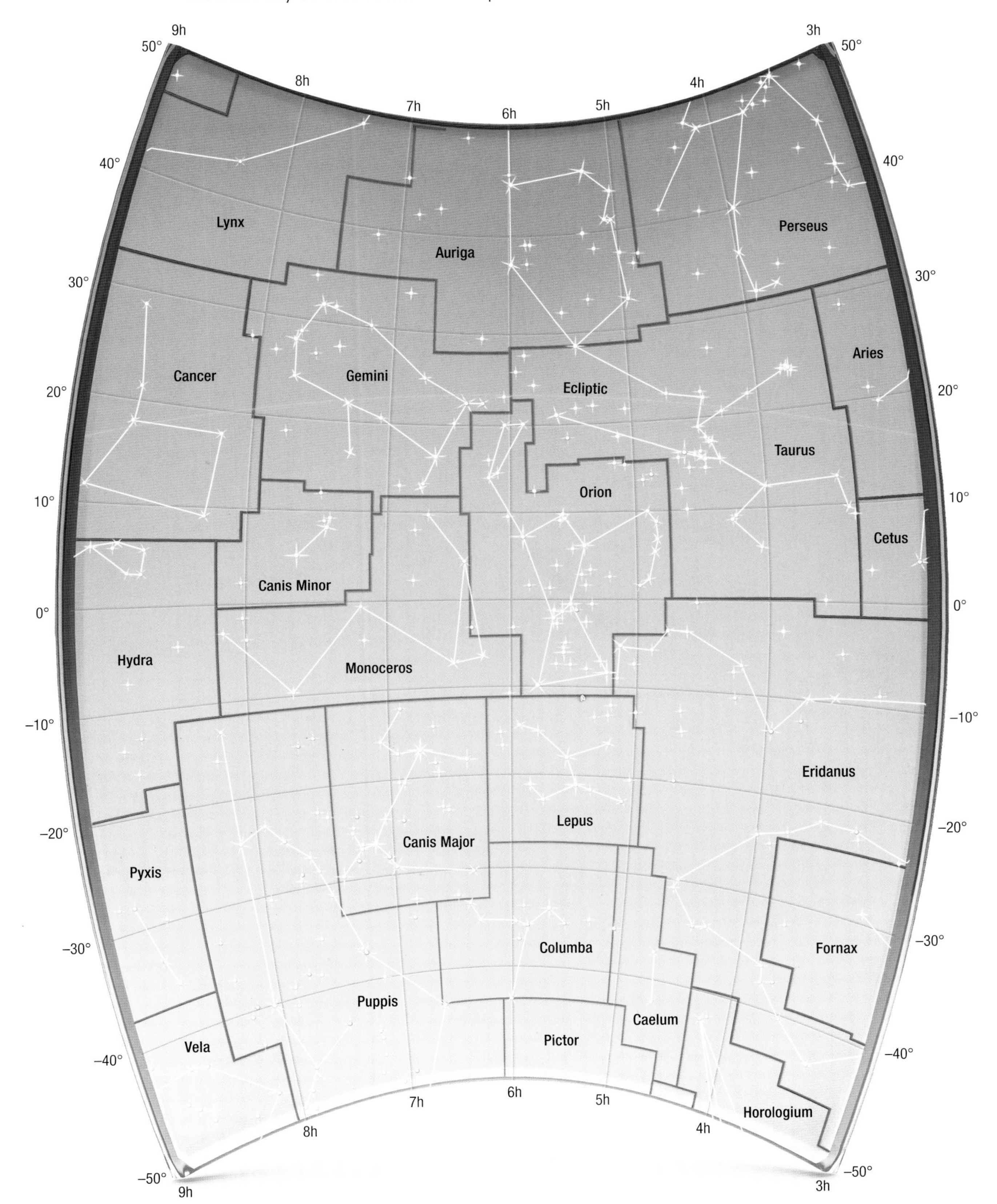

URSA MINOR THE LITTLE BEAR

URSA MINOR CONTAINS THE NORTH CELESTIAL POLE, AND ITS BRIGHTEST STAR, POLARIS, IS THE NORTH POLE STAR. THE CONSTELLATION REPRESENTS A SMALL BEAR, A COMPANION OF URSA MAJOR, THE GREAT BEAR.

Consisting of seven main stars arranged in a saucepan shape, Ursa Minor resembles a small version of the Big Dipper, hence its popular name of the Little Dipper. In Greek mythology, it represents a nymph who nursed the god Zeus as an infant. Polaris, its brightest star, lies very near the north celestial pole and is an easy guide to finding north at night.

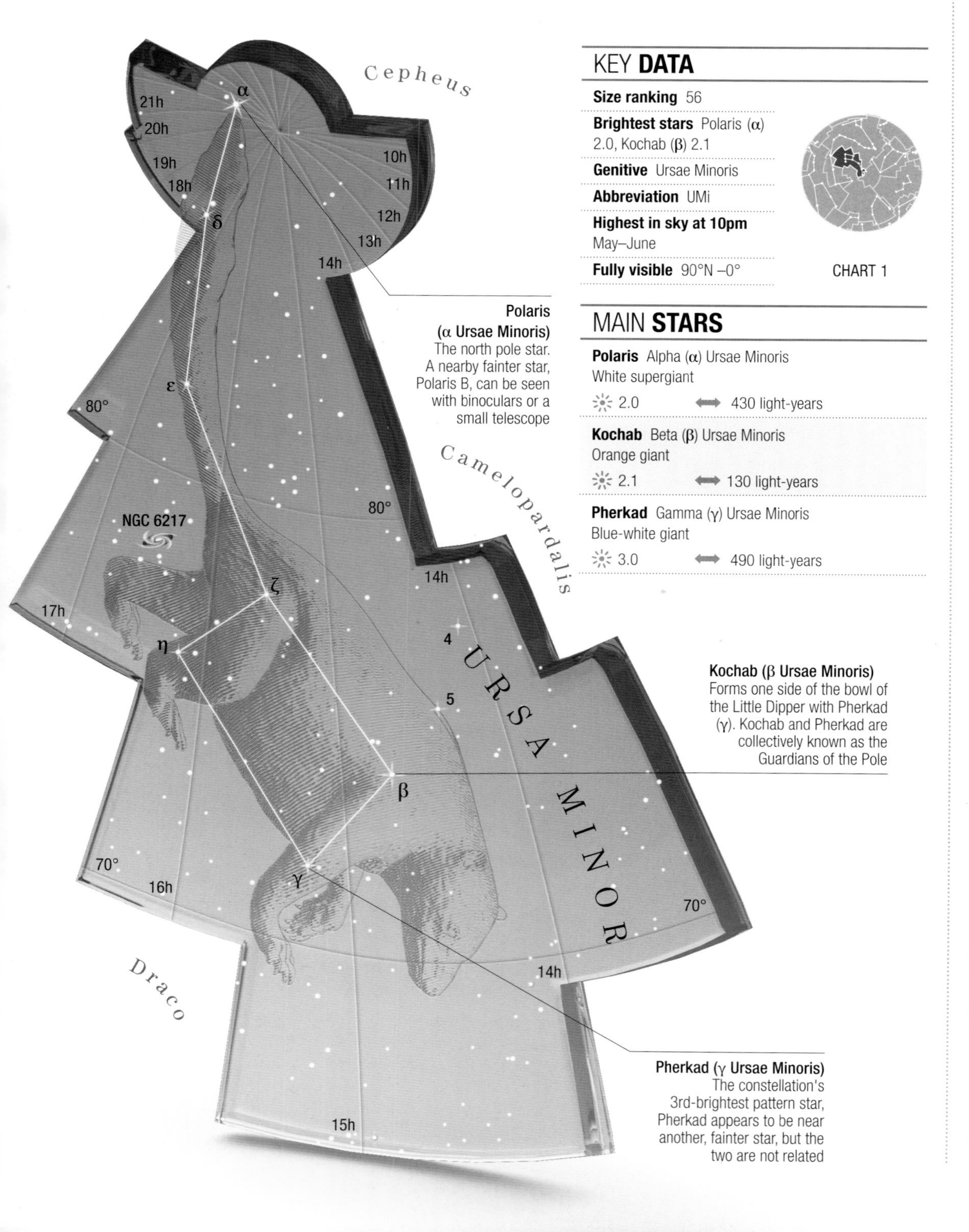
Cepheus
Camelopardalis
URSA MINOR
Draco
NGC 6217
α
β
γ
δ
ε
ζ
η
4
5
21h
20h
19h
18h
17h
16h
15h
14h
13h
12h
11h
10h
80°
70°
Polaris (α Ursae Minoris) The north pole star. A nearby fainter star, Polaris B, can be seen with binoculars or a small telescope
Kochab (β Ursae Minoris) Forms one side of the bowl of the Little Dipper with Pherkad (γ). Kochab and Pherkad are collectively known as the Guardians of the Pole
Pherkad (γ Ursae Minoris) The constellation's 3rd-brightest pattern star, Pherkad appears to be near another, fainter star, but the two are not related
CHART 1

KEY DATA

Size ranking	56
Brightest stars	Polaris (α) 2.0, Kochab (β) 2.1
Genitive	Ursae Minoris
Abbreviation	UMi
Highest in sky at 10pm	May–June
Fully visible	90°N –0°

MAIN STARS

Star	Designation	Type	Magnitude	Distance
Polaris	Alpha (α) Ursae Minoris	White supergiant	2.0	430 light-years
Kochab	Beta (β) Ursae Minoris	Orange giant	2.1	130 light-years
Pherkad	Gamma (γ) Ursae Minoris	Blue-white giant	3.0	490 light-years

CEPHEUS

MYTHICAL KING OF ETHIOPIA

A FAINT NORTHERN CONSTELLATION, CEPHEUS IS SHAPED LIKE A BUILDING WITH A POINTED ROOF. IT REPRESENTS A KING IN GREEK MYTHOLOGY AND CONTAINS THE PROTOTYPE OF THE CEPHEID VARIABLE STARS.

Cepheus was supposedly the King of Ethiopia, a mythical country on the eastern Mediterranean, not the African country we know today. He was the husband of Cassiopeia, who lies next to him in the sky, and the father of Andromeda.

The constellation's most important features are two famous variable stars. Delta Cephei was the first of the pulsating stars known as Cepheid variables to be discovered. In 1784, the English amateur astronomer John Goodricke noted variations in its brightness, which cycles from magnitude 3.5 to 4.4 and back every 5 days 9 hours. It is also a triple star, with one fainter companion visible through a small telescope. Mu Cephei, another variable, is known as the Garnet Star because of its strong red colour. A red supergiant, it ranges between magnitudes 3.4 and 5.1 approximately every two years.

△ **IC 1396**
Situated near the border with Cygnus in the south of Cepheus, IC 1396 is a star cluster surrounded by a large cloud of glowing gas. Seen in silhouette against the bright gas in this image is a dark area called the Elephant's Trunk Nebula, which is a region of gas and dust in which new stars are forming.

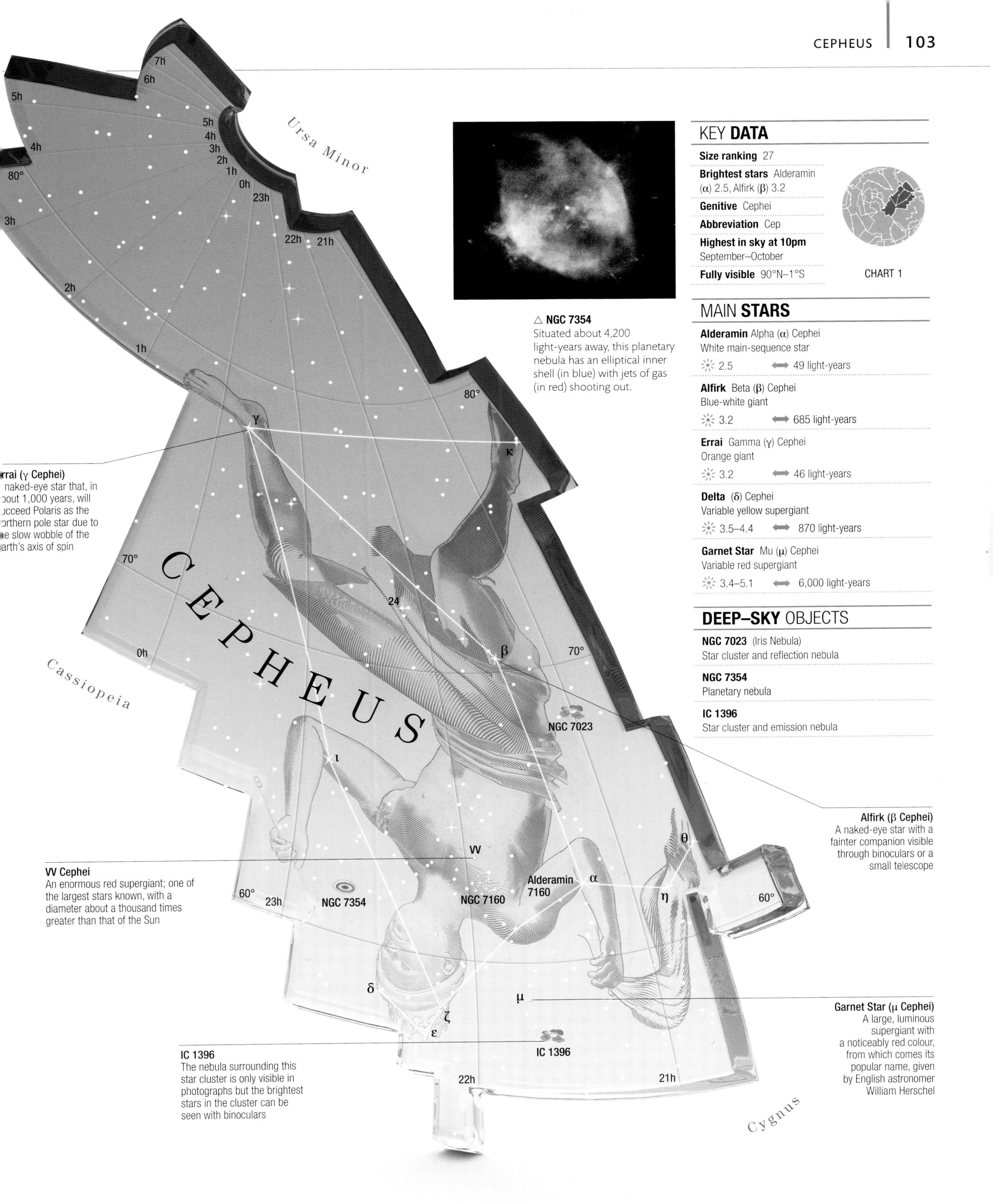

△ **NGC 7354**
Situated about 4,200 light-years away, this planetary nebula has an elliptical inner shell (in blue) with jets of gas (in red) shooting out.

KEY **DATA**

Size ranking 27

Brightest stars Alderamin (α) 2.5, Alfirk (β) 3.2

Genitive Cephei

Abbreviation Cep

Highest in sky at 10pm September–October

Fully visible 90°N–1°S

CHART 1

MAIN **STARS**

Alderamin Alpha (α) Cephei
White main-sequence star
2.5 ⟷ 49 light-years

Alfirk Beta (β) Cephei
Blue-white giant
3.2 ⟷ 685 light-years

Errai Gamma (γ) Cephei
Orange giant
3.2 ⟷ 46 light-years

Delta (δ) Cephei
Variable yellow supergiant
3.5–4.4 ⟷ 870 light-years

Garnet Star Mu (μ) Cephei
Variable red supergiant
3.4–5.1 ⟷ 6,000 light-years

DEEP–SKY OBJECTS

NGC 7023 (Iris Nebula)
Star cluster and reflection nebula

NGC 7354
Planetary nebula

IC 1396
Star cluster and emission nebula

rrai (γ Cephei)
naked-eye star that, in bout 1,000 years, will ucceed Polaris as the orthern pole star due to e slow wobble of the arth's axis of spin

Alfirk (β Cephei)
A naked-eye star with a fainter companion visible through binoculars or a small telescope

VV Cephei
An enormous red supergiant; one of the largest stars known, with a diameter about a thousand times greater than that of the Sun

Garnet Star (μ Cephei)
A large, luminous supergiant with a noticeably red colour, from which comes its popular name, given by English astronomer William Herschel

IC 1396
The nebula surrounding this star cluster is only visible in photographs but the brightest stars in the cluster can be seen with binoculars

LUMINOSITIES

Omega Draconis
6 Suns

Nu[1] Draconis
9 Suns

Delta Draconi
46 Suns

DRACO THE DRAGON

DRACO WINDS NEARLY HALFWAY ROUND THE NORTH CELESTIAL POLE. IT IS MOST EASILY IDENTIFIED BY THE PATTERN OF THE FOUR STARS THAT MARK ITS HEAD.

Draco represents the dragon of Ancient Greek mythology that was killed by Hercules as one of his 12 labours. In the sky, Hercules kneels next to the dragon, with one foot on its head. Despite its large size, Draco is not a particularly prominent constellation. Its brightest star, Gamma – popularly known as Etamin or Eltanin – is of only 2nd magnitude. The constellation contains many double stars divisible by small telescopes or even binoculars, including Nu, a 5th-magnitude pair; Psi, a 5th- and 6th-magnitude pair; 16 and 17 Draconis, both of 5th magnitude; and 40 and 41 Draconis, both of 6th magnitude. Draco's comparatively few notable deep-sky objects include the Cat's Eye Nebula (NGC 6543) and the distorted spiral Tadpole Galaxy (UGC 10214).

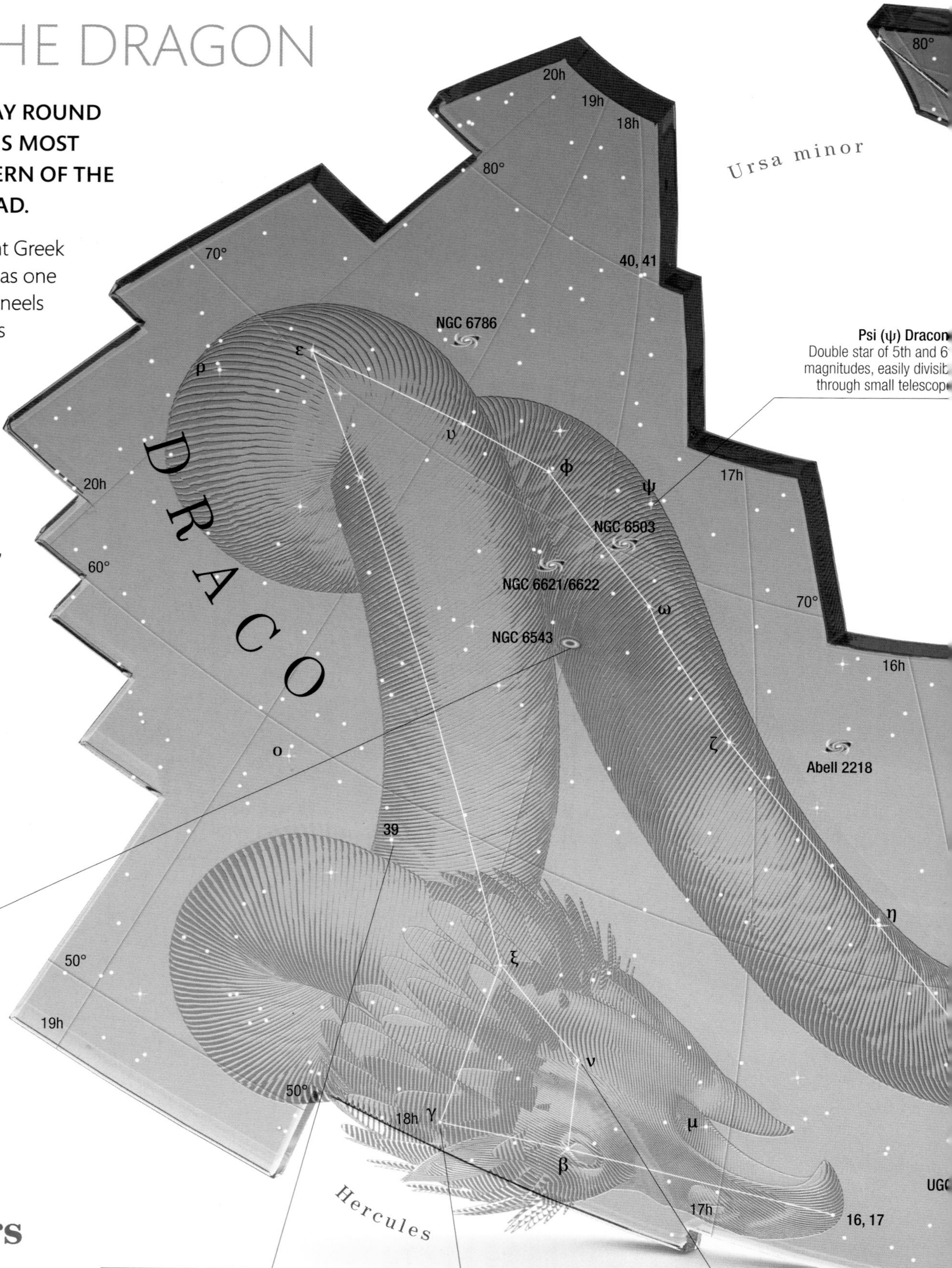

Psi (ψ) Dracon
Double star of 5th and 6
magnitudes, easily divisib
through small telescop

NGC 6543
Planetary nebula, popularly known as the Cat's Eye Nebula, lying about 3,000 light-years away and visible through small telescopes as a bluish disk

39 Draconis
A wide pair of stars, of 5th and 8th magnitudes, divisible through binoculars or small telescopes

Etamin (γ Draconis)
Also called Eltanin, Draco's brightest star, magnitude 2.2. It forms a lozenge shape with Beta (β), Nu (ν), and Xi (ξ), which marks the dragon's head

Nu (ν) Draconi
Widely spaced pair o
matching 5th-magnitude whit
stars, visible with binoculars o
small telescope

Thuban was the north **pole star about 3,000 years ago** but is now far from the pole due to wobbling of the Earth's axis of spin

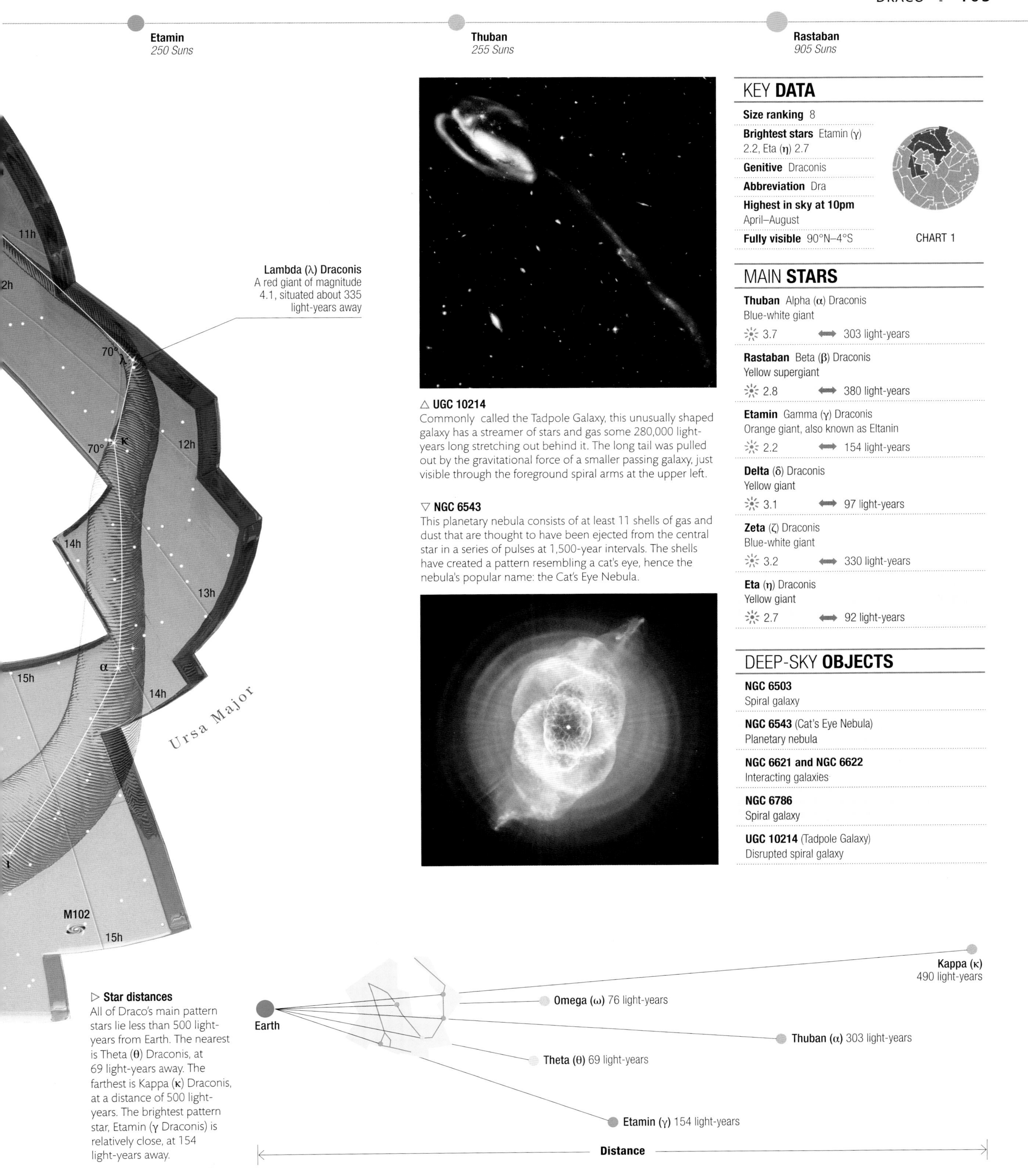

△ **UGC 10214**
Commonly called the Tadpole Galaxy, this unusually shaped galaxy has a streamer of stars and gas some 280,000 light-years long stretching out behind it. The long tail was pulled out by the gravitational force of a smaller passing galaxy, just visible through the foreground spiral arms at the upper left.

▽ **NGC 6543**
This planetary nebula consists of at least 11 shells of gas and dust that are thought to have been ejected from the central star in a series of pulses at 1,500-year intervals. The shells have created a pattern resembling a cat's eye, hence the nebula's popular name: the Cat's Eye Nebula.

▷ **Star distances**
All of Draco's main pattern stars lie less than 500 light-years from Earth. The nearest is Theta (θ) Draconis, at 69 light-years away. The farthest is Kappa (κ) Draconis, at a distance of 500 light-years. The brightest pattern star, Etamin (γ Draconis) is relatively close, at 154 light-years away.

KEY **DATA**

Size ranking 8

Brightest stars Etamin (γ) 2.2, Eta (η) 2.7

Genitive Draconis

Abbreviation Dra

Highest in sky at 10pm April–August

Fully visible 90°N–4°S

MAIN **STARS**

Thuban Alpha (α) Draconis
Blue-white giant
3.7 ⟷ 303 light-years

Rastaban Beta (β) Draconis
Yellow supergiant
2.8 ⟷ 380 light-years

Etamin Gamma (γ) Draconis
Orange giant, also known as Eltanin
2.2 ⟷ 154 light-years

Delta (δ) Draconis
Yellow giant
3.1 ⟷ 97 light-years

Zeta (ζ) Draconis
Blue-white giant
3.2 ⟷ 330 light-years

Eta (η) Draconis
Yellow giant
2.7 ⟷ 92 light-years

DEEP-SKY **OBJECTS**

NGC 6503
Spiral galaxy

NGC 6543 (Cat's Eye Nebula)
Planetary nebula

NGC 6621 and NGC 6622
Interacting galaxies

NGC 6786
Spiral galaxy

UGC 10214 (Tadpole Galaxy)
Disrupted spiral galaxy

LUMINOSITIES

Eta Cassiopeiae *1 Sun* · Caph *30 Suns* · Ruchbah *70 Suns*

CASSIOPEIA

THE VAIN QUEEN

CASSIOPEIA LIES WITHIN THE BAND OF THE MILKY WAY. ITS FIVE MAIN STARS FORM A ZIG-ZAG SHAPE THAT RESEMBLES THE LETTER "W", MAKING THIS CONSTELLATION EASY TO RECOGNIZE IN THE NORTHERN SKY.

Cassiopeia was a vain queen of Greek mythology, the wife of King Cepheus. As punishment for Cassiopeia's vanity, the sea god Poseidon sent a monster to ravage her country's coastline. To rid themselves of the monster, Cassiopeia and Cepheus chained their daughter Andromeda to a rock as a sacrifice. Fortunately, she was rescued from the monster's jaws by the hero Perseus. All the characters in this myth are represented by constellations close together in the night sky.

Cassiopeia contains the remains of two supernova explosions. One, called Tycho's Star, became visible from Earth in 1572. The other occurred about a century later, but went unseen at the time.

The major features of the constellation for users of small telescopes are the beautiful double star Eta Cassiopeiae and several open clusters of stars, notably M52, M103, and NGC 457.

50,000 BCE

50,000 CE

100,000 CE

◁ **Changing shape**
All stars are moving through space, so constellation patterns gradually change with time. These diagrams show the stars that make up Cassiopeia 50,000 years ago (top) and how it will appear in 50,000 and 100,000 years from now (centre and bottom).

IC 1805
Surrounding this star cluster is a cloud of glowing gas called the Heart Nebula, so-named because it resembles the shape of a human heart

◁ **Cassiopeia A supernova remnant**
The strongest radio source in the sky, Cassiopeia A has been identified as the remains of a supernova explosion some 11,000 light-years away. Light from the supernova should have reached Earth in the 1600s. However, there is no record of it having been observed, so it was probably dimmed by the surrounding dust. This image of the exploded star is a composite of observations made at infrared (red), optical (yellow), and X-ray (green and blue) wavelengths.

▷ **Star distances**
One might be forgiven for thinking that the five main stars in Cassiopeia's distinctive "W" formation are relatively close together, but in fact they lie at greatly differing distances from Earth. The farthest away, Gamma Cassiopeiae (the central star in the "W"), is over ten times more distant than the nearest of the five to us, Caph (Beta Cassiopeiae)

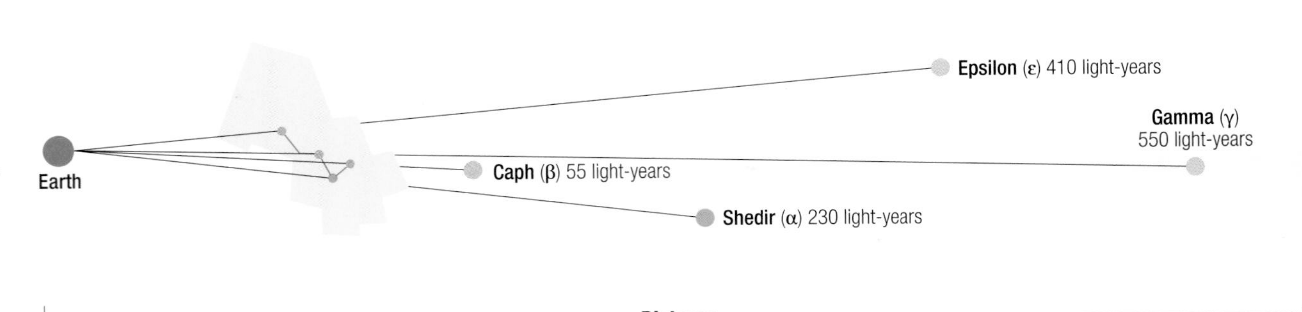

Shedir
540 Suns

Epsilon Cassiopeiae
630 Suns

Gamma Cassiopeiae
3,400 Suns

In November 1572, a **supernova** in Cassiopeia was **as bright as** the planet **Venus** and **visible by day**

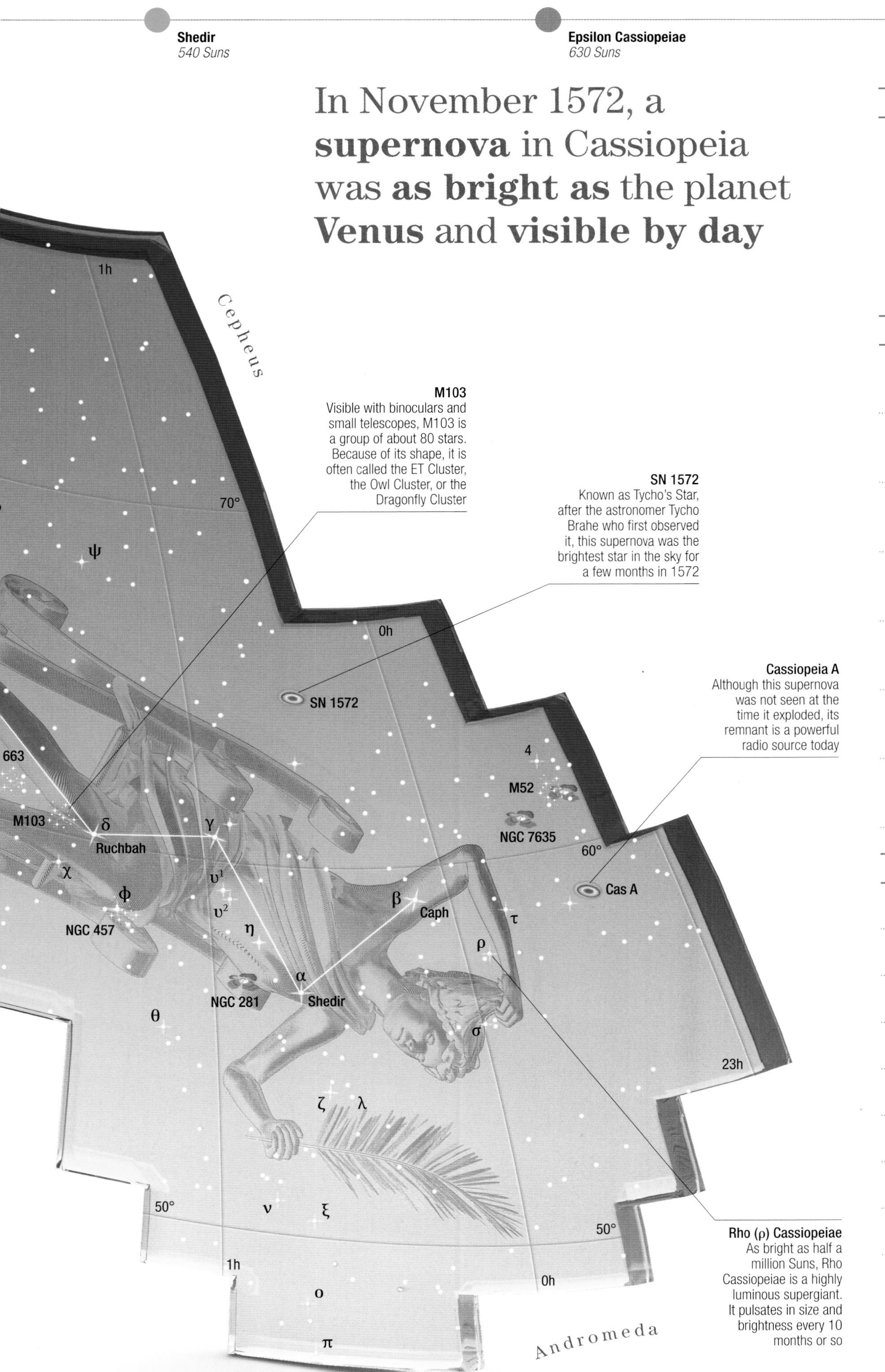

M103
Visible with binoculars and small telescopes, M103 is a group of about 80 stars. Because of its shape, it is often called the ET Cluster, the Owl Cluster, or the Dragonfly Cluster

SN 1572
Known as Tycho's Star, after the astronomer Tycho Brahe who first observed it, this supernova was the brightest star in the sky for a few months in 1572

Cassiopeia A
Although this supernova was not seen at the time it exploded, its remnant is a powerful radio source today

Rho (ρ) Cassiopeiae
As bright as half a million Suns, Rho Cassiopeiae is a highly luminous supergiant. It pulsates in size and brightness every 10 months or so

KEY DATA

Size ranking 25

Brightest stars Shedir (α) 2.2, Gamma (γ) 2.2

Genitive Cassiopeiae

Abbreviation Cas

Highest in sky at 10pm October–December

Fully visible 90°N–12°S

CHART 1

MAIN STARS

Shedir Alpha (α) Cassiopeiae
Orange giant; Schedar is an alternative spelling
2.2 — 230 light-years

Caph Beta (β) Cassiopeiae
White giant
2.3 — 55 light-years

Gamma (γ) Cassiopeiae
Blue-white subgiant
2.4 — 550 light-years

Ruchbah Delta (δ) Cassiopeiae
White subgiant
2.7 — 99 light-years

Epsilon (ε) Cassiopeiae
Blue giant
3.4 — 410 light-years

Eta (η) Cassiopeiae
Yellow main-sequence star
3.4 — 19 light-years

Rho (ρ) Cassiopeiae
Yellow supergiant variable
4.1–6.2 — 12,000 light-years

DEEP-SKY OBJECTS

M52
Bright open cluster of about 100 stars

M103
Small open cluster of about 25 stars

NGC 457
Loose open cluster of about 80 stars

NGC 663
Large open cluster of about 80 stars

NGC 7635
Emission nebula; also known as the Bubble Nebula

IC 1805
Star cluster surrounded by the Heart Nebula

Cassiopeia A
Supernova remnant; strong radio source

SN 1572
Supernova remnant

LYNX

THE LYNX

THIS NORTHERN CONSTELLATION FILLS A BLANK AREA OF SKY BETWEEN URSA MAJOR AND AURIGA. THE LYNX IS DRAWN AROUND A CHAIN OF STARS THAT STRETCHES FROM ITS NOSE TO ITS TAIL.

Johannes Hevelius, the Polish astronomer who defined this constellation in 1687, was renowned for his sharp eyesight. He noted that only those who were as keen-sighted as cats would be able to see it. Most naked-eye observers will see little more than its brightest star Alpha. With a telescope, interesting double and multiple stars can be seen, such as the triple star 19 Lyncis, which consists of two stars of 6th and 7th magnitude and a wider 8th-magnitude companion. Notable deep-sky objects are the distant globular cluster NGC 2419 and the huge star-forming region known as the Lynx Arc.

◁ **The Lynx Arc**
A vast arc of brilliant light about 12 billion light-years away gives a glimpse back in time to a period of intense star formation. The Lynx Arc is the biggest, brightest, and hottest star-forming region known. It is a million times brighter than the better-known Orion Nebula and contains a million blue-white stars, twice as hot as similar stars in the Milky Way.

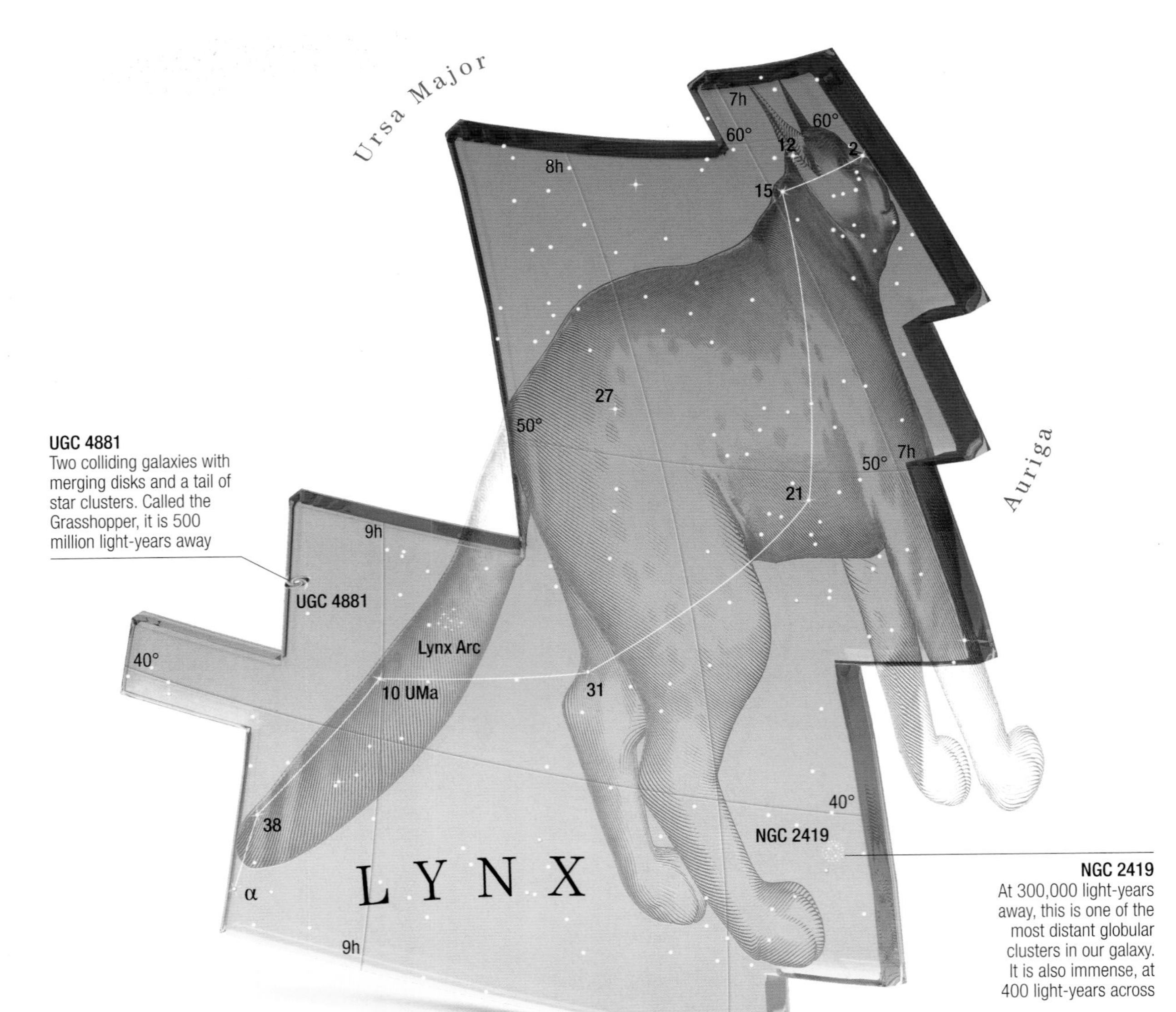

UGC 4881
Two colliding galaxies with merging disks and a tail of star clusters. Called the Grasshopper, it is 500 million light-years away

NGC 2419
At 300,000 light-years away, this is one of the most distant globular clusters in our galaxy. It is also immense, at 400 light-years across

KEY DATA

Size ranking 28
Brightest stars Alpha (α) 3.1, 38 Lyncis (α) 3.8
Genitive Lyncis
Abbreviation Lyn
Highest in sky at 10pm February–March
Fully visible 90°N–28°S

CHART 6

MAIN STARS

Alpha (α) Lyncis
Orange giant
3.1 ⟷ 203 light-years

5 Lyncis
Optical double star
5.2 ⟷ 625 light-years

12 Lyncis
Triple-star system
4.9 ⟷ 215 light-years

19 Lyncis
Triple-star system
5.8 ⟷ 470 light-years

38 Lyncis
Blue-white main-sequence star and double star
3.8 ⟷ 125 light-years

DEEP-SKY OBJECTS

NGC 2419
Globular cluster

UGC 4881
Pair of interacting galaxies; also called the Grasshopper

Lynx Arc
Star-formation region

CAMELOPARDALIS

THE GIRAFFE

OCCUPYING AN AREA OF SKY BETWEEN CASSIOPEIA AND THE "HEAD" OF THE GREAR BEAR (URSA MAJOR), CAMELOPARDALIS LACKS BRIGHT OBJECTS AND IS BEST FOUND BY FIRST LOCATING ITS NEIGHBOURS.

Left blank by the Ancient Greeks, this large and barren region of northern sky contains no stars brighter than 4th magnitude. The gap was eventually filled in 1612 when Dutch theologian and astronomer Petrus Plancius drew a giraffe around some of its stars. Its front legs, body, and back legs fit around an inverted "U" shape of stars. The giraffe's distinctive neck is drawn around no particular stars and stretches up towards Draco. The constellation's most notable feature is a trail of unrelated stars called Kemble's Cascade that lead away from NGC 1502 towards Cassiopeia.

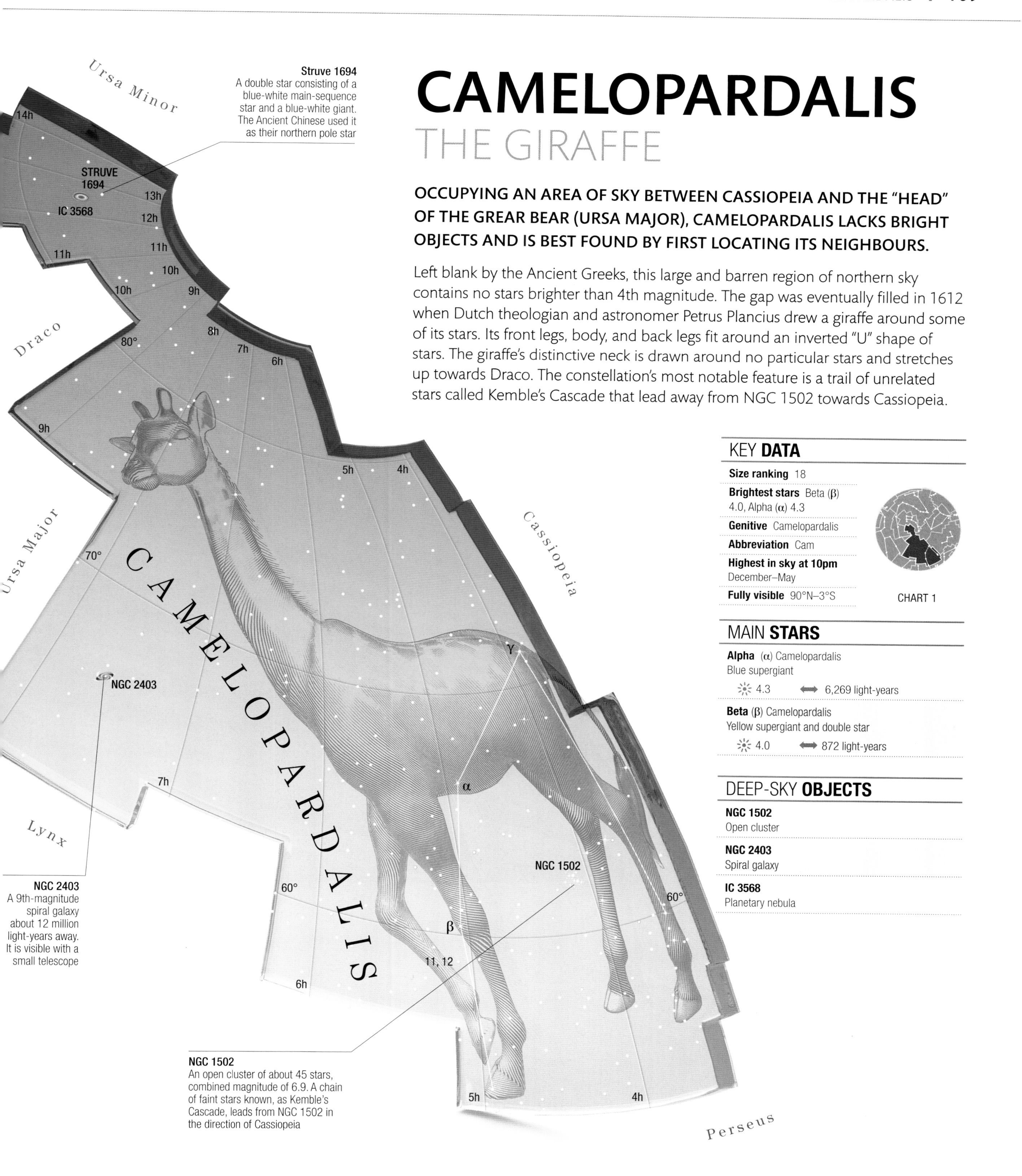
Ursa Minor
Draco
Ursa Major
Lynx
Cassiopeia
Perseus
CAMELOPARDALIS
STRUVE 1694
IC 3568
NGC 2403
NGC 1502
14h
13h
12h
11h
10h
9h
8h
7h
6h
5h
4h
80°
70°
60°
α
β
γ
11, 12
CHART 1
Struve 1694
A double star consisting of a blue-white main-sequence star and a blue-white giant. The Ancient Chinese used it as their northern pole star
NGC 2403
A 9th-magnitude spiral galaxy about 12 million light-years away. It is visible with a small telescope
NGC 1502
An open cluster of about 45 stars, combined magnitude of 6.9. A chain of faint stars known, as Kemble's Cascade, leads from NGC 1502 in the direction of Cassiopeia

KEY DATA

Size ranking 18

Brightest stars Beta (β) 4.0, Alpha (α) 4.3

Genitive Camelopardalis

Abbreviation Cam

Highest in sky at 10pm December–May

Fully visible 90°N–3°S

MAIN STARS

Alpha (α) Camelopardalis
Blue supergiant
4.3 ⟷ 6,269 light-years

Beta (β) Camelopardalis
Yellow supergiant and double star
4.0 ⟷ 872 light-years

DEEP-SKY OBJECTS

NGC 1502
Open cluster

NGC 2403
Spiral galaxy

IC 3568
Planetary nebula

LUMINOSITIES

Xi Ursae Majoris *1.5 Suns*

Megrez *25 Suns*

Merak *60 Suns*

Phad *62 Suns*

URSA MAJOR THE GREAT BEAR

THE THIRD-LARGEST CONSTELLATION, URSA MAJOR IS BEST KNOWN FOR CONTAINING THE PLOUGH (ALSO CALLED THE BIG DIPPER), PROBABLY THE MOST FAMOUS STAR PATTERN IN THE ENTIRE SKY.

Seven stars make up the familiar ladle-shaped pattern known as the Plough: Dubhe, Merak, Phad, Megrez, Alioth, Mizar, and Alkaid. The second star in the handle of the dipper is a wide double. The brighter of the pair is Mizar, and its companion is Alcor. Two stars in the bowl of the dipper, Merak and Dubhe, point towards the north pole star, Polaris, in nearby Ursa Minor, the Little Bear.

Ursa Major also contains several interesting deep-sky objects. These include M101, a face-on spiral also known as the Pinwheel Galaxy; M81 and M82 (also called the Cigar Galaxy), a pair of galaxies that are thought to have had a close encounter about 300 million years ago; and the planetary nebula M97, popularly called the Owl Nebula because of its resemblance to an owl's face.

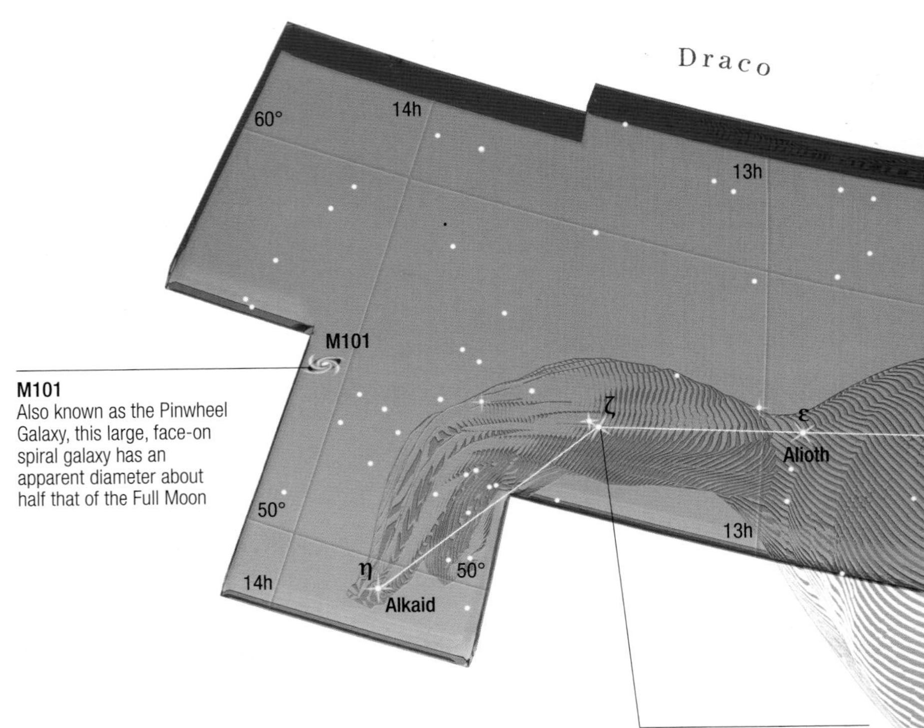

M101
Also known as the Pinwheel Galaxy, this large, face-on spiral galaxy has an apparent diameter about half that of the Full Moon

Mizar (ζ Ursae Majoris)
A 2nd-magnitude star with a 4th-magnitude companion, Alcor, that can just be seen with the naked eye but is easily visible with binoculars

◁ **NGC 3982**
Pink clouds of glowing hydrogen gas stand out along the spiral arms of this face-on spiral galaxy, nearly 70 million light-years away. Like the bright nebulae in our own galaxy, these clouds are areas where stars are being born, while the bluer regions consist of hot young stars. NGC 3982 is about 30,000 light-years wide, nearly one-third of the diameter of the Milky Way.

◁ **M82**
Popularly known as the Cigar Galaxy, this is undergoing a huge surge in star formation as a result of an interaction with its neighbouring galaxy, M81. Plumes of hot, ionized gas (red in this Hubble image) are being blasted out above and below the disk of the Cigar Galaxy. Situated in the northern part of Ursa Major, both galaxies are 12 million light-years from Earth.

▷ **Star distances**
Ursa Major's main pattern stars lie between 29 and 358 light-years away from Earth. The two stars that form the ends of the Plough asterism – Dubhe and Alkaid – are 123 and 104 light-years away, respectively. The other five stars of the asterism – Merak, Phad, Megrez, Alioth, and Mizar – all lie at similar distances (about 80–86 light-years away) and are moving in the same direction through space. They form what is known as the Ursa Major Moving Group, a former open cluster that has drifted apart.

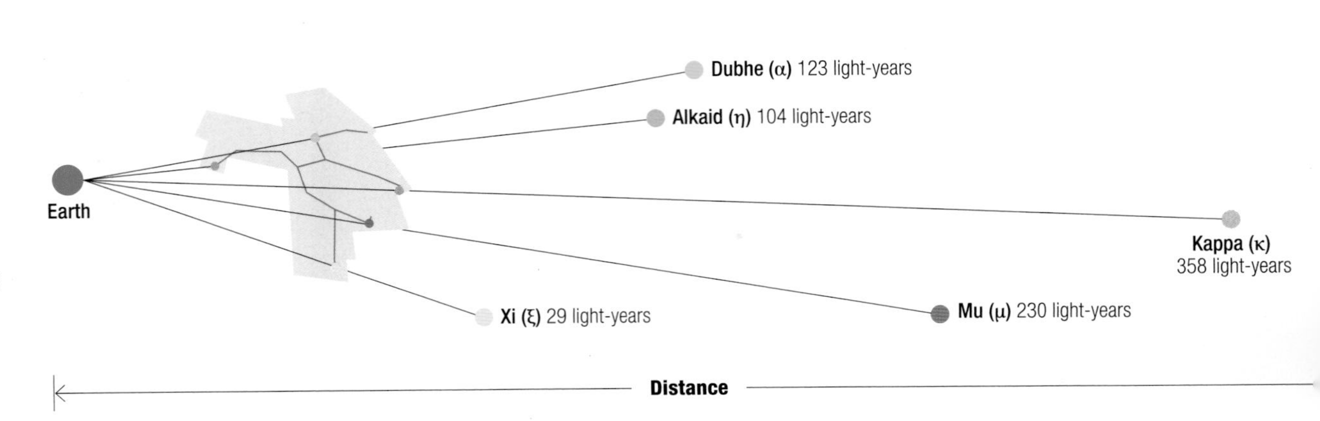

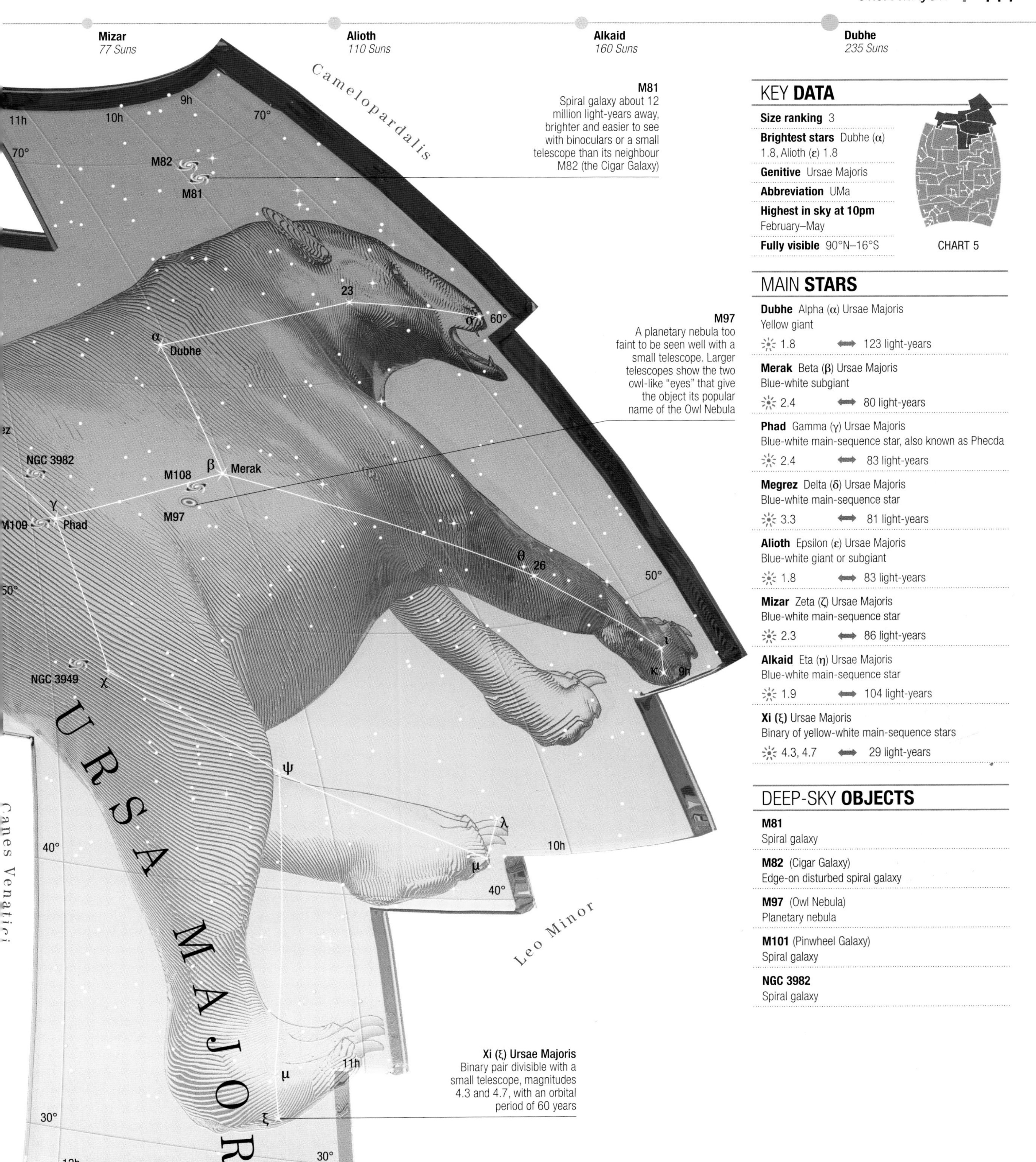

Mizar
77 Suns
Alioth
110 Suns
Alkaid
160 Suns
Dubhe
235 Suns
Camelopardalis
M81
Spiral galaxy about 12 million light-years away, brighter and easier to see with binoculars or a small telescope than its neighbour M82 (the Cigar Galaxy)
M82
M81
M97
A planetary nebula too faint to be seen well with a small telescope. Larger telescopes show the two owl-like "eyes" that give the object its popular name of the Owl Nebula
Dubhe
Merak
M108
M97
NGC 3982
M109
Phad
NGC 3949
URSA MAJOR
Leo Minor
Canes Venatici
Xi (ξ) Ursae Majoris
Binary pair divisible with a small telescope, magnitudes 4.3 and 4.7, with an orbital period of 60 years
CHART 5

KEY DATA

Size ranking 3

Brightest stars Dubhe (α) 1.8, Alioth (ε) 1.8

Genitive Ursae Majoris

Abbreviation UMa

Highest in sky at 10pm February–May

Fully visible 90°N–16°S

MAIN STARS

Dubhe Alpha (α) Ursae Majoris
Yellow giant
1.8 ⟷ 123 light-years

Merak Beta (β) Ursae Majoris
Blue-white subgiant
2.4 ⟷ 80 light-years

Phad Gamma (γ) Ursae Majoris
Blue-white main-sequence star, also known as Phecda
2.4 ⟷ 83 light-years

Megrez Delta (δ) Ursae Majoris
Blue-white main-sequence star
3.3 ⟷ 81 light-years

Alioth Epsilon (ε) Ursae Majoris
Blue-white giant or subgiant
1.8 ⟷ 83 light-years

Mizar Zeta (ζ) Ursae Majoris
Blue-white main-sequence star
2.3 ⟷ 86 light-years

Alkaid Eta (η) Ursae Majoris
Blue-white main-sequence star
1.9 ⟷ 104 light-years

Xi (ξ) Ursae Majoris
Binary of yellow-white main-sequence stars
4.3, 4.7 ⟷ 29 light-years

DEEP-SKY OBJECTS

M81
Spiral galaxy

M82 (Cigar Galaxy)
Edge-on disturbed spiral galaxy

M97 (Owl Nebula)
Planetary nebula

M101 (Pinwheel Galaxy)
Spiral galaxy

NGC 3982
Spiral galaxy

LUMINOSITIES

Beta Canum Venaticorum *1.2 Suns*

RS Canum Venaticorum *13 Suns*

CANES VENATICI
THE HUNTING DOGS

BETWEEN BOÖTES AND URSA MAJOR LIES THE CONSTELLATION CANES VENATICI, REPRESENTING A PAIR OF HUNTING DOGS HELD ON A LEAD BY BOÖTES. SEVERAL REMARKABLE GALAXIES LIE WITHIN ITS BORDERS, MOST NOTABLY M51, POPULARLY KNOWN AS THE WHIRLPOOL.

Not recognized by the Ancient Greeks, the constellation Canes Venatici was introduced in 1687 by Johannes Hevelius, a Polish astronomer who invented several new sky figures. He imagined it as two hounds held on a lead by the adjacent Boötes, the Herdsman.

This constellation has few stars of note. Its brightest star was named Cor Caroli (Charles's Heart) in the 17th century to commemorate King Charles I of England, who was beheaded by the republican parliament in 1649.

Near the constellation's upper border with Ursa Major lies M51 (see pp.114–15), a face-on spiral galaxy. Its spiral structure was first detected in 1845 by an Irish astronomer, Lord Rosse, using a telescope he had built himself at his home at Birr Castle, County Offaly. Rosse's discovery led to speculation that such spiral objects could be separate galaxies far off in space. In the case of M51, the distance is about 30 million light-years. A smaller galaxy, called NGC 5195, lies near the end of one of its arms.

In 1845, Lord Rosse observed M51, the **first spiral galaxy** to be recognized, using what was then the **world's largest telescope**

KEY DATA

Size ranking 38

Brightest stars Alpha (α) 2.9, Beta (β) 4.3

Genitive Canum Venaticorum

Abbreviation CVn

Highest in sky at 10pm April–May

Fully visible 90°N–27°S

CHART 5

MAIN STARS

Cor Caroli Alpha (α) Canum Venaticorum
Blue-white main sequence
2.9 ⟷ 115 light-years

Beta (β) Canum Venaticorum
Yellow main sequence
4.3 ⟷ 28 light-years

La Superba Y Canum Venaticorum
Red giant variable
4.9–7.3 ⟷ 1,000 light-years

RS Canum Venaticorum
Eclipsing binary
7.9–9.1 ⟷ 520 light-years

DEEP-SKY OBJECTS

M3
Globular cluster

M51
Spiral galaxy; also known as the Whirlpool Galaxy

M63
Spiral galaxy; also known as the Sunflower Galaxy

M94
Spiral galaxy

M106
Spiral galaxy

NGC 4244
Edge-on spiral galaxy

NGC 4449
Irregular dwarf galaxy

NGC 4631
Edge-on spiral galaxy; also called the Whale Galaxy

◁ **M106**
This view of spiral galaxy M106 is a composite of images from the Hubble Space Telescope and two amateur astrophotographers, Robert Gendler and Jay GaBany.

▽ **NGC 4449**
The glowing patches in this dwarf galaxy are bursts of star formation, most probably triggered by an interaction or merger with one or more smaller galaxies.

Cor Caroli
75 Suns

La Superba
608 Suns

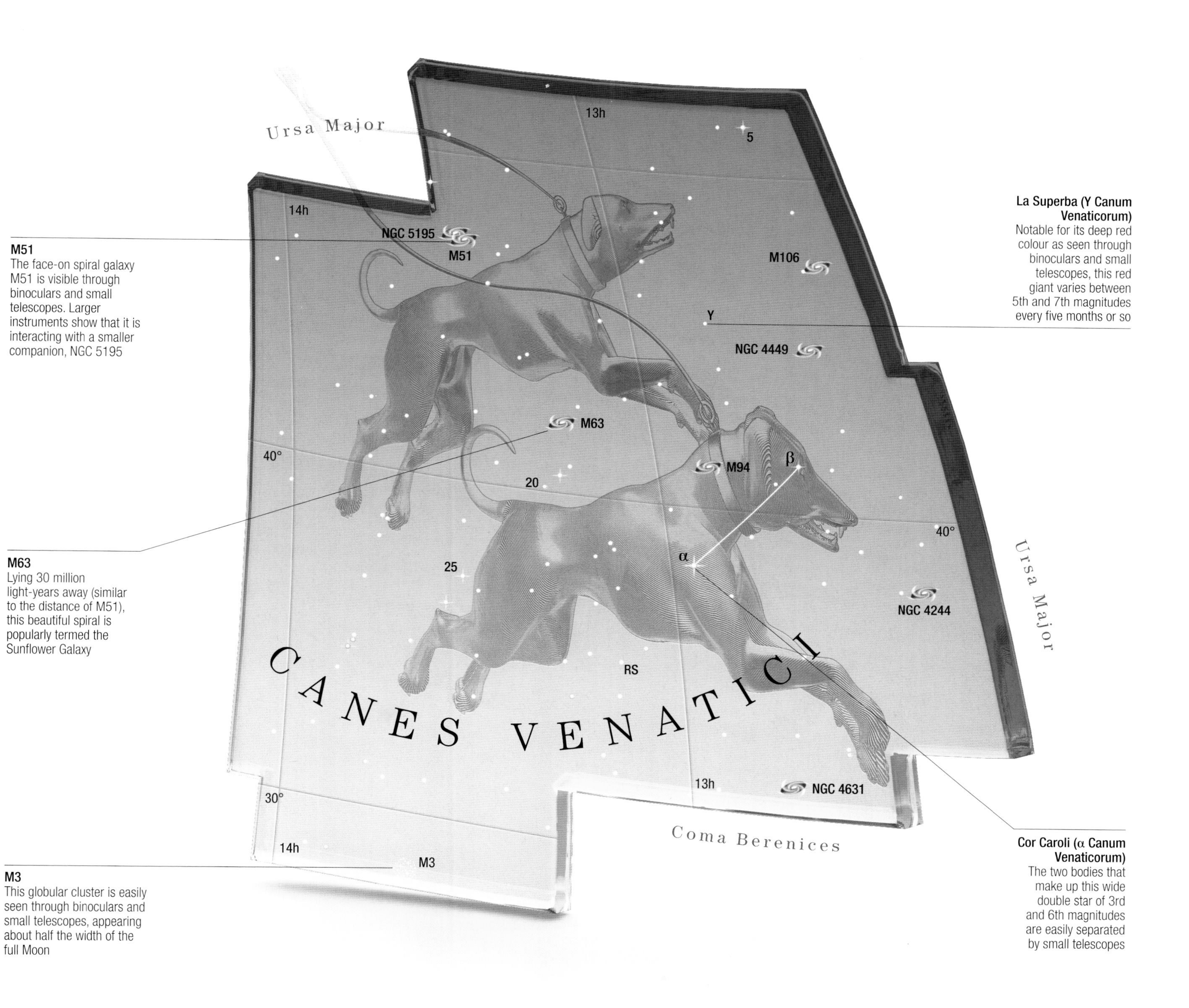

M51
The face-on spiral galaxy M51 is visible through binoculars and small telescopes. Larger instruments show that it is interacting with a smaller companion, NGC 5195

M63
Lying 30 million light-years away (similar to the distance of M51), this beautiful spiral is popularly termed the Sunflower Galaxy

M3
This globular cluster is easily seen through binoculars and small telescopes, appearing about half the width of the full Moon

La Superba (Y Canum Venaticorum)
Notable for its deep red colour as seen through binoculars and small telescopes, this red giant varies between 5th and 7th magnitudes every five months or so

Cor Caroli (α Canum Venaticorum)
The two bodies that make up this wide double star of 3rd and 6th magnitudes are easily separated by small telescopes

▷ **Star distances**
Canes Venatici contains numerous celestial objects of interest, but the constellation figure is made up of only two pattern stars. The brighter of the two, Cor Caroli, lies more than four times farther than the fainter Beta Canum Venaticorum.

Earth

Beta (β)
28 light-years

Cor Caroli (α)
115 light-years

Distance

THE WHIRLPOOL GALAXY

1 Grand spiral
Long lines of stars and dust-laced gas wind round the centre of the M51, known as the Whirlpool Galaxy. The arms are star-forming factories where hydrogen gas is compressed and new stars are born. The young hot stars make the arms look bluish, and cause clouds of hydrogen to glow pink. The small galaxy (NGC 5195) at right is passing behind the Whirlpool, triggering star formation as it glides by.

2 Galaxy core
When seen imaged in X-rays, the galaxy's core shines brightly. This image, taken by the Chandra X-ray Observatory, reveals vast clouds of multi-million degree gas at either side of it. The cloud at upper left of the bright central core is 1,500 light-years across. The gas is heated by a high-velocity jet of material accelerating away from a supermassive black hole within the galaxy's nucleus.

3 Inside the core
This Hubble Space Telescope image takes us to the very heart of the galaxy – the active galactic nucleus (AGN) of its central core. The dark "X" silhouetted against the bright nucleus marks the exact location of a black hole, but hides it and its disk of infalling hot gas from view. The broad line of the X is a dust ring 100 light-years across lying at right angles to the galaxy's disk.

4 X-ray view
More than 400 X-ray sources are revealed in this image of the Whirlpool by the space-based Chandra X-ray Observatory, which took 11 hours of observation to create. Most are X-ray binary star systems in which a neutron star, or more rarely a stellar black hole, captures material from an orbiting companion star. The infalling material heats to millions of degrees, producing a luminous X-ray source.

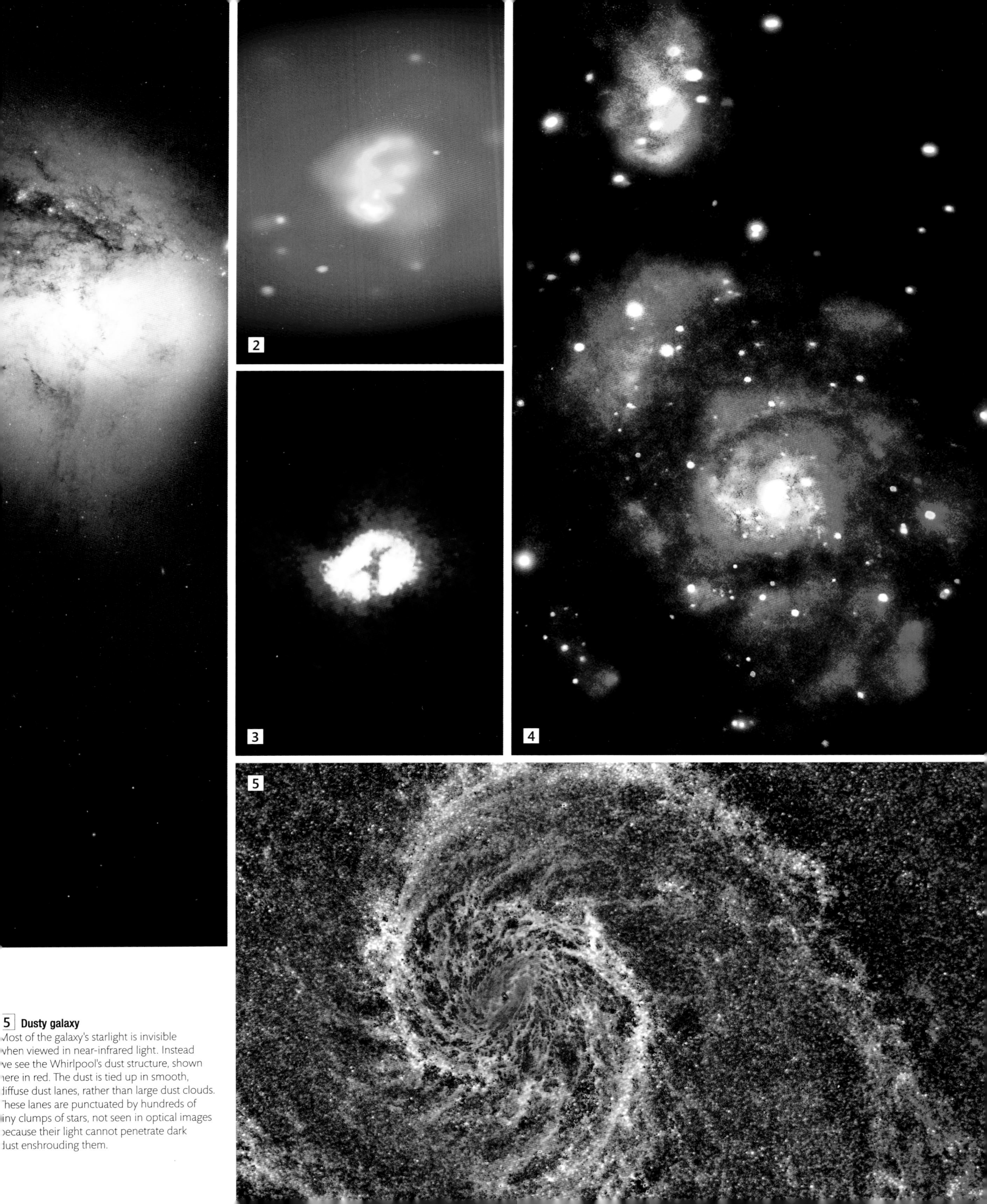

5 Dusty galaxy
Most of the galaxy's starlight is invisible when viewed in near-infrared light. Instead we see the Whirlpool's dust structure, shown here in red. The dust is tied up in smooth, diffuse dust lanes, rather than large dust clouds. These lanes are punctuated by hundreds of tiny clumps of stars, not seen in optical images because their light cannot penetrate dark dust enshrouding them.

BOÖTES
THE HERDSMAN

A DISTINCTIVE KITE-SHAPED PATTERN IN THE NORTHERN SKY, BOÖTES IS HOME TO ARCTURUS, ONE OF THE BRIGHTEST AND CLOSEST STARS TO US.

A large constellation, Boötes extends from Draco and Ursa Major in the north to Virgo in the south. Myths differ about what exactly Boötes represents but he is often taken to be a herdsman who is driving away two bears – represented by the constellations Ursa Major and Ursa Minor - with the aid of his dogs, represented by the adjacent constellation Canes Venatici. Boötes' brightest star is Arcturus, which is "bear guard" or "bear keeper" in Greek. It is the brightest star north of the celestial equator. Boötes is also noted for its double stars. The best is Izar, one of the most beautiful doubles in the sky. The Quadrantid meteor shower, named after Quadrans, an obsolete constellation that once took up part of Boötes, radiates from this area of sky every January.

Arcturus emits over **100 times more energy** than the **Sun** even though it is only slightly more massive

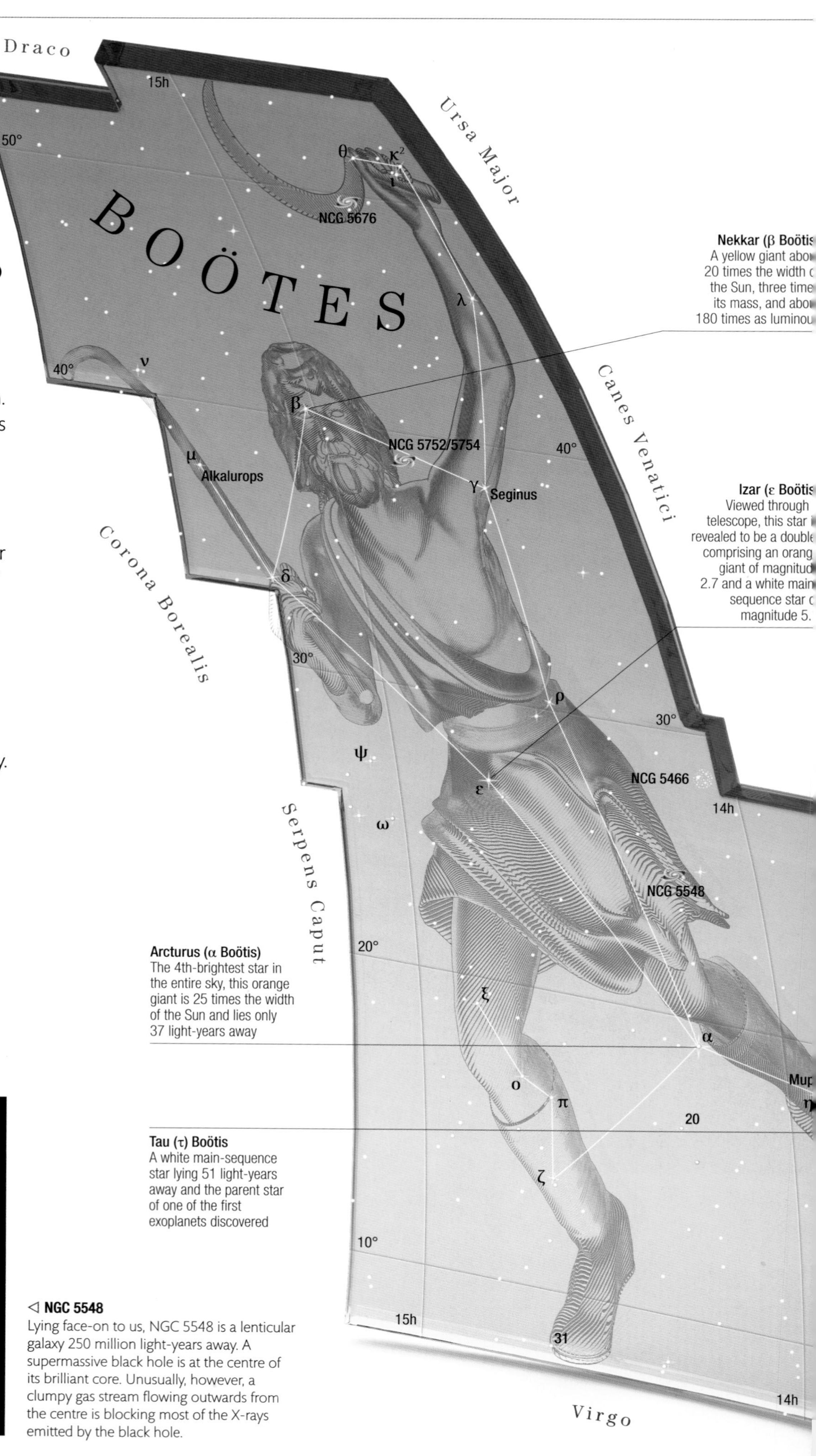
Draco
Ursa Major
Canes Venatici
Corona Borealis
Serpens Caput
Virgo
BOÖTES
15h
50°
40°
30°
20°
10°
14h
NGC 5676
NCG 5752/5754
Alkalurops
Seginus
NCG 5466
NCG 5548
Mup
20
31
Nekkar (β Boötis)
A yellow giant about 20 times the width of the Sun, three times its mass, and about 180 times as luminous
Izar (ε Boötis)
Viewed through a telescope, this star is revealed to be a double comprising an orange giant of magnitude 2.7 and a white main-sequence star of magnitude 5.
Arcturus (α Boötis)
The 4th-brightest star in the entire sky, this orange giant is 25 times the width of the Sun and lies only 37 light-years away
Tau (τ) Boötis
A white main-sequence star lying 51 light-years away and the parent star of one of the first exoplanets discovered

◁ **NGC 5548**
Lying face-on to us, NGC 5548 is a lenticular galaxy 250 million light-years away. A supermassive black hole is at the centre of its brilliant core. Unusually, however, a clumpy gas stream flowing outwards from the centre is blocking most of the X-rays emitted by the black hole.

KEY DATA

Size ranking 13

Brightest stars Arcturus (α) -0.1, Izar (ε) 2.4

Genitive Boötis

Abbreviation Boo

Highest in sky at 10pm May–June

Fully visible 90°N–35°S

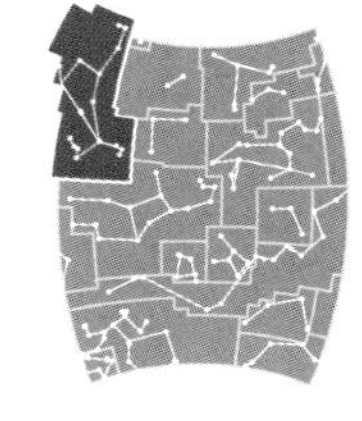

CHART 5

MAIN STARS

Arcturus Alpha (α) Boötis
Orange giant
-0.1 ⟷ 37 light-years

Nekkar Beta (β) Boötis
Yellow giant
3.5 ⟷ 225 light-years

Seginus Gamma (γ) Boötis
White giant; also a variable star
3.0 ⟷ 87 light-years

Delta (δ) Boötis
Yellow giant; also a double star
3.5 ⟷ 122 light-years

Izar Epsilon (ε) Boötis
Orange giant; also a double star
2.4 ⟷ 202 light-years

Muphrid Eta (η) Boötis
Yellow subgiant
2.7 ⟷ 37 light-years

Alkalurops Mu (μ) Boötis
White main-sequence star; also a triple star
4.3 ⟷ 113 light-years

DEEP-SKY OBJECTS

NGC 5248
Spiral galaxy

NGC 5466
Globular cluster

NGC 5548
Lenticular galaxy; also a Seyfert galaxy

NGC 5676
Spiral galaxy

NGC 5752 and NGC 5754
Pair of interacting galaxies

CORONA BOREALIS
THE NORTHERN CROWN

A HORSESHOE-SHAPED PATTERN OF STARS REPRESENTING A MAGNIFICENT CROWN, CORONA BOREALIS IS A SMALL BUT DISTINCTIVE CONSTELLATION IN THE NORTHERN SKY.

One of the original 48 constellations of Ancient Greece, Corona Borealis represents the jewel-studded crown worn by the mythical Princess Ariadne of Crete at her wedding to the god Dionysus. Newly married, Dionysus tossed the crown into the sky, where its jewels became stars. The crown shape is drawn around seven linked stars. Found between Boötes and Hercules, it is easily spotted despite the relative faintness of its stars. Corona Borealis contains interesting double stars and variables. It is also host to several galaxy clusters, including SDS J1531+3414 and Abell 2065. The latter contains more than 400 galaxies but is 1.5 billion light-years away and is too faint to be visible with most amateur telescopes.

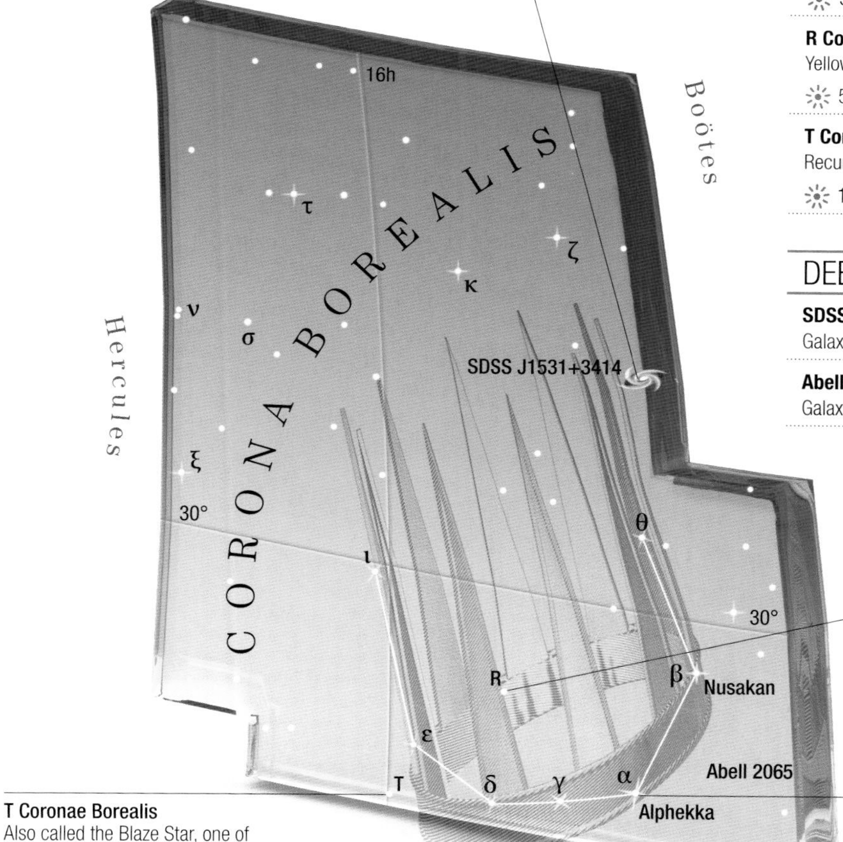

SDSS J1531+3414
A dense cluster of mainly giant elliptical galaxies, with a few spiral and irregular galaxies

R Coronae Borealis
A yellow supergiant usually just visible to the naked eye but which diminishes in brightness every few years to about magnitude 14

T Coronae Borealis
Also called the Blaze Star, one of the brightest and most reliable recurrent novae, brightening from about magnitude 10 to about 2 every few decades

Alphekka (α Coronae Borealis)
An eclipsing binary varying in brightness between magnitudes 2.1 and 2.3 in a 17.4-day cycle

KEY DATA

Size ranking 73

Brightest stars Alphekka (α) 2.1–2.3, Nusakan (β) 3.7

Genitive Coronae Borealis

Abbreviation CrB

Highest in sky at 10pm June

Fully visible 90°N–50°S

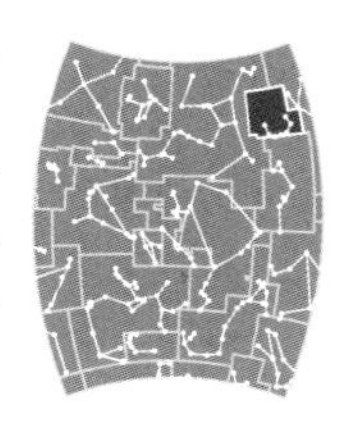

CHART 4

MAIN STARS

Alphekka Alpha (α) Coronae Borealis
White main-sequence star; also an eclipsing binary
2.1–2.3 ⟷ 75 light-years

Nusakan Beta (β) Coronae Borealis
White main-sequence star; also a binary
3.7 ⟷ 112 light-years

Gamma (γ) Coronae Borealis
White main-sequence star
3.8 ⟷ 146 light-years

Zeta (ζ) Coronae Borealis
Blue-white main-sequence star; also a double star
4.9 ⟷ 470 light-years

Nu (ν) Coronae Borealis
Red giant; also a double star
5.2 ⟷ 640 light-years

Sigma (σ) Coronae Borealis
White main-sequence star; also a double star
5.6 ⟷ 69 light-years

R Coronae Borealis
Yellow supergiant; also a variable star
5.7 ⟷ 81,500 light-years

T Coronae Borealis
Recurrent nova, also known as the Blaze Star
10.2 ⟷ 3,470 light-years

DEEP-SKY OBJECTS

SDSS J1531+3414
Galaxy cluster

Abell 2065
Galaxy cluster

LUMINOSITIES

Mu Herculis
3 Suns

Zeta Herculis
8 Suns

Delta Herculis
26 Suns

HERCULES
THE STRONGMAN

HERCULES IS A LARGE BUT NOT PARTICULARLY PROMINENT CONSTELLATION LYING BETWEEN LYRA AND BOÖTES. ITS MOST NOTABLE FEATURES ARE GLOBULAR STAR CLUSTERS, INCLUDING M13, GENERALLY REGARDED AS THE FINEST IN NORTHERN SKIES.

Hercules is oriented with his feet pointing north and his head in the south. He represents the strongman of Ancient Greek mythology who was ordered to undertake 12 epic labours. Among them was slaying a dragon, and in the sky Hercules is visualized with his left foot over the dragon's head, represented by the constellation Draco to the north. Hercules's head is marked by the star called Rasalgethi, which is a red giant of variable brightness. Although Rasalgethi is labelled Alpha Herculis, the constellation's brightest star is Beta Herculis, also known as Kornephoros.

Four of the constellation's pattern stars (Epsilon, Zeta, Eta, and Pi Herculis) form a quadrangular shape called the Keystone. The Keystone marks the lower body of Hercules. On one side of the Keystone lies the bright globular cluster M13, which is nearly 150 light-years across and contains more than one-quarter of a million stars.

Hercules also contains several attractive double stars that can be separated with the use of a small telescopes, notably Rho Herculis, 95 Herculis, and a relatively bright and nearby white dwarf, 110 Herculis.

△ **M13**
The brightest globular cluster in the northern sky, M13 lies about 25,000 light-years away and contains an estimated 300,000 stars. It is just visible to the naked eye; with small telescopes, details such as chains of stars are visible.

◁ **Hercules A**
Jets of gas a million light-years long shoot out from Hercules A, an elliptical galaxy about two billion light-years away. Although invisible at visible-light wavelengths, the jets can be detected at radio wavelengths and can be seen clearly in this combined visible-light and radio-wave image. The jets are thought to be powered by a black hole with a mass of about 2.5 billion Suns at the galaxy's centre.

113

Aquila

20°

110 Herculis
White dwarf about 63 light-years away. With a magnitude of 4.2, it is visible with the naked eye

▷ **Star distances**
The nearest of Hercules's main pattern stars, Mu (μ) Herculis, is only 27 light-years away while the farthest is Theta (θ) Herculis, at about 758 light-years. Coincidentally, these are also the least and most luminous of the pattern stars. Mu emits as much energy as about three Suns whereas Theta emits the equivalent of about 1,330 Suns.

Earth

Theta (θ) 758 light-years

Pi (π) 377 light-years

Mu (μ) 27 light-years

Delta (δ) 75 light-years

Beta (β) 139 light-years

Distance

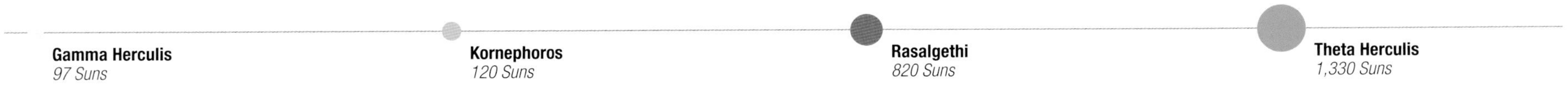
Gamma Herculis
97 Suns
Kornephoros
120 Suns
Rasalgethi
820 Suns
Theta Herculis
1,330 Suns

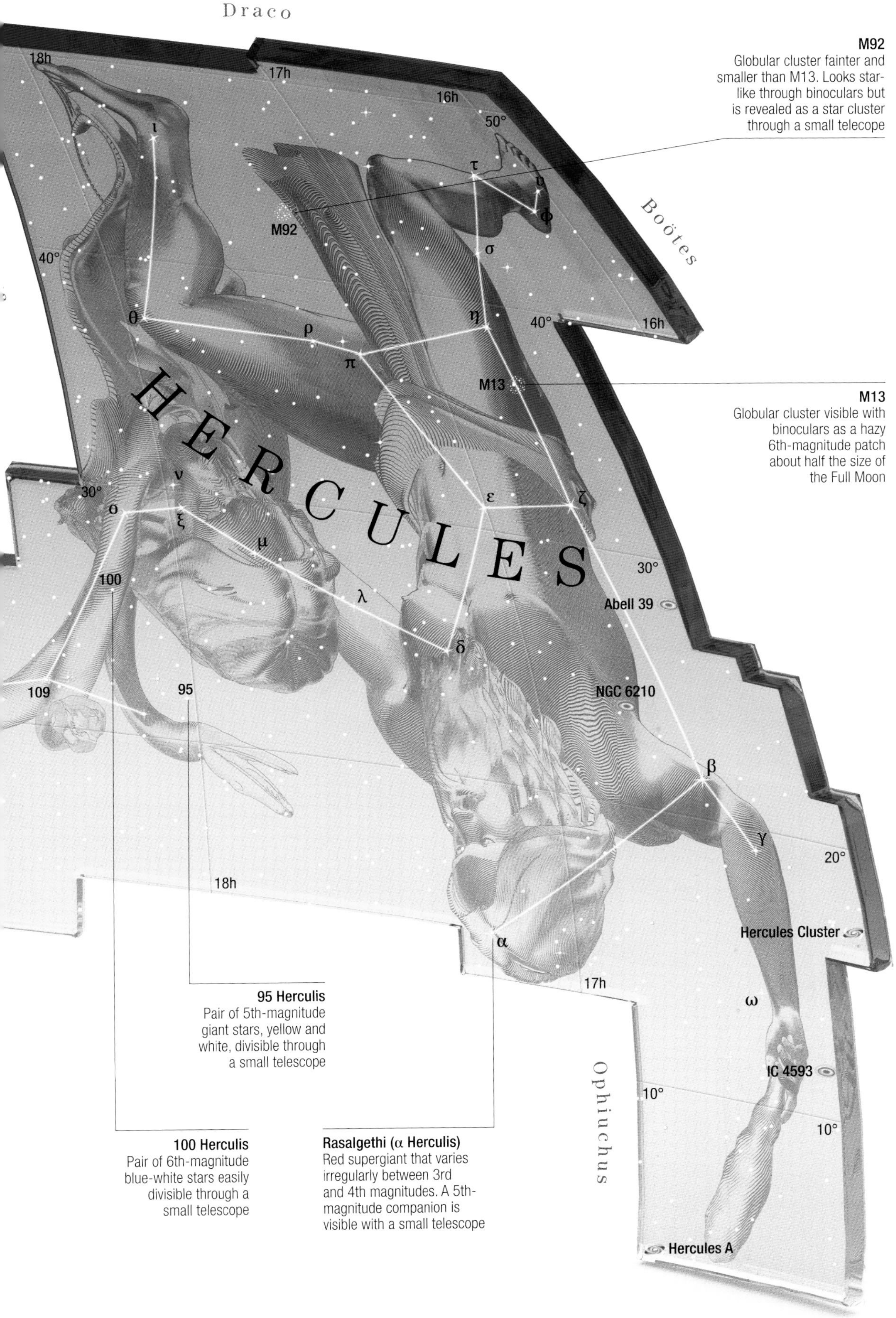
Draco
Boötes
Ophiuchus
HERCULES
18h
17h
16h
50°
40°
30°
20°
10°
M92
Globular cluster fainter and smaller than M13. Looks star-like through binoculars but is revealed as a star cluster through a small telecope
M13
Globular cluster visible with binoculars as a hazy 6th-magnitude patch about half the size of the Full Moon
Abell 39
NGC 6210
Hercules Cluster
IC 4593
Hercules A
109
100
95
95 Herculis
Pair of 5th-magnitude giant stars, yellow and white, divisible through a small telescope
100 Herculis
Pair of 6th-magnitude blue-white stars easily divisible through a small telescope
Rasalgethi (α Herculis)
Red supergiant that varies irregularly between 3rd and 4th magnitudes. A 5th-magnitude companion is visible with a small telescope

KEY DATA

Size ranking 5

Brightest stars Kornephoros (β) 2.8, Zeta (ζ) 2.8

Genitive Herculis

Abbreviation Her

Highest in sky at 10pm June–July

Fully visible 90°N–38°S

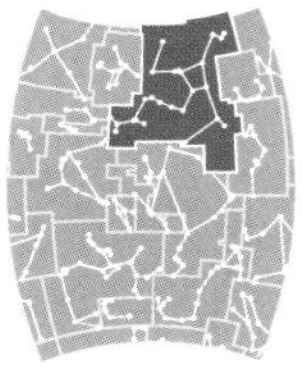

CHART 4

MAIN STARS

Rasalgethi Alpha (α) Herculis
Variable red supergiant
2.7–4.0 ⟷ 360 light-years

Kornephoros Beta (β) Herculis
Yellow giant
2.8 ⟷ 139 light-years

Gamma (γ) Herculis
White giant
3.8 ⟷ 193 light-years

Delta (δ) Herculis
Blue-white supergiant
3.1 ⟷ 75 light-years

Zeta (ζ) Herculis
Yellow-white supergiant
2.8 ⟷ 35 light-years

Eta (η) Herculis
Yellow giant
3.5 ⟷ 109 light-years

Pi (π) Herculis
Orange giant
3.2 ⟷ 377 light-years

DEEP-SKY OBJECTS

M13
Globular cluster

M92
Globular cluster

NGC 6210
Planetary nebula

IC 4539
Planetary nebula

Abell 39
Planetary nebula

Hercules Cluster
Cluster of about 200 galaxies

LUMINOSITIES

Epsilon Lyrae *29 Suns*

Vega *50 Suns*

Delta¹ Lyrae *470 Suns*

LYRA THE LYRE

THIS PROMINENT CONSTELLATION IN THE NORTHERN SKY CONTAINS THE FIFTH-BRIGHTEST STAR VISIBLE FROM EARTH, VEGA, ALONG WITH SEVERAL INTERESTING DOUBLE STARS AND A FAMOUS PLANETARY NEBULA.

The constellation Lyra is said to represent the lyre, or harp, played by the legendary Greek musician Orpheus. However, Arab astronomers visualized it as an eagle or vulture, and the name of its brightest star, Vega, comes from an Arabic phrase meaning "swooping eagle (or vulture)".

Near the brilliant Vega lies Epsilon Lyrae, a celebrated quadruple star about 160 light-years away. Telescopes show that each star in this "double" is in fact a close pair – hence its popular name, the Double Double. Another famous double is Beta Lyrae. In this case, the brighter component is also an eclipsing binary (see p.43), which varies between magnitudes 3.3 and 4.4 every 12.9 days.

Delta Lyrae is an unrelated pair of red and blue-white stars, of 4th and 6th magnitudes, which is easily divided with binoculars. Zeta Lyrae, another pair of 4th and 6th magnitudes, can be separated with binoculars or small telescopes. Between Beta and Gamma Lyrae lies the Ring Nebula, M57, a beautiful planetary nebula shaped like a smoke ring (see pp.122–23).

▽ **Structure of M57 (Ring Nebula)**
From our viewpoint on Earth, the planetary nebula M57 looks like a smoke ring around a central star. Seen side-on, though, it would look more like the diagram above. A doughnut-shaped ring of gas with denser knots in it is expanding away from the central star's equator, while fainter lobes of thinner gas extend from the star's poles (top and bottom).

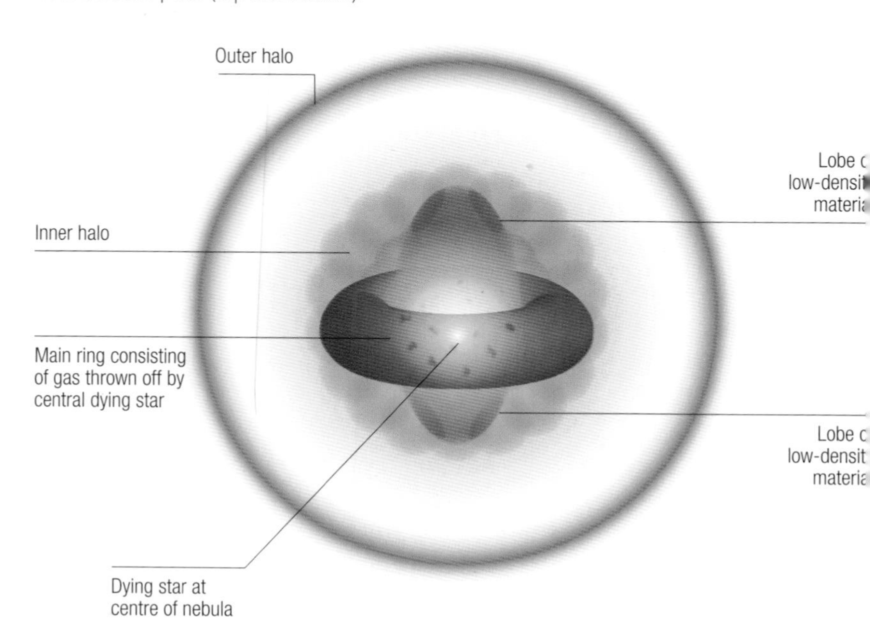

Epsilon Lyrae, the **Double Double**, is a remarkable family of four stars, all **linked by gravity**

▷ **NGC 6745**
A collision between two galaxies has produced this strangely shaped object, which resembles a bird's head. The main part of the "head", seen here in this Hubble Space Telescope image, is a spiral galaxy. It has been highly distorted by its encounter with a smaller elliptical galaxy, which is just visible in the bottom right corner, at the end of the "beak". The blue-white patches at the top and right of the spiral are areas of star formation triggered by the collision.

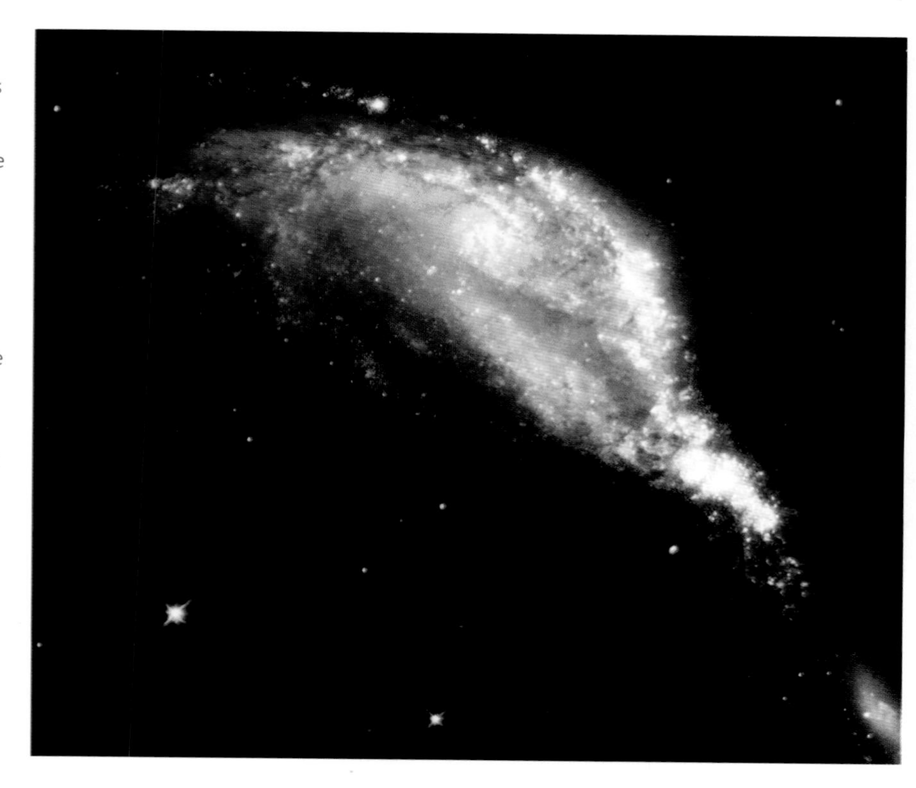

△ **M56**
This globular cluster is a ball of ancient stars just over 30,000 light-years from the Sun. It requires a fair-sized telescope to be seen well. Studies of the chemical composition and age of the stars in the cluster suggest that it was once part of an older dwarf galaxy that subsequently merged with the Milky Way.

Delta² Lyrae
910 Suns

Sulafat
1,580 Suns

Sheliak
2,960 Suns

R Lyrae
The red giant variable R Lyrae ranges in brightness between magnitudes 3.9 and 5.0 every six or seven weeks

Epsilon (ε) Lyrae
This four-star family appears in binoculars as a wide double of 5th-magnitude stars. Telescopes can further divide each star into a close pair

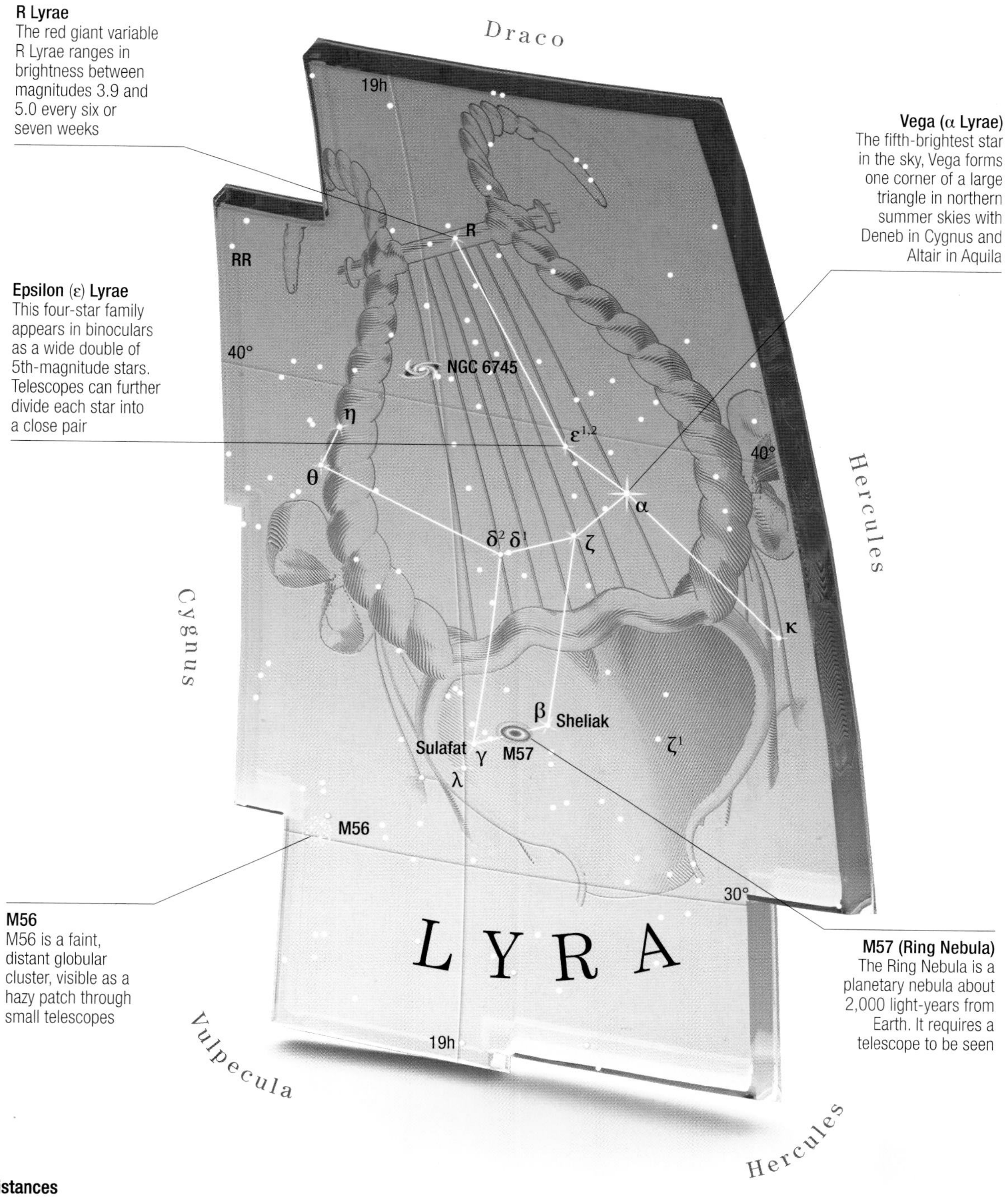

M56
M56 is a faint, distant globular cluster, visible as a hazy patch through small telescopes

Vega (α Lyrae)
The fifth-brightest star in the sky, Vega forms one corner of a large triangle in northern summer skies with Deneb in Cygnus and Altair in Aquila

M57 (Ring Nebula)
The Ring Nebula is a planetary nebula about 2,000 light-years from Earth. It requires a telescope to be seen

KEY DATA

Size ranking 52

Brightest stars Vega (α) 0.0, Sulafat (γ) 3.3

Genitive Lyrae

Abbreviation Lyr

Highest in sky at 10pm July–August

Fully visible 90°N–42°S

CHART 4

MAIN STARS

Vega Alpha (α) Lyrae
Blue-white main sequence
0.0 | 25 light-years

Sheliak Beta (β) Lyrae
Blue-white giant; eclipsing binary
3.3–4.4 | 960 light-years

Sulafat Gamma (γ) Lyrae
Blue-white giant
3.3 | 620 light-years

Delta¹ (δ¹) Lyrae
Blue-white main sequence
5.6 | 990 light-years

Delta² (δ²) Lyrae
Red giant
4.3 | 740 light-years

Epsilon¹ (ε¹) Lyrae
Blue-white main sequence
4.7 | 160 light-years

Epsilon² (ε²) Lyrae
Blue-white main sequence
4.6 | 155 light-years

Zeta (ζ) Lyrae
Blue-white main sequence
4.4 | 155 light-years

R Lyrae
Red giant variable
3.9–5.0 | 300 light-years

RR Lyrae
White giant variable
7.1–8.1 | 940 light-years

DEEP-SKY OBJECTS

M56
Globular cluster, 8th magnitude

M57 (Ring Nebula)
Planetary nebula, 9th magnitude

NGC 6745
Pair of colliding galaxies

▽ **Star distances**
The closest of the pattern stars that make up Lyra is the also one of the brightest stars in the night sky. Vega is relatively close at 25 light-years, with the rest of the pattern stars all over 100 light-years distant. The most distant of the pattern stars is Eta Lyrae, which is nearly 1,400 light-years from Earth.

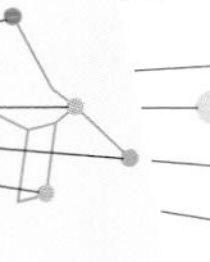

R Lyrae 300 light-years

Eta (η) 1,390 light-years

Vega (α) 25 light-years

Kappa (κ) 250 light-years

Earth

Sheliak (β) 960 light-years

Distance

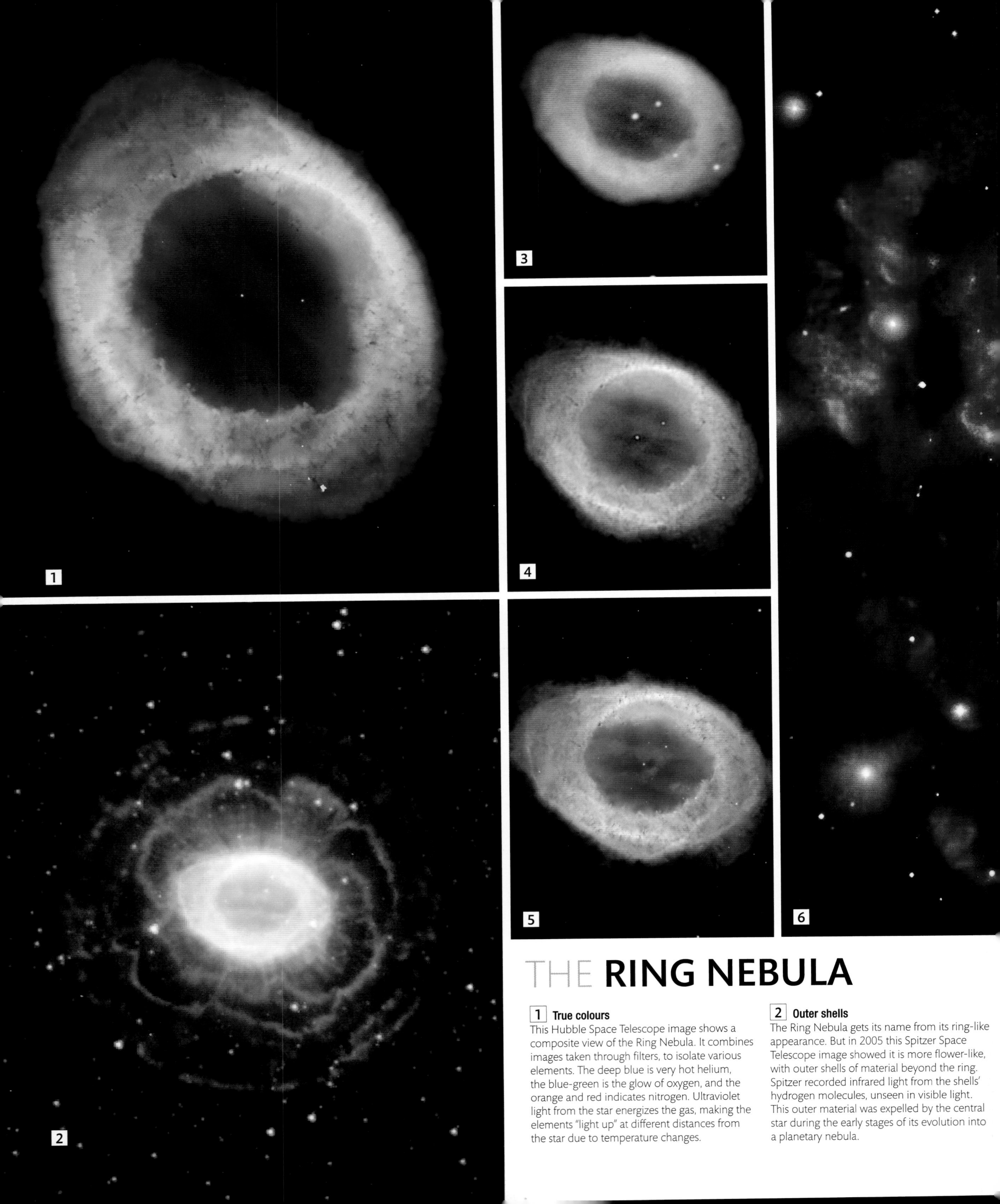

THE RING NEBULA

1 True colours
This Hubble Space Telescope image shows a composite view of the Ring Nebula. It combines images taken through filters, to isolate various elements. The deep blue is very hot helium, the blue-green is the glow of oxygen, and the orange and red indicates nitrogen. Ultraviolet light from the star energizes the gas, making the elements "light up" at different distances from the star due to temperature changes.

2 Outer shells
The Ring Nebula gets its name from its ring-like appearance. But in 2005 this Spitzer Space Telescope image showed it is more flower-like, with outer shells of material beyond the ring. Spitzer recorded infrared light from the shells' hydrogen molecules, unseen in visible light. This outer material was expelled by the central star during the early stages of its evolution into a planetary nebula.

3 Captured on film
This image shows the Ring Nebula as captured on film in 1973, nearly 200 years after it was discovered. The nebula was discovered in 1779, independently by French astronomers Antoine Darquier de Pellepoix and Charles Messier. The picture was taken using the 4m (13ft) telescope at Kitt Peak National Observatory, USA. A decade would pass before astronomical images were routinely recorded digitally.

4 False-colour details
Taken three decades after picture 3 (above), this view of the Ring Nebula was also recorded at Kitt Peak National Observatory, this time using the 3.5-metre (11½ft) telescope. It combines separate images taken through different coloured filters. A red one highlights hydrogen and nitrogen, and a green one isolates oxygen. The use of these false-colour filters help to bring out greater detail in the nebula's shells.

5 Shape and structure
The nebula's shape is more complicated than appears at first glance. Its overall shape is similar to that of a barrel, appearing round because the "barrel" is positioned end-on to us (see p120). Its blue centre is in the shape of a rugby ball and this protrudes out of opposite sides of an orange-red doughnut ring of material. Dark knots of dense gas are embedded on the inner edge of the rim.

6 Dying star
Data from telescopes based in space and on the ground combine to give a complete picture of the Ring Nebula. The tiny white dot at the centre is a white dwarf, the central remains of the star that blew off the surrounding material thousands of years ago. The nebula's ring shape is just under one light-year across, and the whole nebula is getting larger, expanding at 69,000 km (43,000 miles) per hour.

LUMINOSITIES

Mu Cygni
7 Suns

Gienah
44 Suns

Delta Cygni
160 Suns

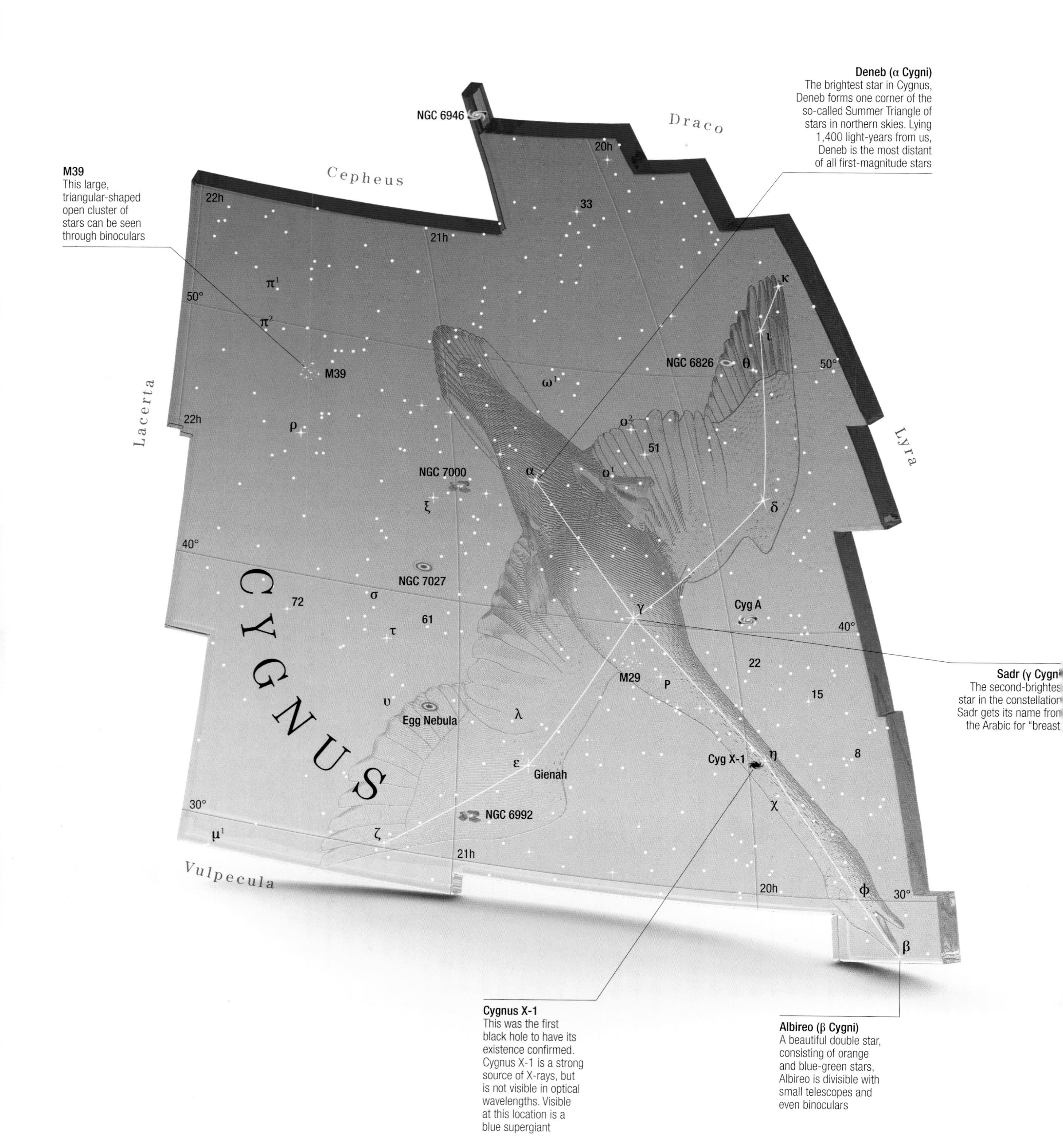

M39
This large, triangular-shaped open cluster of stars can be seen through binoculars

Deneb (α Cygni)
The brightest star in Cygnus, Deneb forms one corner of the so-called Summer Triangle of stars in northern skies. Lying 1,400 light-years from us, Deneb is the most distant of all first-magnitude stars

Sadr (γ Cygni)
The second-brightest star in the constellation, Sadr gets its name from the Arabic for "breast"

Cygnus X-1
This was the first black hole to have its existence confirmed. Cygnus X-1 is a strong source of X-rays, but is not visible in optical wavelengths. Visible at this location is a blue supergiant

Albireo (β Cygni)
A beautiful double star, consisting of orange and blue-green stars, Albireo is divisible with small telescopes and even binoculars

Albireo
930 Suns

Sadr
35,250 Suns

Deneb
51,620 Suns

CYGNUS THE SWAN

SOMETIMES REFERRED TO AS THE NORTHERN CROSS BECAUSE OF ITS DISTINCTIVE SHAPE, CYGNUS IS PROMINENT IN NORTHERN SKIES. THE MILKY WAY'S HAZY BAND RUNS THROUGH IT, DIVIDED INTO TWO STREAMS BY A DARK LANE OF DUST AND GAS CALLED THE CYGNUS RIFT.

Cygnus is an ancient Greek constellation representing a swan. Myths tell how the god Zeus turned himself into a swan to pursue the beautiful Queen Leda of Sparta, and the constellation commemorates his disguise. Deneb, the brightest star in Cygnus, lies in the swan's tail, while the bird's long neck extends along the Milky Way to the star Albireo, a true binary star (see pp.40–41) that marks its beak. Other stars suggest the swan's outstretched wings.

Near Deneb lies a cloud of glowing gas named the North America Nebula, since it looks very much like that continent in shape. Difficult to see with smaller instruments, the nebula shows up best on long-exposure photographs.

Between Epsilon Cygni and the border with Vulpecula lies NGC 6992, another nebula best seen on photographs. Known as the Cygnus Loop or Veil Nebula, it is the remains of a star that exploded some 5,000 years ago.

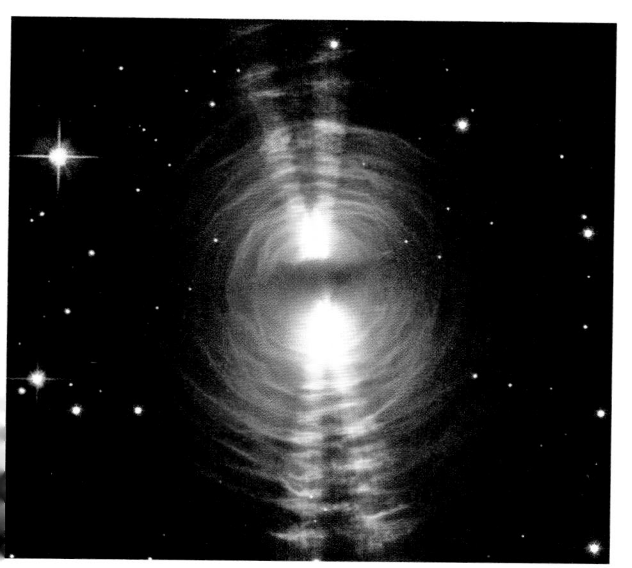

◁ **Egg Nebula**
Starlight shining through thin shells of dust creates beautiful patterns in this false-coloured Hubble Space Telescope image of the planetary nebula. A thicker inner dust belt blocks out light from the central star.

△ **Cygnus A**
One of the strongest radio sources in the sky, this galaxy has a central supermassive black hole, from which jets of gas (coloured red) are thrown out. Hot X-ray-emitting gas is shown in blue.

About **6,000 light-years** from Earth, Cygnus X-1 is a **black hole** that orbits a **blue supergiant** every 5.6 days

▷ **Star distances**
The constellation of Cygnus is the 16th-largest constellation and spans vast distances in space when the distances to its pattern stars are taken into account. The tail (Deneb) of the Swan is over 1,000 light-years from its beak (Albireo). Sadr, in its chest, is farther away still.

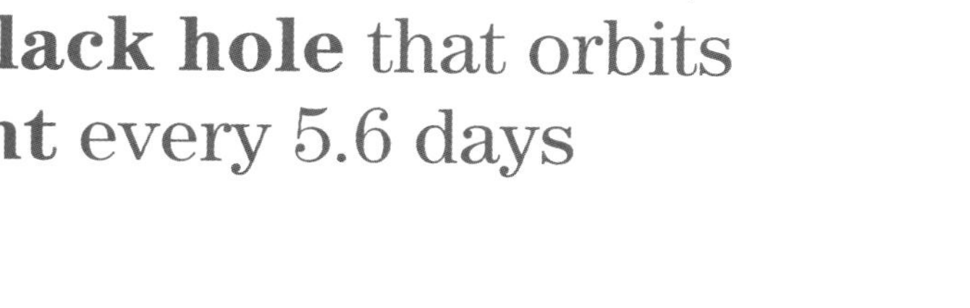

KEY **DATA**

Size ranking 16

Brightest stars Deneb (α) 1.25, Sadr (γ) 2.2

Genitive Cygni

Abbreviation Cyg

Highest in sky at 10pm August–September

Fully visible 58°N–83°S

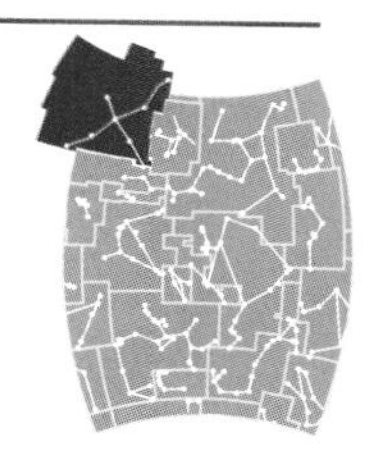

CHART 4

MAIN **STARS**

Deneb Alpha (α) Cygni
Blue-white supergiant; brightest star in Cygnus
1.25 ⟷ 1,400 light-years

Albireo Beta (β) Cygni
Wide double star; colours are orange and blue-green
3.1, 5.1 ⟷ 400 light-years

Sadr Gamma (γ) Cygni
White supergiant in the middle of the northern cross
2.2 ⟷ 1,800 light-years

Delta (δ) Cygni
Binary star; period 920 years
2.8 ⟷ 165 light-years

Gienah Epsilon (ε) Cygni
Orange giant
2.5 ⟷ 73 light-years

Zeta (ζ) Cygni
Yellow giant; spectroscopic binary
3.2 ⟷ 145 light-years

Mu (μ) Cygni
Binary star; period 790 years
4.5 ⟷ 720 light-years

DEEP-SKY **OBJECTS**

M39
Open cluster of around 30 stars

NGC 6826 (Blinking Planetary)
Planetary nebula

NGC 6992 (Cygnus Loop / Veil Nebula)
Supernova remnant

NGC 7000 (North America Nebula)
Emission nebula

Cygnus X-1
X-ray binary system containing a black hole

Egg Nebula
Planetary nebula

Cygnus A
Radio galaxy

LUMINOSITIES

Upsilon Andromedae *4 Suns*

Delta Andromedae *45 Suns*

Alpheratz *115 Suns*

ANDROMEDA
THE CAPTIVE PRINCESS

ANDROMEDA LIES NEXT TO ONE CORNER OF THE SQUARE OF PEGASUS. THIS CONSTELLATION CONTAINS THE NEAREST LARGE GALAXY TO THE MILKY WAY, A VAST SPIRAL KNOWN AS M31.

Andromeda was a mythical princess, the daughter of Queen Cassiopeia and King Cepheus. In one of the most famous Greek myths, the gods ordered Andromeda to be sacrificed to the sea monster Cetus to atone for her mother's vanity, but she was rescued from the monster's jaws by the hero Perseus. Andromeda and her rescuer are commemorated in the sky by adjoining constellations.

The most important object in this constellation is the Andromeda Galaxy, also known as M31, a huge spiral of stars similar to our own Milky Way but even larger. The Andromeda Galaxy is just visible to the unaided eye on clear, dark nights near the star Nu Andromedae. It is easy to see with binoculars or a small telescope, appearing as an elongated smudge. Lying 2.5 million light-years away, M31 is the most distant object visible to the naked eye.

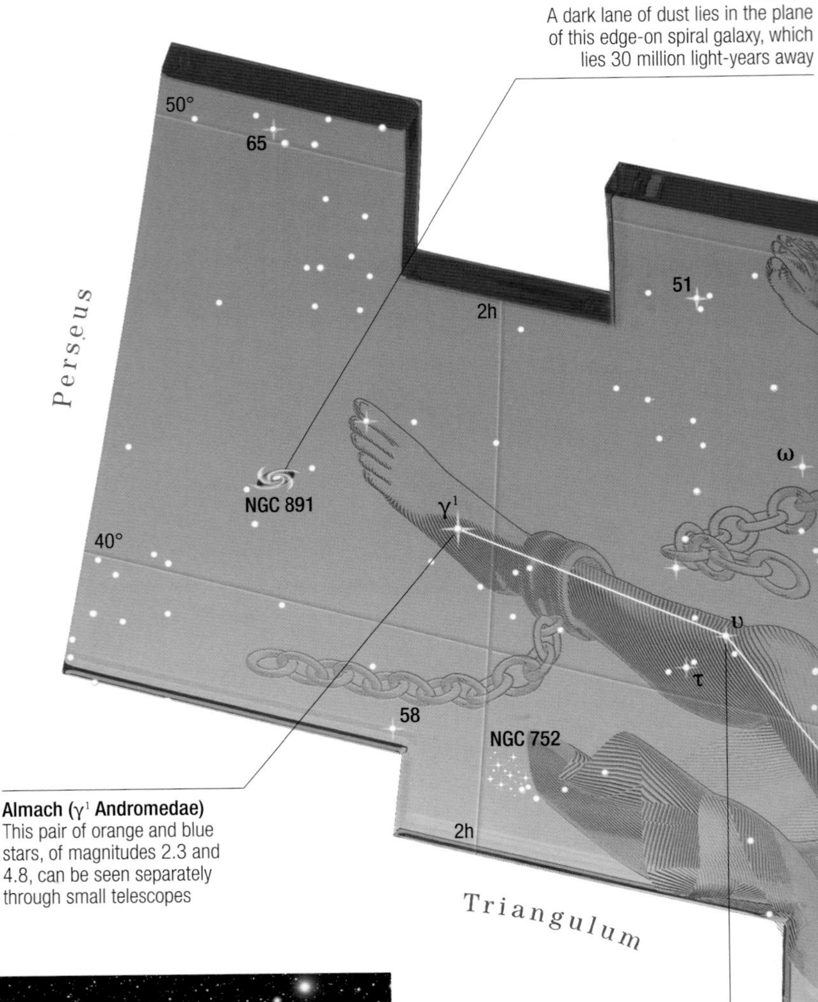

NGC 891
A dark lane of dust lies in the plane of this edge-on spiral galaxy, which lies 30 million light-years away

Almach (γ[1] Andromedae)
This pair of orange and blue stars, of magnitudes 2.3 and 4.8, can be seen separately through small telescopes

Upsilon (υ) Andromedae
This was the first star found to be orbited by more than one planet. Four planets are now known in its system

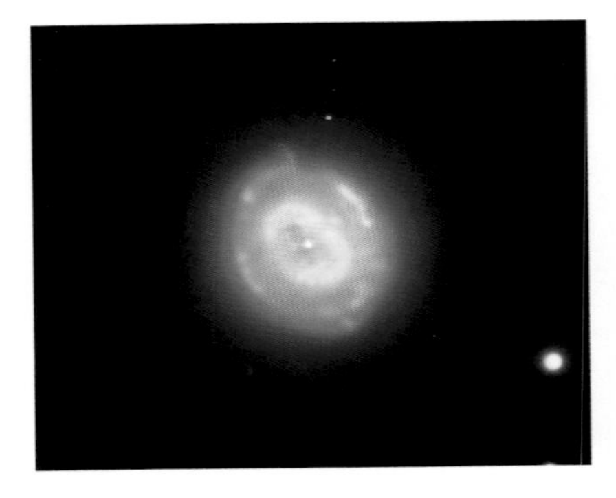

△ **NGC 7662**
Popularly known as the Blue Snowball, this 9th-magnitude planetary nebula appears as an elliptical blue-green patch through small telescopes. Photographs reveal the central star.

◁ **M31**
The great spiral M31 is tilted at an angle to us, so that it appears elliptical. Two smaller companion galaxies are also visible in this picture – M110 below it and M32 on its upper edge.

▷ **Star distances**
The nearest of Andromeda's pattern stars is Upsilon (υ) Andromedae at 44 light-years away. The brightest star, Alpheraz (α Andromedae) is just over double that distance, at 97 light-years away. The most distant pattern star is Phi, which is about 700 light-years from Earth.

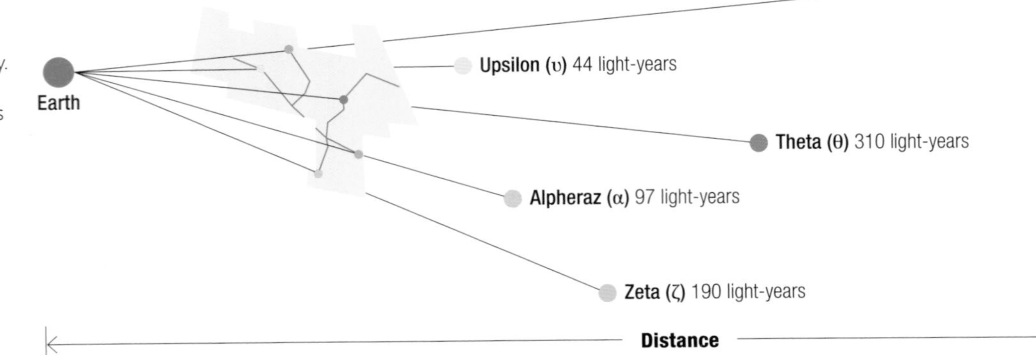

Mirach
475 Suns

Pi Andromedae
540 Suns

Omicron Andromedae
1,380 Suns

Almach
1,830 Suns

M31
The Andromeda Galaxy and its companions appear in the sky near the star Nu (ν) Andromedae, but in reality they are nearly 2 million light-years farther away

Omicron (ο) Andromedae
Studies have shown that the components of this binary are binaries themselves, making it a quad-star system

Alpheratz (α Andromedae)
Marking Andromeda's head is Alpheratz. With Mirach, it is the joint-brightest star in the constellation, at magnitude 2.1

KEY DATA

Size ranking 19

Brightest stars Alpheratz (α) 2.1, Mirach (β) 2.1

Genitive Andromedae

Abbreviation And

Highest in sky at 10pm October–November

Fully visible 90°N–7°S

CHART 3

MAIN STARS

Alpheratz Alpha (α) Andromedae
Blue-white star; also known as Sirrah
2.1 ⟷ 97 light-years

Mirach Beta (β) Andromedae
Red giant
2.1 ⟷ 200 light-years

Almach Gamma (γ[1]) Andromedae
Double star for small telescopes; also called Almaak
2.3, 5.8 ⟷ 360 light-years

Delta (δ) Andromedae
Orange giant
3.3 ⟷ 105 light-years

Omicron (ο) Andromedae
Blue-white giant
3.6 ⟷ 700 light-years

Pi (π) Andromedae
Double star for small telescopes
4.3, 9.0 ⟷ 600 light-years

Upsilon (υ) Andromedae
Yellow-white main-sequence star with planets
4.1 ⟷ 44 light-years

DEEP-SKY OBJECTS

M31
Large spiral galaxy 2.5 million light-years away

M32
Small elliptical galaxy; companion of M31

M110
Small elliptical galaxy; companion of M31

NGC 752
Large open cluster visible with binoculars

NGC 891
Edge-on spiral galaxy

NGC 7662
Planetary nebula, popularly termed the Blue Snowball

In a few billion years' time, the **Milky Way** and the **Andromeda Galaxy** will **collide** and **merge** to create a **super-galaxy**

TRIANGULUM
THE TRIANGLE

THOUGH THIS ELONGATED TRIANGLE IS DRAWN AROUND THREE INSIGNIFICANT STARS, IT IS EASY TO SPOT BECAUSE OF ITS COMPACT SIZE. THE NEARBY SPIRAL GALAXY M33 IS THIS NORTHERN CONSTELLATION'S FINEST SIGHT.

More than 2,000 years ago this three-sided star pattern was imagined as the Greek capital letter Delta, the Nile river delta, and the island of Sicily. A fourth option – an isosceles triangle – prevailed and this is how the constellation is seen today. It is home to the spiral galaxy M33, better known as the Triangulum Galaxy. It is the third-largest member of our Local Group of galaxies and, at 2.7 million light-years away, is one of the closest.

KEY DATA

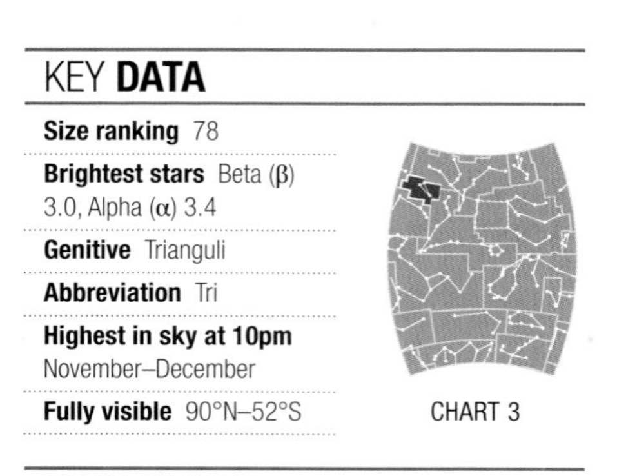

Size ranking 78

Brightest stars Beta (β) 3.0, Alpha (α) 3.4

Genitive Trianguli

Abbreviation Tri

Highest in sky at 10pm November–December

Fully visible 90°N–52°S

CHART 3

MAIN STARS

Alpha (α) Trianguli
White giant or subgiant
3.4 — 63 light-years

Beta (β) Trianguli
White subgiant
3.0 — 130 light-years

Gamma (γ) Trianguli.
White main-sequence star
4.0 — 112 light-years

6 Trianguli
Yellow giant, and a double star
5.0 — 290 light-years

R Trianguli
Red giant star, also a variable star
6.8 — 960 light-years

DEEP-SKY OBJECTS

M33 (Triangulum Galaxy)
Spiral galaxy, also known as NGC 598

NGC 604
Star-formation nebula in M33

NGC 784
Barred spiral galaxy

NGC 925
Barred spiral galaxy

△ **NGC 604**
The huge billowing cloud of hydrogen in M33 is 1,500 light-years across and a centre of star formation. Its red glow is a product of the ultraviolet energy released by hundreds of young, bright stars.

▷ **M33**
Galaxy M33 lies almost face-on to Earth and its arms are, in fact, a series of separate patches. The galaxy is one of the most distant objects visible to the naked eye, as long as conditions are good.

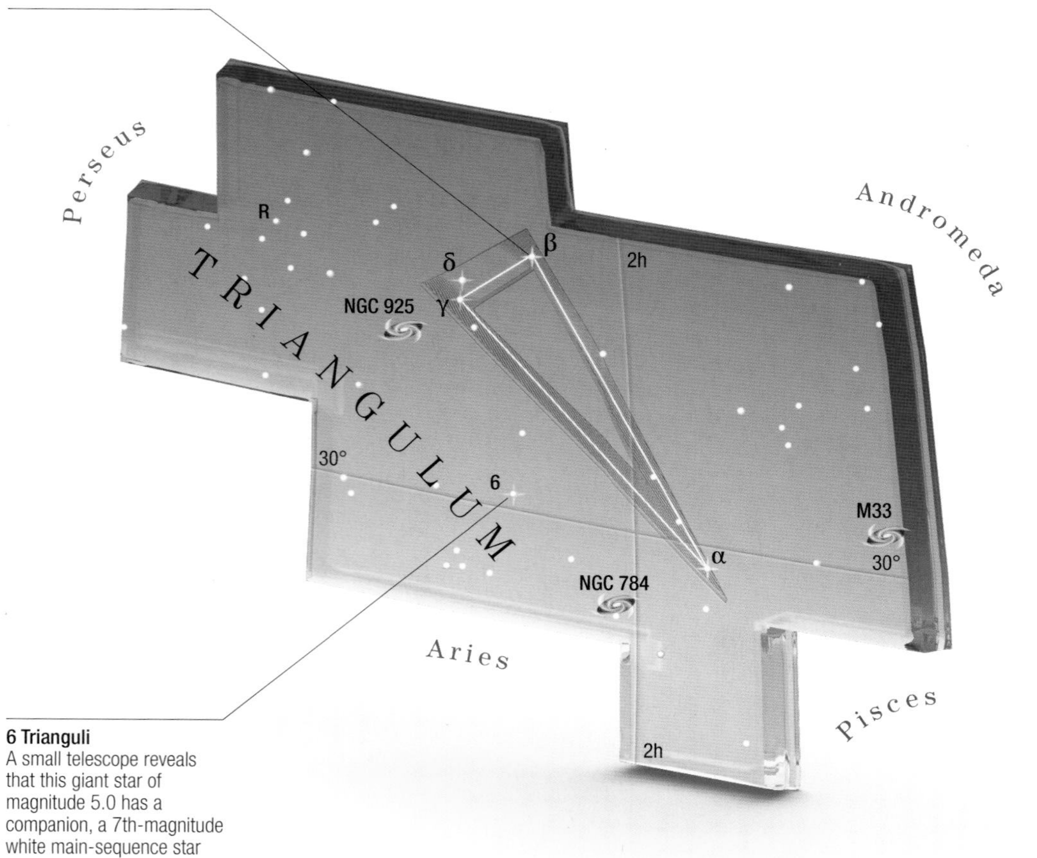

Beta (β) Trianguli
At magnitude 3.0 this is the brightest star in Triangulum. It is a white star about four times the width of the Sun

6 Trianguli
A small telescope reveals that this giant star of magnitude 5.0 has a companion, a 7th-magnitude white main-sequence star

LACERTA
THE LIZARD

A SMALL, OBSCURE CONSTELLATION, LACERTA HAS A ZIGZAG PATTERN THAT RESEMBLES A SCURRYING LIZARD. IT LIES IN THE NORTHERN PATH OF THE MILKY WAY, SANDWICHED BETWEEN THE CONSTELLATIONS OF ANDROMEDA AND CYGNUS. ITS MAIN STARS ARE IN THE REPTILE'S HEAD.

Lacerta was named in 1687 when the Polish astronomer Johannes Hevelius first described it. It was one of 11 new constellations he devised to fill gaps in the northern sky; seven of these are still in use today. Lacerta's stars are faint and none have names, but it has been the site of occasional nova explosions. Among the constellation's dense Milky Way star clouds are few significant deep-sky objects. However, one object of note is BL Lacertae, the prototype of a strange class of galaxy with active nuclei, known as B Lac objects or blazars. These are a type of quasar that emit energetic jets directly towards Earth, giving them the appearance of a star.

KEY DATA

Size ranking 68
Brightest stars Alpha (α) 3.8, 1 Lacertae 4.1
Genitive Lacertae
Abbreviation Lac
Highest in sky at 10pm September–October
Fully visible 90°N–33°S

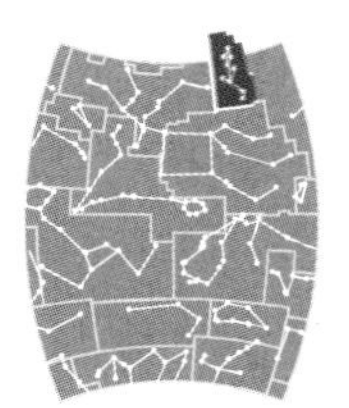

CHART 3

MAIN STARS

Alpha (α) Lacertae
Blue-white main-sequence star
3.8 — 103 light-years

Beta (β) Lacertae
Yellow giant
4.4 — 170 light-years

1 Lacertae
Orange giant
4.1 — 621 light-years

DEEP-SKY OBJECTS

NGC 7243
Open star cluster

BL Lacertae
Blazar and the prototype of B Lac objects

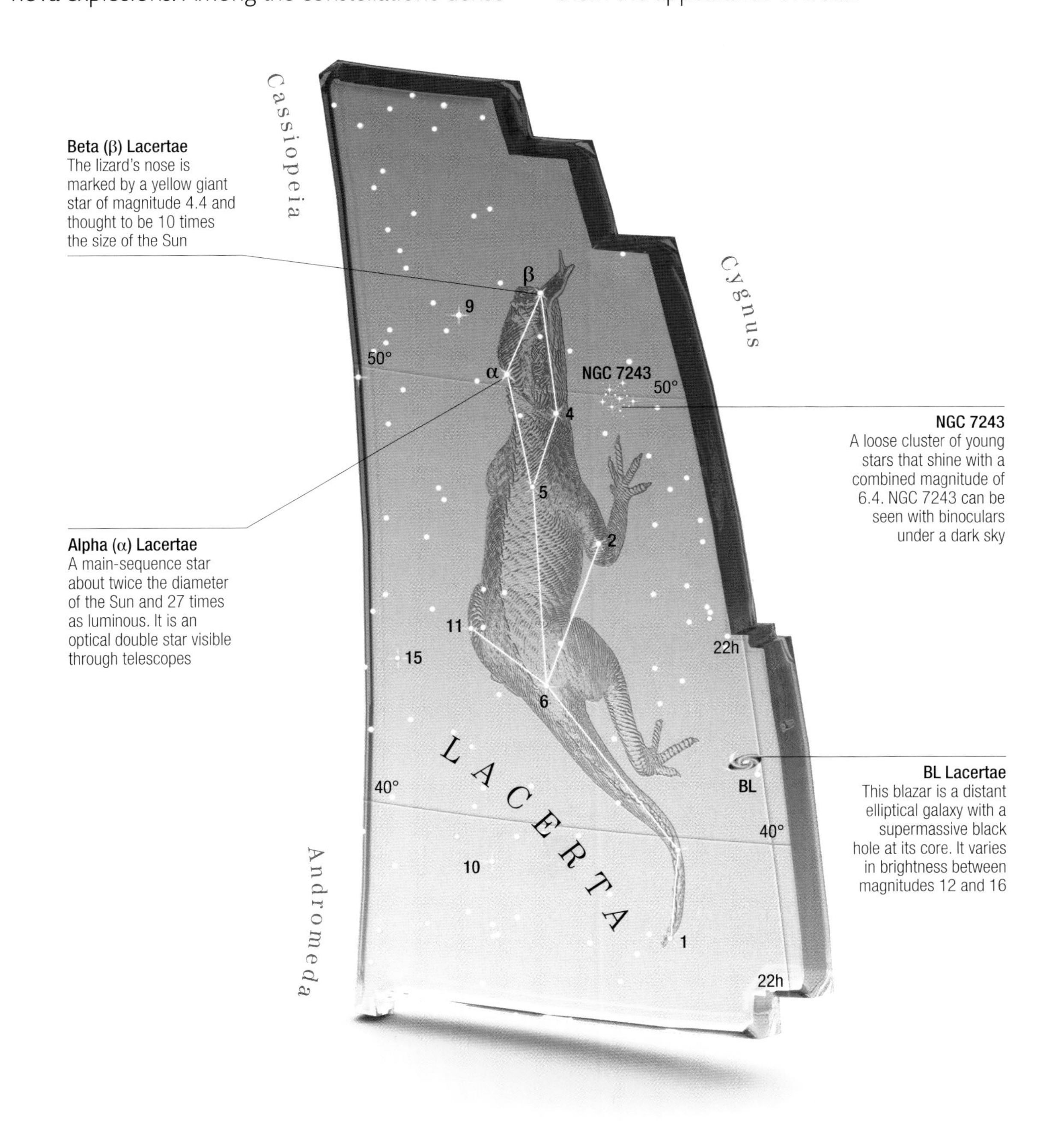

Beta (β) Lacertae
The lizard's nose is marked by a yellow giant star of magnitude 4.4 and thought to be 10 times the size of the Sun

Alpha (α) Lacertae
A main-sequence star about twice the diameter of the Sun and 27 times as luminous. It is an optical double star visible through telescopes

NGC 7243
A loose cluster of young stars that shine with a combined magnitude of 6.4. NGC 7243 can be seen with binoculars under a dark sky

BL Lacertae
This blazar is a distant elliptical galaxy with a supermassive black hole at its core. It varies in brightness between magnitudes 12 and 16

△ **NGC 7243**
This group of blue-white stars is about 2,800 light-years away. The young blue-white stars stand out against a rich starfield of yellow and red stars. Seen through a small telescope, there are about 120 stars spread across an area equal to that of the Full Moon. Their loose scattering has made it uncertain whether or not they form a true star cluster.

LUMINOSITIES

16 Persei *25 Suns*

Algol *95 Suns*

Gamma Persei *330 Suns*

PERSEUS
THE VICTORIOUS HERO

PERSEUS IS A PROMINENT CONSTELLATION OF THE NORTHERN SKY. IT LIES IN THE MILKY WAY BETWEEN ANDROMEDA AND AURIGA, NORTH OF TAURUS THE BULL. PERSEUS FEATURES A TWIN CLUSTER OF STARS AND A FAMOUS VARIABLE STAR, ALGOL.

In Greek myth, Perseus was sent to bring back the head of Medusa the Gorgon, whose gaze turned people to stone. He cut off Medusa's head and was returning home with it when he saw Princess Andromeda chained to a rock as a sacrifice to the sea monster Cetus. Perseus killed Cetus, freed Andromeda, and took her as his bride. The constellations Perseus and Andromeda lie side-by-side in the night sky.

Perseus is represented holding Medusa's severed head, marked by the variable star Algol in his left hand. Algol is a binary, in which the fainter star eclipses the brighter one every 2.9 days, causing it to fade for 10 hours.

Perseus is the source of the annual Perseid meteor shower, which radiates from the north of the constellation, near the border with Cassiopeia, in mid-August each year.

In 1901, **Nova Persei erupted** to become one of the **brightest stars** in the sky for several days before gradually fading away

▷ **NGC 869 and NGC 884**
Known as the Double Cluster, these twin open clusters are visible to the naked eye as a brighter patch in the Milky Way near the border with Cassiopeia. NGC 869 (left in the image) is the brighter and richer of the pair. They lie about 7,000 light-years away.

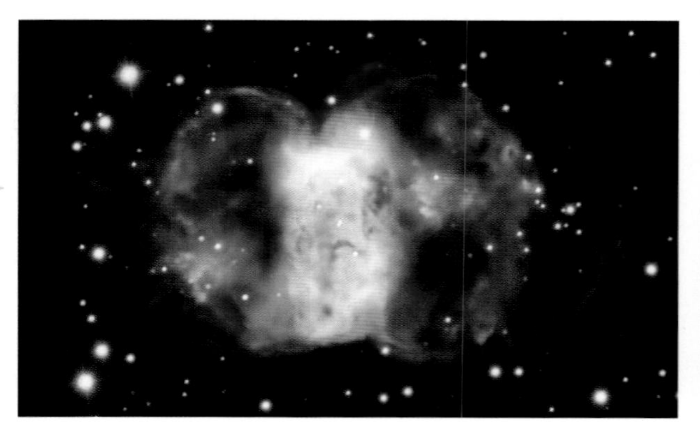

△ **M76**
The double-lobed shape of this planetary nebula gives it its popular name: the Little Dumbbell. At 10th magnitude, M76 is the faintest object in Charles Messier's catalogue of deep-sky objects.

△ **GK Persei**
The explosion of Nova Persei threw off a glowing shell of hot gas, resulting in the nova remnant known as GK Persei (also sometimes called the Firework Nebula).

KEY DATA

Size ranking 24

Brightest stars Mirphak (α) 1.8, Algol (β) 2.1–3.4

Genitive Persei

Abbreviation Per

Highest in sky at 10pm November–December

Fully visible 58°N–83°S

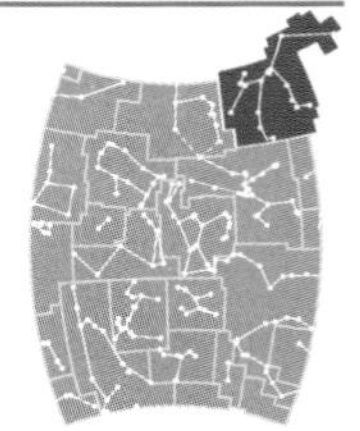

CHART 6

MAIN STARS

Mirphak Alpha (α) Persei
Also spelled Mirfak, a white supergiant
1.8 ⟷ 500 light-years

Algol Beta (β) Persei
An eclipsing binary star with variable brightness
2.1–3.4 ⟷ 90 light-years

Gamma (γ) Persei
A yellow giant
2.9 ⟷ 240 light-years

Delta (δ) Persei
A blue giant
3.0 ⟷ 520 light-years

Epsilon (ε) Persei
A blue giant
2.9 ⟷ 640 light-years

Zeta (ζ) Persei
A blue supergiant and Perseus's third-brightest star
2.9 ⟷ 750 light-years

Rho (ρ) Persei
A variable red giant
3.3–4.0 ⟷ 310 light-years

DEEP-SKY OBJECTS

Alpha Persei Cluster
A scattered cluster of stars around Mirphak

GK Persei
The Nova Persei remnant, also called the Firework Nebula

M34
A large open cluster of about 60 stars

M76
A planetary nebula, known as the Little Dumbbell

NGC 869 and NGC 884
Twin open clusters, known as the Double Cluster

NGC 1499
An emission nebula, known as the California Nebula

Perseus A (NGC 1275)
A supergiant elliptical galaxy

Rho Persei *360 Suns*

Epsilon Persei *2,310 Suns*

Zeta Persei *3,380 Suns*

Mirphak *4,040 Suns*

Mirphak (α Persei)
At magnitude 1.8, Mirphak is the brightest star in Perseus. It is surrounded by a loose cluster of dimmer stars

NGC 1528
An open cluster, NGC 1528 was discovered in 1790 by the British astronomer William Herschel. Consisting of about 160 stars, the cluster can be seen with binoculars

M34
Lying about 1,400 light-years away, M34 is a large open cluster of about 60 stars

Algol (β Persei)
An eclipsing binary star, Algol fades from magnitude 2.1 to 3.4 for 10 hours every 2.9 days

Zeta (ζ) Persei
_ying about 750 light-years away, Zeta Persei is a blue supergiant that is about 3,400 times more luminous than the Sun

Camelopardalis
Cassiopeia
Andromeda
Triangulum
Taurus
PERSEUS
NGC 869
NGC 884
NGC 1528
M76
GK Persei
M34
Perseus A
NGC 1499
NGC 1342
3h 4h 2h 50° 40°

Star distances
At about 90 light-years away, Algol is the nearest of Perseus's main pattern stars and the nearer of the constellation's two brightest stars. Mirphak is more than five times farther away, at 500 light-years, but is the brighter of the two. Xi is the most distant of the constellation's main pattern stars, situated about 1,240 light-years away.

Earth
Mirphak (α) 500 light-years
Kappa (κ) 115 light-years
Algol (β) 90 light-years
Xi (ξ) 1,240 light-years
Zeta (ζ) 750 light-years
Distance

LEO MINOR
THE LITTLE LION

A SMALL AND FAINT CONSTELLATION IN THE NORTHEN SKY REPRESENTING A LION CUB, LEO MINOR WAS INVENTED IN THE LATE 17TH CENTURY BY JOHANNES HEVELIUS.

Leo Minor is not one of the constellations known to the Ancient Greeks but was introduced in 1687 by the Polish astronomer Johannes Hevelius. Squeezed into a gap between Leo and Ursa Major, it is easily overlooked. It has no star labelled Alpha, although there is a Beta. This is due to an error by the English astronomer Francis Baily, who omitted to label its brightest member, 46 Leonis Minoris, when cataloguing its stars in 1845. The constellation's most famous object is Hanny's Voorwerp ("Hanny's Object" in Dutch), an unusual cloud of gas discovered by Dutch amateur astronomer Hanny van Arkel.

KEY DATA

Size ranking 64

Brightest stars 46 Leonis Minoris 3.8, Beta (β) 4.2

Genitive Leonis Minoris

Abbreviation LMi

Highest in sky at 10pm March–April

Fully visible 90°N–48°S

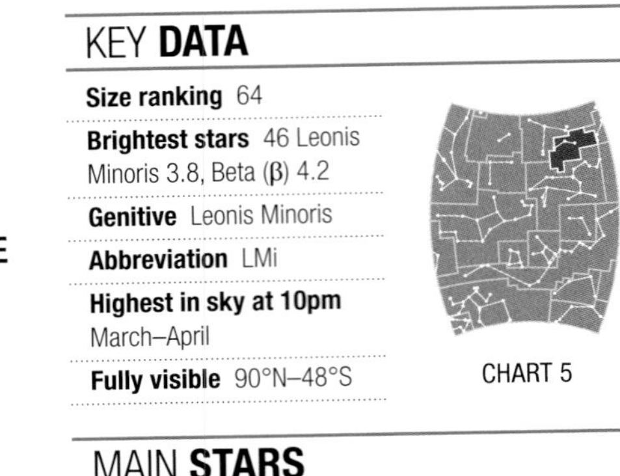

CHART 5

MAIN STARS

Beta (β) Leonis Minoris
Yellow-orange giant
4.2 ⟷ 154 light-years

46 Leonis Minoris
Orange giant
3.8 ⟷ 95 light-years

DEEP-SKY OBJECTS

IC 2497 and **Hanny's Voorwerp**
Active galaxy with nearby gas cloud

NGC 3021
Spiral galaxy

46 Leonis Minoris
The brightest star in Leo Minor; an orange giant with a diameter about 8.5 times that of the Sun and a mass of about 1.5 Suns
Beta (β) Leonis Minoris
The 2nd-brightest star in Leo Minor and the only star in the constellation with a Greek letter
10h
40°
40°
Lynx
Ursa Major
β
46
21
11h
IC 2497/Hanny's Voorwerp
NGC 3021
LEO MINOR
30°
30°
11h

▷ **IC 2497 and Hanny's Voorwerp**
The irregular object coloured green in this Hubble image is Hanny's Voorwerp, a cloud of gas lit up by radiation from an old quasar in the galaxy IC 2497 seen above it.

The main star **Capella** is the **brightest naked-eye star** with the **same colour as the Sun**

Lynx
40°

AURIGA
THE CHARIOTEER

A LARGE AND PROMINENT CONSTELLATION OF THE NORTHERN SKY, AURIGA CONTAINS THE SIXTH-BRIGHTEST STAR IN THE SKY, CAPELLA.

Auriga represents a charioteer of Greek legend although the chariot itself is not part of the constellation. Auriga's brightest star, Capella, represents a goat carried by the charioteer. Another notable star is the white supergiant Epsilon Aurigae, which is an eclipsing binary with an exceptionally long period of 27 years. Three open clusters, M36, M37, and M38, can be seen with binoculars. All lie about 4,000 light-years away. M37 is the largest but M36 is the easiest to spot. The star marking the charioteer's right foot was once shared with Taurus. When borders for all the constellations were officially decided in 1930, this star was allocated to Taurus, hence its present-day name of Beta Tauri.

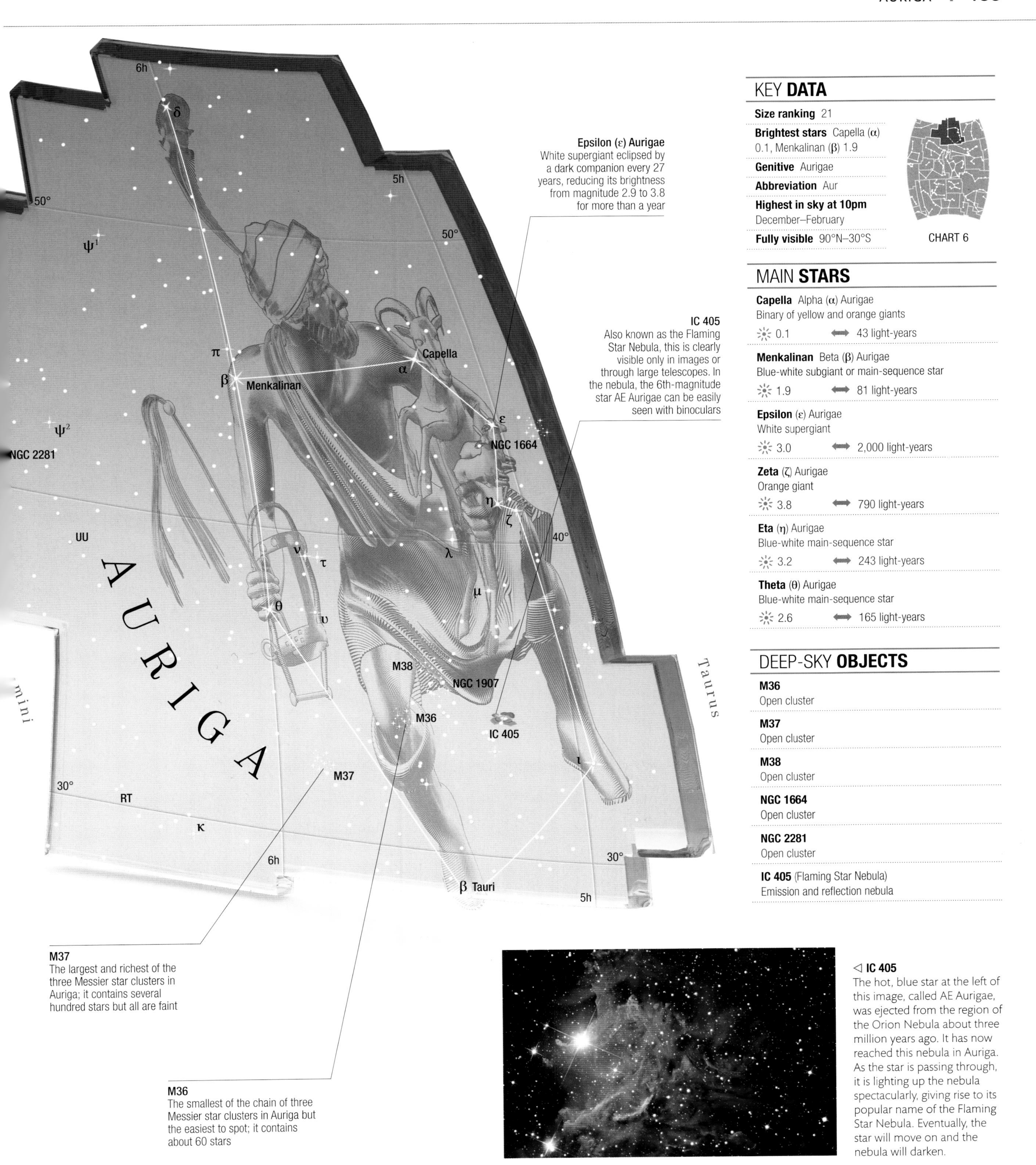
Epsilon (ε) Aurigae
White supergiant eclipsed by a dark companion every 27 years, reducing its brightness from magnitude 2.9 to 3.8 for more than a year
IC 405
Also known as the Flaming Star Nebula, this is clearly visible only in images or through large telescopes. In the nebula, the 6th-magnitude star AE Aurigae can be easily seen with binoculars
M37
The largest and richest of the three Messier star clusters in Auriga; it contains several hundred stars but all are faint
M36
The smallest of the chain of three Messier star clusters in Auriga but the easiest to spot; it contains about 60 stars
AURIGA
6h
5h
50°
40°
30°
δ
ψ¹
ψ²
π
β
Menkalinan
α
Capella
ε
NGC 1664
η
ζ
λ
ν
τ
θ
υ
μ
M38
NGC 1907
M36
IC 405
M37
ι
UU
RT
κ
NGC 2281
β Tauri
Taurus
mini
CHART 6

KEY DATA

Size ranking 21

Brightest stars Capella (α) 0.1, Menkalinan (β) 1.9

Genitive Aurigae

Abbreviation Aur

Highest in sky at 10pm December–February

Fully visible 90°N–30°S

MAIN STARS

Capella Alpha (α) Aurigae
Binary of yellow and orange giants
0.1 — 43 light-years

Menkalinan Beta (β) Aurigae
Blue-white subgiant or main-sequence star
1.9 — 81 light-years

Epsilon (ε) Aurigae
White supergiant
3.0 — 2,000 light-years

Zeta (ζ) Aurigae
Orange giant
3.8 — 790 light-years

Eta (η) Aurigae
Blue-white main-sequence star
3.2 — 243 light-years

Theta (θ) Aurigae
Blue-white main-sequence star
2.6 — 165 light-years

DEEP-SKY OBJECTS

M36
Open cluster

M37
Open cluster

M38
Open cluster

NGC 1664
Open cluster

NGC 2281
Open cluster

IC 405 (Flaming Star Nebula)
Emission and reflection nebula

◁ **IC 405**
The hot, blue star at the left of this image, called AE Aurigae, was ejected from the region of the Orion Nebula about three million years ago. It has now reached this nebula in Auriga. As the star is passing through, it is lighting up the nebula spectacularly, giving rise to its popular name of the Flaming Star Nebula. Eventually, the star will move on and the nebula will darken.

LUMINOSITIES

Iota Leonis
12 Suns

Denebola
15 Suns

LEO THE LION

LEO IS A LARGE CONSTELLATION OF THE ZODIAC, EASY TO RECOGNIZE BECAUSE OF ITS RESEMBLANCE TO A CROUCHING LION. A SICKLE SHAPE FORMED BY SIX STARS OUTLINES THE HEAD AND CHEST OF THE LION.

Leo is said to represent the lion killed by Hercules, the strong man of Greek myth, as the first of his 12 labours. The Sickle, delineating the lion's head and chest, looks like a back-to-front question mark. At the base of the Sickle is Leo's brightest star, Regulus – the heart of the lion.

In the middle of the Sickle lies Gamma Leonis, popularly known as Algieba. This binary has two yellow-orange giants that orbit each other every 550 years. A 5th-magnitude star nearby, 40 Leonis, is unrelated. Zeta Leonis, of 3rd magnitude, has two fainter companions visible through binoculars, but the three stars are not gravitationally bound.

A number of spiral galaxies can be seen through small telescopes under the lion's body. Of these galaxies M65, M66, M95, and M96 are the most prominent.

Every November, Earth moves through a stream of particles left by the comet Tempel–Tuttle, and the Leonid meteors radiate from the region of the Sickle. Rates are usually low, but storms occasionally occur, as in 1833.

Shooting stars were said to have fallen from the sky **"like snowflakes"** in the Leonid **meteor storm** of 1833

▷ **NGC 3521**
Patches of star formation give a mottled look to the arms of this spiral galaxy, 35 million light-years away. This image is from the European Southern Observatory's Very Large Telescope in Chile.

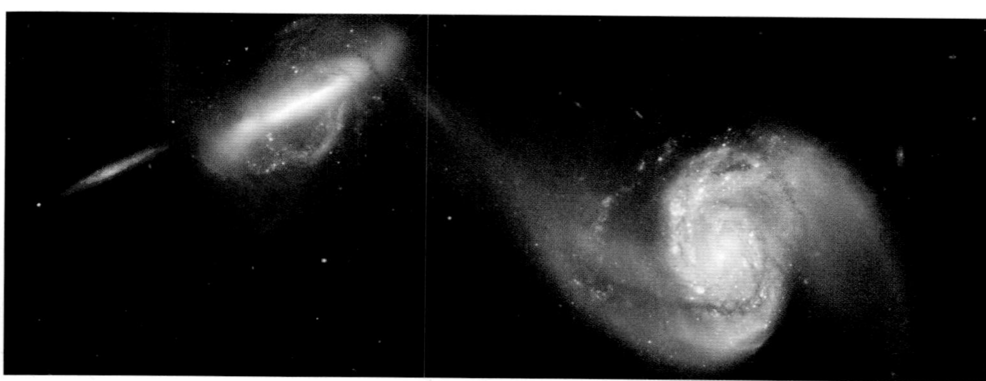

▽ **NGC 3808 and NGC 3808A**
Two spiral galaxies intertwine in this image from the Hubble Space Telescope. Stars, gas, and dust flow from NGC 3808 (right) and coil around its smaller companion, NGC 3808A (seen edge-on).

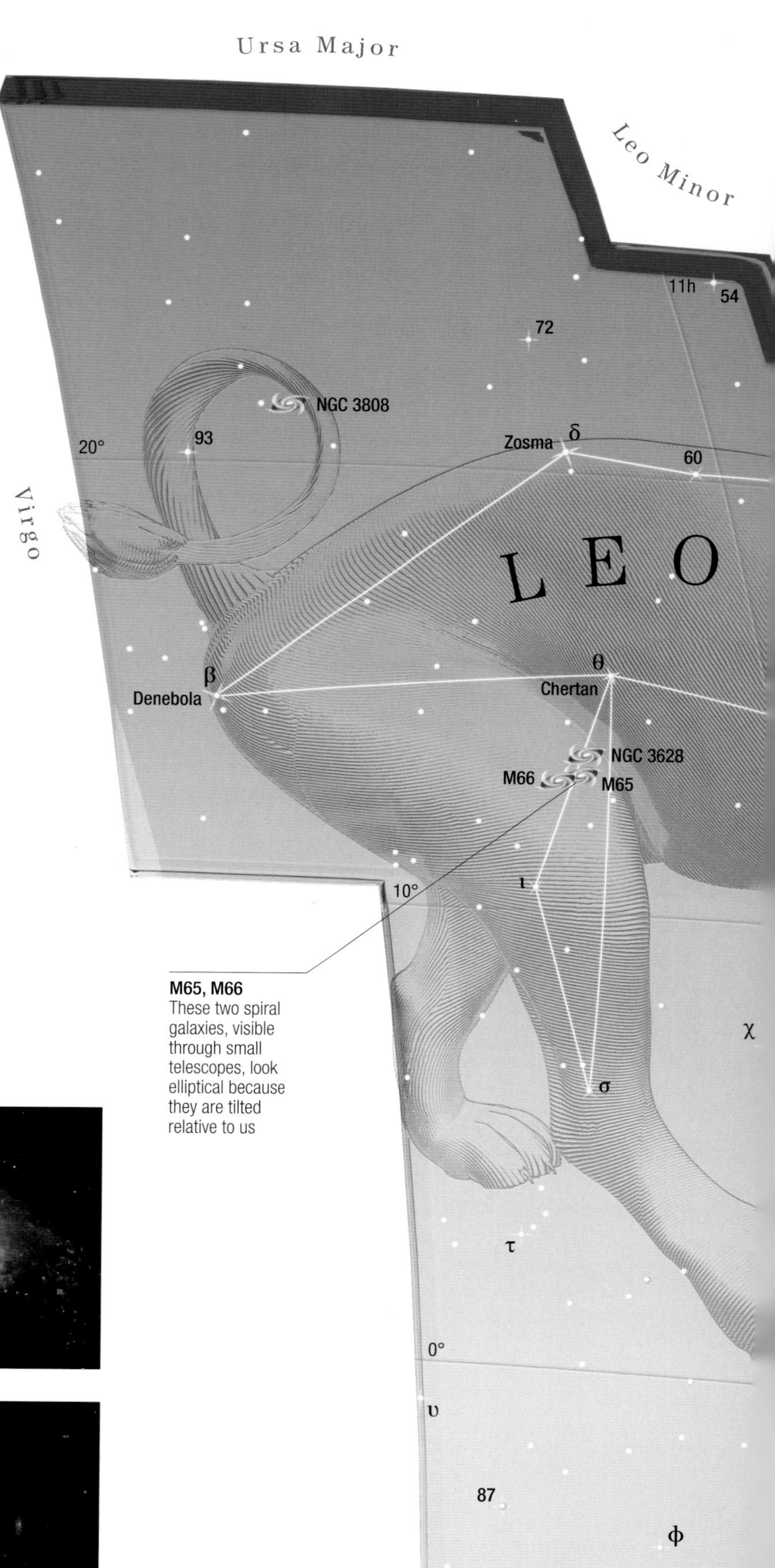

M65, M66
These two spiral galaxies, visible through small telescopes, look elliptical because they are tilted relative to us

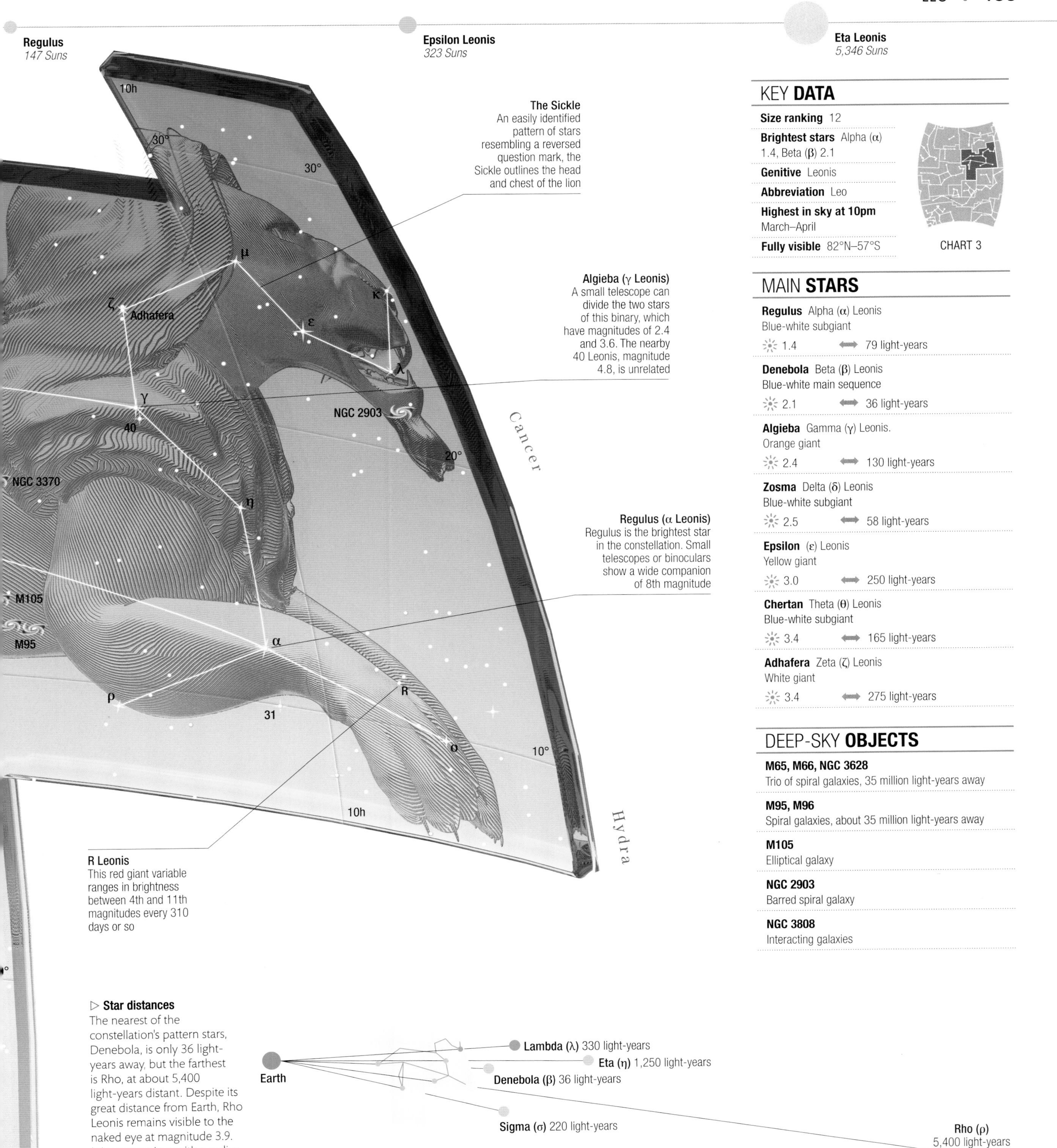

The Sickle
An easily identified pattern of stars resembling a reversed question mark, the Sickle outlines the head and chest of the lion

Algieba (γ Leonis)
A small telescope can divide the two stars of this binary, which have magnitudes of 2.4 and 3.6. The nearby 40 Leonis, magnitude 4.8, is unrelated

Regulus (α Leonis)
Regulus is the brightest star in the constellation. Small telescopes or binoculars show a wide companion of 8th magnitude

R Leonis
This red giant variable ranges in brightness between 4th and 11th magnitudes every 310 days or so

KEY DATA

Size ranking 12

Brightest stars Alpha (α) 1.4, Beta (β) 2.1

Genitive Leonis

Abbreviation Leo

Highest in sky at 10pm March–April

Fully visible 82°N–57°S

CHART 3

MAIN STARS

Regulus Alpha (α) Leonis
Blue-white subgiant
1.4 ⟷ 79 light-years

Denebola Beta (β) Leonis
Blue-white main sequence
2.1 ⟷ 36 light-years

Algieba Gamma (γ) Leonis.
Orange giant
2.4 ⟷ 130 light-years

Zosma Delta (δ) Leonis
Blue-white subgiant
2.5 ⟷ 58 light-years

Epsilon (ε) Leonis
Yellow giant
3.0 ⟷ 250 light-years

Chertan Theta (θ) Leonis
Blue-white subgiant
3.4 ⟷ 165 light-years

Adhafera Zeta (ζ) Leonis
White giant
3.4 ⟷ 275 light-years

DEEP-SKY OBJECTS

M65, M66, NGC 3628
Trio of spiral galaxies, 35 million light-years away

M95, M96
Spiral galaxies, about 35 million light-years away

M105
Elliptical galaxy

NGC 2903
Barred spiral galaxy

NGC 3808
Interacting galaxies

▷ **Star distances**
The nearest of the constellation's pattern stars, Denebola, is only 36 light-years away, but the farthest is Rho, at about 5,400 light-years distant. Despite its great distance from Earth, Rho Leonis remains visible to the naked eye at magnitude 3.9. It is a supergiant with a radius about 37 times that of our Sun.

Earth
Lambda (λ) 330 light-years
Eta (η) 1,250 light-years
Denebola (β) 36 light-years
Sigma (σ) 220 light-years
Rho (ρ) 5,400 light-years
Distance

LUMINOSITIES

Zavijava
4 Suns

Porrima
10 Suns

△ **Sombrero Galaxy**
Also known as M104, this edge-on spiral galaxy resembles a sombrero hat, its brim edged by a dark lane of dust. Situated on the border with the constellation Corvus, the Sombrero Galaxy is about 30 million light-years away.

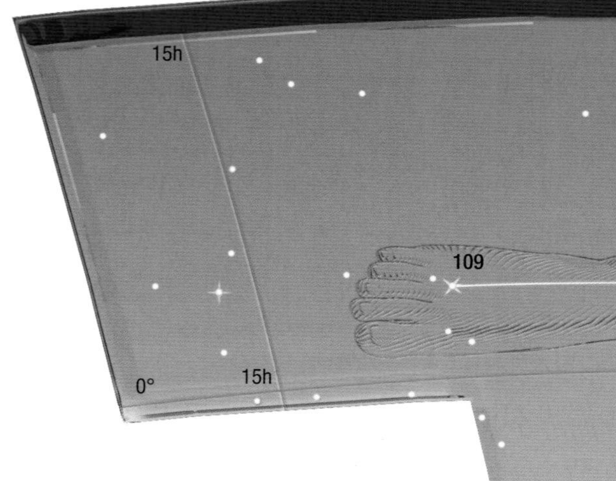

VIRGO THE VIRGIN

VIRGO IS THE LARGEST CONSTELLATION IN THE ZODIAC, AND ALSO THE SECOND LARGEST IN THE ENTIRE SKY. IT CONTAINS THE MAJOR CLUSTER OF GALAXIES CLOSEST TO US, AS WELL AS THE BRIGHTEST QUASAR.

Virgo had several identities in Ancient Greek mythology. In one story, she represented Demeter, the corn goddess, and was depicted holding an ear of grain marked by the constellation's brightest star, Spica. Usually, though, she was equated with Dike, the goddess of justice, and the adjoining constellation Libra was visualized as her scales of justice.

The constellation is shaped like a sloping letter Y, with Spica at the base. In the bowl of the Y is situated the Virgo Cluster, some 55 million light-years away. This cluster of galaxies has more than 2,000 members, the brightest of which can be seen through a small telescope. So large is the cluster that it spills over Virgo's northern border into the adjacent constellation of Coma Berenices. At the heart of the Virgo Cluster is the giant elliptical galaxy M87. The brightest quasar (see pp.60–61) as seen from Earth, 3C 273, also lies in Virgo but is over 50 times farther away than the Virgo Cluster.

Virgo's **M87** is one of the **most massive local galaxies**, with a mass of almost **3 trillion Suns**

▷ **Star distances**
The nearest of Virgo's pattern stars is Zavijava (β Virginis) at 36 light-years away. The brightest star, Spica (α Virginis), is much more distant, at about 250 light-years away. The most distant pattern star is Nu (ν) Virginis, which is about 1,170 light-years from Earth.

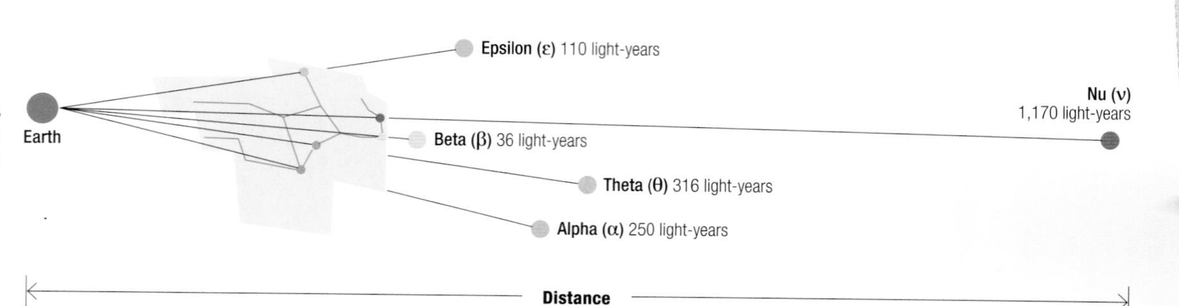

Vindemiatrix
70 Suns

Delta Virginis
140 Suns

Spica
2,070 Suns

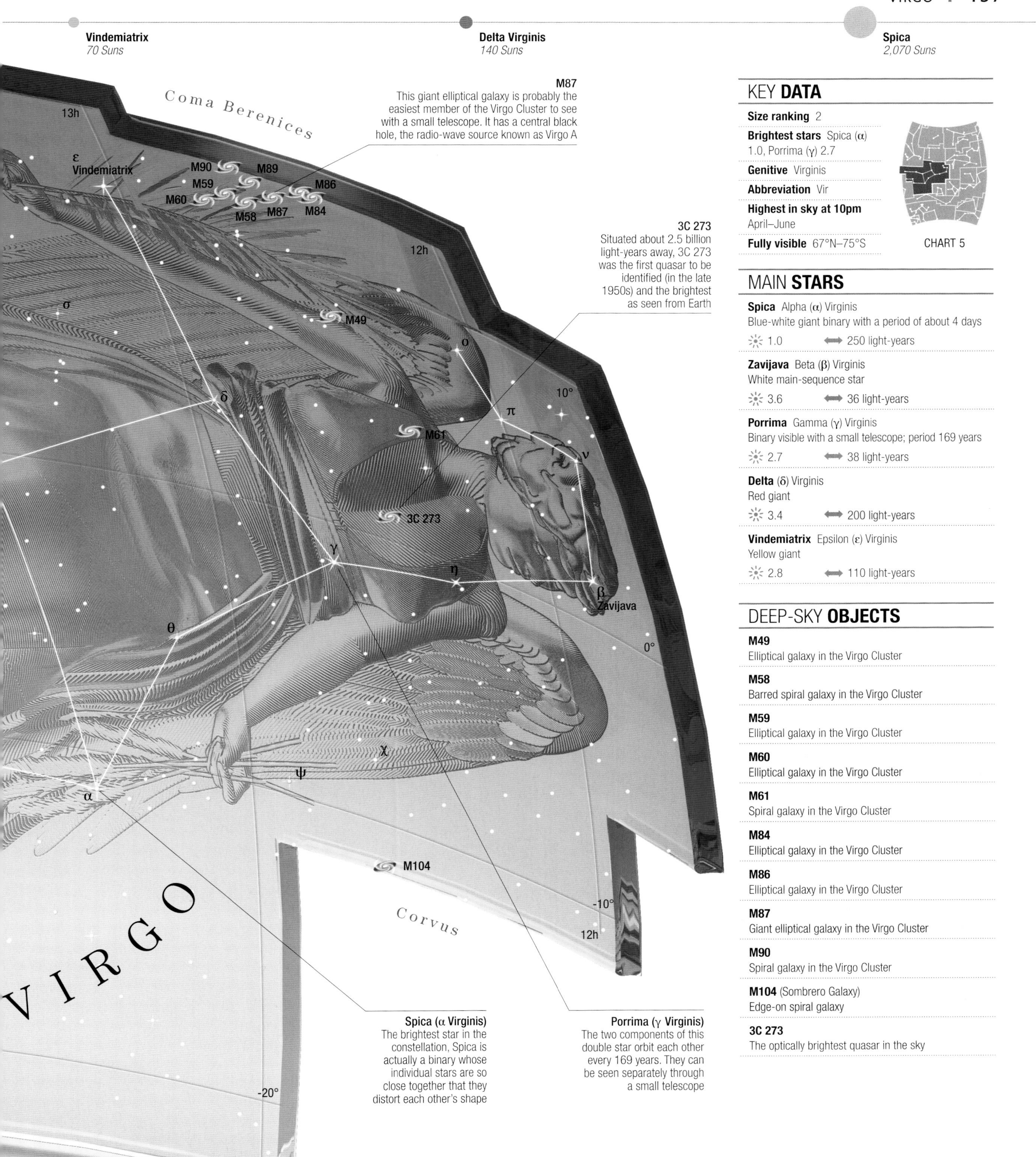

M87
This giant elliptical galaxy is probably the easiest member of the Virgo Cluster to see with a small telescope. It has a central black hole, the radio-wave source known as Virgo A

3C 273
Situated about 2.5 billion light-years away, 3C 273 was the first quasar to be identified (in the late 1950s) and the brightest as seen from Earth

Spica (α Virginis)
The brightest star in the constellation, Spica is actually a binary whose individual stars are so close together that they distort each other's shape

Porrima (γ Virginis)
The two components of this double star orbit each other every 169 years. They can be seen separately through a small telescope

KEY **DATA**

Size ranking 2

Brightest stars Spica (α) 1.0, Porrima (γ) 2.7

Genitive Virginis

Abbreviation Vir

Highest in sky at 10pm April–June

Fully visible 67°N–75°S

CHART 5

MAIN **STARS**

Spica Alpha (α) Virginis
Blue-white giant binary with a period of about 4 days
1.0 ⟷ 250 light-years

Zavijava Beta (β) Virginis
White main-sequence star
3.6 ⟷ 36 light-years

Porrima Gamma (γ) Virginis
Binary visible with a small telescope; period 169 years
2.7 ⟷ 38 light-years

Delta (δ) Virginis
Red giant
3.4 ⟷ 200 light-years

Vindemiatrix Epsilon (ε) Virginis
Yellow giant
2.8 ⟷ 110 light-years

DEEP-SKY **OBJECTS**

M49
Elliptical galaxy in the Virgo Cluster

M58
Barred spiral galaxy in the Virgo Cluster

M59
Elliptical galaxy in the Virgo Cluster

M60
Elliptical galaxy in the Virgo Cluster

M61
Spiral galaxy in the Virgo Cluster

M84
Elliptical galaxy in the Virgo Cluster

M86
Elliptical galaxy in the Virgo Cluster

M87
Giant elliptical galaxy in the Virgo Cluster

M90
Spiral galaxy in the Virgo Cluster

M104 (Sombrero Galaxy)
Edge-on spiral galaxy

3C 273
The optically brightest quasar in the sky

COMA BERENICES

BERENICE'S HAIR

NAMED AFTER THE ANCIENT EGYPTIAN QUEEN BERENICES II, COMA BERENICES IS EASILY LOCATED BETWEEN THE MORE PROMINENT CONSTELLATIONS OF LEO AND BOÖTES. IT CONTAINS CLUSTERS OF BOTH STARS AND GALAXIES.

Considered part of Leo until 1536, Coma Berenices was first shown as a separate constellation on a globe by German cartographer Caspar Vopel. The constellation has no stars brighter than 4th magnitude, but plenty of interesting deep-sky objects. Galaxies such as M85, M88, M99, and M100 near the southern border with Virgo are part of the Virgo Cluster about 50 million light-years away. Others belong to the Coma Cluster, which is six times more distant. Melotte 111 is one of the closest open star clusters. More than 20 of its stars are visible to the naked eye.

△ **M64**
The brightest galaxy in Coma Berenices, M64 is nicknamed the Black Eye Galaxy because of the lane of dark dust near its bright core. It lies 17 million light-years from Earth.

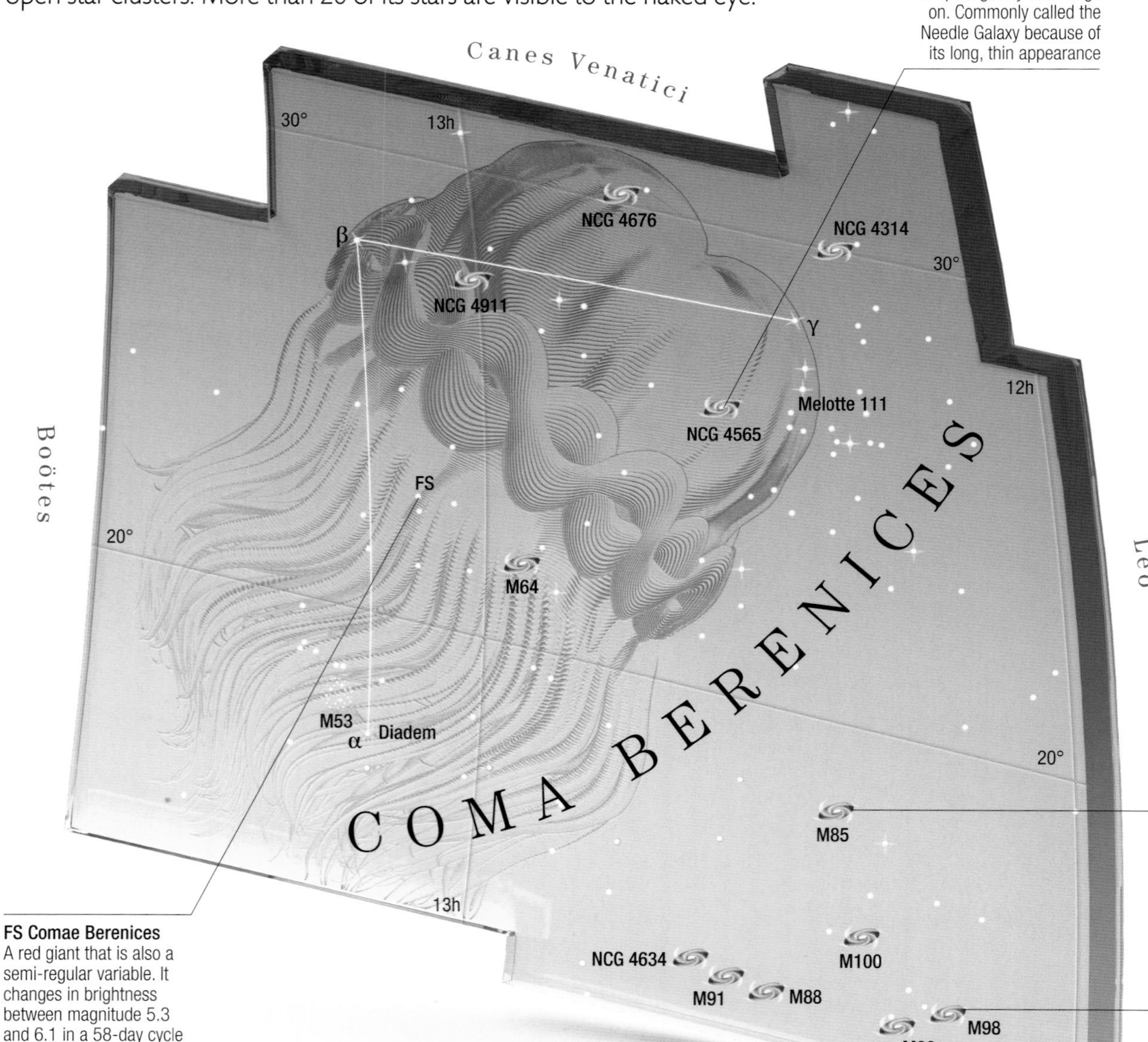

NGC 4565
A spiral galaxy seen edge on. Commonly called the Needle Galaxy because of its long, thin appearance

FS Comae Berenices
A red giant that is also a semi-regular variable. It changes in brightness between magnitude 5.3 and 6.1 in a 58-day cycle

M85
A lenticular galaxy about 125,000 light-years wide and 60 million light-years from Earth

M98
A spiral galaxy seen nearly edge-on, 44 million light-years away. Part of the Virgo Cluster and discovered in 1791 on the same day as M99 and M100

KEY DATA

Size ranking 42

Brightest stars Beta (β) 4.2, Diadem (α) 4.3

Genitive Comae Berenices

Abbreviation Com

Highest in sky at 10pm April–May

Fully visible 90°N–56°S

CHART 5

MAIN STARS

Diadem Alpha (α) Comae Berenices
Binary star consisting of two main-sequence stars
4.3 ⟷ 58 light-years

Beta (β) Comae Berenices
Yellow main-sequence star
4.2 ⟷ 30 light-years

Gamma (γ) Comae Berenices
Orange giant
4.3 ⟷ 167 light-years

FS Comae Berenices
Red giant and semi-regular variable star
5.6 ⟷ 736 light-years

DEEP-SKY OBJECTS

Melotte 111 (Coma Star Cluster)
Open cluster

M53 (NGC 5024)
Globular cluster

M64 (Black Eye Galaxy, NGC 4826)
Spiral galaxy

M85 (NGC 4382)
Lenticular galaxy

M88 (NGC 4501)
Spiral galaxy

M91 (NGC 4548)
Barred spiral galaxy

M99 (NGC 4254)
Spiral galaxy

M100 (NGC 4321)
Spiral galaxy

NGC 4565 (Needle Galaxy)
Spiral galaxy

LIBRA
THE SCALES

OCCUPYING AN AREA OF SKY ONCE SEEN AS PART OF SCORPIUS, LIBRA REPRESENTS THE SCALES OF JUSTICE. IT IS THE LEAST CONSPICUOUS ZODIAC CONSTELLATION AND THE ONLY ONE TO DEPICT AN INANIMATE OBJECT.

The Ancient Greeks called this part of the sky Chelae Scorpionis, the scorpion's claws, but by the 5th century BCE, the Romans were describing it as a balance. Today, Libra is characterized as the scales of justice held aloft by neighbouring Virgo. The names of its brightest stars reflect Libra's past: Zubenelgenubi is Arabic for southern claw, and Zubeneschamali, northern claw.

Lying just south of the celestial equator, the faint constellation of Libra is best found by locating its brighter neighbours. Libra's variable star Delta changes between the 5th and 6th magnitudes in a 2-day, 8-hour cycle, and Iota is a multiple star.

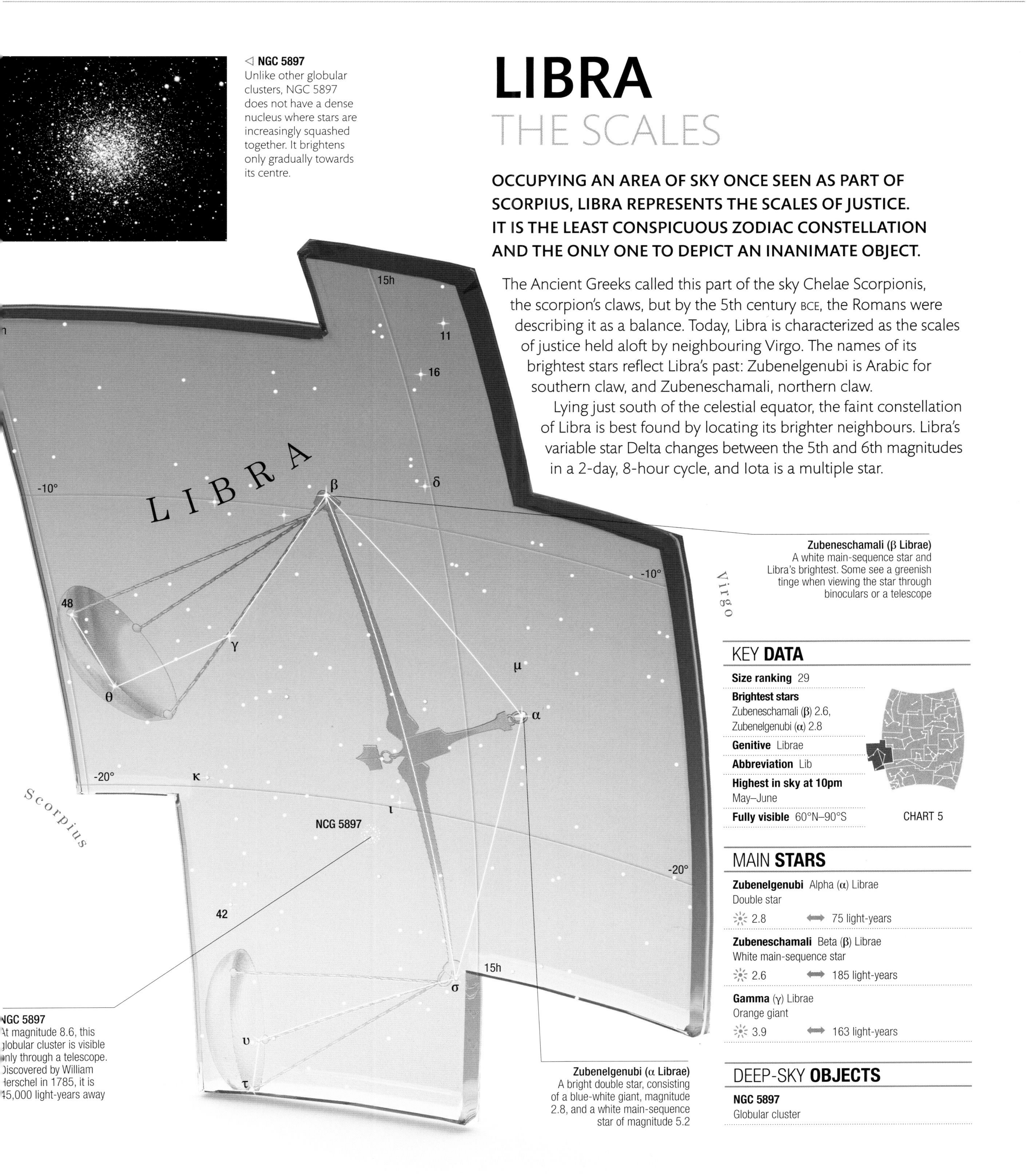

◁ **NGC 5897**
Unlike other globular clusters, NGC 5897 does not have a dense nucleus where stars are increasingly squashed together. It brightens only gradually towards its centre.

Zubeneschamali (β Librae)
A white main-sequence star and Libra's brightest. Some see a greenish tinge when viewing the star through binoculars or a telescope

NGC 5897
At magnitude 8.6, this globular cluster is visible only through a telescope. Discovered by William Herschel in 1785, it is 45,000 light-years away

Zubenelgenubi (α Librae)
A bright double star, consisting of a blue-white giant, magnitude 2.8, and a white main-sequence star of magnitude 5.2

KEY DATA

Size ranking 29

Brightest stars Zubeneschamali (β) 2.6, Zubenelgenubi (α) 2.8

Genitive Librae

Abbreviation Lib

Highest in sky at 10pm May–June

Fully visible 60°N–90°S

CHART 5

MAIN STARS

Zubenelgenubi Alpha (α) Librae
Double star
2.8 ⟷ 75 light-years

Zubeneschamali Beta (β) Librae
White main-sequence star
2.6 ⟷ 185 light-years

Gamma (γ) Librae
Orange giant
3.9 ⟷ 163 light-years

DEEP-SKY OBJECTS

NGC 5897
Globular cluster

LUMINOSITIES

Epsilon Scorpii
40 Suns

Graffias
1,265 Suns

Lesath
2,260 Suns

Antares is over **800 times** the diameter of **the Sun**, so a phone call would take **more than an hour** to get from **one side** of it to **the other**

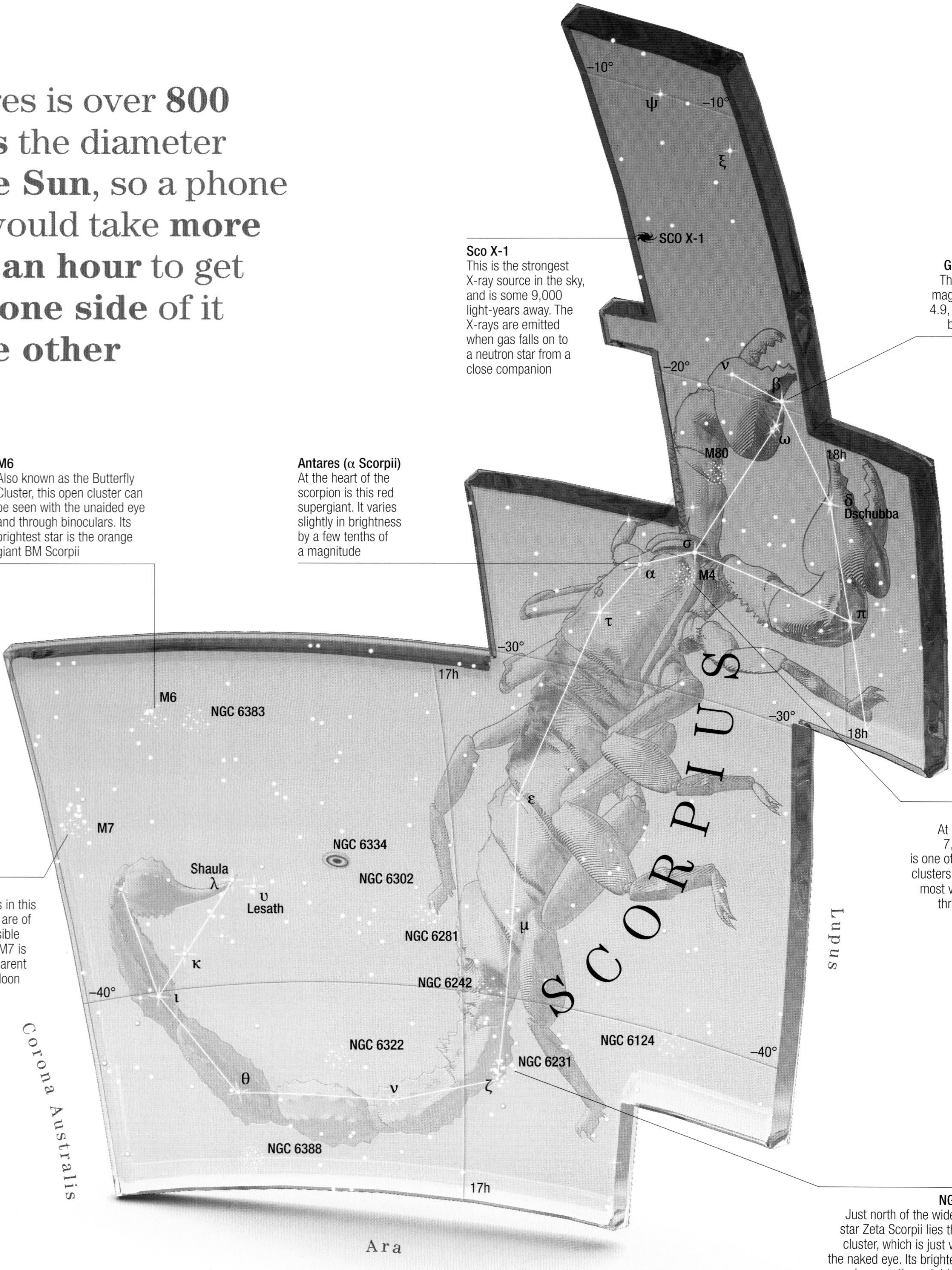

Sco X-1
This is the strongest X-ray source in the sky, and is some 9,000 light-years away. The X-rays are emitted when gas falls on to a neutron star from a close companion

Graffias (β Scorpii)
This double star, with magnitudes of 2.6 and 4.9, is easily separated by small telescopes

M6
Also known as the Butterfly Cluster, this open cluster can be seen with the unaided eye and through binoculars. Its brightest star is the orange giant BM Scorpii

Antares (α Scorpii)
At the heart of the scorpion is this red supergiant. It varies slightly in brightness by a few tenths of a magnitude

M4
At a distance of around 7,000 light-years, this is one of the closest globular clusters. Large but faint, it is most visible on dark nights through binoculars or small telescopes

M7
The brightest stars in this large open cluster are of 6th magnitude. Visible to the naked eye, M7 is over twice the apparent width of the Full Moon

NGC 6231
Just north of the wide double star Zeta Scorpii lies this open cluster, which is just visible to the naked eye. Its brightest stars can be seen through binoculars

Dschubba
2,400 Suns

Shaula
6,000 Suns

Antares
9,450 Suns

SCORPIUS
THE SCORPION

A PROMINENT CONSTELLATION OF THE ZODIAC SOUTH OF THE CELESTIAL EQUATOR, SCORPIUS IS IDENTIFIED BY ITS DISTINCTIVE HOOK SHAPE, WHICH MARKS THE SCORPION'S TAIL. RICH MILKY WAY STAR FIELDS LIE HERE, TOWARDS THE CENTRE OF OUR GALAXY.

Scorpius represents the scorpion that stung Orion the Hunter to death. Myth tells how the adversaries were placed on opposite sides of the sky so that as the scorpion rises, Orion sets. The constellation was once also called Scorpio, but astronomers no longer use this name.

Scorpius's brightest star, the red supergiant Antares, marks the scorpion's heart. An arc of stars leading south from Antares gives the impression of a scorpion's tail. At the tail's end is the constellation's second-brightest star, Shaula (from the Arabic for "stinger"). The tail, situated in a dense area towards the Milky Way's centre, is dotted with star clusters.

Many of the brightest stars in Scorpius and its adjoining constellations lie about 500 light-years away. They are all members of an area of recent star formation called the Scorpius–Centaurus Association. Antares is its brightest member.

◁ **NGC 6302**
NGC 6302, also known as the Bug or Butterfly Nebula, is a complex planetary nebula. Gas flows away from the central star in two directions, forming the "wings" seen in this image from the Hubble Space Telescope.

△ **M80**
Bright red giants are identifiable by their colour in this Hubble image of the dense globular cluster M80, around 28,000 light-years away. Red giants are stars like the Sun that are nearing the ends of their life.

KEY DATA

Size ranking 33

Brightest stars Antares (α) 0.9, Shaula (λ) 1.6

Genitive Scorpii

Abbreviation Sco

Highest in sky at 10pm June–July

Fully visible 44°N–90°S

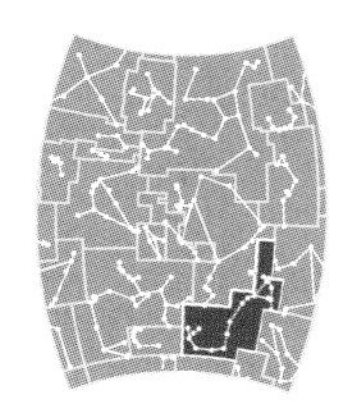

CHART 4

MAIN STARS

Antares Alpha (α) Scorpii
Variable red supergiant
0.9 ⟷ 550 light-years

Graffias Beta (β) Scorpii
Blue-white main sequence star
2.6 ⟷ 400 light-years

Dschubba Delta (δ) Scorpii
Blue-white subgiant
2.3 ⟷ 500 light-years

Epsilon (ε) Scorpii
Orange giant
2.3 ⟷ 64 light-years

Theta (θ) Scorpii
White giant
1.9 ⟷ 300 light-years

Shaula Lambda (λ) Scorpii
Blue-white subgiant
1.6 ⟷ 570 light-years

Lesath Upsilon (υ) Scorpii
Blue-white subgiant
2.7 ⟷ 580 light-years

DEEP-SKY OBJECTS

M4
Globular cluster

M6 (Butterfly Cluster)
Open cluster

M7
Open cluster

M80
Globular cluster

NGC 6302 (Bug Nebula)
Planetary nebula, also known as the Butterfly Nebula

▷ **Star distances**
The nearest of Scorpius's main pattern stars is Epsilon (ε) Scorpii, at 64 light-years away. The most distant is Zeta[1] (ζ[1]) Scorpii, at about 2,570 light-years from Earth. (The other member of the double star Zeta Scorpii, Zeta[2], is only about 130 light-years away.) Many of the stars in Scorpius are members of the Scorpius–Centaurus Association (a group of young stars formed at about the same time) and are at similar distances from Earth: about 500 light-years away.

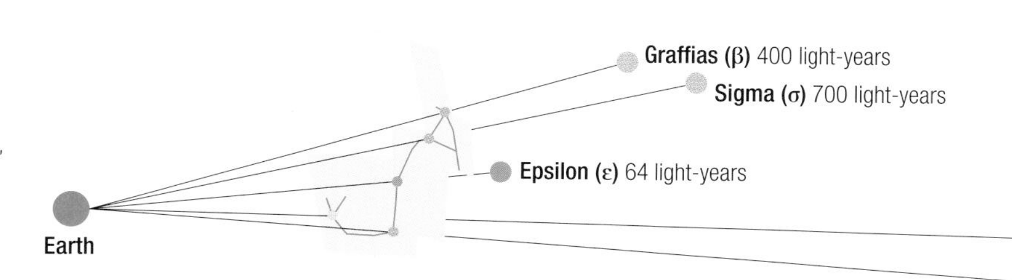

Iota[1] (ι[1]) 1,930 light-years

Zeta[1] (ζ[1]) 2,570 light-years

Distance

LUMINOSITIES

Gamma Serpentis
3 Suns

Eta Serpentis
15 Suns

SERPENS
THE SERPENT

UNIQUELY AMONG THE CONSTELLATIONS, SERPENS IS DIVIDED IN TWO, WITH ITS HEAD ON ONE SIDE OF OPHIUCHUS AND ITS TAIL ON THE OTHER. THE TWO PARTS COUNT AS A SINGLE CONSTELLATION.

Serpens represents a snake or serpent held by Ophiuchus, who grasps its head in his left hand and its tail in his right. The head section of the constellation is known as Serpens Caput, while the tail is Serpens Cauda.

The constellation's brightest star is 3rd-magnitude Alpha Serpentis. Its popular name, Unukalhai, is from the Arabic for "serpent's neck", which is where it lies. In the serpent's head is Beta Serpentis, which has a 7th-magnitude unrelated companion visible with binoculars. In the serpent's tail is M16, a star cluster surrounded by the Eagle Nebula, made famous by the "pillars of creation" Hubble image. Also in the tail, Alya (Theta Serpentis) is a 5th-magnitude double divisible through a small telescope. Nearby is IC 4765, an open cluster visible through binoculars.

◁ **Seyfert's Sextet**
A group of galaxies that consists of four interacting galaxies and two other members. The small spiral at the centre of this image from the Hubble Space Telescope is not part of the interaction but a background object in the same line of sight by chance. The sixth member is not a galaxy at all but a long tail of stars torn from one of the galaxies.

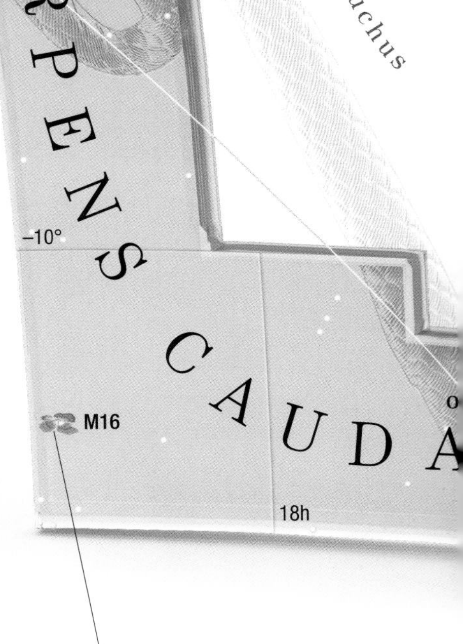

IC 4756
Open cluster visible through binoculars, appearing larger than the Full Moon. Its brightest stars are of 8th magnitude

Alya (θ Serpentis)
Pair of 5th-magnitude stars easily divisible through a small telescope

M16
Open cluster visible through binoculars and small telescopes; appears hazy because it is embedded in the Eagle Nebula

◁ **Pillars of Creation**
This iconic image taken by the Hubble Space telescope shows columns of gas and dust in the Eagle Nebula. About 4 light-years tall, the pillars are a site of new star formation. At the same time, they are also being eroded by ultraviolet light from other hot, newborn stars nearby.

▽ **Hoag's Object**
This unusual galaxy consists of a ring of hot, blue stars encircling a core of yellow, older stars. It is thought that the ring may be the remains of another galaxy that passed too close and was shredded.

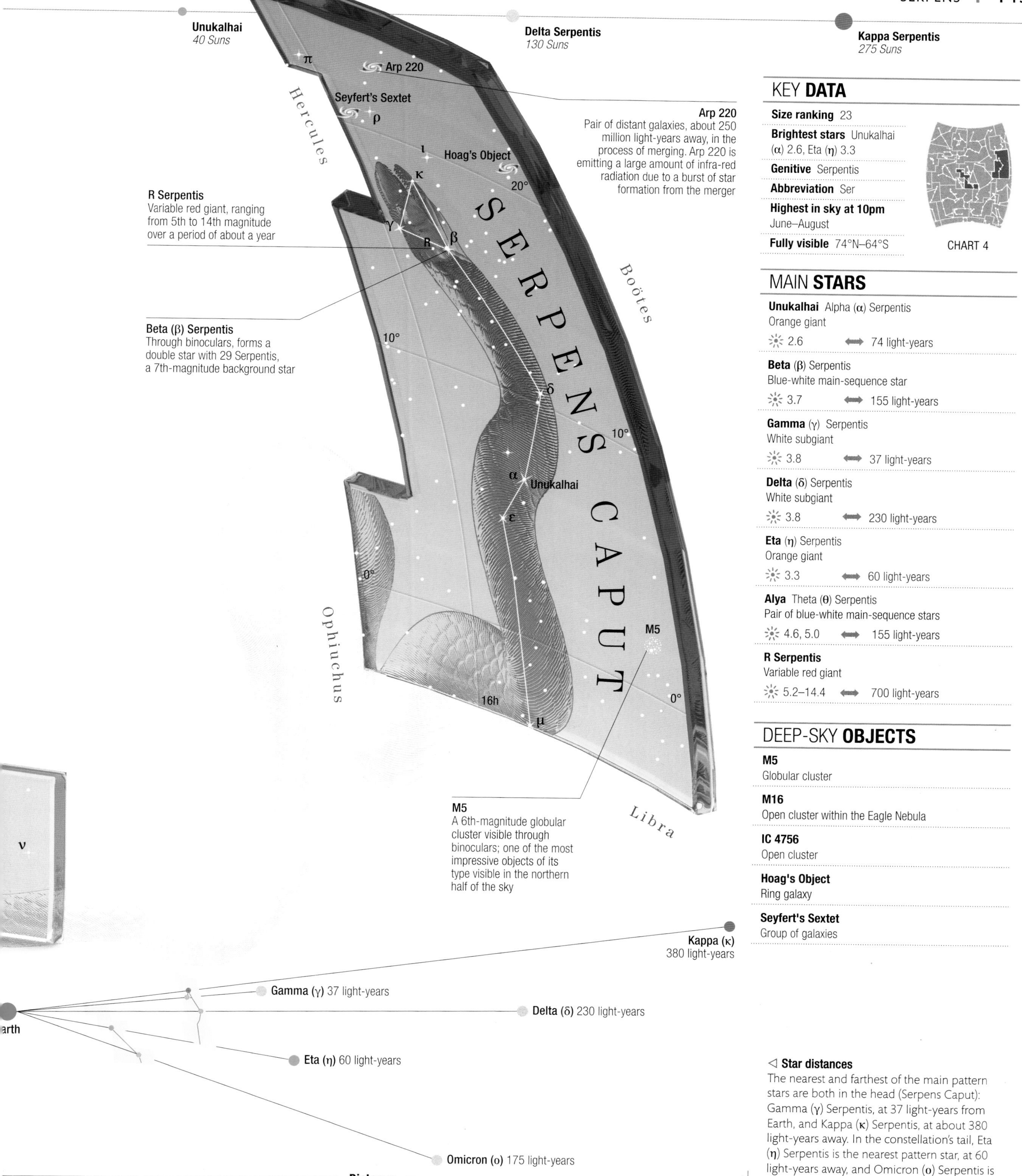

Unukalhai
40 Suns
Delta Serpentis
130 Suns
Kappa Serpentis
275 Suns
Arp 220
Seyfert's Sextet
Hoag's Object
Hercules
Boötes
Ophiuchus
Libra
SERPENS CAPUT
Unukalhai
M5
20°
10°
0°
16h
Arp 220
Pair of distant galaxies, about 250 million light-years away, in the process of merging. Arp 220 is emitting a large amount of infra-red radiation due to a burst of star formation from the merger
R Serpentis
Variable red giant, ranging from 5th to 14th magnitude over a period of about a year
Beta (β) Serpentis
Through binoculars, forms a double star with 29 Serpentis, a 7th-magnitude background star
M5
A 6th-magnitude globular cluster visible through binoculars; one of the most impressive objects of its type visible in the northern half of the sky
Kappa (κ) 380 light-years
Gamma (γ) 37 light-years
Delta (δ) 230 light-years
Eta (η) 60 light-years
Omicron (ο) 175 light-years
Earth
Distance

KEY DATA

Size ranking 23

Brightest stars Unukalhai (α) 2.6, Eta (η) 3.3

Genitive Serpentis

Abbreviation Ser

Highest in sky at 10pm June–August

Fully visible 74°N–64°S

CHART 4

MAIN STARS

Unukalhai Alpha (α) Serpentis
Orange giant
2.6 — 74 light-years

Beta (β) Serpentis
Blue-white main-sequence star
3.7 — 155 light-years

Gamma (γ) Serpentis
White subgiant
3.8 — 37 light-years

Delta (δ) Serpentis
White subgiant
3.8 — 230 light-years

Eta (η) Serpentis
Orange giant
3.3 — 60 light-years

Alya Theta (θ) Serpentis
Pair of blue-white main-sequence stars
4.6, 5.0 — 155 light-years

R Serpentis
Variable red giant
5.2–14.4 — 700 light-years

DEEP-SKY OBJECTS

M5
Globular cluster

M16
Open cluster within the Eagle Nebula

IC 4756
Open cluster

Hoag's Object
Ring galaxy

Seyfert's Sextet
Group of galaxies

◁ **Star distances**
The nearest and farthest of the main pattern stars are both in the head (Serpens Caput): Gamma (γ) Serpentis, at 37 light-years from Earth, and Kappa (κ) Serpentis, at about 380 light-years away. In the constellation's tail, Eta (η) Serpentis is the nearest pattern star, at 60 light-years away, and Omicron (ο) Serpentis is the farthest, at about 175 light-years distant.

LUMINOSITIES

Rasalhague
28 Suns

Cebalrai
43 Suns

Sabik
68 Su

OPHIUCHUS
THE SERPENT HOLDER

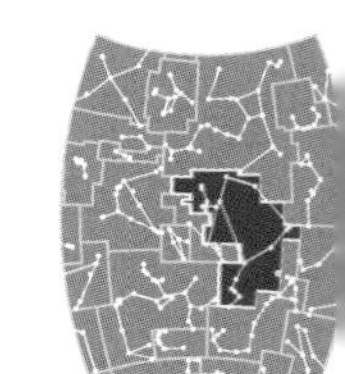

OPHIUCHUS IS A LARGE CONSTELLATION THAT LIES ON THE CELESTIAL EQUATOR. IT EXTENDS FROM HERCULES IN THE NORTH TO SCORPIUS AND SAGITTARIUS IN THE SOUTH.

Ophiuchus represents a legendary healer called Aesculapius, who was reputed to be able to revive the dead. In the sky, he is depicted holding a snake (a traditional symbol of healing) in the form of the constellation Serpens.

Although large, the constellation is not particularly prominent. Its brightest star is second-magnitude Rasalhague (Alpha Ophiuchi), which marks Ophiuchus's head. Its most celebrated star is Barnard's Star, a faint (10th-magnitude) red dwarf a mere 5.9 light-years away. Ophiuchus contains numerous globular clusters, of which M10 and M12 are the easiest to see through a small telescope. The Sun passes through Ophiuchus in the first half of December each year but it is not regarded by many as a traditional constellation of the zodiac (see pp.92–93).

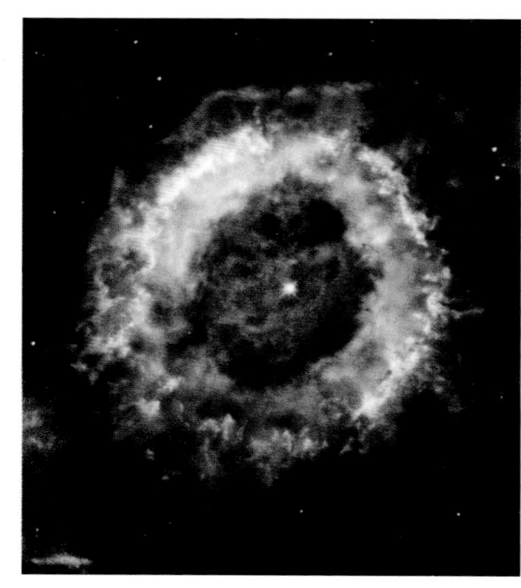

△ **NGC 6369**
Seen here through the Hubble Space Telescope, this planetary nebula, popularly known as the Little Ghost, consists of a ring of gas about a light-year across, illuminated by ultraviolet light from the central core.

KEY DATA

Size ranking 11

Brightest stars Rasalhague (α) 2.1, Sabik (η) 2.4

Genitive Ophiuchi

Abbreviation Oph

Highest in sky at 10pm June–July

Fully visible 59°N–75°S

CHART 4

MAIN STARS

Rasalhague Alpha (α) Ophiuchi
Blue-white giant
2.1 ⟷ 49 light-years

Cebalrai Beta (β) Ophiuchi
Orange giant
2.8 ⟷ 82 light-years

Delta (δ) Ophiuchi
Red giant
2.8 ⟷ 170 light-years

Zeta (ζ) Ophiuchi
Blue-white subdwarf
2.6 ⟷ 365 light-years

Sabik Eta (η) Ophiuchi
Blue-white subdwarf
2.4 ⟷ 88 light-years

Barnard's Star
Red dwarf
9.5 ⟷ 5.9 light-years

DEEP-SKY OBJECTS

Kepler's Star
Remains of a supernova seen in October 1604

M10
Globular cluster

M12
Globular cluster

NGC 6369
Planetary nebula, also known as the Little Ghost

NGC 6633
Open cluster

IC 4665
Open cluster

Pipe Nebula
Dark nebula

Twin Jet Nebula (Minkowski 2–9)
Bipolar planetary nebula

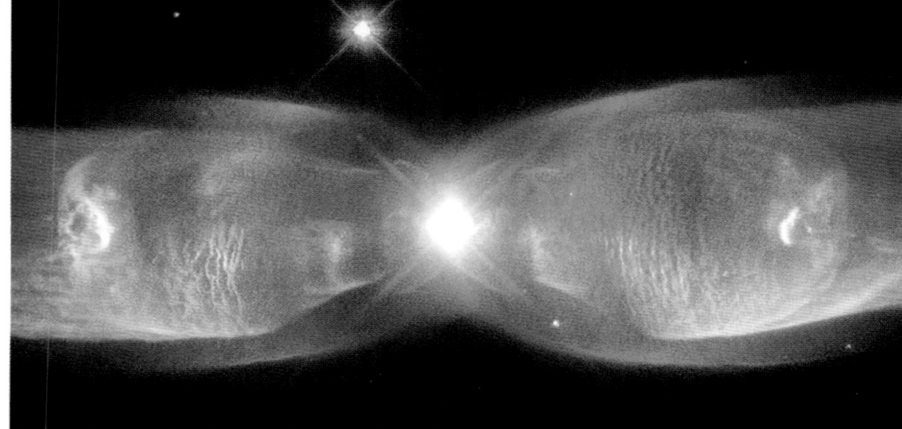

▷ **Twin Jet Nebula**
Two lobes of shimmering gas stream outwards at speeds greater than a million km per hour (620,000 miles per hour) from a central binary star, creating the butterfly-like shape seen in this Hubble Space Telescope image.

▽ **Star distances**
The closest pattern star, at about 20 light-years from Earth, is the binary pair 36 Ophiuchi. The most distant is 67 Ophiuchi, which is 60 times further away, at about 1,230 light-years from Earth.

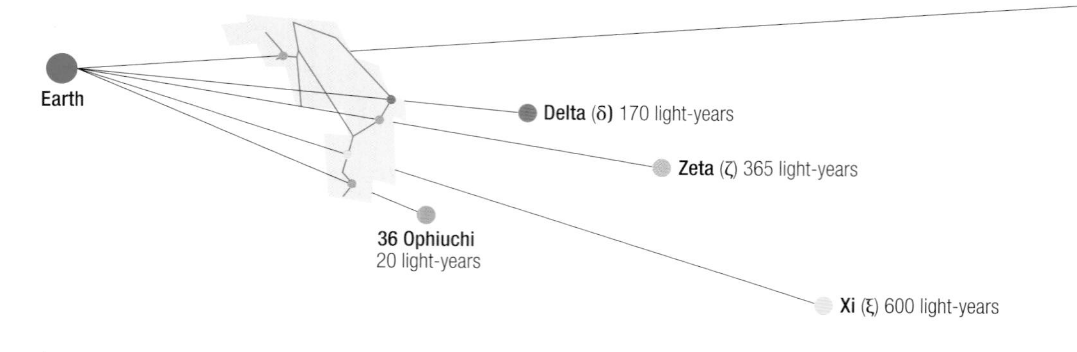

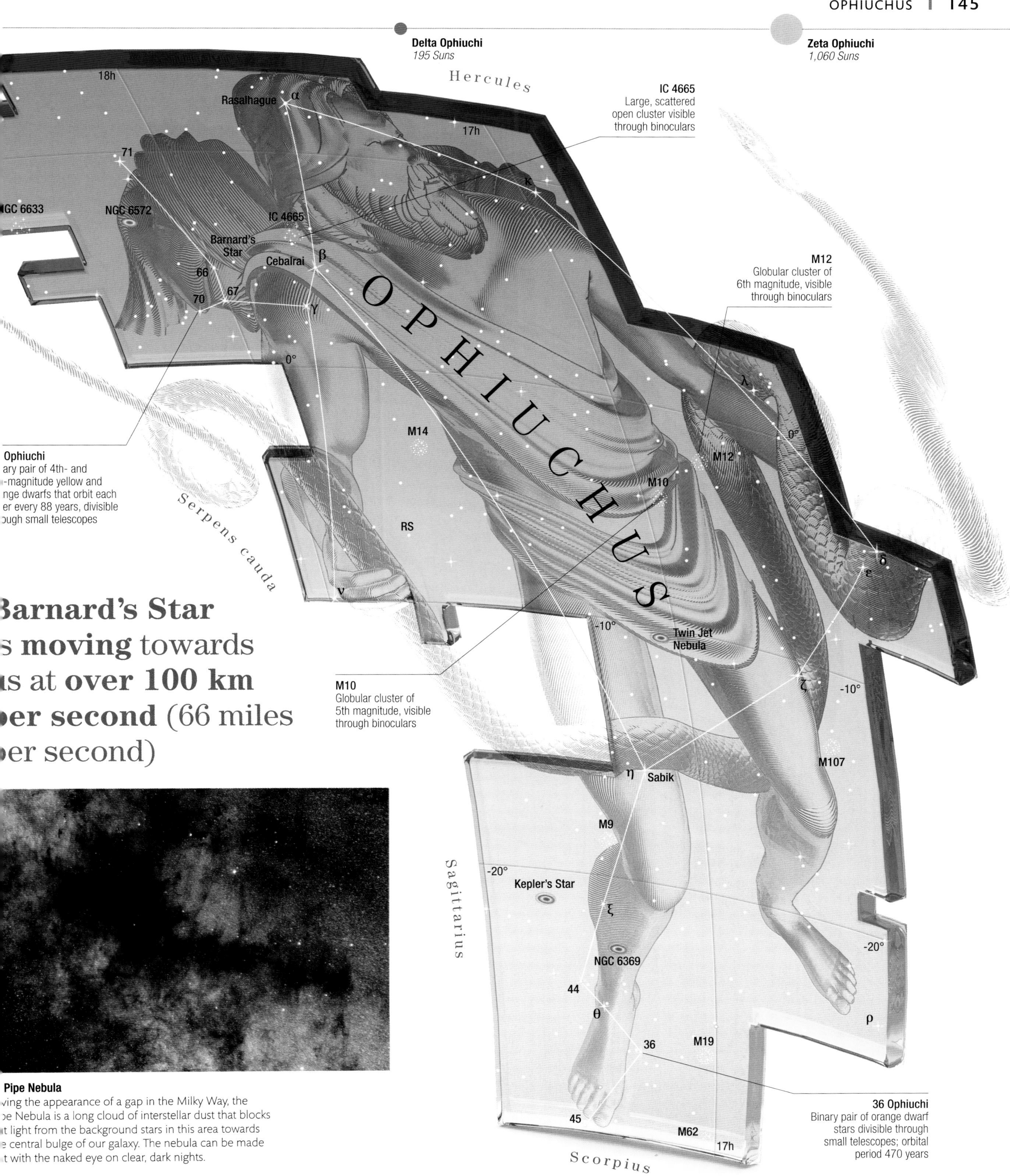

Barnard's Star s **moving** towards us at **over 100 km** **per second** (66 miles per second)

Pipe Nebula
ving the appearance of a gap in the Milky Way, the pe Nebula is a long cloud of interstellar dust that blocks t light from the background stars in this area towards e central bulge of our galaxy. The nebula can be made t with the naked eye on clear, dark nights.

AQUILA THE EAGLE

THE PATTERN MADE BY THE STARS IN AQUILA CAN EASILY BE IMAGINED AS AN EAGLE SOARING IN THE SKY. AQUILA IS A SPECIAL EAGLE ASSOCIATED WITH THE GREEK GOD ZEUS.

One of the original 48 constellations, Aquila straddles the celestial equator in a rich region of the Milky Way. Aquila could be the eagle that carried Zeus's thunderbolts, or alternatively, it could be Zeus in the form of an eagle, which enabled him to carry Ganymede to Mount Olympus to serve the gods. The eagle appears to swoop down towards adjacent Aquarius, which is identified with Ganymede.

Aquila can best be found by spotting its brightest star, Altair (Alpha Aquilae), whose name is Arabic for flying eagle. It is the 12th-brightest star in the entire night sky and at only 17 light-years away, also one of the closest bright stars. With Deneb (in Cygnus) and Vega (in Lyra) it forms the Summer Triangle of northern skies. The supergiant Eta Aquilae is one of the brightest naked-eye Cepheid variables. Its magnitude changes from 3.5 to 4.4 in a 7.2 day cycle.

KEY DATA

Size ranking 22

Brightest stars Altair (α) 0.8, Tarazed (γ) 2.7

Genitive Aquilae

Abbreviation Aql

Highest in sky at 10pm July–August

Fully visible 78°N–71°S

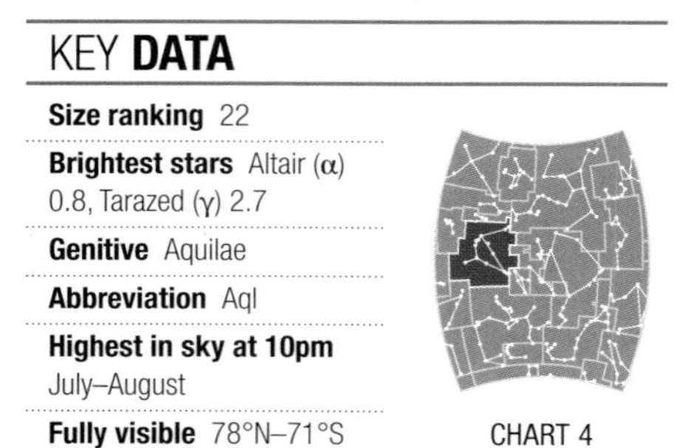

CHART 4

MAIN STARS

Altair Alpha (α) Aquilae
White main-sequence star
0.8 — 17 light-years

Alshain Beta (β) Aquilae
Yellow subgiant
3.7 — 45 light-years

Tarazed Gamma (γ) Aquilae
Orange giant
2.7 — 395 light-years

Zeta (ζ) Aquilae
White main-sequence star
3.0 — 83 light-years

DEEP-SKY OBJECTS

NGC 6709
Open star cluster

NGC 6751
Planetary nebula

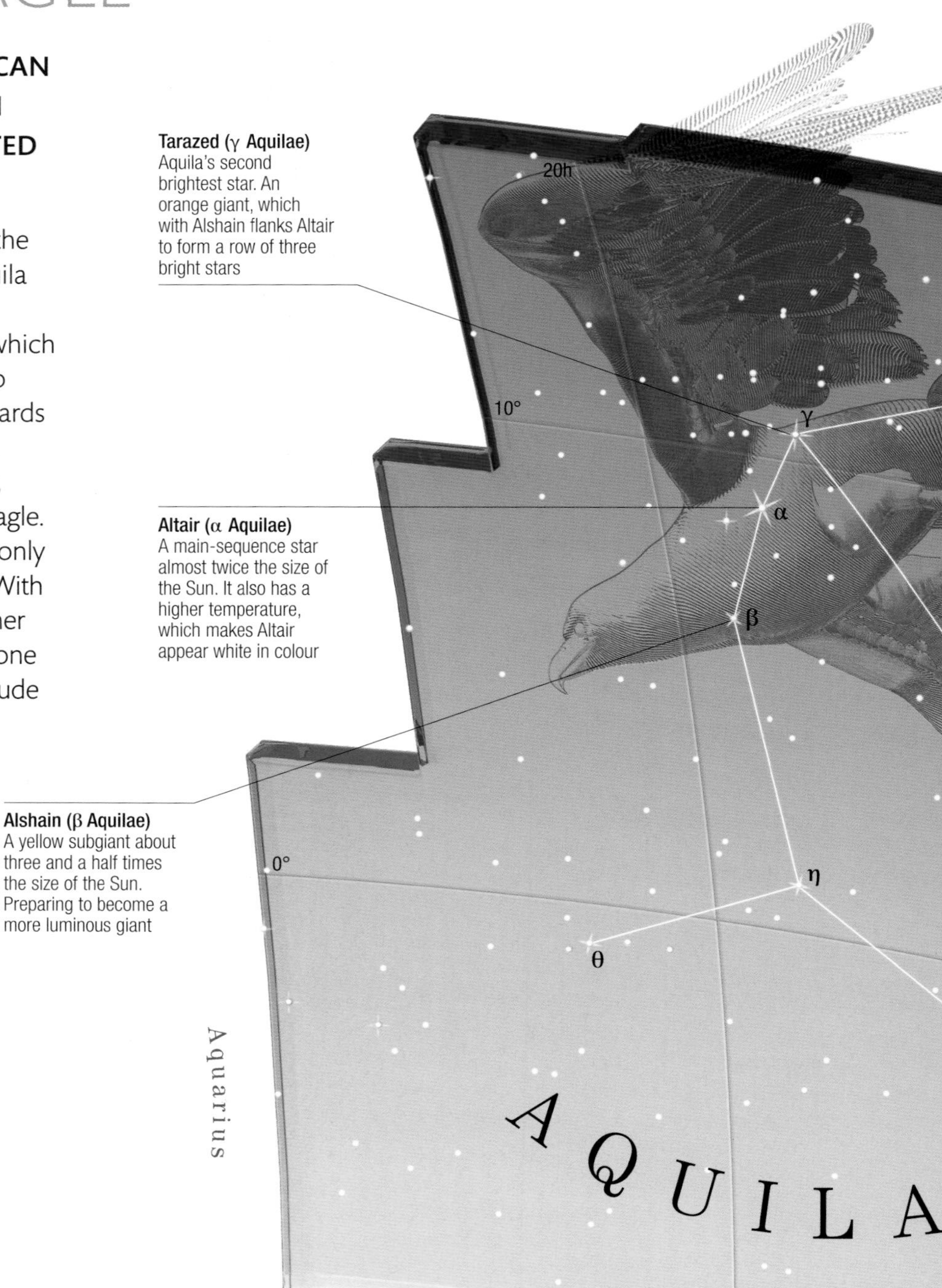

Tarazed (γ Aquilae) Aquila's second brightest star. An orange giant, which with Alshain flanks Altair to form a row of three bright stars

Altair (α Aquilae) A main-sequence star almost twice the size of the Sun. It also has a higher temperature, which makes Altair appear white in colour

Alshain (β Aquilae) A yellow subgiant about three and a half times the size of the Sun. Preparing to become a more luminous giant

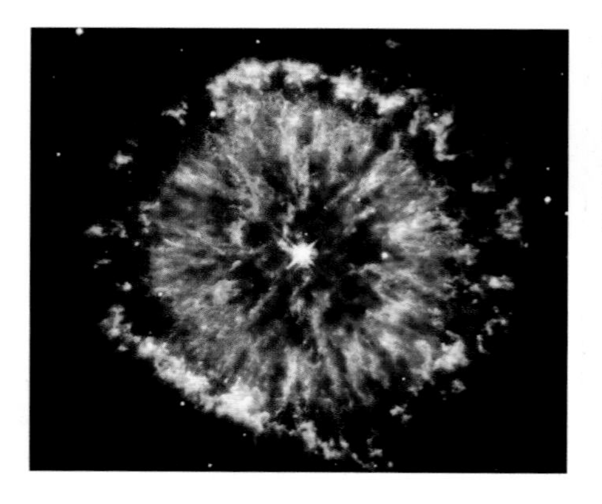

◁ **NGC 6751** This planetary nebula, with a magnitude of 12.5, is about 6,500 light-years from Earth. At its centre is a white dwarf of magnitude 15.5.

SCUTUM THE SHIELD

THE FIFTH-SMALLEST CONSTELLATION IN THE SKY, SCUTUM LIES IN A BRIGHT AREA OF THE MILKY WAY BETWEEN SAGITTARIUS AND THE PROMINENT STAR ALTAIR IN AQUILA.

The constellation Scutum was defined by the Polish astronomer Johannes Hevelius in 1684. He devised it in honour of his patron, King John III Sobiesci of Poland; its original name being Scutum Sobiescianum – Sobiesci's Shield. Just south of the celestial equator, its brightest stars are only fourth magnitude, none are named, and two, Delta and R Scuti are interesting variables. It is, however, crossed by a bright, star-rich region of the Milky Way. This includes the Scutum Star Cloud, the brightest part of the Milky Way outside Sagittarius. The Star Cloud is home to the Wild Duck Cluster, which contains around 3,000 stars. Scutum's brightest star is Alpha Scuti, 132 times more luminous than the Sun.

KEY DATA

Size ranking 84

Brightest stars Alpha (α) 3.8, Beta (β) 4.2

Genitive Scuti

Abbreviation Sct

Highest in sky at 10pm July–August

Fully visible 74°N–90°S

CHART 4

MAIN STARS

Alpha (α) Scuti
Orange giant
3.8 — 199 light-years

DEEP SKY OBJECTS

M11 (Wild Duck Cluster, NGC 6705)
Open star cluster

M26
Open star cluster

Scutum Star Cloud
Star-rich region of the Milky Way

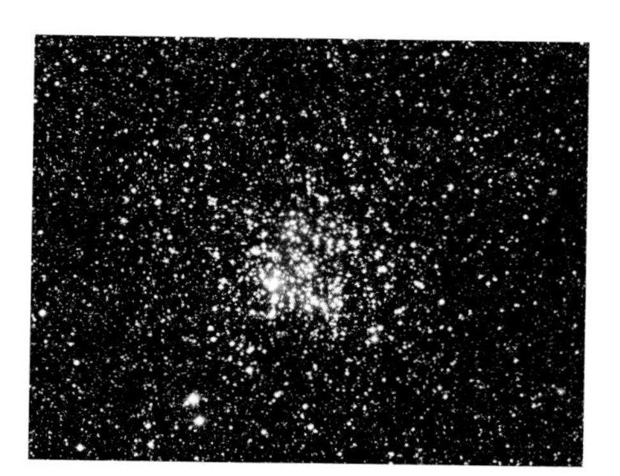

△ **M11**
Also known as the Wild Duck Cluster, this relatively compact open cluster of about 3,000 stars lies approximately 6,000 light-years away. About 20 light-years across and roughly 250 million years old, it is visible to the naked eye but can be seen in more detail with binoculars or a telescope.

Zeta (ζ) Aquilae
A white main-sequence star marking the eagle's tail. It is about twice the size and mass of the Sun and 40 times its luminosity

NGC 6709
A loose cluster of about 60 young stars. At magnitude 6.7, this cluster is just beyond visibility by the naked eye

NGC 6751
A planetary nebula of magnitude 12.5 with a central white dwarf of magnitude 15.5

R Scuti
This orange supergiant is a pulsating variable. It changes in magnitude from 4.5 to 8.8, in a cycle lasting 144 days

Beta (β) Scuti
A yellow giant of magnitude 4.2. It is 690 light-years away, 64 times the size of the Sun, and 1,760 times its luminosity

Alpha (α) Scuti
The brightest star in Scutum, of magnitude 3.8. This orange giant is about 21 times the width of the Sun

M26
A tight, open star cluster of magnitude 8.9. About 5,000 light-years away, its stars are 90 million years old

VULPECULA
THE FOX

A FAINT NORTHERN CONSTELLATION NEAR THE HEAD OF CYGNUS, THE SWAN, VULPECULA WAS INTRODUCED AT THE END OF THE 17TH CENTURY BY THE POLISH ASTRONOMER JOHANNES HEVELIUS. IT CONTAINS A FAMOUS PLANETARY NEBULA, THE DUMBBELL.

Johannes Hevelius originally called this constellation *Vulpecula cum Ansere*, the fox and goose, but modern astronomers have simplified the name to just Vulpecula. It consists of a scattering of stars of 4th magnitude and fainter in the Milky Way south of Cygnus. On its southern border with Sagitta lies a grouping called Brocchi's Cluster. Through binoculars, this group appears as a line of six stars with a protruding hook, reminiscent of a coat hanger, which gives rise to its popular name, the Coathanger. However, it is not a true cluster, because all its stars are at different distances from us. Another celebrated object in Vulpecula is M27, a planetary nebula popularly known as the Dumbbell from its supposed resemblance to a barbell used for weight training.

KEY DATA

Size ranking 55

Brightest stars Alpha (α) 4.5, 13 Vulpeculae 4.6

Genitive Vulpeculae

Abbreviation Vul

Highest in sky at 10pm August–September

Fully visible 90°N–61°S

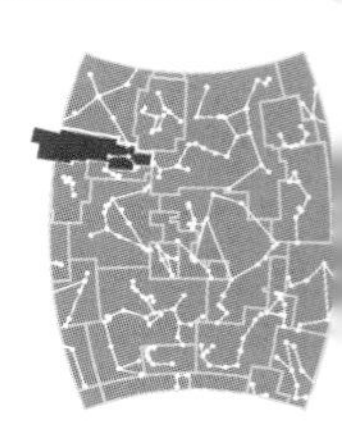

CHART 4

MAIN STARS

Alpha (α) **Vulpeculae**
Red giant
4.5 ⟷ 297 light-years

T Vulpeculae
Variable yellow-white supergiant
5.4–6.1 ⟷ 1,200 light-years

DEEP-SKY OBJECTS

M27 (Dumbbell Nebula)
Planetary nebula about 1,200 light-years away

Brocchi's Cluster (Collinder 399; "the Coathanger")
Grouping of 10 unrelated stars

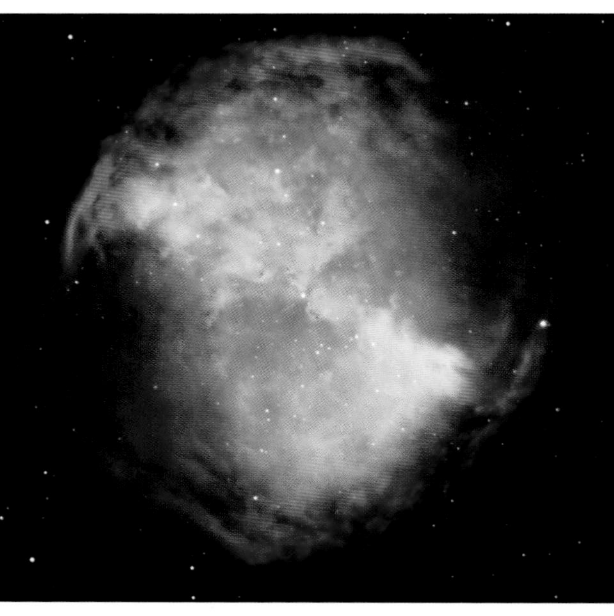

◁ **Dumbbell Nebula**
A well-known planetary nebula also known as M27, the Dumbbell lies about 1,200 light-years away. It consists of gas ejected from a dying star, the exposed core of which can be seen as a faint white dot at the centre of this image.

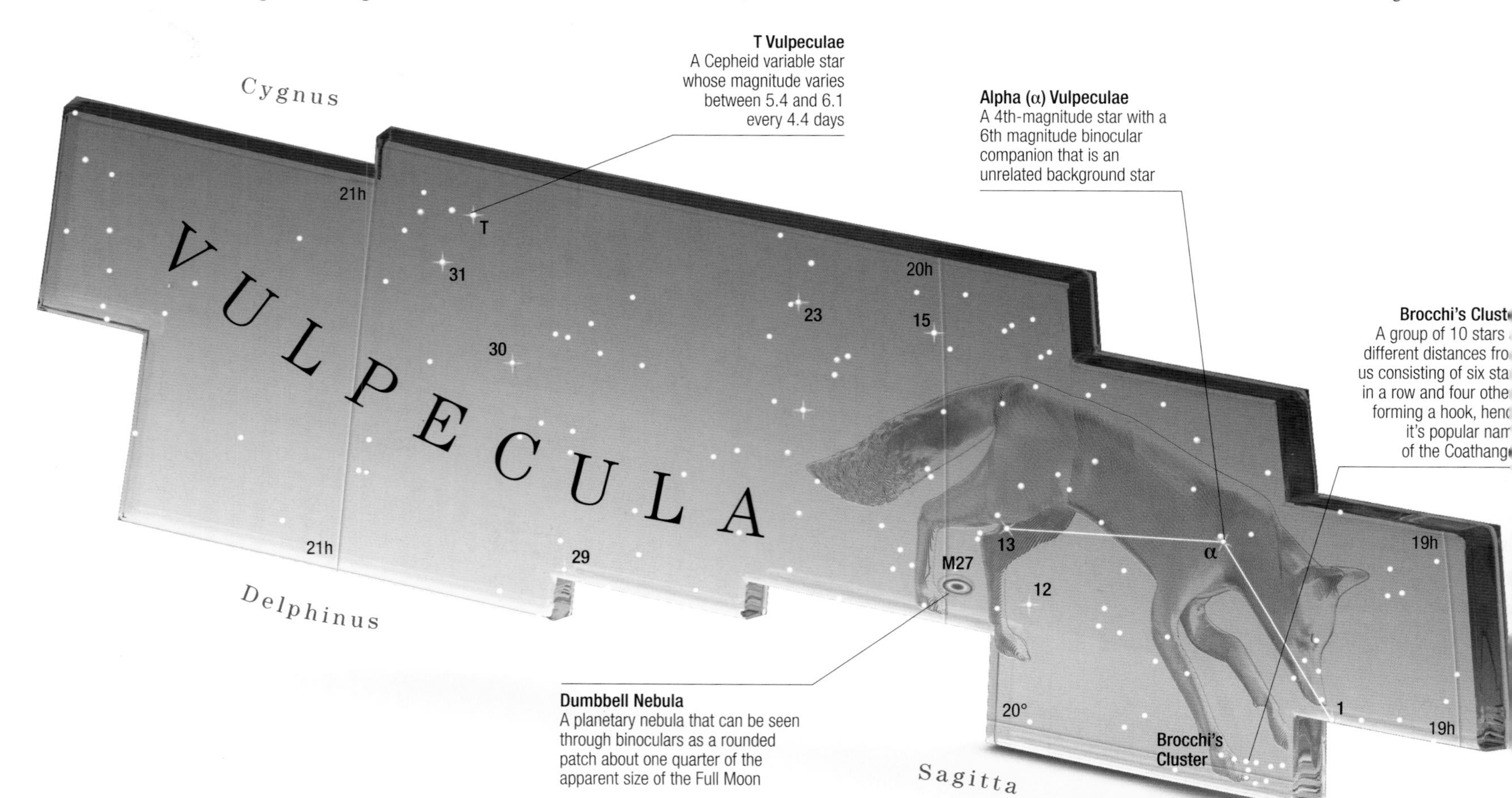
VULPECULA
Cygnus
Delphinus
Sagitta
21h
21h
20h
19h
19h
20°
T
31
30
23
15
29
M27
13
12
α
1
Brocchi's Cluster
T Vulpeculae
A Cepheid variable star whose magnitude varies between 5.4 and 6.1 every 4.4 days
Alpha (α) Vulpeculae
A 4th-magnitude star with a 6th magnitude binocular companion that is an unrelated background star
Dumbbell Nebula
A planetary nebula that can be seen through binoculars as a rounded patch about one quarter of the apparent size of the Full Moon

Brocchi's Clust
A group of 10 stars different distances fro us consisting of six sta in a row and four othe forming a hook, henc it's popular nam of the Coathang

SAGITTA
THE ARROW

THE THIRD-SMALLEST CONSTELLATION IN THE SKY, SAGITTA LIES IN THE BAND OF THE MILKY WAY BETWEEN AQUILA AND VULPECULA.

Sagitta is one of the original 48 constellations known to the Ancient Greeks. Its four brightest stars, all of 4th magnitude, suggest the shape of an arrow. The main object of interest for users of binoculars or small telescopes is M71. It was long considered to be a rich open cluster but is now classified as a globular cluster, even though it lacks the dense central concentration typical of most globulars. Other notable objects include WZ Sagittae, a dwarf nova star system that undergoes periodic outbursts of energy, and the Necklace Nebula, a planetary nebula with a ring of bright "knots" that resembles a necklace.

▷ **M71**
This Hubble Space Telescope image shows a brilliant splash of stars in the heart of the globular cluster M71 in Sagitta. M71 lies roughly 13,000 light-years away and is about 27 light-years in diameter.

KEY DATA

Size ranking	86
Brightest stars	Gamma (γ) 3.5, Delta (δ) 3.8
Genitive	Sagittae
Abbreviation	Sge
Highest in sky at 10pm	August
Fully visible	90°N–69°S

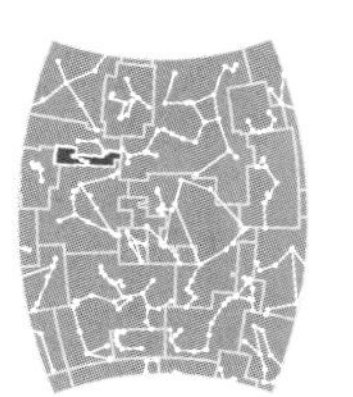

CHART 4

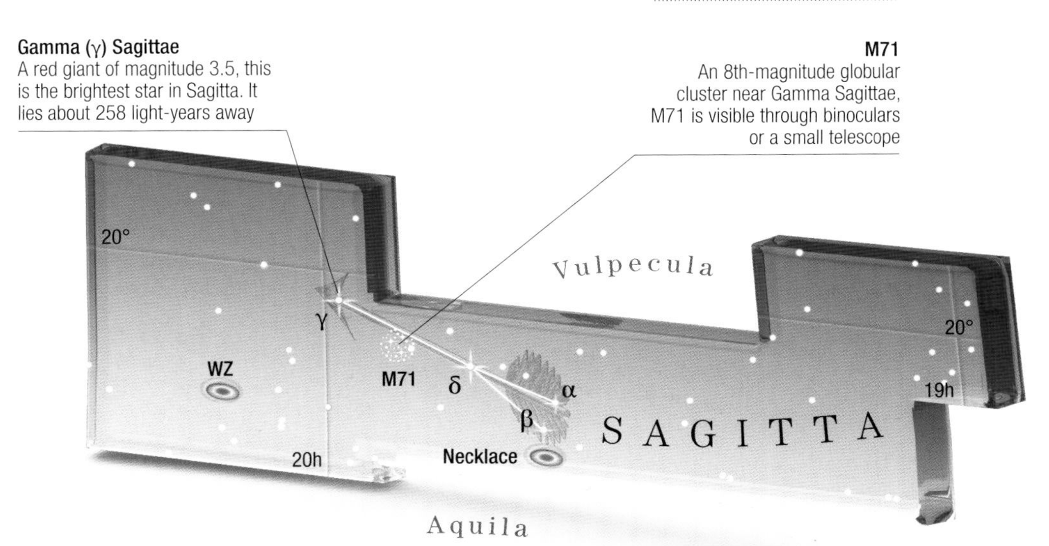
Gamma (γ) Sagittae
A red giant of magnitude 3.5, this is the brightest star in Sagitta. It lies about 258 light-years away
M71
An 8th-magnitude globular cluster near Gamma Sagittae, M71 is visible through binoculars or a small telescope
20°
Vulpecula
20°
γ
WZ
M71
δ
α
β
19h
SAGITTA
20h
Necklace
Aquila

DELPHINUS
THE DOLPHIN

A SMALL NORTHERN CONSTELLATION REPRESENTING A DOLPHIN, DELPHINUS LIES ON THE EDGE OF THE MILKY WAY BETWEEN PEGASUS AND AQUILA.

It is easy to visualize a leaping dolphin among the stars of Delphinus. In Greek mythology, dolphins were the messengers of Poseidon, the sea god. Alpha and Beta Delphini, the constellation's brightest stars, bear the odd names Sualocin and Rotanev. In reverse, they spell Nicolaus Venator, the Latinized form of Niccolò Cacciatore, an Italian astronomer who seemingly named the stars after himself in the early 19th century. Among other objects of interest are Gamma Delphini, a wide double star easily divided with binoculars, the faint globular cluster NGC 6934, and a pair of colliding galaxies known as ZW II 96.

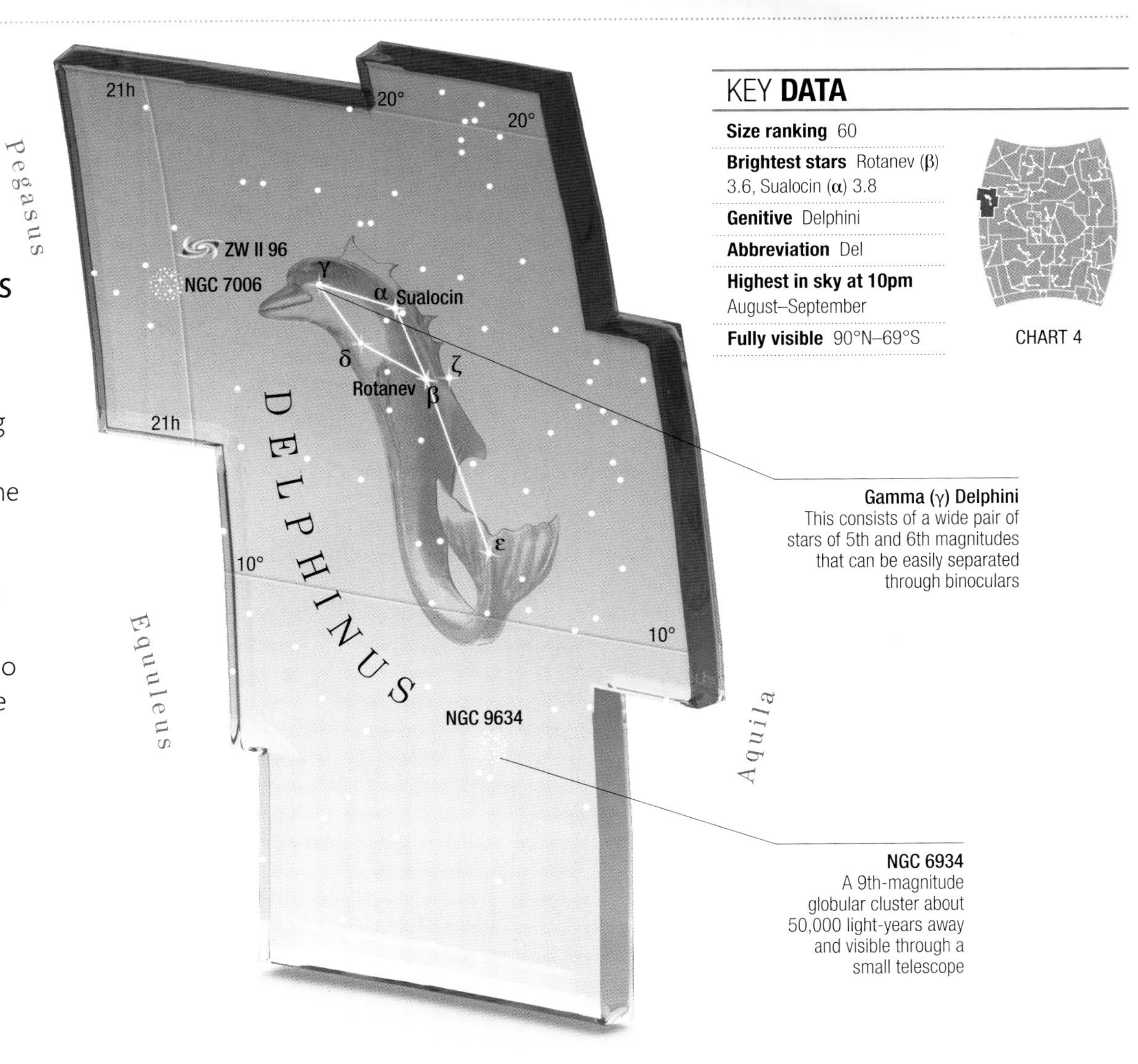
21h
20°
20°
Pegasus
ZW II 96
NGC 7006
γ
α Sualocin
δ
ζ
Rotanev β
21h
DELPHINUS
ε
10°
10°
Equuleus
NGC 9634
Aquila
Gamma (γ) Delphini
This consists of a wide pair of stars of 5th and 6th magnitudes that can be easily separated through binoculars
NGC 6934
A 9th-magnitude globular cluster about 50,000 light-years away and visible through a small telescope

KEY DATA

Size ranking	60
Brightest stars	Rotanev (β) 3.6, Sualocin (α) 3.8
Genitive	Delphini
Abbreviation	Del
Highest in sky at 10pm	August–September
Fully visible	90°N–69°S

CHART 4

PEGASUS
THE WINGED HORSE

ONE OF THE ORIGINAL 48 GREEK CONSTELLATIONS, PEGASUS REPRESENTS THE FLYING HORSE RIDDEN BY BELLEROPHON. THE GREAT SQUARE OF PEGASUS IS FORMED OF STARS IN THE HORSE'S BODY.

Pegasus is the seventh-largest constellation and occupies an area of the sky to the north of Aquarius and Pisces. It represents the head and forequarters of the flying horse. The bright stars Markab, Scheat, and Algenib mark three of the four corners of the Great Square of Pegasus. A star in the neighbouring constellation of Andromeda completes it.

KEY DATA

Size ranking 7

Brightest stars Scheat (β) 2.3–2.7, Enif (ε) 2.4

Genitive Pegasi

Abbreviation Peg

Highest in sky at 10pm September–October

Fully visible 90°N–53°S

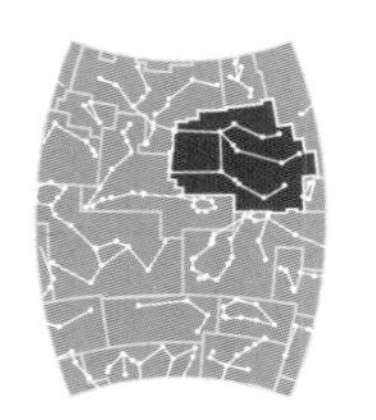

CHART 3

MAIN STARS

Markab Alpha (α) Pegasi
Blue-white giant
2.5 — 133 light-years

Scheat Beta (β) Pegasi
Red giant with variable brightness
2.3–2.7 — 196 light-years

Algenib Gamma (γ) Pegasi
Blue-white subgiant
2.8 — 391 light-years

Enif Epsilon (ε) Pegasi
Orange-yellow supergiant
2.4 — 121 light-years

Matar Eta (η) Pegasi
Binary star
3.0 — 214 light-years

DEEP-SKY OBJECTS

M15
Globular cluster

NGC 7331
Spiral galaxy

Stephan's Quintet
Group of five galaxies

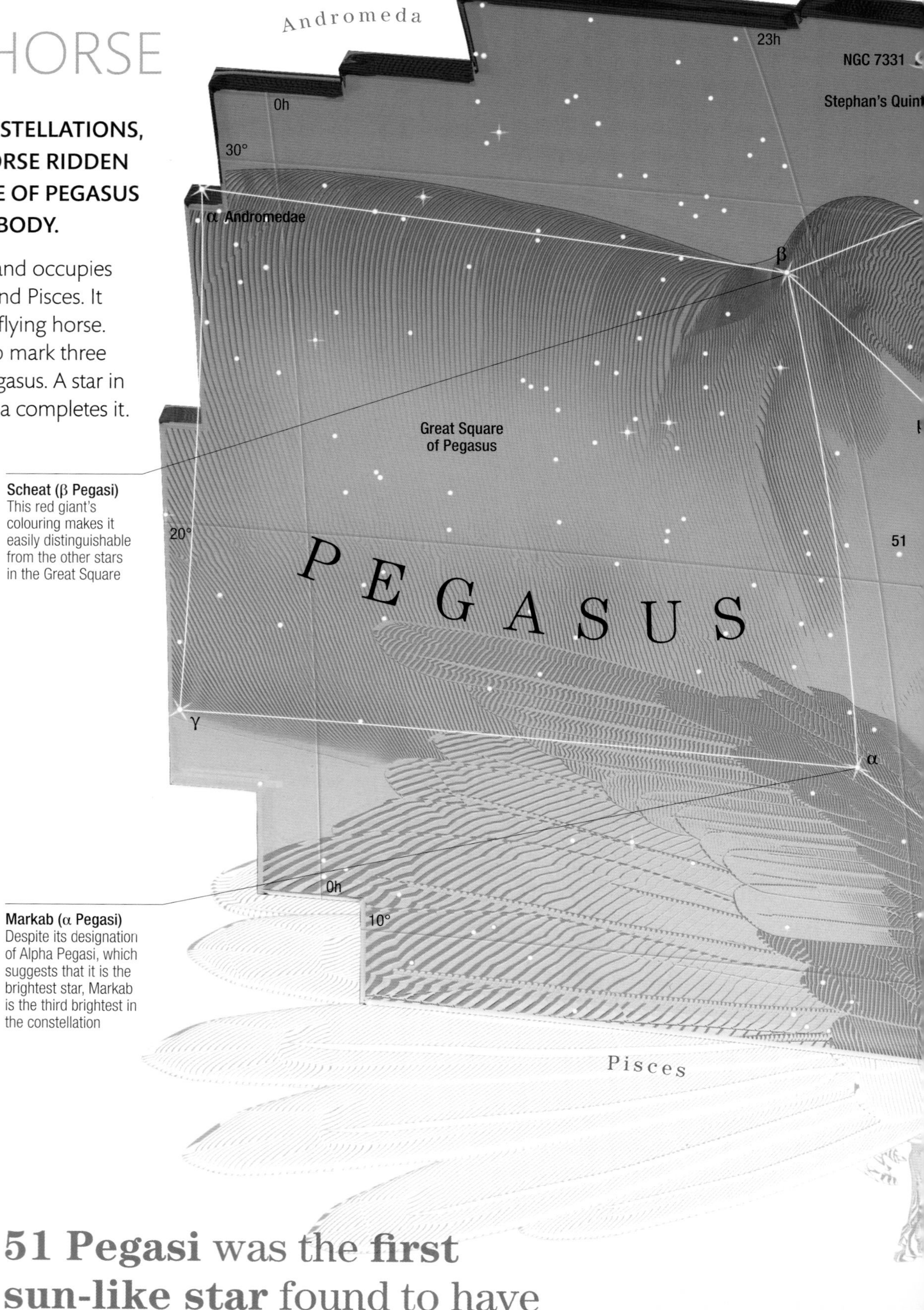

Scheat (β Pegasi) This red giant's colouring makes it easily distinguishable from the other stars in the Great Square

Markab (α Pegasi) Despite its designation of Alpha Pegasi, which suggests that it is the brightest star, Markab is the third brightest in the constellation

51 Pegasi was the **first sun-like star** found to have an **exoplanet**

EQUULEUS

THE FOAL

THE SECOND-SMALLEST CONSTELLATION AND WITH NO BRIGHT STARS, EQUULEUS DEPICTS A FOAL'S HEAD LYING NEXT TO THE LARGER HEAD OF PEGASUS.

Equuleus has been a companion to Pegasus in the sky since ancient times. One Greek myth suggests it could be Celeris, the offspring or brother of the winged horse. It is a faint constellation that is easily overlooked. The double star Gamma Equulei has components that are readily separated with binoculars.

KEY **DATA**

Size ranking	87
Brightest stars	Kitalpha (α) 3.9, Delta (δ) 4.4
Genitive	Equulei
Abbreviation	Equ
Highest in sky at 10pm	September
Fully visible	90°N–77°S

CHART 3

△ **NGC 7331**
A spiral galaxy seen edge-on from Earth, NGC 7331 is often used as an example of how our Milky Way galaxy might look from the outside.

M15
One of the densest known globular clusters, M15 is also one of the finest of the northern sky and easily spotted through binoculars

Enif (ε Pegasi)
From the Arabic for "nose", the bright star Enif marks the horse's nose and is easily visible with the naked eye

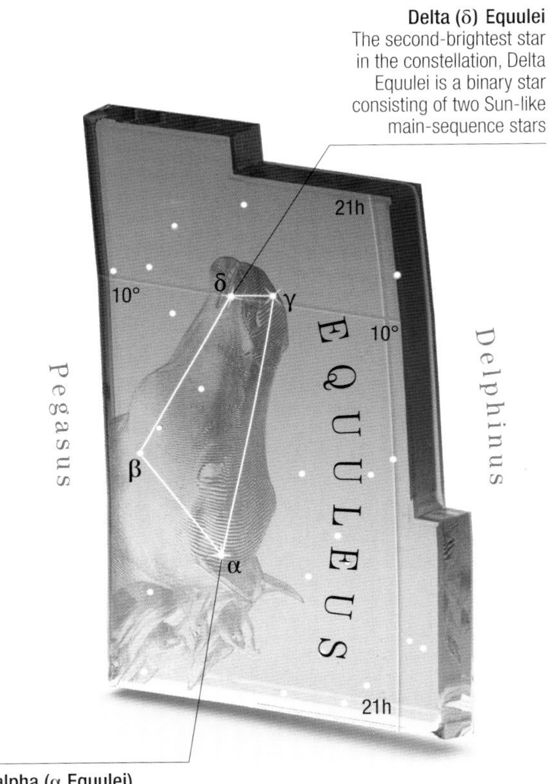

Delta (δ) Equulei
The second-brightest star in the constellation, Delta Equulei is a binary star consisting of two Sun-like main-sequence stars

Kitalpha (α Equulei)
A yellow giant, 190 light-years distant and 75 times more luminous than the Sun; its name comes from the Arabic for "piece of horse"

LUMINOSITIES

Omega[1] Aquarii *17 Suns*

Zeta Aquarii *24 Suns*

Sadachbia *65 Suns*

AQUARIUS
THE WATER CARRIER

THIS CONSTELLATION DEPICTS A YOUNG MAN POURING WATER FROM A JAR. IT CONTAINS TWO FAMOUS PLANETARY NEBULAE: HELIX AND SATURN.

Aquarius represents a young shepherd boy called Ganymede, who in Greek mythology was taken up to the heavens by Zeus to serve as a waiter to the gods on Mount Olympus. In the sky, he is visualized as pouring water out of a jar. The water jar is marked by a Y-shaped group of four stars in the north of the constellation – Gamma, Pi, Zeta, and Eta Aquarii. The flow of water from the jar is suggested by a stream of stars cascading southwards to the constellation's border with Piscis Austrinus, the Southern Fish.

The Eta Aquarid meteor shower – caused by dust from Halley's Comet entering the atmosphere as Earth crosses the comet's path – radiates from the area of the Water Jar asterism in early May each year. At the peak of the shower, as many as 35 meteors can be seen per hour.

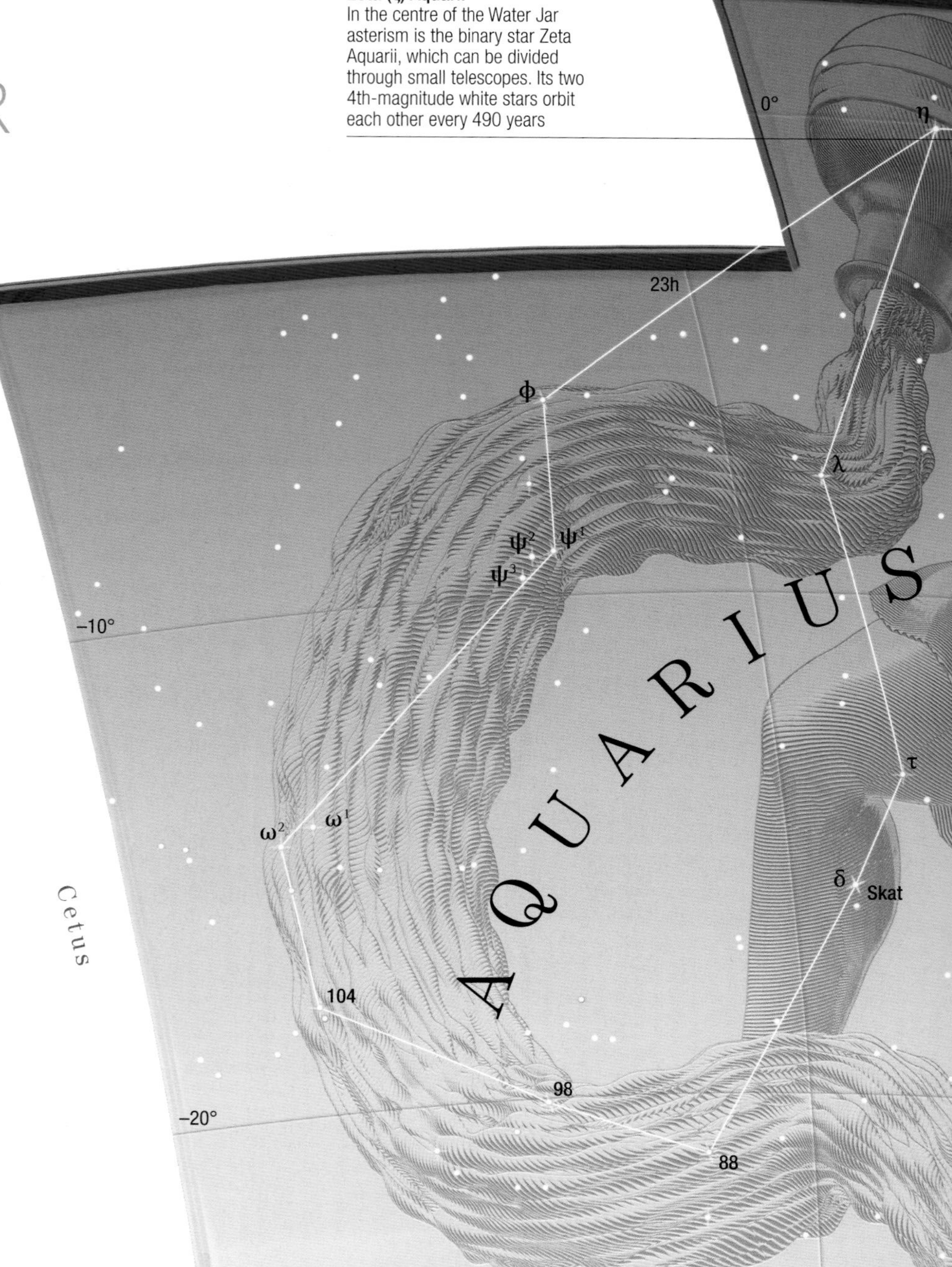

Zeta (ζ) Aquarii
In the centre of the Water Jar asterism is the binary star Zeta Aquarii, which can be divided through small telescopes. Its two 4th-magnitude white stars orbit each other every 490 years

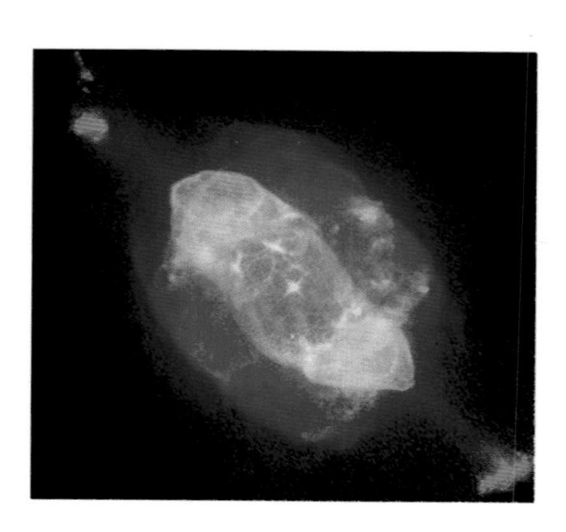

◁ **NGC 7009 (Saturn Nebula)**
This planetary nebula, seen here through the Hubble Space Telescope, gets its popular name from the "handles" at each end that look like the rings of Saturn. It lies 1,400 light-years away.

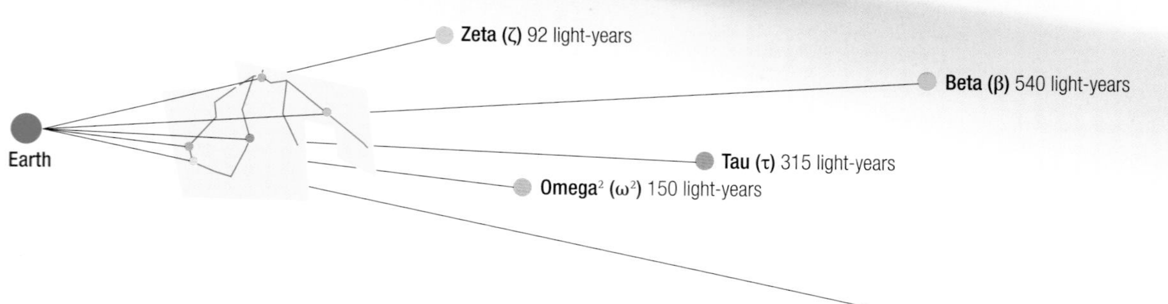

▷ **Star distances**
Most of the main pattern stars of Aquarius are relatively close to Earth, lying between about 100 and 300 light-years away. The nearest main pattern star is Zeta (ζ) Aquarii, which is the central star of the Water Jar asterism and lies about 92 light-years away. The most distant of Aquarius's pattern stars is 104 Aquarii, which is about 840 light-years away.

Skat
105 Suns

Sadalmelik
1,480 Suns

Sadalsuud
1,635 Suns

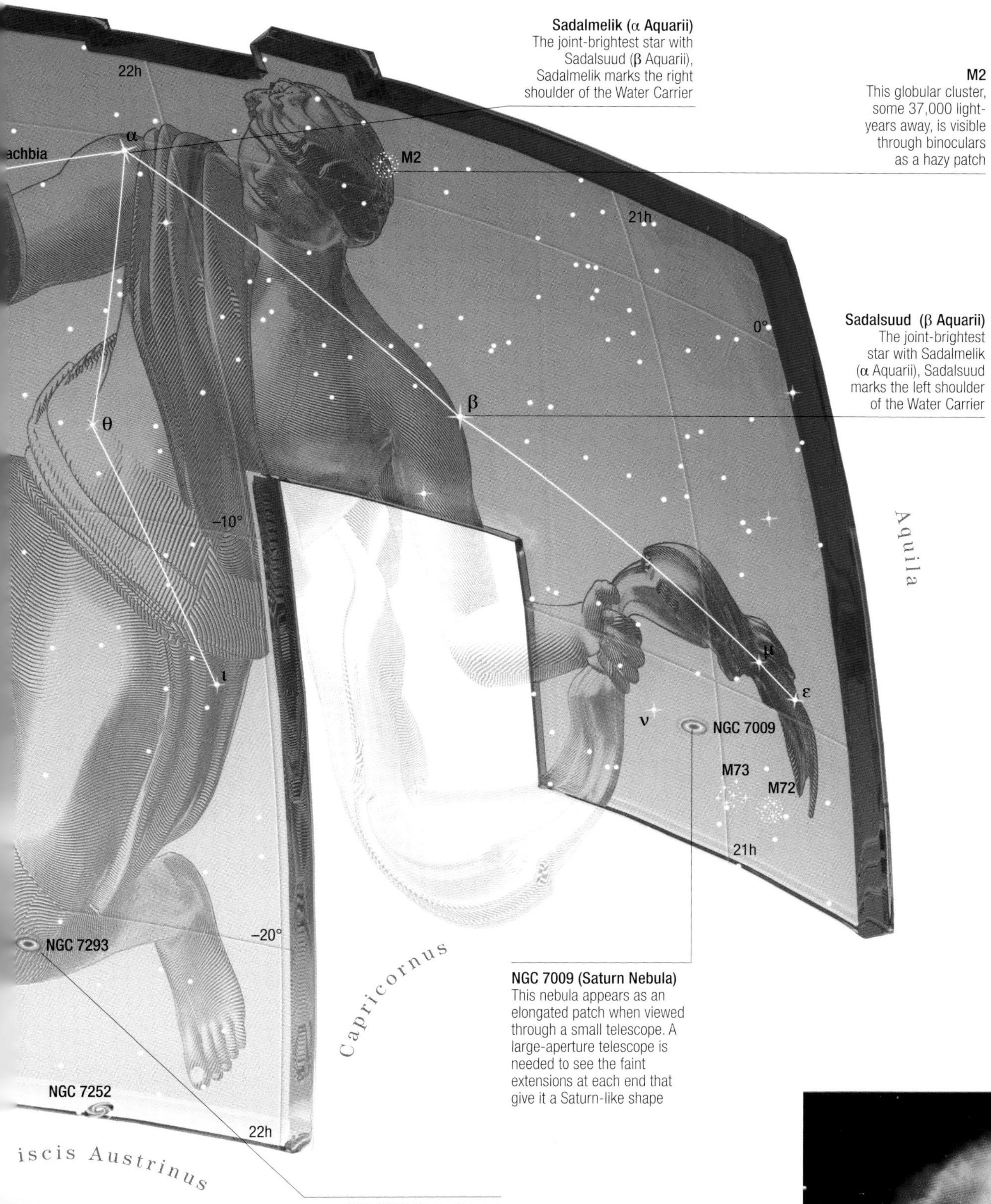

Sadalmelik (α Aquarii)
The joint-brightest star with Sadalsuud (β Aquarii), Sadalmelik marks the right shoulder of the Water Carrier

M2
This globular cluster, some 37,000 light-years away, is visible through binoculars as a hazy patch

Sadalsuud (β Aquarii)
The joint-brightest star with Sadalmelik (α Aquarii), Sadalsuud marks the left shoulder of the Water Carrier

NGC 7009 (Saturn Nebula)
This nebula appears as an elongated patch when viewed through a small telescope. A large-aperture telescope is needed to see the faint extensions at each end that give it a Saturn-like shape

NGC 7293 (Helix Nebula)
Visible through binoculars and small telescopes as a large, pale patch nearly half the apparent width of the Full Moon, NGC 7293 is the largest planetary nebula as seen from Earth

KEY DATA

Size ranking 10
Brightest stars Sadalmelik (α) 2.9, Sadalsuud (β) 2.9
Genitive Aquarii
Abbreviation Aqr
Highest in sky at 10pm June–July
Fully visible 65°N–86°S

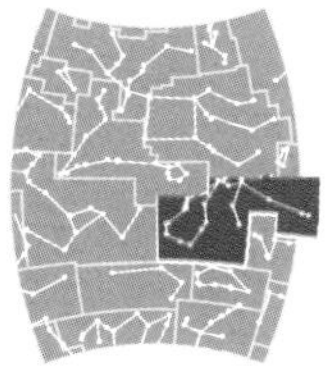

CHART 3

MAIN STARS

Sadalmelik Alpha (α) Aquarii
Yellow supergiant
2.9 ⟷ 525 light-years

Sadalsuud Beta (β) Aquarii
Yellow supergiant
2.9 ⟷ 540 light-years

Sadachbia Gamma (γ) Aquarii
Blue-white main sequence star
3.8 ⟷ 165 light-years

Skat Delta (δ) Aquarii
Blue-white main sequence star
3.3 ⟷ 160 light-years

Zeta (ζ) Aquarii
White giant
3.7 ⟷ 92 light-years

DEEP-SKY OBJECTS

M2
6th-magnitude globular cluster

M72
Globular cluster

M73
Small group of faint, unrelated stars

NGC 7009 (Saturn Nebula)
Planetary nebula similar in size to Saturn

NGC 7252 (Atoms for Peace Galaxy)
Colliding galaxies

NGC 7293 (Helix Nebula)
Large planetary nebula

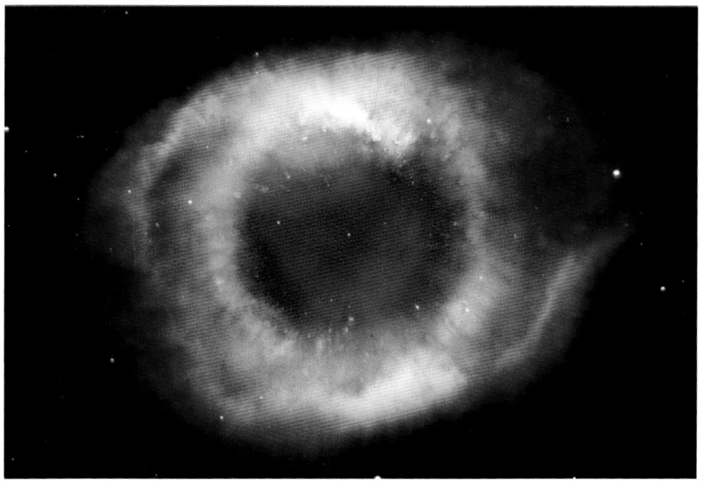

▷ **NGC 7293 (Helix Nebula)**
The Helix is the closest planetary nebula to the Sun, located about 650 light-years away. It consists of a cloud of gas about three light-years across, which surrounds a central white dwarf star.

LUMINOSITIES

PISCES THE FISH

ONE OF THE 48 CONSTELLATIONS OF CLASSICAL TIMES, PISCES IS A ZODIAC CONSTELLATION THAT REPRESENTS TWO FISH. ITS MOST DISTINCTIVE FEATURE IS A RING OF STARS CALLED THE CIRCLET.

A faint constellation lodged between Aquarius and Aries, Pisces can be found by looking south of the Great Square of Pegasus and locating a ring of stars. Named the Circlet, this ring marks the body of a fish. Pisces' second fish faces the opposite direction but the two are tied together by "ribbons". The star Alrescha marks the knot joining the two ribbons. According to Greek myth, the fish are linked to the goddess Aphrodite and her son Eros. Pisces is probably best known for containing the point where the Sun crosses the celestial equator into the northern hemisphere. Called the First Point of Aries, or vernal (spring) equinox, this point is used to measure celestial coordinates (see pp.90–91).

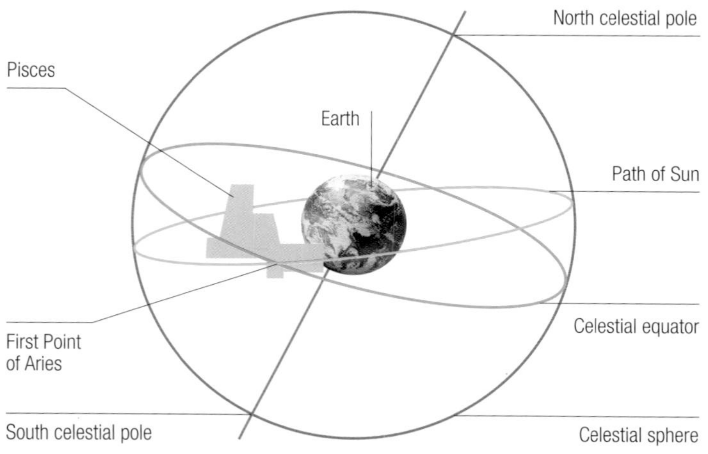

△ **The First Point of Aries**
Pisces contains the vernal (spring) equinox, also known as the First Point of Aries. This is the point at which the Sun crosses the celestial equator from south to north in March each year and is the point from which right ascension is measured. Originally in the constellation Aries, the vernal equinox has shifted position over time because of precession, the slow wobble of the Earth on its axis.

▷ **Star distances**
The nearest and farthest of Pisces' main pattern stars both form part of the Circlet: Iota (ι) Piscium at 45 light-years away, and TX Piscium, which is 20 times farther away, at about 900 light-years from Earth. These two stars are also the least and most luminous of Pisces' pattern stars, with Iota emitting about four times as much energy as the Sun gives off and TX Piscium about 690 times as much.

Psi¹ 275 light-years
Eta (η) 350 light-years
Earth
Epsilon (ε) 180 light-years
Iota (ι) 45 light-years
TX Piscium
900 light-years
Distance

Gamma Piscium *52 Suns* — **Alrescha** *55 Suns* — **Eta Piscium** *355 Suns* — **TX Piscium** *690 Suns*

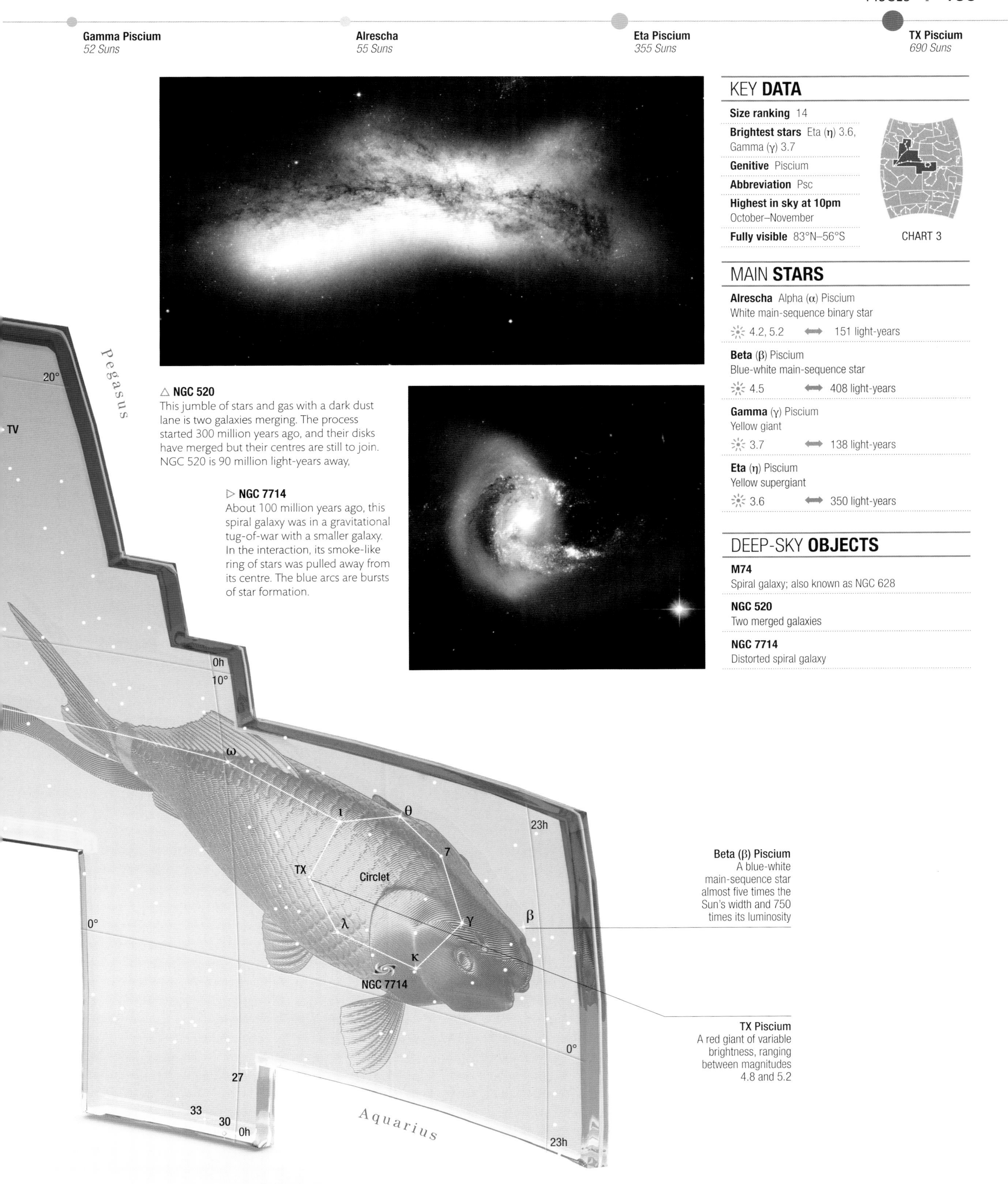

△ **NGC 520**
This jumble of stars and gas with a dark dust lane is two galaxies merging. The process started 300 million years ago, and their disks have merged but their centres are still to join. NGC 520 is 90 million light-years away,

▷ **NGC 7714**
About 100 million years ago, this spiral galaxy was in a gravitational tug-of-war with a smaller galaxy. In the interaction, its smoke-like ring of stars was pulled away from its centre. The blue arcs are bursts of star formation.

Beta (β) Piscium
A blue-white main-sequence star almost five times the Sun's width and 750 times its luminosity

TX Piscium
A red giant of variable brightness, ranging between magnitudes 4.8 and 5.2

KEY DATA

Size ranking 14

Brightest stars Eta (η) 3.6, Gamma (γ) 3.7

Genitive Piscium

Abbreviation Psc

Highest in sky at 10pm October–November

Fully visible 83°N–56°S

MAIN STARS

Alrescha Alpha (α) Piscium
White main-sequence binary star
4.2, 5.2 ⟷ 151 light-years

Beta (β) Piscium
Blue-white main-sequence star
4.5 ⟷ 408 light-years

Gamma (γ) Piscium
Yellow giant
3.7 ⟷ 138 light-years

Eta (η) Piscium
Yellow supergiant
3.6 ⟷ 350 light-years

DEEP-SKY OBJECTS

M74
Spiral galaxy; also known as NGC 628

NGC 520
Two merged galaxies

NGC 7714
Distorted spiral galaxy

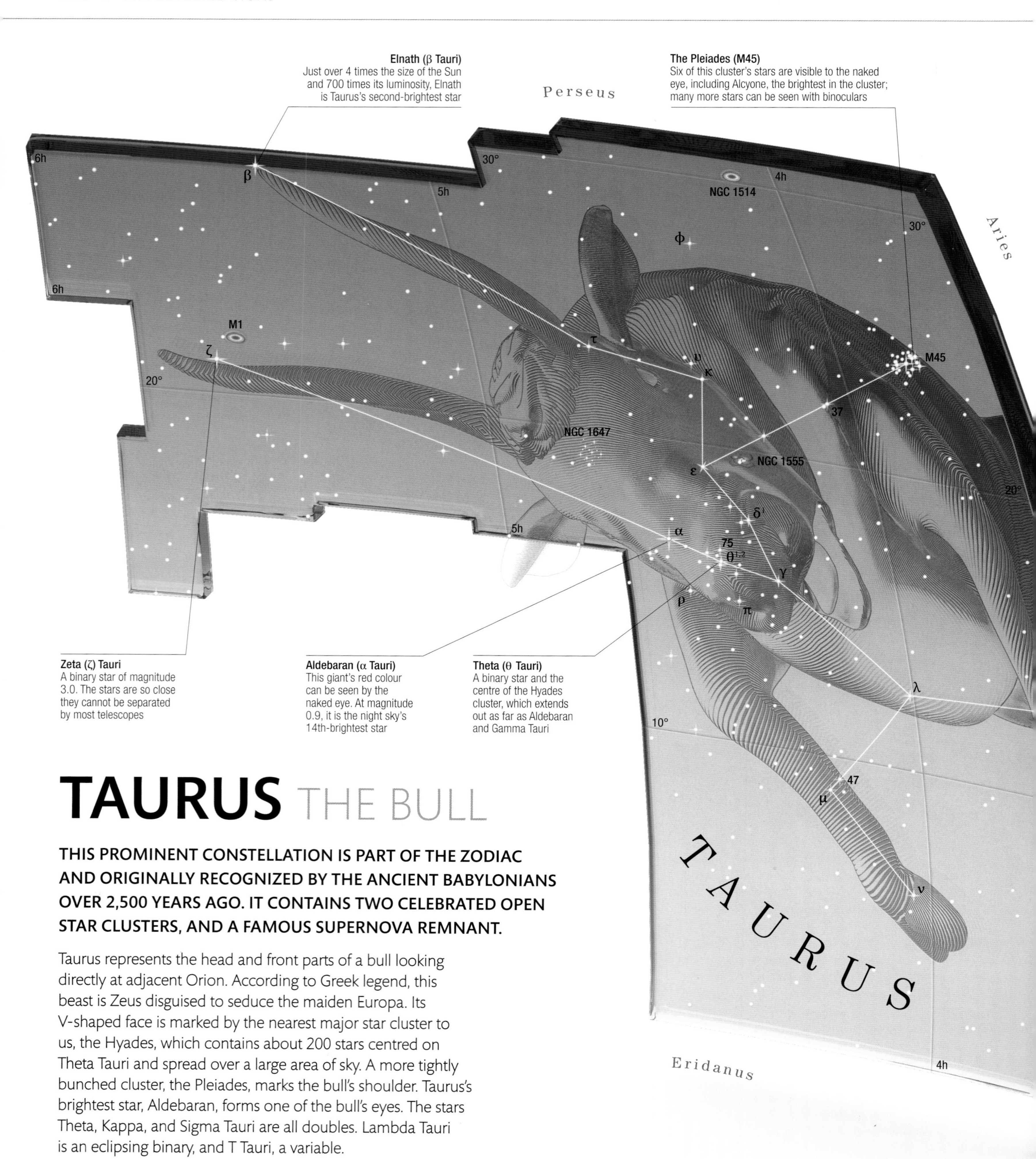
Elnath (β Tauri)
Just over 4 times the size of the Sun and 700 times its luminosity, Elnath is Taurus's second-brightest star
Perseus
The Pleiades (M45)
Six of this cluster's stars are visible to the naked eye, including Alcyone, the brightest in the cluster; many more stars can be seen with binoculars
Aries
6h
5h
4h
30°
20°
10°
NGC 1514
M1
M45
37
NGC 1647
NGC 1555
75
47
Zeta (ζ) Tauri
A binary star of magnitude 3.0. The stars are so close they cannot be separated by most telescopes
Aldebaran (α Tauri)
This giant's red colour can be seen by the naked eye. At magnitude 0.9, it is the night sky's 14th-brightest star
Theta (θ Tauri)
A binary star and the centre of the Hyades cluster, which extends out as far as Aldebaran and Gamma Tauri
TAURUS
Eridanus

TAURUS THE BULL

THIS PROMINENT CONSTELLATION IS PART OF THE ZODIAC AND ORIGINALLY RECOGNIZED BY THE ANCIENT BABYLONIANS OVER 2,500 YEARS AGO. IT CONTAINS TWO CELEBRATED OPEN STAR CLUSTERS, AND A FAMOUS SUPERNOVA REMNANT.

Taurus represents the head and front parts of a bull looking directly at adjacent Orion. According to Greek legend, this beast is Zeus disguised to seduce the maiden Europa. Its V-shaped face is marked by the nearest major star cluster to us, the Hyades, which contains about 200 stars centred on Theta Tauri and spread over a large area of sky. A more tightly bunched cluster, the Pleiades, marks the bull's shoulder. Taurus's brightest star, Aldebaran, forms one of the bull's eyes. The stars Theta, Kappa, and Sigma Tauri are all doubles. Lambda Tauri is an eclipsing binary, and T Tauri, a variable.

KEY DATA

Size ranking 17

Brightest stars Aldebaran (α) 0.9, Alnath (β) 1.7

Genitive Tauri

Abbreviation Tau

Highest in sky at 10pm December–January

Fully visible 88°N–58°S

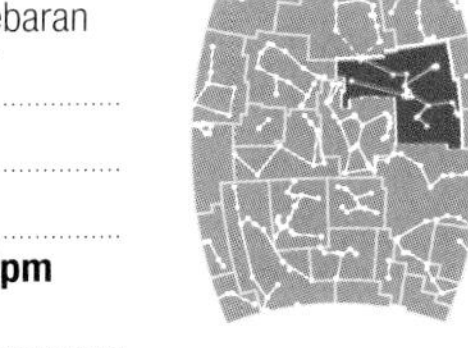

CHART 6

MAIN STARS

Aldebaran Alpha (α) Tauri
Red giant
0.9 ⟷ 67 light-years

Elnath Beta (β) Tauri
Blue-white giant
1.7 ⟷ 134 light-years

Zeta (ζ) Tauri
Binary star
3.0 ⟷ 445 light-years

Alcyone Eta (η) Tauri
Blue-white giant in the Pleiades cluster
2.9 ⟷ 403 light-years

DEEP-SKY OBJECTS

Hyades
Open star cluster centred on Theta (θ) Tauri

M45 (Pleiades)
Open star cluster

M1 (Crab Nebula)
Supernova remnant

NGC 1514
Planetary nebula

NGC 1555 (Hind's Variable Nebula)
Reflection nebula

△ **M1**
Popularly called the Crab Nebula, this is the remnant of a supernova that exploded in 1054 and was bright enough to be seen in the daytime. It is about 10 light-years across and still expanding as filaments of gas rush outwards from the site of the explosion. A neutron star, the central remnant of the original star is in its centre.

ARIES THE RAM

ONE OF THE ZODIAC CONSTELLATIONS, ARIES REPRESENTS A CROUCHING RAM WITH ITS HEAD TURNED TOWARDS TAURUS. ITS PATTERN IS RELATIVELY FAINT AND HARD TO IDENTIFY.

According to Greek legend, Aries is the ram whose golden fleece was sought by Jason and the Argonauts. Its most prominent part is a bent line made from three bright stars that mark the ram's head. More than 2,000 years ago Aries contained the vernal equinox, the point where the Sun crosses the celestial equator, south to north. The point, also known as the "First Point of Aries" defines zero hours right ascension. Today, it is in neighbouring Pisces.

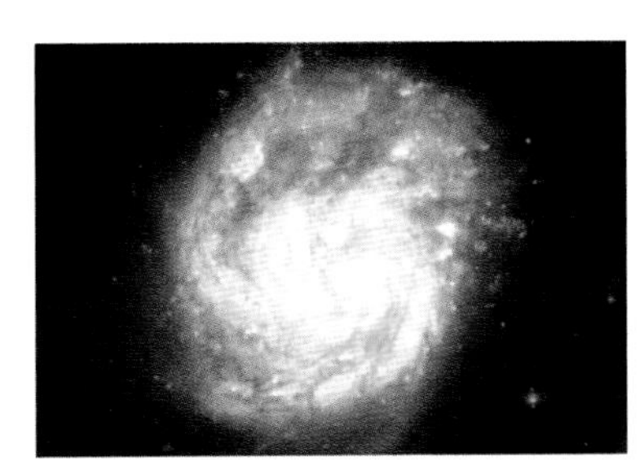

◁ **NGC 695**
This spiral galaxy, about 450 million light-years away, lies face-on to us. Its spiral arms are not well defined and very loosely wound. Knots of star formation are tangled in a mesh of dust and gas, which gives the whole galaxy a peculiar appearance.

KEY DATA

Size ranking 39

Brightest stars Hamal (α) 2.0, Sheratan (β) 2.7

Genitive Arietis

Abbreviation Ari

Highest in sky at 10pm November–December

Fully visible 90°N–58°S

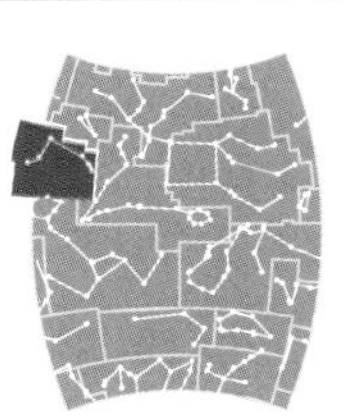

CHART 3

MAIN STARS

Hamal Alpha (α) Arietis
Yellow-orange giant
2.0 ⟷ 66 light-years

Sheratan Beta (β) Arietis
Binary star
2.7 ⟷ 59 light-years

Mesartim Gamma (γ) Arietis
Binary star
3.9 ⟷ 164 light-years

Lambda (λ) Arietis
Binary star
4.8 ⟷ 129 light-years

DEEP-SKY OBJECTS

NGC 695
Spiral galaxy

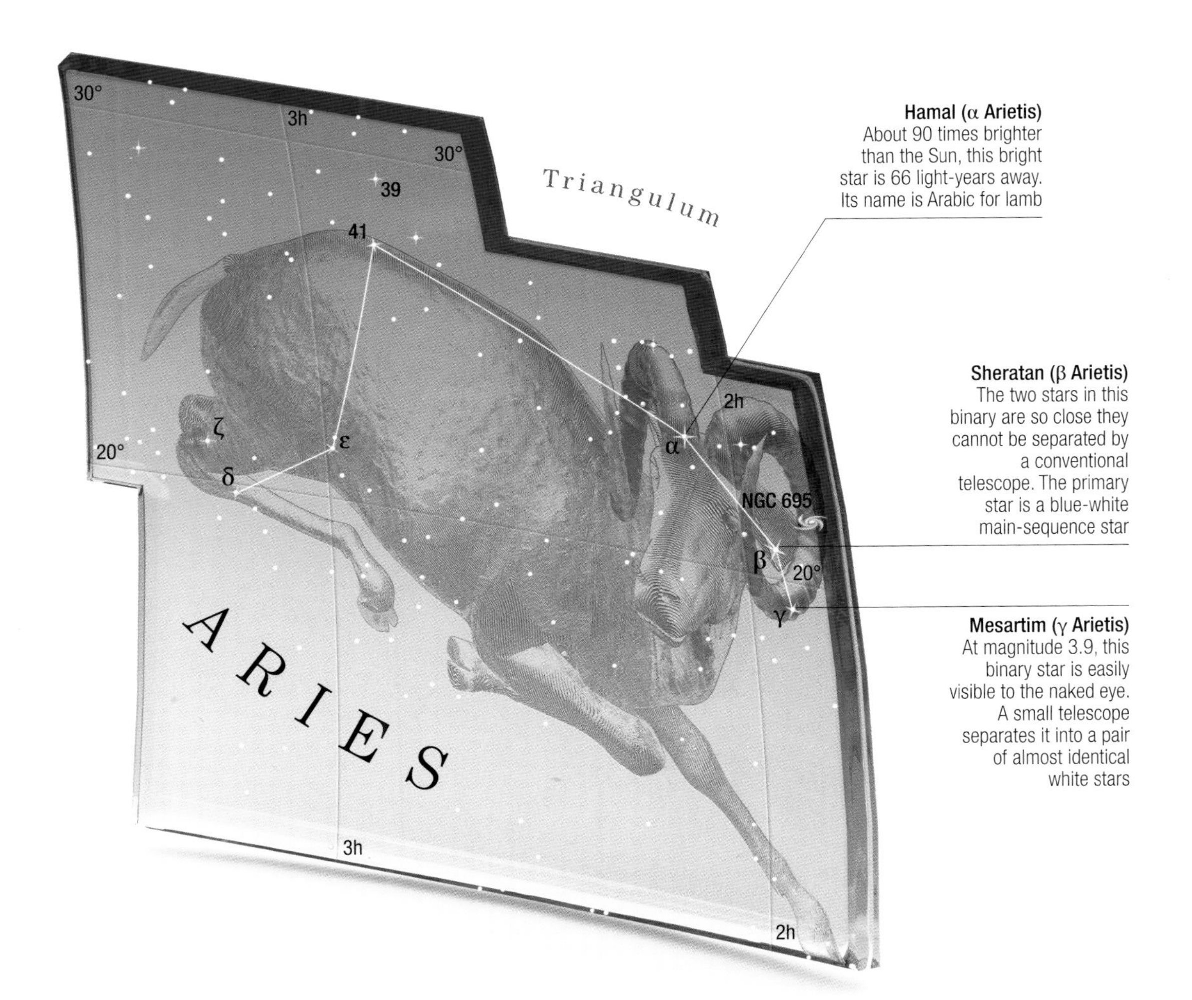

Hamal (α Arietis)
About 90 times brighter than the Sun, this bright star is 66 light-years away. Its name is Arabic for lamb

Sheratan (β Arietis)
The two stars in this binary are so close they cannot be separated by a conventional telescope. The primary star is a blue-white main-sequence star

Mesartim (γ Arietis)
At magnitude 3.9, this binary star is easily visible to the naked eye. A small telescope separates it into a pair of almost identical white stars

LUMINOSITIES

△ **Arp 256**
Lying about 350 million light-years away, these two galaxies, collectively known as ARP 256, are at an early stage of merging. Their interaction has disrupted their shapes and triggered star formation.

CETUS
THE SEA MONSTER

STRADDLING THE CELESTIAL EQUATOR, CETUS IS THE FOURTH-LARGEST CONSTELLATION IN THE SKY. IT IS NOT PARTICULARLY PROMINENT, SO CAN BE CHALLENGING TO IDENTIFY.

One of the original 48 Greek constellations, this sea monster was killed by Perseus as it was about to savage Andromeda, chained to cliffs as a sacrifice to the monster. It is often represented as a strange hybrid, with a large head, a land mammal's front legs and body, and a sea serpent's tail.

The constellation has several notable stars and deep-sky objects. Mira is one of the most prominent variables in the sky. Its brightness changes as it undergoes a long, regular cycle of pulsations. In contrast, UV Ceti is a red dwarf flare star whose brightness increases dramatically without warning. Tau Ceti is a Sun-like star only 12 light-years away and is orbited by five exoplanets.

Menkar (α Ceti)
The second-brightest star, Menkar, is about 89 times the size of the Sun and 2.3 times its mass. It has a wide but unrelated companion of magnitude 5.6, visible with binoculars

M77
Lying 47 million light-years away, M77 is the closest Seyfert galaxy to us. It is also the brightest, due to a central supermassive black hole with 15 million times the mass of the Sun

Mira (ο Ceti)
Appearing distinctly red, Mira varies in brightness as it fluctuates in size. It changes from magnitude 2.0 to 10 over 322 days, going from a naked-eye object to a telescopic one

◁ **Star distances**
Cetus's main pattern stars are situated between 12 and about 650 light-years away. The nearest, Tau (τ) Ceti, is one of the closest yellow main-sequence, Sun-like stars to us. Delta (δ) Ceti, is a blue giant that, as well as being the most distant of Cetus's pattern stars, is also the most luminous, with a luminosity of more than 800 Suns.

Mira	Gamma Ceti	Deneb Kaitos	Menkar	Delta Ceti
19 Suns	*21 Suns*	*115 Suns*	*490 Suns*	*805 Suns*

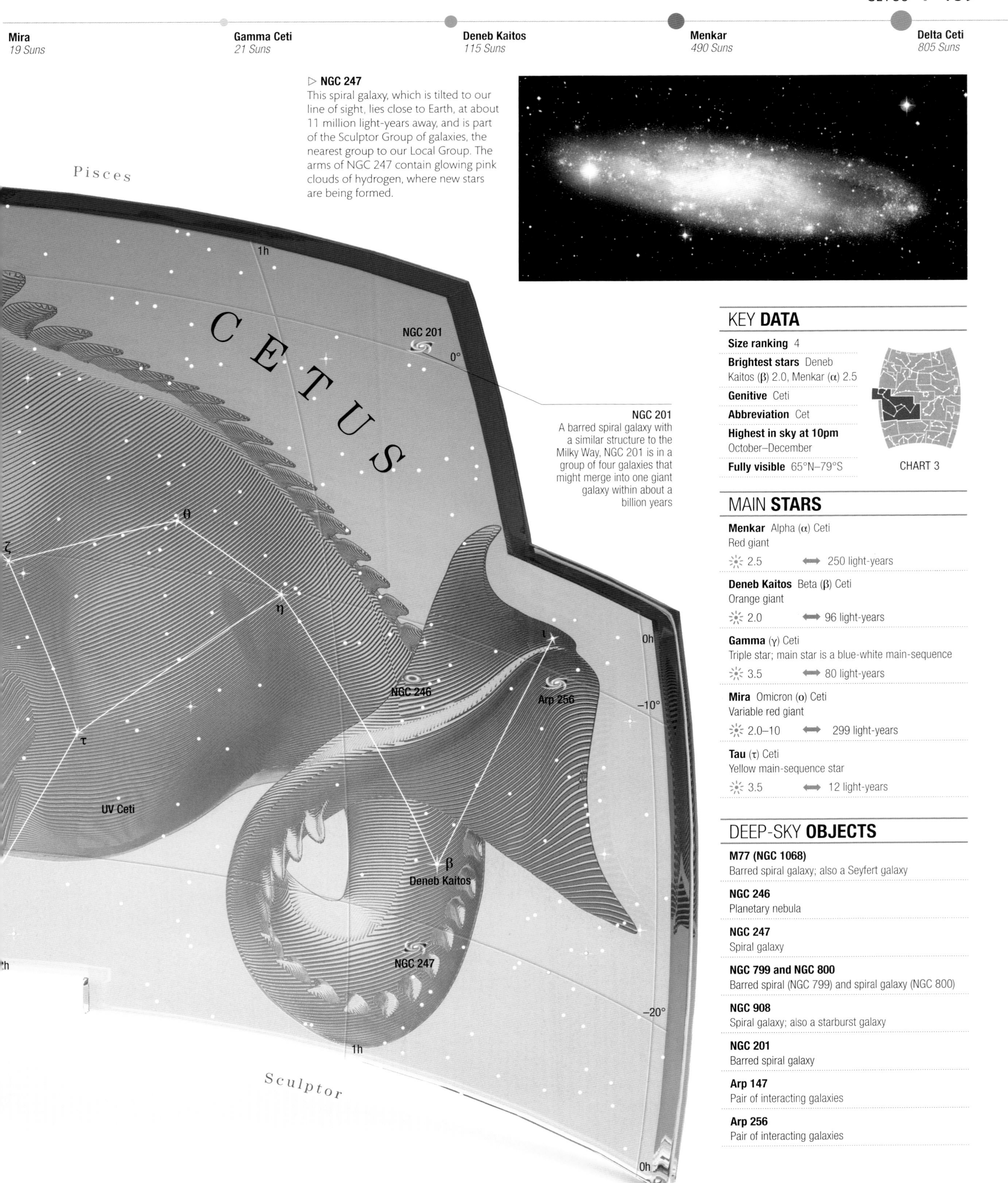

▷ **NGC 247**
This spiral galaxy, which is tilted to our line of sight, lies close to Earth, at about 11 million light-years away, and is part of the Sculptor Group of galaxies, the nearest group to our Local Group. The arms of NGC 247 contain glowing pink clouds of hydrogen, where new stars are being formed.

NGC 201
A barred spiral galaxy with a similar structure to the Milky Way, NGC 201 is in a group of four galaxies that might merge into one giant galaxy within about a billion years

KEY **DATA**

Size ranking 4
Brightest stars Deneb Kaitos (β) 2.0, Menkar (α) 2.5
Genitive Ceti
Abbreviation Cet
Highest in sky at 10pm October–December
Fully visible 65°N–79°S

CHART 3

MAIN **STARS**

Menkar Alpha (α) Ceti
Red giant
2.5 ⟷ 250 light-years

Deneb Kaitos Beta (β) Ceti
Orange giant
2.0 ⟷ 96 light-years

Gamma (γ) Ceti
Triple star; main star is a blue-white main-sequence
3.5 ⟷ 80 light-years

Mira Omicron (ο) Ceti
Variable red giant
2.0–10 ⟷ 299 light-years

Tau (τ) Ceti
Yellow main-sequence star
3.5 ⟷ 12 light-years

DEEP-SKY **OBJECTS**

M77 (NGC 1068)
Barred spiral galaxy; also a Seyfert galaxy

NGC 246
Planetary nebula

NGC 247
Spiral galaxy

NGC 799 and NGC 800
Barred spiral (NGC 799) and spiral galaxy (NGC 800)

NGC 908
Spiral galaxy; also a starburst galaxy

NGC 201
Barred spiral galaxy

Arp 147
Pair of interacting galaxies

Arp 256
Pair of interacting galaxies

LUMINOSITIES

Epsilon Eridani *0.3 Suns*

Cursa *51 Suns*

Acamar *150 Suns*

ERIDANUS THE RIVER

ERIDANUS MEANDERS FROM ORION DEEP INTO THE SOUTHERN SKY, ENDING AT THE BRIGHT STAR ACHERNAR. IN ANCIENT GREEK TIMES, THE CONSTELLATION DID NOT END SO FAR SOUTH, BUT IT WAS EXTENDED WHEN EUROPEAN NAVIGATORS BEGAN TO CHART THE SOUTHERN STARS.

Eridanus represents the mythical river that Phaethon, the son of the Sun-god Helios, fell into while attempting to drive his father's Sun-chariot across the sky.

Ancient Greek astronomers traced the celestial river as far south as the star we know as Acamar (Theta Eridani). Eridanus was later extended farther south so that it now ends at the star we call Achernar (Alpha Eridani). The names Acamar and Achernar both come from the Arabic meaning "river's end". Present-day Eridanus has the greatest north-to-south span of any constellation, nearly 60 degrees.

Eridanus contains several notable celestial objects, including the flattened, fast-spinning star, Achernar, as well as the classic barred spiral galaxy NGC 1300 (see pp.52–53), the face-on spiral NGC 1309, and the galaxy NGC 1291, which is surrounded by an outer ring in which new stars are being formed.

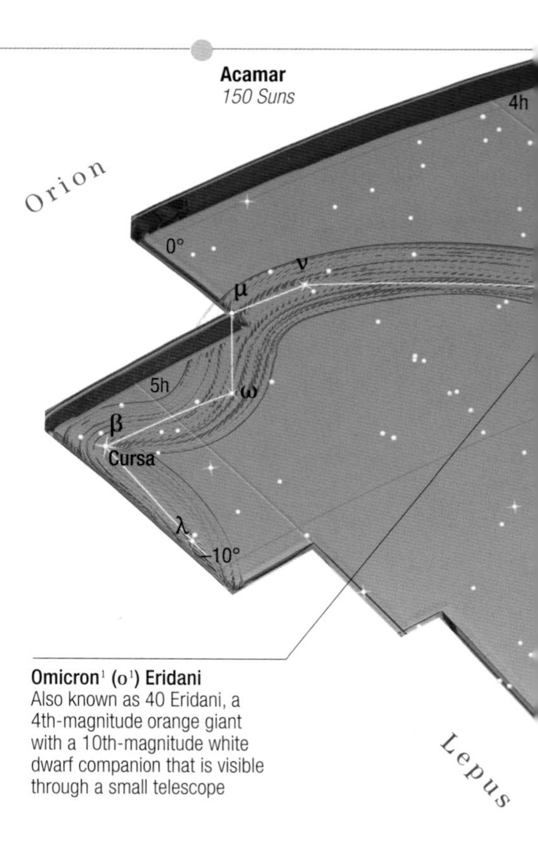

Omicron[1] (ο[1]) Eridani
Also known as 40 Eridani, a 4th-magnitude orange giant with a 10th-magnitude white dwarf companion that is visible through a small telescope

Achernar, the brightest star in Eridanus, is the **least spherical star** known

△ **NGC 1309**
Lying about 100 million light-years away, NGC 1309 is a spiral galaxy that is about three-quarters as wide as our own Milky Way. In this Hubble image, bright areas of active star formation can be seen in the arms, with brown dust lanes spiralling out from the pale yellow nucleus containing older stars.

◁ **NGC 1291**
A ring of newborn stars encircles the galaxy NGC 1291 in this false-colour image taken at infrared wavelengths by NASA's Spitzer Space Telescope. In this image, young stars are shown in red, and the older stars at the galaxy's centre are shown in blue. When galaxies like this are young, star formation is concentrated near their centres, but as gas at the galaxy's centre is used up, star formation moves to the outer regions, as has occurred here.

▷ **Star distances**
Eridanus's main pattern stars lie between about 10 light-years and 810 light-years from Earth. The nearest and farthest stars are both in the northern part of the constellation: Epsilon (ε) Eridani at 10.5 light-years from Earth and Lambda (λ) Eridani at about 810 light-years away.

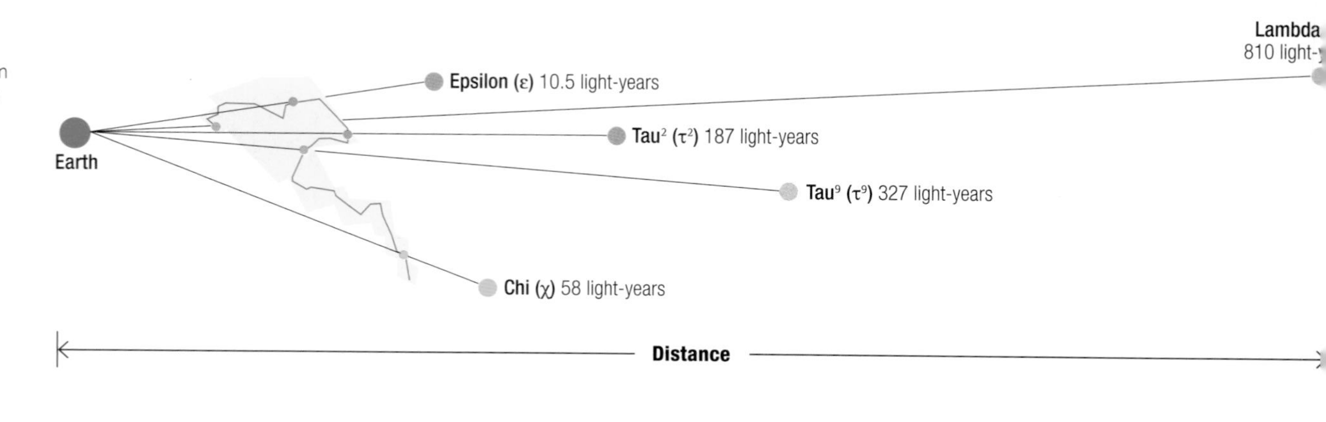

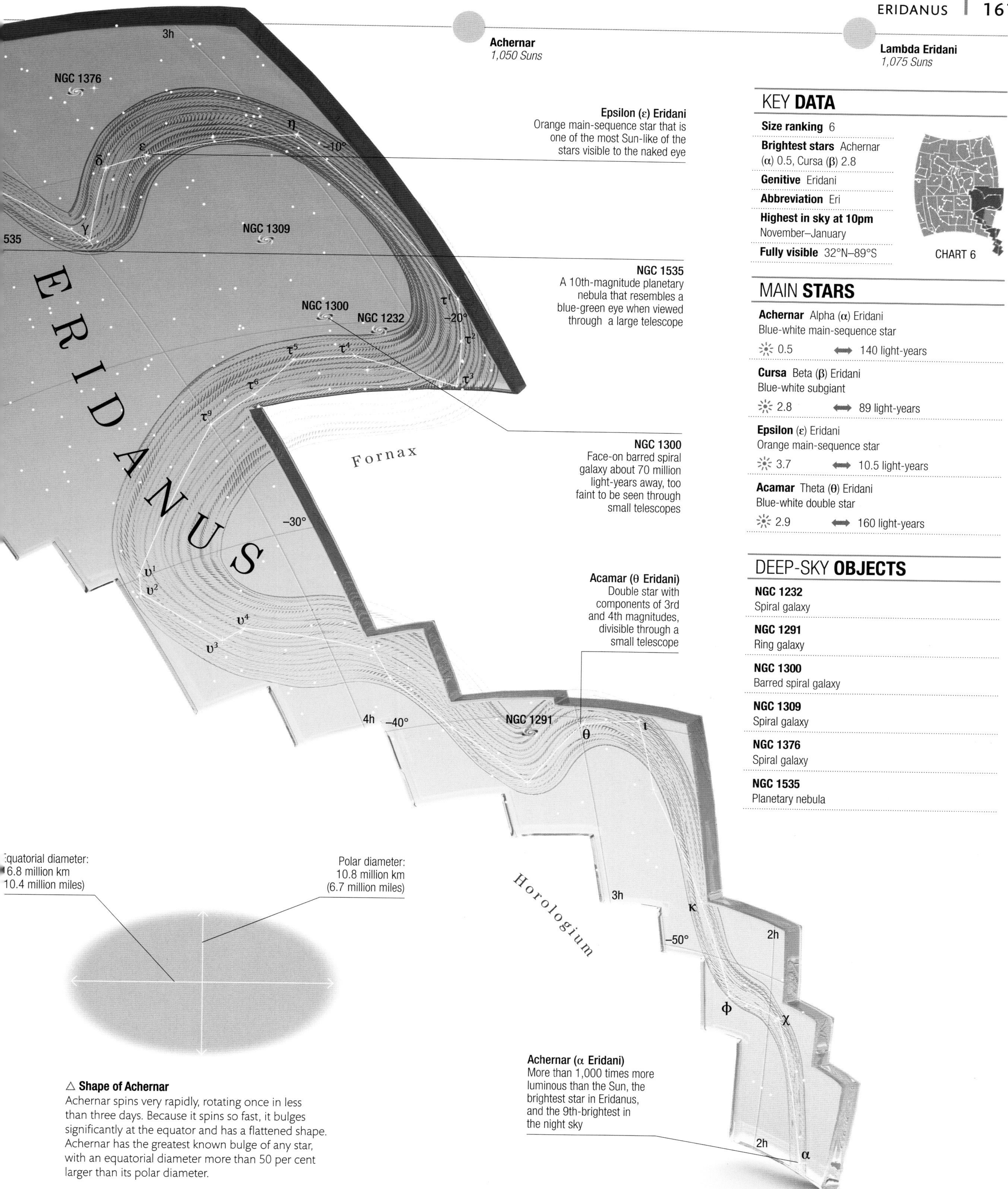

△ **Shape of Achernar**
Achernar spins very rapidly, rotating once in less than three days. Because it spins so fast, it bulges significantly at the equator and has a flattened shape. Achernar has the greatest known bulge of any star, with an equatorial diameter more than 50 per cent larger than its polar diameter.

KEY **DATA**

Size ranking 6

Brightest stars Achernar (α) 0.5, Cursa (β) 2.8

Genitive Eridani

Abbreviation Eri

Highest in sky at 10pm November–January

Fully visible 32°N–89°S

CHART 6

MAIN **STARS**

Achernar Alpha (α) Eridani
Blue-white main-sequence star
0.5 ⟷ 140 light-years

Cursa Beta (β) Eridani
Blue-white subgiant
2.8 ⟷ 89 light-years

Epsilon (ε) Eridani
Orange main-sequence star
3.7 ⟷ 10.5 light-years

Acamar Theta (θ) Eridani
Blue-white double star
2.9 ⟷ 160 light-years

DEEP-SKY **OBJECTS**

NGC 1232
Spiral galaxy

NGC 1291
Ring galaxy

NGC 1300
Barred spiral galaxy

NGC 1309
Spiral galaxy

NGC 1376
Spiral galaxy

NGC 1535
Planetary nebula

LUMINOSITIES

Chi^1 *1 Sun*

Mintaka *4,945 Suns*

Alnitak *8,940 Suns*

ORION THE HUNTER

FORMED FROM AN EASILY RECOGNIZABLE PATTERN OF STARS, ORION IS FAMILIAR TO MOST SKYWATCHERS. IT CONTAINS SEVERAL BRIGHT STARS AND THE ORION NEBULA, ONE OF THE MOST BEAUTIFUL SIGHTS IN THE NIGHT SKY.

Orion is an ancient constellation that represents a hunter or warrior in Greek myth. Orion was the son of Poseidon, the god of the sea, and was a hunter of great prowess. Despite his hunting skill, he was killed by a mere scorpion, possibly in retribution for his boastfulness. In the sky, the scorpion is depicted by the constellation Scorpius, and as Orion sets below the horizon, Scorpius rises and pursues him across the sky. Close to Orion's heels are Canis Major and Canis Minor, representing his hunting dogs.

The two brightest stars in Orion provide a striking colour contrast: the red supergiant Betelgeuse marks the hunter's shoulder, while the blue supergiant Rigel is positioned in one of this feet. Many of the constellation's highlights are near the line of stars that represents Orion's Belt. The belt is easy to find because it is made up of three equally spaced bright stars – Alnitak, Alnilam, and Mintaka – which form an almost perfectly straight line. Just below the belt is a complex of stars and nebulae that represent the hunter's sword. This area includes a vast area of star formation called the Orion Nebula (M42), the largest and closest nebula of its kind in the night sky. Other nebulae lie nearby, including the Horsehead Nebula, a dark nebula silhoutted against the bright emission nebula, known as IC 434.

When Betelgeuse **explodes,** it will release **more energy** in an **instant** than the **Sun** will produce in its **lifetime**

▷ **Betelgeuse**
The supergiant star Betelgeuse is more than 500 times bigger than the Sun. If it was placed at the centre of our Solar System, it would engulf the Sun and all the planets out as far as Jupiter. Betelgeuse is both relatively young and highly unstable, varying erratically in brightness. At some point in the next million years, it will probably explode in a supernova.

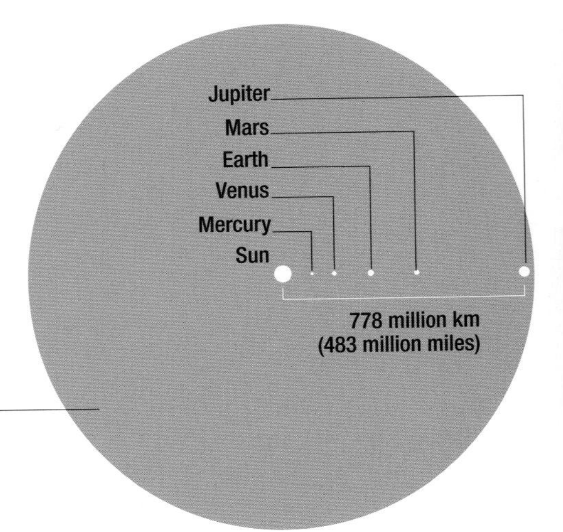

▷ **M42**
Better known as the Orion Nebula, this star-forming region is about 24 light-years across. It is shown here as bright pink because radiation from hot young stars causes hydrogen gas to glow pink. The nebula is embedded in a much larger dark cloud; dust in this cloud is coloured dull pink.

▽ **The Trapezium**
The stars forming in the Orion Nebula include a group called the Trapezium. As well as the four stars seen here, the system includes two fainter members.

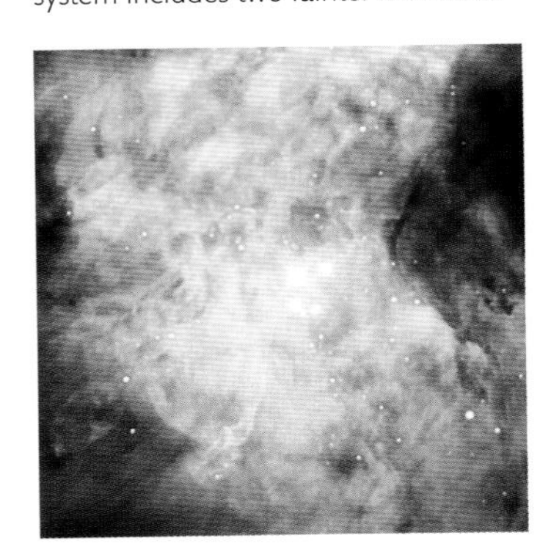

▷ **Star distances**
At just under 500 light-years away, Betelgeuse is the closer of Orion's two brightest stars. Rigel is much more distant – about 860 light-years away – but most of the time, Rigel looks brighter than Betelgeuse because it emits far more light. The three stars that form the line of Orion's Belt are widely spread out in space, with Alnilam being the most distant. In fact, Alnilam is the farthest of all Orion's main pattern stars, at nearly 2,000 light-years away. The nearest pattern star is Pi^3 Orionis, which is only 26 light-years from Earth

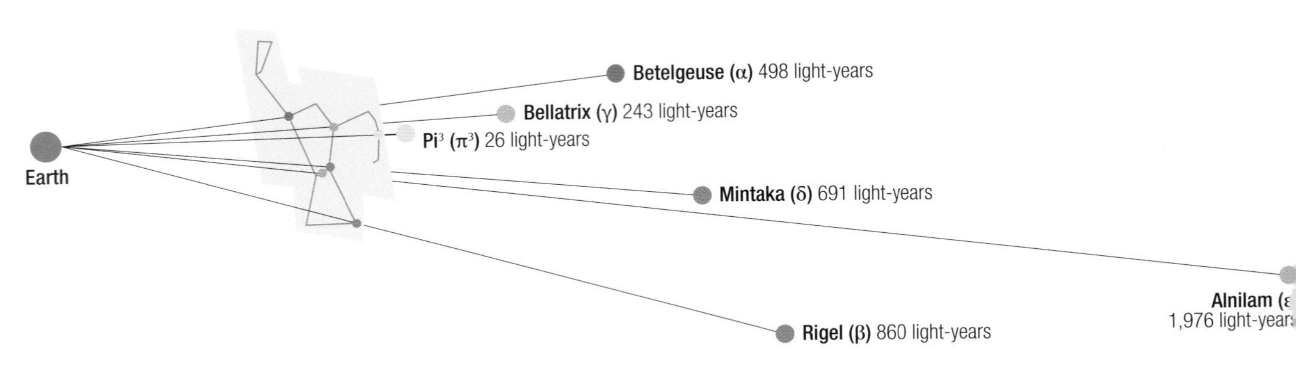

Betelgeuse
13,415 Suns

Rigel
51,665 Suns

Alnilam
67,480 Suns

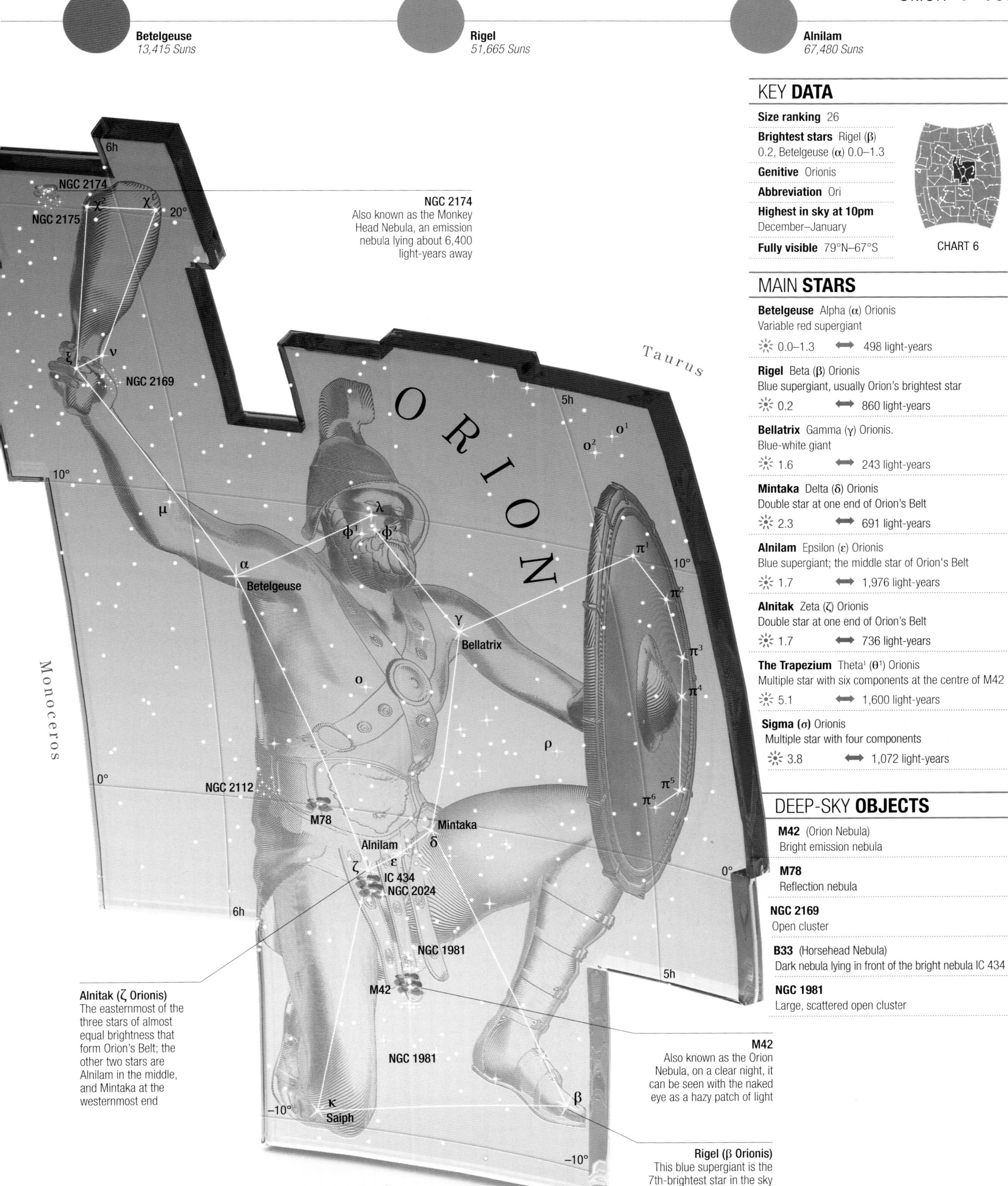

NGC 2174
Also known as the Monkey Head Nebula, an emission nebula lying about 6,400 light-years away

Alnitak (ζ Orionis)
The easternmost of the three stars of almost equal brightness that form Orion's Belt; the other two stars are Alnilam in the middle, and Mintaka at the westernmost end

M42
Also known as the Orion Nebula, on a clear night, it can be seen with the naked eye as a hazy patch of light

Rigel (β Orionis)
This blue supergiant is the 7th-brightest star in the sky

KEY DATA

Size ranking 26
Brightest stars Rigel (β) 0.2, Betelgeuse (α) 0.0–1.3
Genitive Orionis
Abbreviation Ori
Highest in sky at 10pm December–January
Fully visible 79°N–67°S

CHART 6

MAIN STARS

Betelgeuse Alpha (α) Orionis
Variable red supergiant
0.0–1.3 ⟷ 498 light-years

Rigel Beta (β) Orionis
Blue supergiant, usually Orion's brightest star
0.2 ⟷ 860 light-years

Bellatrix Gamma (γ) Orionis.
Blue-white giant
1.6 ⟷ 243 light-years

Mintaka Delta (δ) Orionis
Double star at one end of Orion's Belt
2.3 ⟷ 691 light-years

Alnilam Epsilon (ε) Orionis
Blue supergiant; the middle star of Orion's Belt
1.7 ⟷ 1,976 light-years

Alnitak Zeta (ζ) Orionis
Double star at one end of Orion's Belt
1.7 ⟷ 736 light-years

The Trapezium Theta[1] (θ^1) Orionis
Multiple star with six components at the centre of M42
5.1 ⟷ 1,600 light-years

Sigma (σ) Orionis
Multiple star with four components
3.8 ⟷ 1,072 light-years

DEEP-SKY OBJECTS

M42 (Orion Nebula)
Bright emission nebula

M78
Reflection nebula

NGC 2169
Open cluster

B33 (Horsehead Nebula)
Dark nebula lying in front of the bright nebula IC 434

NGC 1981
Large, scattered open cluster

VIEWS OF THE **ORION NEBULA**

1 Orion's sword

The Orion Nebula is one of the most observed and photographed objects in the night sky. To the naked eye it is a just a fuzzy patch marking the sword of Orion. However, photographs transform it into a colourful maelstrom of star-birth. This wide-field view shows all of the massive star-formation region. It was taken by the VISTA infrared telescope at the European Paranal Observatory in Chile.

2 Heart of the nebula

The heart of the Orion Nebula stellary nursery contains thousands of young stars and developing protostars. The infant stars have blown away much of the dust and gas in which they formed, carving a huge cavity in the cloud, seen here in red. The brilliant starlight of the region at the top of the cavity comes from a tight open cluster of young stars, known as the Trapezium (see p.162).

3 Infrared composite

This infrared view combines data from two space telescopes, Spitzer and Herschel. It shows a region of the nebula about 10 light-years across, with the Trapezium to the left of the image. In infrared light, it is dust rather than the gas and stars that shines brightest. The red regions show cold dust, condensed into clumps around stars in the process of forming. Blue indicates warmer dust, heated by fully formed hot, young stars.

4 High-temperature gas

A high-temperature gas cloud, shown here in blue, is revealed when the nebula is imaged in X-ray light by the space telescope XMM-Newton. It appears to fill the nebula's huge cavi which is visible in optical and infrared views. Th cloud was produced in a violent collision whe wind from a massive star was heated to millions of degrees as it slammed into surrounding gas The bright yellow patch is the Trapezium cluste

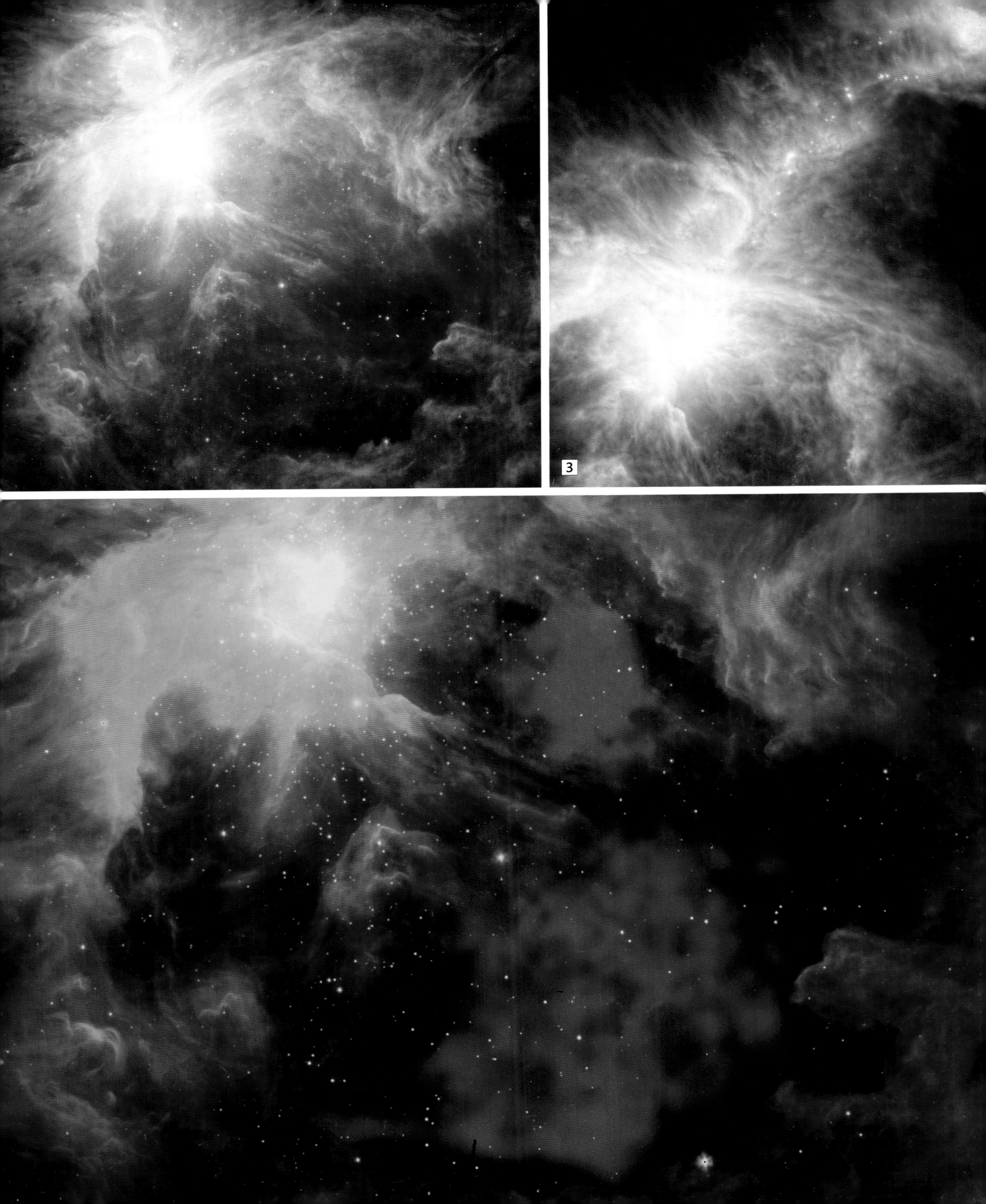
3

LUMINOSITIES

Wasat
12 Suns

Pollux
32 Suns

Castor
49 Suns

GEMINI THE TWINS

GEMINI IS A PROMINENT CONSTELLATION OF THE ZODIAC, REPRESENTING THE MYTHOLOGICAL TWINS CASTOR AND POLLUX. ITS TWO BRIGHTEST STARS MARK THE HEADS OF THE TWINS. ITEMS OF INTEREST INCLUDE A BRIGHT STAR CLUSTER AND AN UNUSUAL-LOOKING PLANETARY NEBULA.

In Greek mythology, Castor and Pollux were the sons of Queen Leda of Sparta. Pollux was said to have been fathered by the god Zeus and was immortal, while Castor's father was Leda's husband, King Tyndareus, and he was mortal. The twins joined the crew of the *Argo* and went in search of the Golden Fleece, in one of the great epics of Ancient Greek mythology.

Overall, Gemini is rectangular in shape. One of its two main stars, Castor, is a remarkable multi-star system (see diagram, below right), the two brightest members of which can be seen separately through a small telescope. Although Castor is labelled Alpha Geminorum, it is not the brightest star in Gemini, which is Pollux (Beta Geminorum), the constellation's other main star.

One of the year's richest meteor showers, the Geminids, appears to radiate from a point near Castor around 13 December each year. Up to 100 meteors an hour can be seen. Unlike most meteor showers, the parent body is not a comet but an asteroid, Phaethon.

KEY DATA

Size ranking 30

Brightest stars Pollux (β) 1.1, Castor (α) 1.6

Genitive Geminorum

Abbreviation Gem

Highest in sky at 10pm January–February

Fully visible 90°N–55°S

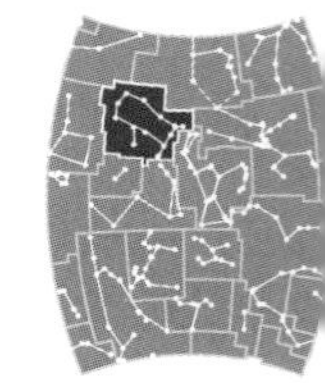

CHART 6

MAIN STARS

Castor Alpha (α) Geminorum
Blue-white multiple star
1.6 ⟷ 51 light-years

Pollux Beta (β) Geminorum
Orange giant
1.1 ⟷ 34 light-years

Alhena Gamma (γ) Geminorum
Blue-white subgiant
1.9 ⟷ 110 light-years

Wasat Delta (δ) Geminorum
White main-sequence star
3.5 ⟷ 60 light-years

Mebsuta Epsilon (ε) Geminorum
Yellow supergiant
3.0 ⟷ 845 light-years

DEEP-SKY OBJECTS

M35
Large, bright open cluster of about 200 stars

NGC 2392 (Eskimo Nebula)
Planetary nebula, also called the Clown Face Nebula

▽ **M35**
This large cluster of about 200 stars is easily visible through binoculars near the border where Gemini meets Taurus. Telescopes also show the fainter and more tightly bunched cluster NGC 2158, seen at bottom left of this image. M35 is approximately 2,800 light-years away, while NGC 2158 is around 10,000 light-years more distant.

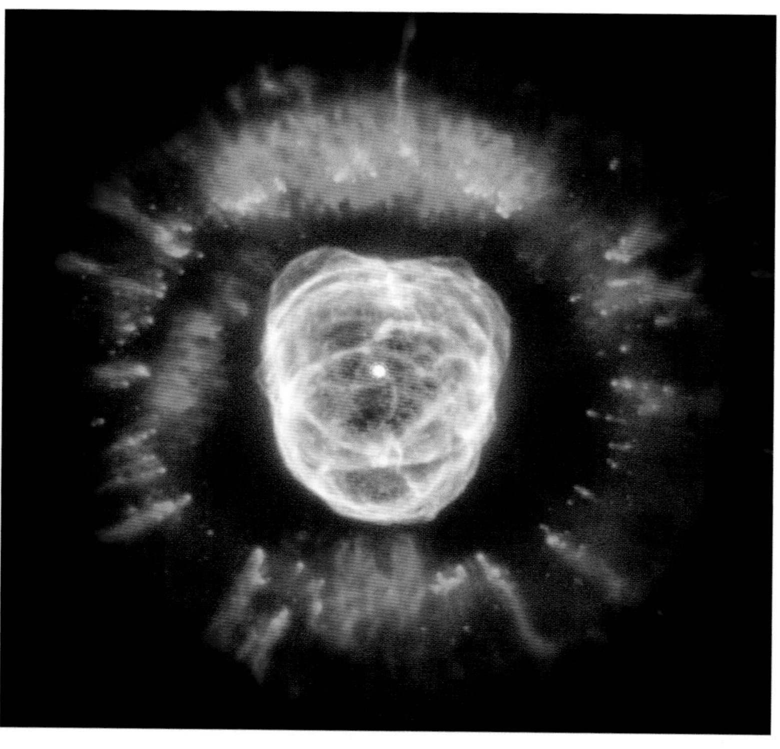

△ **NGC 2392**
This unusual-looking planetary nebula gets its popular name from its resemblance to a face surrounded by a furry parka hood. The "hood" is really a ring of gas streaming away from the central star, creating the appearance of a disk when seen through a small telescope. The Eskimo Nebula is about 5,000 light-years away.

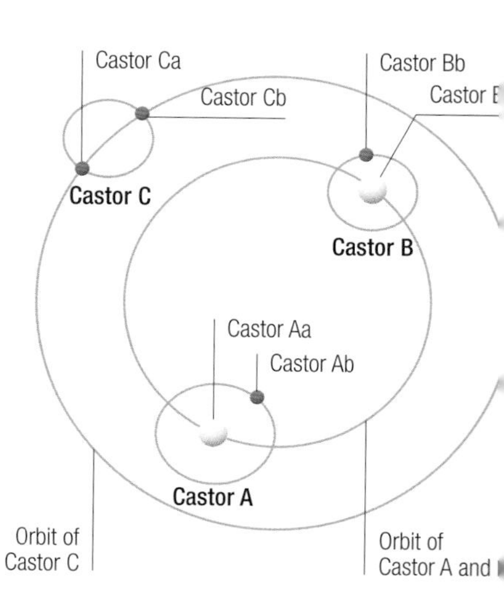

△ **Castor multi-star system**
Through a small telescope, Castor appears as a double star, Castor A and B, with components that orbit one another every 460 years. Each of these stars is itself a binary. To complicate the picture further, Castor A and B also have a faint red dwarf companion, known as Castor C, which is an eclipsing binary, completing a remarkable system of six stars all linked by gravity.

Stars labelled **Alpha (α)** are **not always** the **brightest** in their constellations – an example is **Castor**, which is **dimmer than Pollux**

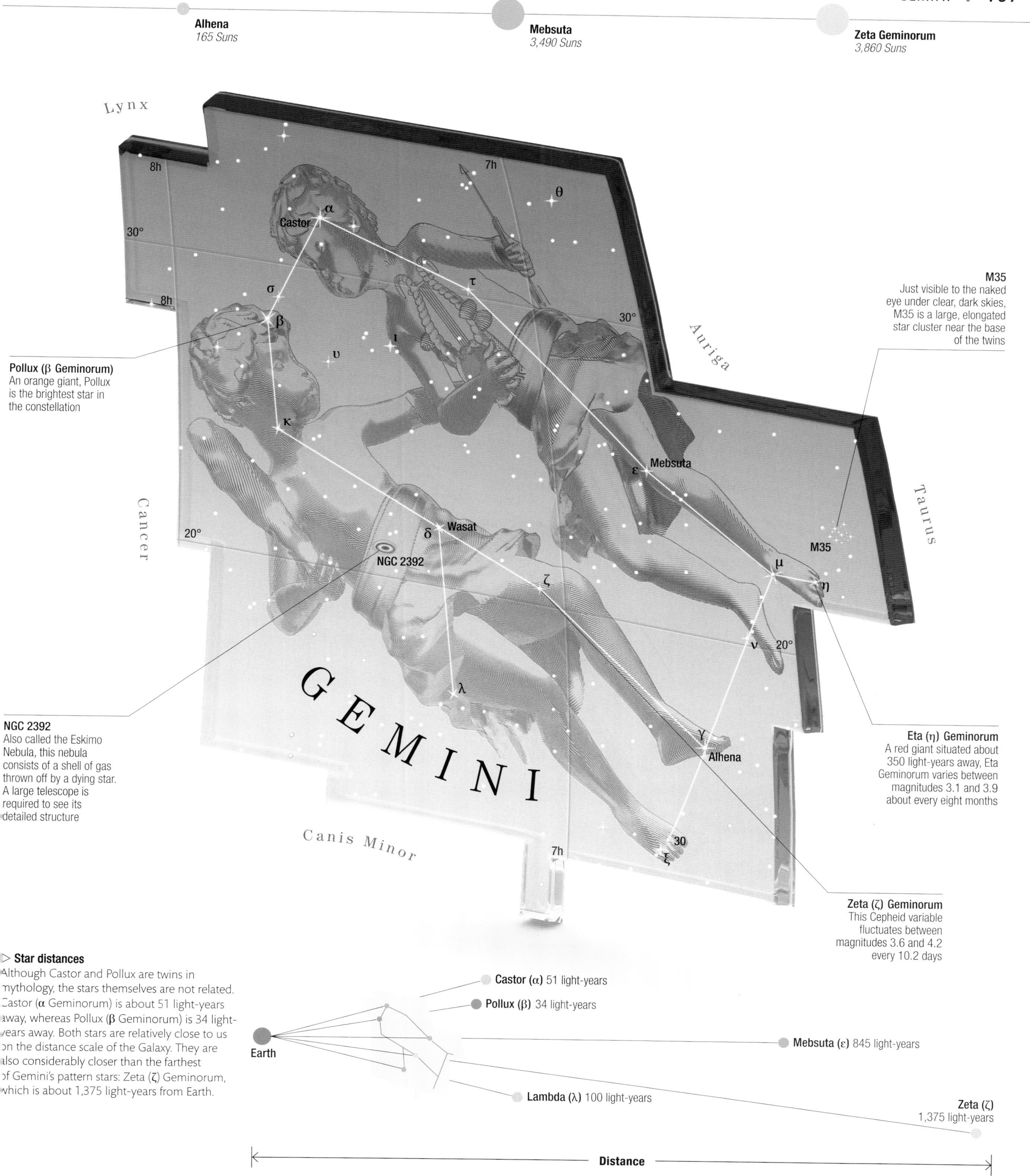

Alhena
165 Suns
Mebsuta
3,490 Suns
Zeta Geminorum
3,860 Suns
Lynx
8h
7h
θ
α
Castor
30°
σ
τ
8h
β
30°
Auriga
ι
υ
M35
Just visible to the naked eye under clear, dark skies, M35 is a large, elongated star cluster near the base of the twins
Pollux (β Geminorum)
An orange giant, Pollux is the brightest star in the constellation
κ
ε
Mebsuta
Cancer
Taurus
δ
Wasat
20°
M35
NGC 2392
μ
ζ
η
ν
20°
λ
GEMINI
γ
Alhena
NGC 2392
Also called the Eskimo Nebula, this nebula consists of a shell of gas thrown off by a dying star. A large telescope is required to see its detailed structure
Eta (η) Geminorum
A red giant situated about 350 light-years away, Eta Geminorum varies between magnitudes 3.1 and 3.9 about every eight months
Canis Minor
30
7h
ξ
Zeta (ζ) Geminorum
This Cepheid variable fluctuates between magnitudes 3.6 and 4.2 every 10.2 days
Castor (α) 51 light-years
Pollux (β) 34 light-years
Earth
Mebsuta (ε) 845 light-years
Lambda (λ) 100 light-years
Zeta (ζ)
1,375 light-years
Distance

▷ **Star distances**

Although Castor and Pollux are twins in mythology, the stars themselves are not related. Castor (α Geminorum) is about 51 light-years away, whereas Pollux (β Geminorum) is 34 light-years away. Both stars are relatively close to us on the distance scale of the Galaxy. They are also considerably closer than the farthest of Gemini's pattern stars: Zeta (ζ) Geminorum, which is about 1,375 light-years from Earth.

CANCER
THE CRAB

ALTHOUGH THE FAINTEST OF THE ZODIAC CONSTELLATIONS, CANCER IS EASY TO FIND BETWEEN THE BRIGHTER STARS OF LEO AND GEMINI.

Cancer represents the crab that attacked the hero Hercules as he fought the multi-headed monster Hydra. A square of stars form the body of the "crab", and the stars Alpha and Iota mark its claws. Alpha takes its name Acubens from the Arabic for claw. The names of the stars Assellus Borealis and Assellus Australis, the northern and southern ass, come from a different legend featuring a donkey. These stars lie either side of the cluster M44 (also known as the Beehive Cluster), which is said to represent the donkey's manger. None of Cancer's stars is particularly bright, and the region is relatively barren. Iota is a yellow giant with a companion detectable with binoculars, and Acubens and Zeta are multiple stars.

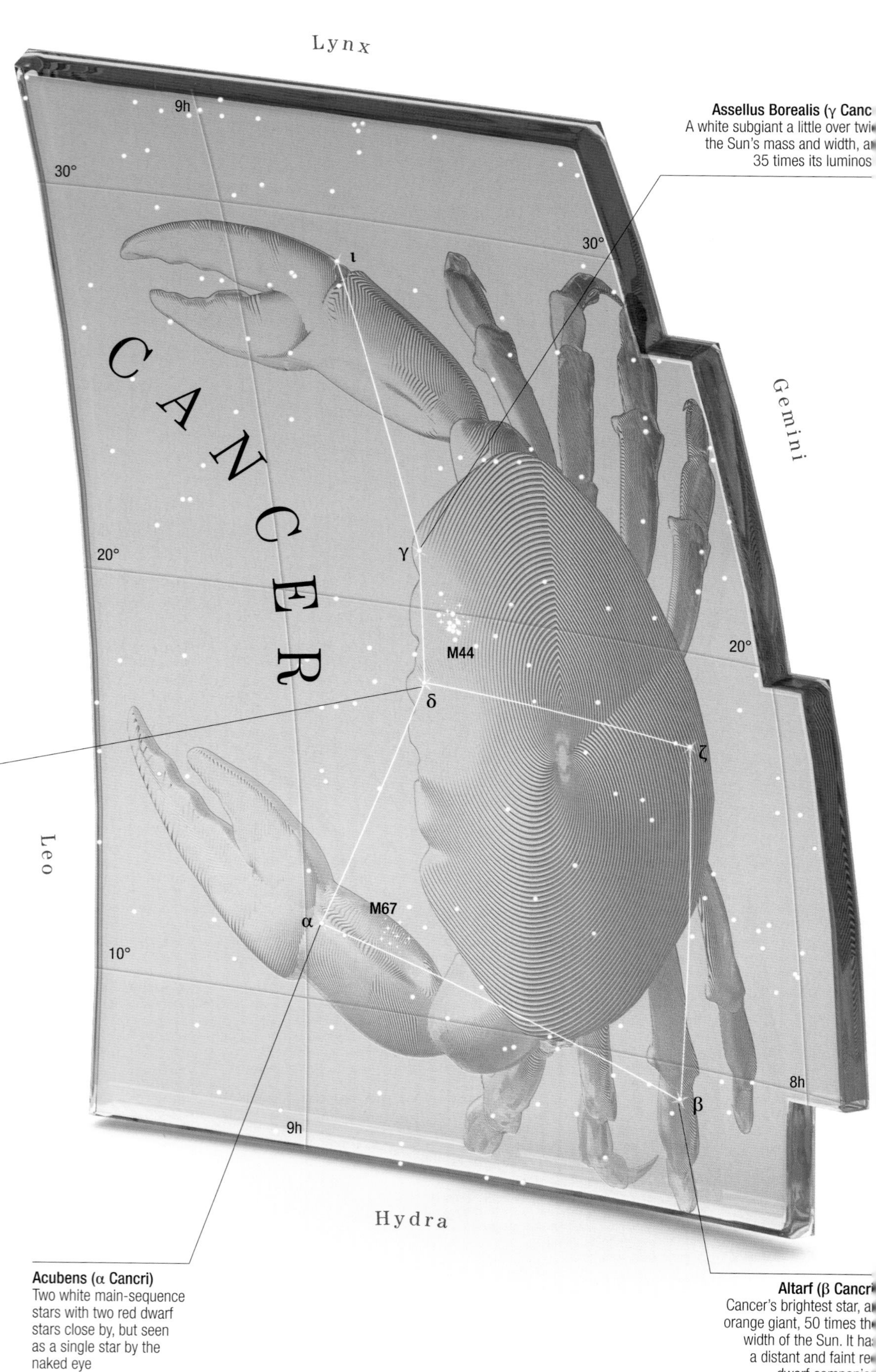

Assellus Borealis (γ Canc
A white subgiant a little over twi
the Sun's mass and width, a
35 times its luminos

Assellus Australis (δ Cancri)
An orange giant of magnitude 3.9, it is 11 times the Sun's width and more than 50 times as luminous

Acubens (α Cancri)
Two white main-sequence stars with two red dwarf stars close by, but seen as a single star by the naked eye

Altarf (β Cancr
Cancer's brightest star, a
orange giant, 50 times th
width of the Sun. It ha
a distant and faint re
dwarf compani

△ **M44**
Only 600 million years old, M44 (also known as the Beehive Cluster or Praesepe) is a relatively loose group of young stars spread across an area of sky three times the size of the Full Moon. It is visible to the naked eye as a starry swarm, and binoculars reveal individual stars of 6th magnitude or fainter. At 590 light-years away, it is one of the closest open clusters to Earth.

KEY DATA

Size ranking 31

Brightest stars Altarf (β) 3.5, Assellus Australis (δ) 3.9

Genitive Cancri

Abbreviation CnC

Highest in sky at 10pm February–March

Fully visible 90°N–57°S

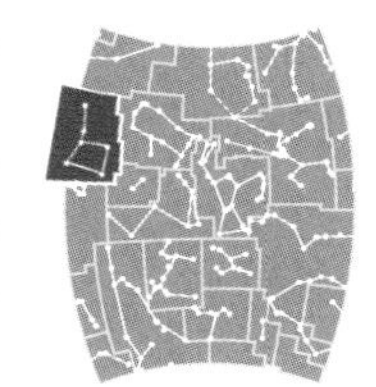

CHART 6

MAIN STARS

Acubens Alpha (α) Cancri
White main-sequence star and multiple star
4.3 ⟷ 188 light-years

Altarf Beta (β) Cancri
Orange giant and binary star
3.5 ⟷ 303 light-years

Assellus Borealis Gamma (γ) Cancri.
White subgiant
4.7 ⟷ 181 light-years

Assellus Australis Delta (δ) Cancri
Orange giant
3.9 ⟷ 131 light-years

DEEP-SKY OBJECTS

M44 (Beehive Cluster, Praesepe)
Open cluster

M67
Open cluster

M67
he open star cluster M67 is about 5 billion years ld, making it one of the oldest open clusters nown. It consists of more than 100 stars with the ame chemical composition as the Sun and red iants. Smaller, denser, and 2,600 light-years away, is more distant than M44 (the Beehive Cluster) ut also covers the width of a Full Moon. It can be een using binoculars.

CANIS MINOR

THE LITTLE DOG

ONE OF THE ORIGINAL CONSTELLATIONS DESCRIBED BY THE ASTRONOMERS OF ANCIENT GREECE, CANIS MINOR IS SMALL BUT EASILY SPOTTED BECAUSE OF ITS BRILLIANT STAR PROCYON.

Canis Minor is the smaller of Orion's two hunting dogs. The little dog is drawn around the constellation's brightest stars Procyon and Gomeisa. Located almost on the celestial equator, the constellation has little of interest other than Procyon, the eighth brightest star in the night sky. Meaning "before the dog" in Greek, the star is so-named because in Mediterranean latitudes it rises shortly before the more brilliant Dog Star, Sirius (in Canis Major). Procyon and Sirius are about the same distance from Earth and their differing brightness therefore indicates a true difference in their luminosity. Procyon, like Sirius, is a binary with a white dwarf companion, Procyon B, visible with a very large telescope. Procyon also marks one corner of the Winter Triangle.

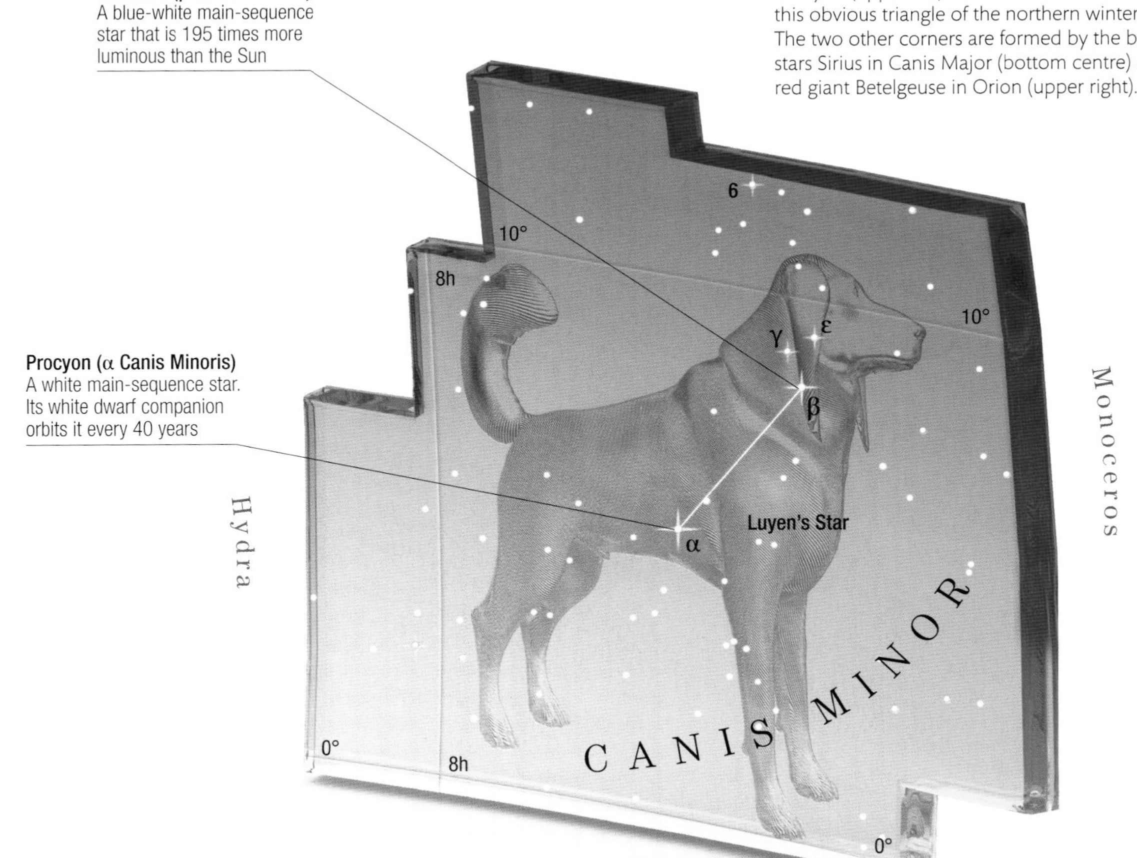

Gomeisa (β Canis Minoris)
A blue-white main-sequence star that is 195 times more luminous than the Sun
Procyon (α Canis Minoris)
A white main-sequence star. Its white dwarf companion orbits it every 40 years
Hydra
Monoceros
Luyen's Star
CANIS MINOR
γ
ε
β
α
6
10°
8h
0°

KEY DATA

Size ranking 71

Brightest stars Procyon (α) 0.4, Gomeisa (β) 2.9

Genitive Canis Minoris

Abbreviation CMi

Highest in sky at 10pm February

Fully visible 89°N–77°S

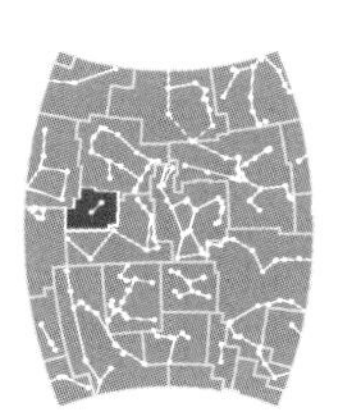

CHART 6

MAIN STARS

Procyon Alpha (α) Canis Minoris
White main-sequence star and binary star
0.4 ⟷ 11 light-years

Gomeisa Beta (β) Canis Minoris
Blue-white main-sequence star
2.9 ⟷ 162 light-years

△ **The Winter Triangle**
Procyon (upper left) marks one of the corners of this obvious triangle of the northern winter sky. The two other corners are formed by the brilliant stars Sirius in Canis Major (bottom centre) and the red giant Betelgeuse in Orion (upper right).

LUMINOSITIES

Alpha Monocerotis *48 Suns*

Delta Monocerotis *265 Suns*

Gamma Monocerotis *515 Suns*

MONOCEROS
THE UNICORN

A LARGE BUT NOT PROMINENT CONSTELLATION, MONOCEROS HAS NO BRIGHT INDIVIDUAL STARS. HOWEVER, IT DOES CONTAIN MANY NOTABLE MULTIPLE STARS AND DEEP-SKY OBJECTS, SUCH AS STAR CLUSTERS AND NEBULAE.

Monoceros is situated between Hydra and Orion, with Canis Major to the south and Canis Minor to the north. Beta Monocerotis is one of the finest triples in the sky. Its three 5th-magnitude stars can be separated with small telescopes. Delta Monocerotis is a wide, unrelated pair of stars, visible with binoculars. Epsilon Monocerotis, of 4th magnitude, has a fainter, unrelated companion visible through small telescopes.

Because it lies in the band of the Milky Way, Monoceros has many star clusters and nebulae. Among its binocular-friendly features are the open cluster M50 in the south of the constellation, and NGC 2264 in the north. Long-exposure images show a faint nebulosity around NGC 2264, including a dark dust lane, the Cone Nebula. Another notable deep-sky object is the Rosette Nebula, surrounding the elongated star cluster NGC 2244.

NGC 2264
This open cluster, about 2,500 light-years away, can be observed with binoculars. Through a small telescope it looks triangular in shape. Its brightest member is the 5th-magnitude star S Monocerotis

NGC 2244
At the core of the Rosette Nebula, about 5,500 light-years away, is this elongated star cluster, which is visible through binoculars. The Rosette Nebula itself (NGC 2237) is three to four times larger than NGC 2244

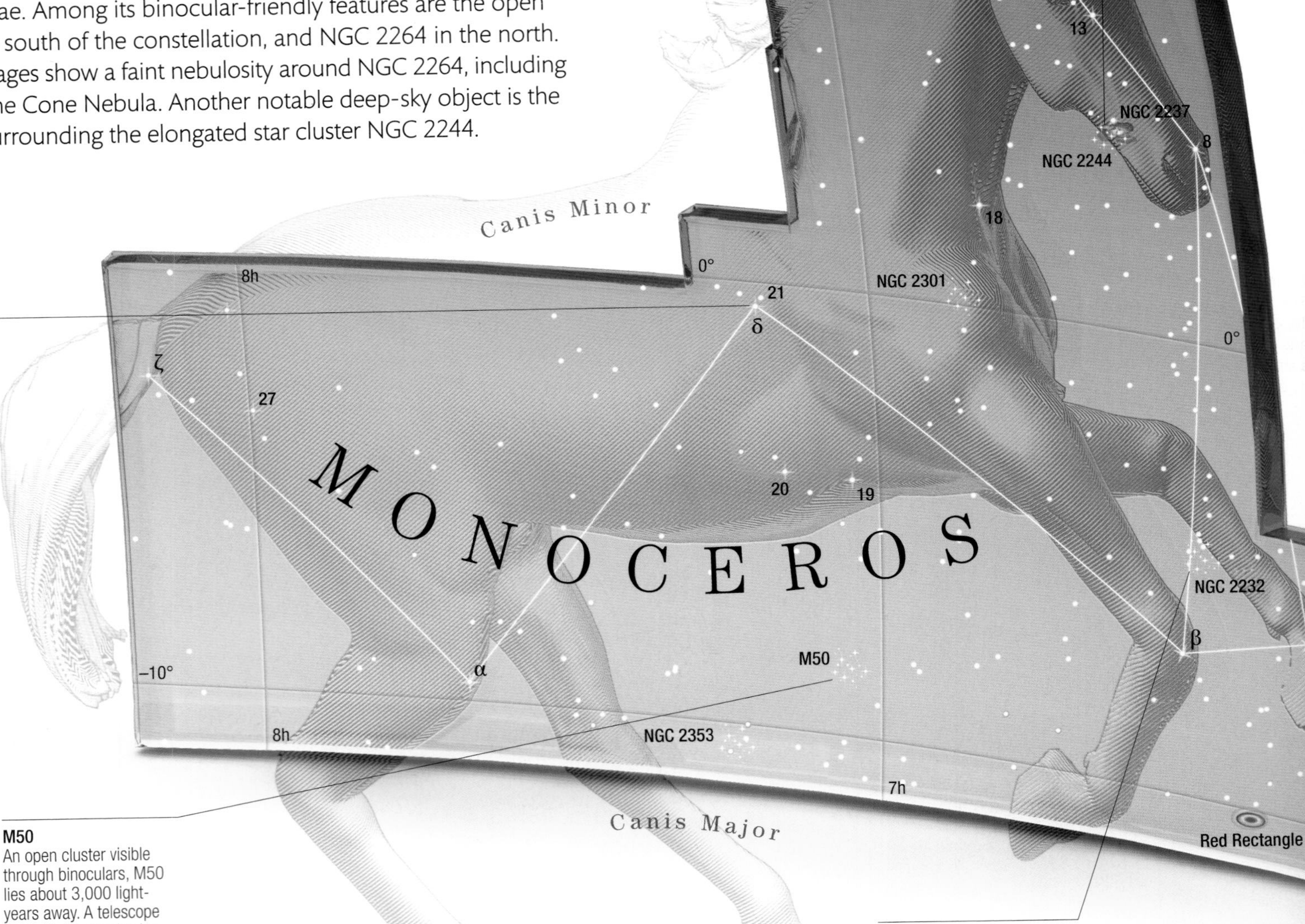

Delta (δ) Monocerotis
This 4th-magnitude star has an unrelated companion, called 21 Monocerotis. The companion star is closer to us than Delta Monocerotis and is visible with binoculars or even sharp eyesight

M50
An open cluster visible through binoculars, M50 lies about 3,000 light-years away. A telescope is needed to resolve its individual stars, which are of 8th magnitude and fainter

NGC 2232
Visible through binoculars, this scattered cluster is about 1,300 light-years from Earth. Its brightest member is the 5th-magnitude star 10 Monocerotis

Beta Monocerotis
1,175 Suns

Zeta Monocerotis
1,655 Suns

13 Monocerotis
142,000,000 Suns

△ **NGC 2237**
The flowery pink gases of NGC 2237 (the Rosette Nebula) surround the star cluster NGC 2244. The stars in the cluster have been born from the nebula and now light up the surrounding gas. The cluster can easily be seen through binoculars, but the faint nebula, larger in apparent diameter than the Full Moon, shows up well only on photographs with large telescopes, as here.

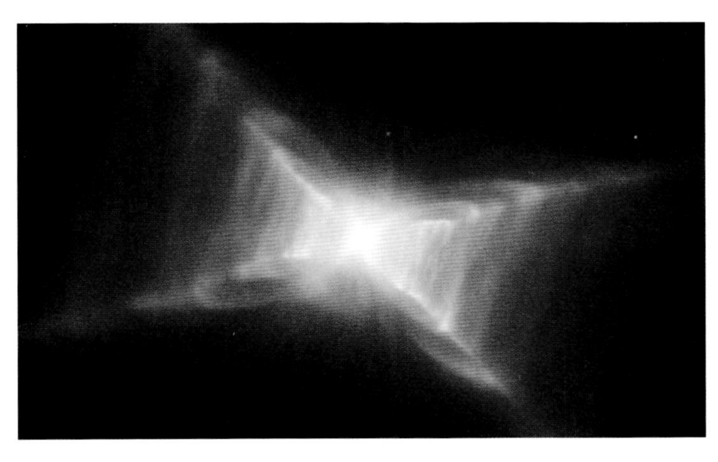

▷ **Red Rectangle**
The Red Rectangle, seen here through the Hubble Space Telescope, is an unusual planetary nebula in which gas and dust flowing out from the central star has produced a striking X-shaped structure.

KEY DATA

Size ranking 35

Brightest stars Alpha (α) 3.9, Gamma (γ) 4.0

Genitive Monocerotis

Abbreviation Mon

Highest in sky at 10pm January–February

Fully visible 78°N–78°S

CHART 6

MAIN STARS

Alpha (α) Monocerotis
Yellow giant
3.9 ⟷ 148 light-years

Beta (β) Monocerotis
Triple star; all three are blue-white main sequence stars
3.7 ⟷ 680 light-years

Gamma (γ) Monocerotis
Orange giant
4.0 ⟷ 500 light-years

Delta (δ) Monocerotis
Blue-white main sequence star
4.2 ⟷ 385 light-years

DEEP-SKY OBJECTS

M50
Open cluster of about 80 stars

NGC 2237 (Rosette Nebula)
Nebulosity around cluster NGC 2244

NGC 2264
Open cluster of about 40 stars

Red Rectangle
Planetary nebula about 2,300 light-years away

Monoceros was **introduced in 1612** by the Dutch cartographer **Petrus Plancius**

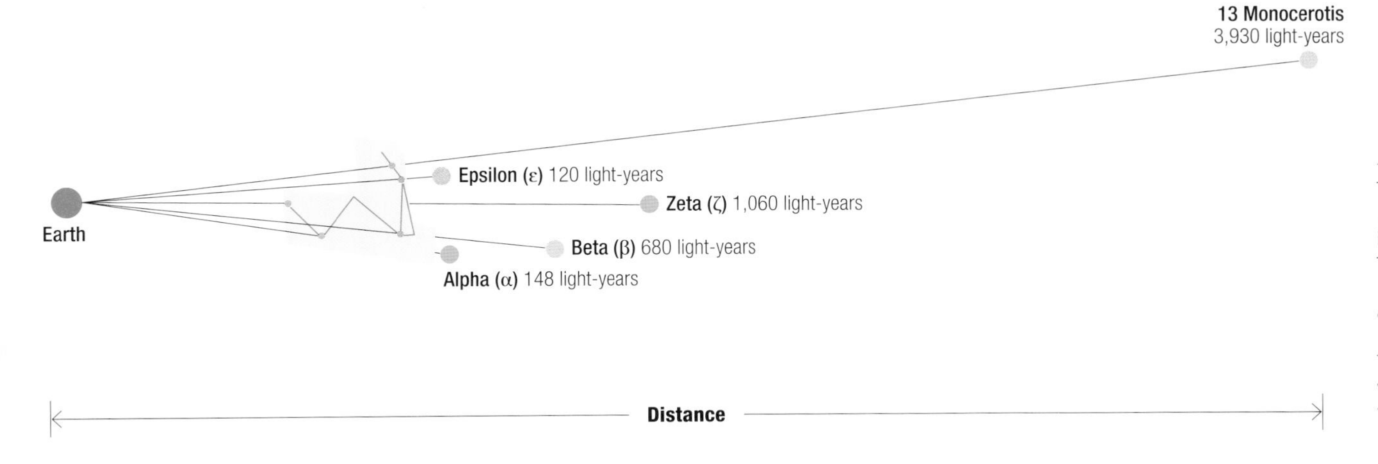

◁ **Star distances**
The main pattern stars of Monoceros vary considerably in their distances from Earth. The nearest is Epsilon (ε) Monocerotis (also sometimes called 8 Monocerotis), which is 120 light-years from Earth. The farthest is 13 Monocerotis, which is more than 3,900 light-years away.

LUMINOSITIES

54 Hydrae *7 Suns* | R Hydrae *37 Suns* | Pi Hydrae *42 Suns*

HYDRA
THE WATER SNAKE

THE LARGEST OF ALL CONSTELLATIONS, HYDRA REPRESENTS A MONSTER SLAIN BY THE GREEK HERO HERCULES. THE SIX STARS MARKING THE SERPENT'S HEAD ARE THE EASIEST TO PICK OUT.

Although the monster confronted by Hercules in the Ancient Greek myths had nine heads, Hydra is depicted in the sky with just a single head. The stars depicting the head are in the northern celestial hemisphere, to the south of Cancer, while most of the body and tail are in the southern celestial hemisphere.

The constellation's brightest star, Alphard, marks the monster's heart. It sits in an otherwise empty looking patch of sky and derives its common name from the Arabic for "the solitary one". The two main objects to look out for in Hydra are the spiral galaxy M83 and the planetary nebula NGC 3242.

The **largest** of the 88 constellations, Hydra stretches more than a **quarter** of the way around the sky

KEY DATA

Size ranking 1

Brightest stars Alphard (α) 2.0, Gamma (γ) 3.0

Genitive Hydrae

Abbreviation Hya

Highest in sky at 10pm February–June

Fully visible 54°N–83°S

CHART 5

MAIN STARS

Alphard Alpha (α) Hydrae
Orange giant
2.0 ⟷ 180 light-years

Gamma (γ) Hydrae
Yellow giant
3.0 ⟷ 145 light-years

Epsilon (γ) Hydrae
Quad star system
3.4 ⟷ 130 light-years

R Hydrae
Mira-type variable
5.0 ⟷ 405 light-years

DEEP-SKY OBJECTS

M68
Globular cluster

M83 (Southern Pinwheel)
Spiral galaxy

NGC 3242 (Ghost of Jupiter)
Planetary nebula, sometimes also called the Eye Nebula

△ **M83**
Also known as the Southern Pinwheel, this spiral galaxy is similar in structure to the Milky Way but is a far more active area of star formation and death. The blue and magenta areas are sites of star birth. Also visible in this image are many supernova remnants, as well as thousands of star clusters and hundreds of thousands of individual stars.

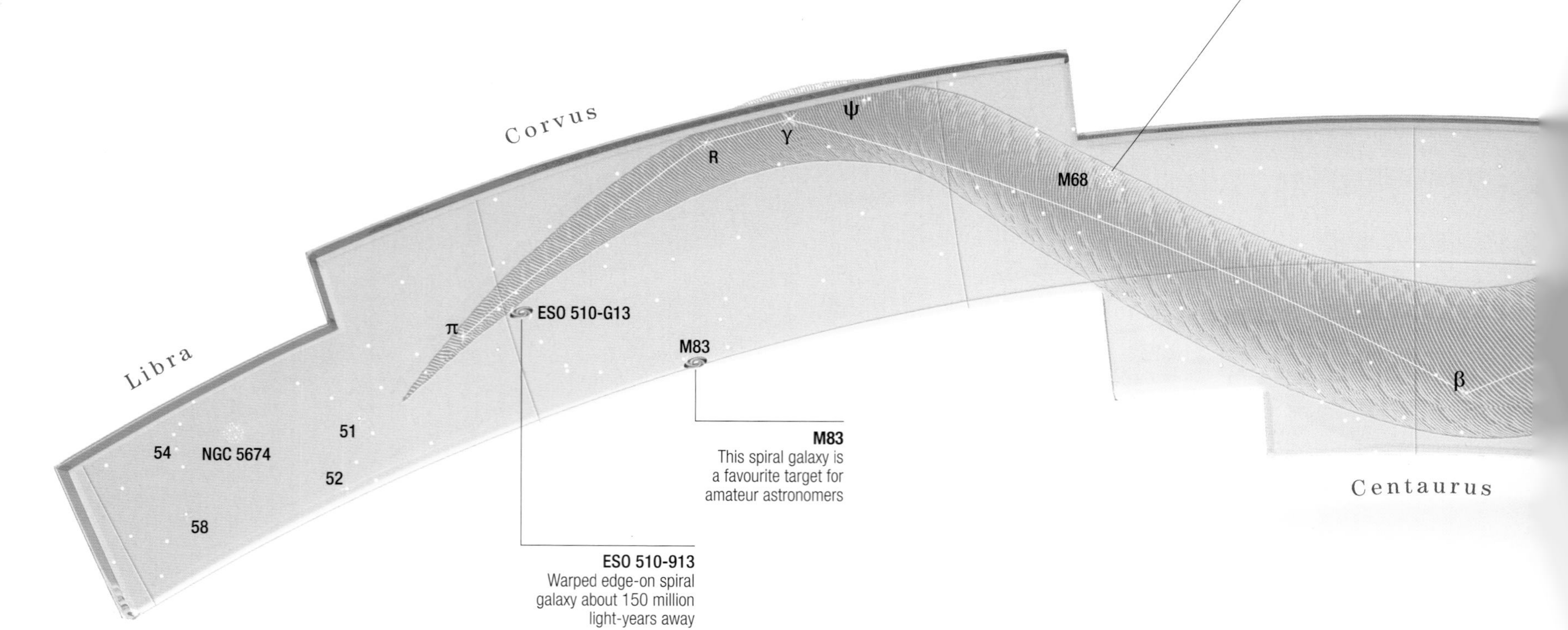

M68
The globular cluster M68 looks like a blurred star when seen with binoculars or a small telescope

M83
This spiral galaxy is a favourite target for amateur astronomers

ESO 510-913
Warped edge-on spiral galaxy about 150 million light-years away

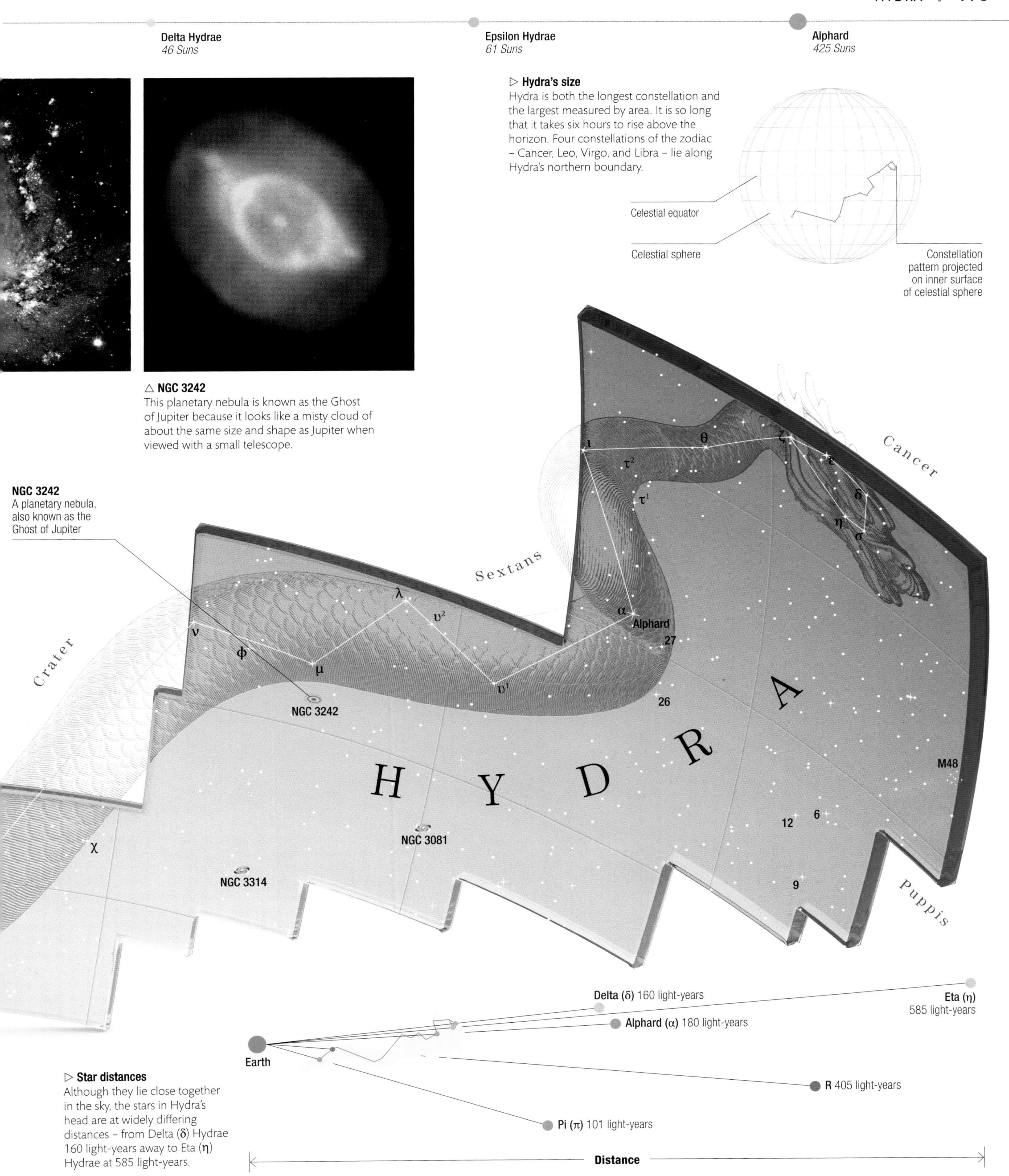

△ **NGC 3242**
This planetary nebula is known as the Ghost of Jupiter because it looks like a misty cloud of about the same size and shape as Jupiter when viewed with a small telescope.

▷ **Hydra's size**
Hydra is both the longest constellation and the largest measured by area. It is so long that it takes six hours to rise above the horizon. Four constellations of the zodiac – Cancer, Leo, Virgo, and Libra – lie along Hydra's northern boundary.

▷ **Star distances**
Although they lie close together in the sky, the stars in Hydra's head are at widely differing distances – from Delta (δ) Hydrae 160 light-years away to Eta (η) Hydrae at 585 light-years.

SEXTANS THE SEXTANT

THIS FAINT CONSTELLATION LIES DIRECTLY ON THE CELESTIAL EQUATOR. IT CAN BE FOUND CLOSE TO THE STAR REGULUS IN LEO.

Just three stars define the Sextant, a constellation identified by the Polish astronomer Johannes Hevelius in 1687. It represents an instrument used on board ship for position-finding. Sextans' stars are relatively dim, its brightest being only magnitude 4.5, and none are named. Its galaxies are best viewed through large telescopes. NGC 3115, called the Spindle Galaxy because it appears spindle-shaped in the sky, is magnitude 8.5 and just visible through binoculars in good conditions. Two unrelated stars of 6th magnitude, 17 and 18 Sextantis are close by and also only visible through binoculars.

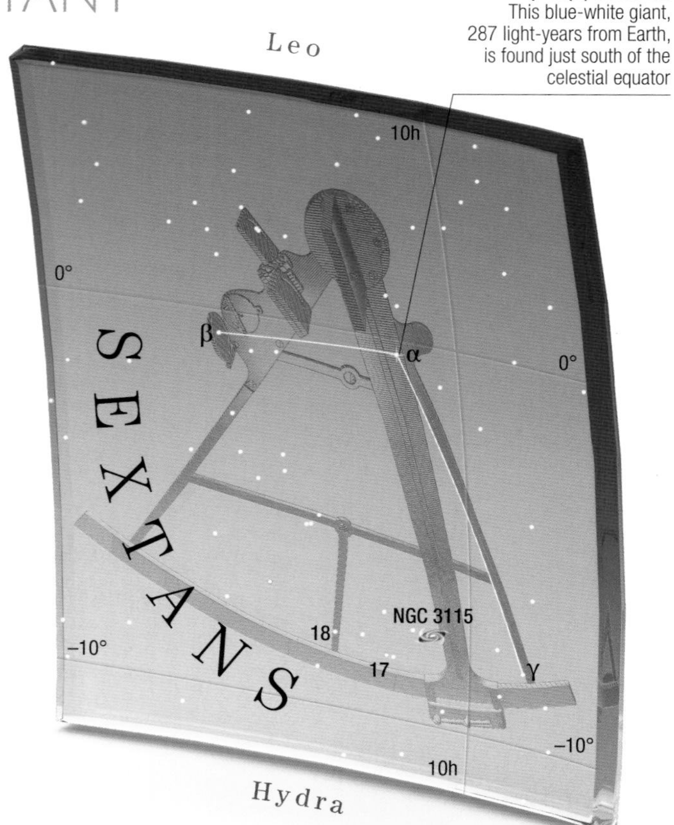

KEY DATA

Size ranking 47

Brightest stars Alpha (α) 4.5, Gamma (γ) 5.1

Genitive Sextantis

Abbreviation Sex

Highest in sky at 10pm March–April

Fully visible 78°N–83°S

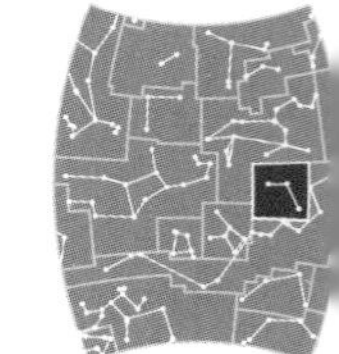
CHART 5

△ **NGC 3115**
This huge lenticular galaxy is seen edge-on from Earth. Its central bulge of stars is clearly visible. A supermassive black hole is hidden from view, deep inside. Also known as the Spindle Galaxy, the galaxy is about 30 million light-years away. It is not to be confused with the Spindle Galaxy in Draco.

CORVUS THE CROW

THIS CONSTELLATION'S SHAPE IS DEFINED BY ITS FOUR BRIGHTEST STARS, WHICH FORM THE BODY OF THE CROW.

Corvus is the sacred bird of the Greek god Apollo. Its story is linked with that of neighbouring Crater (the cup) and Hydra (the water snake). Dispatched by Apollo to collect water in a cup, the crow returned with neither but with a water snake instead. The constellation is best found by looking south-west of the star Spica in Virgo. One corner of Corvus's rectangle shape is marked by Delta Corvi. This double star consists of a bright 3rd-magnitude blue star orbited by a dimmer star. Corvus contains the Antennae Galaxies, one of the nearest and youngest pairs of colliding galaxies.

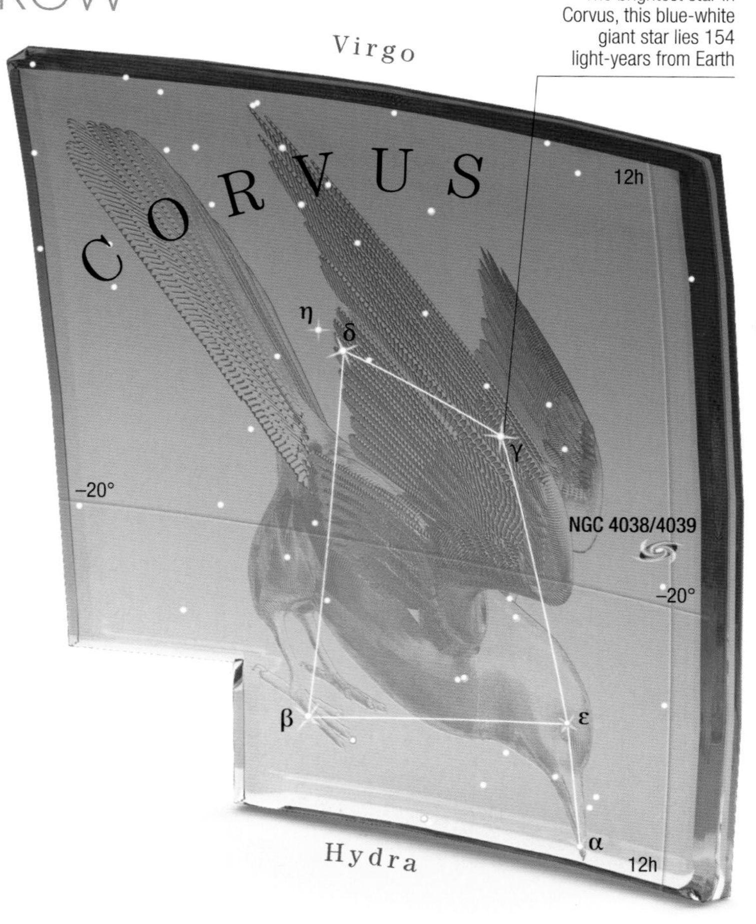

KEY DATA

Size ranking 70

Brightest stars Gienah (γ) 2.6, Beta (β) 2.6

Genitive Corvi

Abbreviation Crv

Highest in sky at 10pm April–May

Fully visible 65°N–90°S

CHART 5

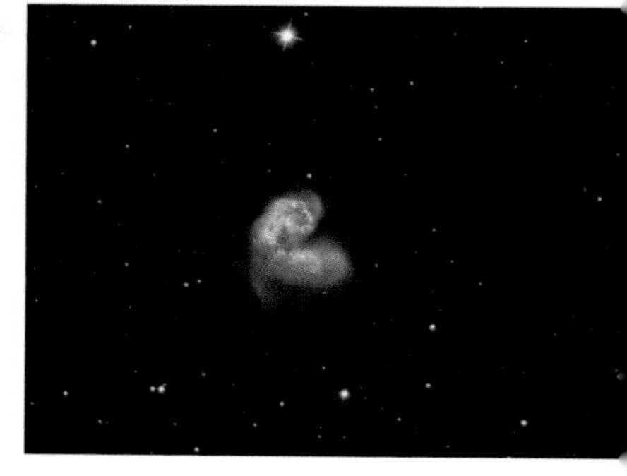

△ **NGC 4038 and NGC 4039**
The faint tails of stars, gas, and dust that extend from NGC 4038 and NGC 4039 give the galaxies their popular name of the Antennae. The tails formed when the galaxies started to interact a few hundred million years ago. The galaxies' collision led to the creation of huge star-forming regions surrounded by glowing hydrogen gas.

CRATER THE CUP

REPRESENTING THE DRINKING CUP OF THE GREEK GOD APOLLO, CRATER IS USUALLY DEPICTED AS A DOUBLE-HANDLED CHALICE. FAINT AND INDISTINCT, THIS CONSTELLATION MAY BE MORE EASILY LOCATED IF IMAGINED AS A LARGE BOW TIE IN THE SKY.

One of the original 48 constellations from Greek mythology, Crater's story links it with its neighbouring constellations Corvus (the crow) and Hydra (the water snake). Apollo is said to have placed the three together in the sky. He was angered that the crow not only was slow to return from a water-collecting trip, but then lied that the water snake had prevented him from collecting any water.

Crater has no brilliant stars, the brightest being Delta, at magnitude 3.6. A large telescope is needed to see any deep-sky objects, such as the barred spiral galaxy NGC 3981. This and galaxies NGC 3511 and NGC 3887 were discovered by British astronomer William Hershel in the mid-1780s. Much more distant, about 6 billion light-years from Earth, is the quasar RXJ 1131.

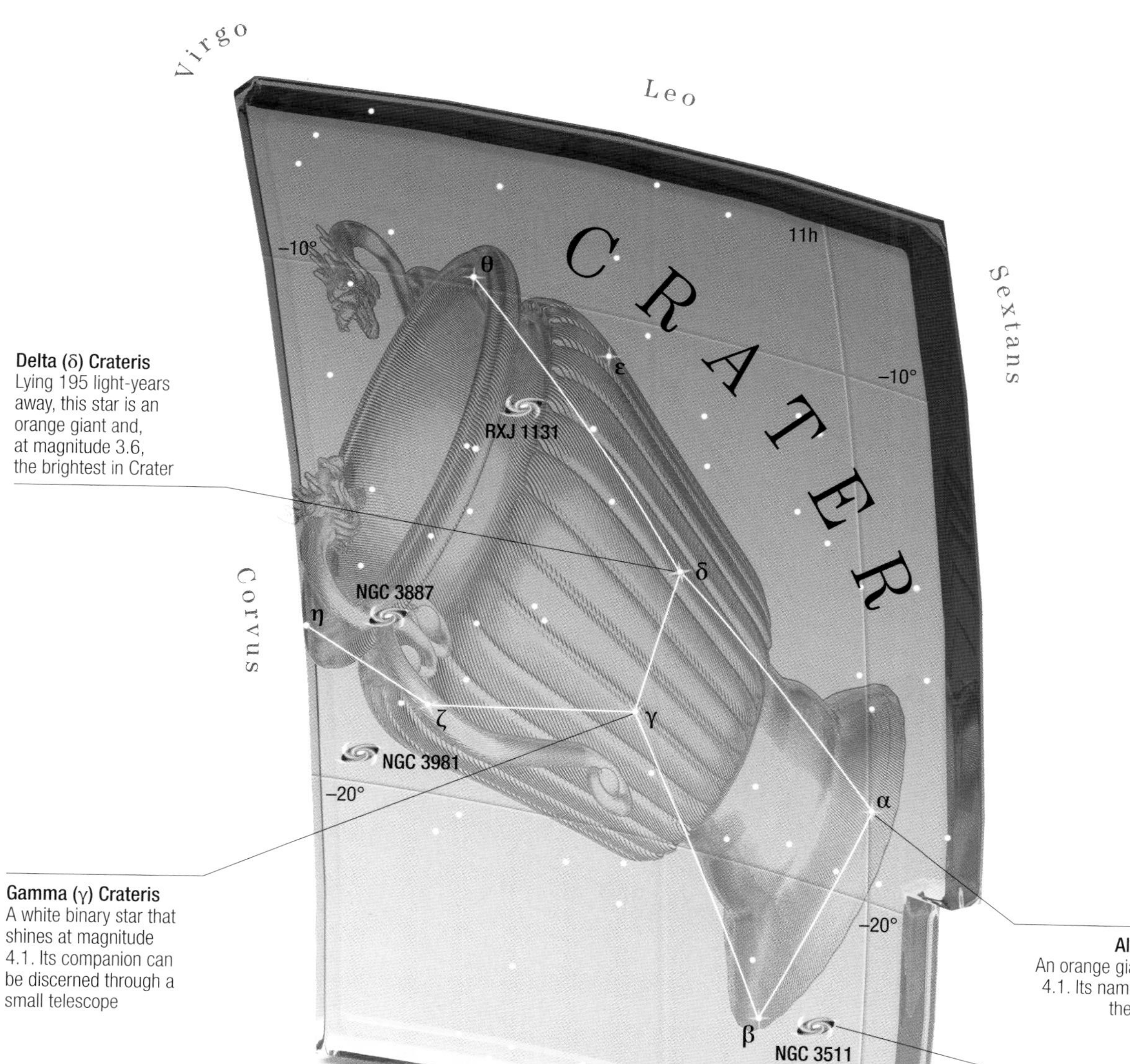

Delta (δ) Crateris
Lying 195 light-years away, this star is an orange giant and, at magnitude 3.6, the brightest in Crater

Gamma (γ) Crateris
A white binary star that shines at magnitude 4.1. Its companion can be discerned through a small telescope

Alkes (α Crateris)
An orange giant of magnitude 4.1. Its name is derived from the Arabic for "cup"

NGC 3511
A barred spiral galaxy tilted almost edge-on to Earth

KEY DATA

Size ranking 53
Brightest stars Delta (δ) 3.6, Alkes (α) 4.1
Genitive Crateris
Abbreviation Crt
Highest in sky at 10pm April
Fully visible 65°N–90°S

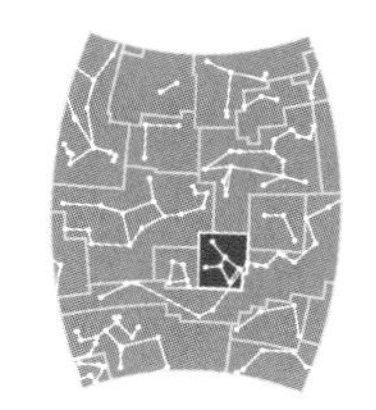
CHART 5

MAIN STARS

Alkes Alpha (α) Crateris
Orange giant
4.1 ⟷ 159 light-years

Delta (δ) Crateris
Orange giant
3.6 ⟷ 195 light-years

DEEP-SKY OBJECTS

NGC 3511
Barred spiral galaxy

NGC 3887
Barred spiral galaxy

NGC 3981
Barred spiral galaxy

RXJ 1131
Quasar powered by a supermassive black hole

△ **RXJ 1131**
The four pink dots in this image are the quasar RXJ 1131. The multiple images are the result of the quasar's light being bent by an elliptical galaxy. That galaxy, seen in the centre, is on the same line of sight as RXJ 1131 but is much closer to us.

LUMINOSITIES

Rigil Kentaurus *1.5 Suns*

Theta Centauri *42 Suns*

Gamma Centauri *183 Suns*

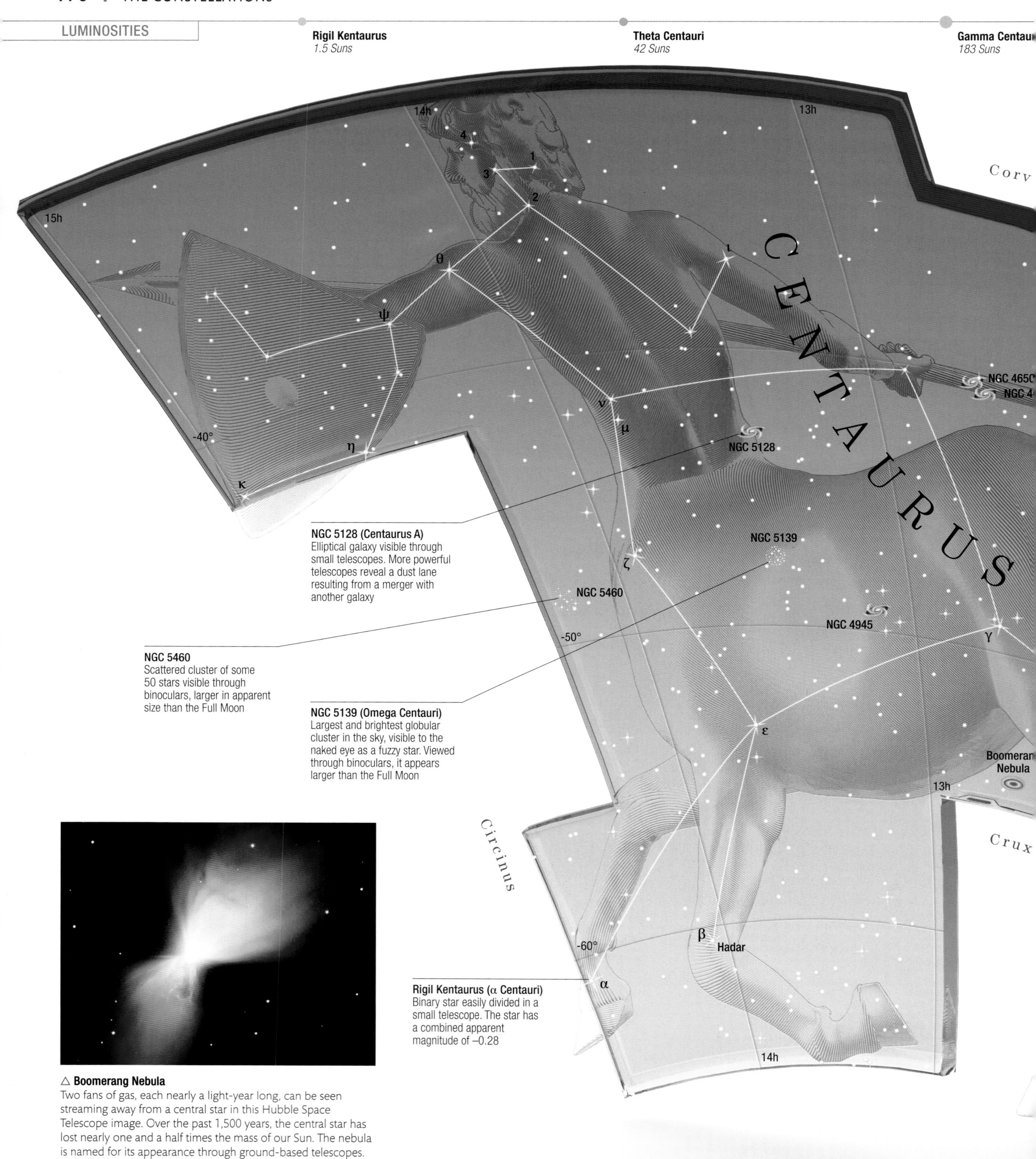

CENTAURUS
Corv
Crux
Circinus
15h
14h
13h
-40°
-50°
-60°
θ
ψ
η
κ
ι
ν
μ
ζ
γ
ε
β
α
1
2
3
4
Hadar
NGC 5128
NGC 5139
NGC 5460
NGC 4945
NGC 465
Boomerang Nebula
NGC 5128 (Centaurus A)
Elliptical galaxy visible through small telescopes. More powerful telescopes reveal a dust lane resulting from a merger with another galaxy
NGC 5460
Scattered cluster of some 50 stars visible through binoculars, larger in apparent size than the Full Moon
NGC 5139 (Omega Centauri)
Largest and brightest globular cluster in the sky, visible to the naked eye as a fuzzy star. Viewed through binoculars, it appears larger than the Full Moon
Rigil Kentaurus (α Centauri)
Binary star easily divided in a small telescope. The star has a combined apparent magnitude of –0.28

△ **Boomerang Nebula**
Two fans of gas, each nearly a light-year long, can be seen streaming away from a central star in this Hubble Space Telescope image. Over the past 1,500 years, the central star has lost nearly one and a half times the mass of our Sun. The nebula is named for its appearance through ground-based telescopes.

Eta Centauri
895 Suns

Epsilon Centauri
1,815 Suns

Hadar
7,170 Suns

CENTAURUS
THE CENTAUR

A PROMINENT SOUTHERN CONSTELLATION THAT CONTAINS THE CLOSEST STAR TO THE SUN, AS WELL AS THE BRIGHTEST GLOBULAR CLUSTER VISIBLE FROM EARTH.

Centaurus is one of the 48 constellations known to the ancient Greeks. It represents Chiron, a wise centaur who taught the gods and mythical heroes of ancient Greece in his cave on Mount Pelion.

Rigil Kentaurus (Alpha Centauri) appears to the naked eye as the third-brightest star in the sky, outshone only by Sirius and Canopus. A telescope splits it into a pair of golden-yellow stars that form a true binary, orbiting each other every 80 years. It is the closest star to the Sun visible to the naked eye. But there is also a third member of the system, only visible with a telescope – a red dwarf called Proxima Centauri, over one-tenth of a light-year closer to the Sun than the other two, making it the closest star of all. In the heart of the constellation lies NGC 5139 (Omega Centauri), a globular cluster so bright that it was at first catalogued as a star. NGC 5128 (Centaurus A) to its north is thought to be the result of a merger between an elliptical and a spiral galaxy.

NGC 3918
Planetary nebula visible through small telescopes as a rounded blue disk, hence its popular name: Blue Planetary Nebula

▽ Star distances
Centaurus contains the closest star to the Sun, Proxima Centauri, at a distance of only 4.2 light-years. Epsilon, one of the stars of the greatest magnitude in the constellation, is 100 times farther away, and Omicron[1], the most distant star, is almost 1,400 times further away.

Earth
Mu (μ) 505 light-years
Zeta (ξ) 382 light-years
Epsilon (ε) 430 light-years
Proxima Centauri 4.2 light years
Omicron[1] (o[1]) 5,720 light-years
Distance

KEY DATA

Size ranking 9
Brightest stars Rigil Kentaurus (α) -0.1, Hadar (β) 0.6
Genitive Centauri
Abbreviation Cen
Highest in sky at 10pm April–June
Fully visible 25°N–90°S

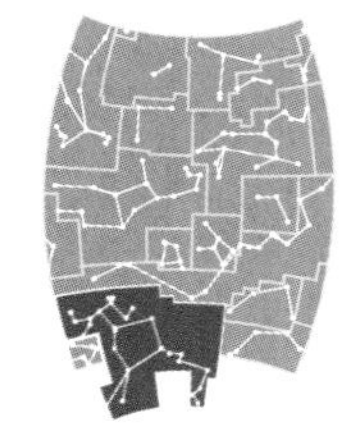
CHART 5

MAIN STARS

Rigil Kentaurus Alpha (α) Centauri
Pair of yellow and orange main-sequence stars
-0.28 ⟷ 4.4 light-years

Hadar Beta (β) Centauri
Blue-white giant
0.6 ⟷ 390 light-years

Gamma (γ) Centauri
Blue-white subgiant
2.2 ⟷ 130 light-years

Epsilon (ε) Centauri
Blue-white giant
2.3 ⟷ 430 light-years

Eta (η) Centauri
Blue-white main-sequence star
2.3 ⟷ 305 light-years

Theta (θ) Centauri
Orange giant
2.1 ⟷ 59 light-years

DEEP-SKY OBJECTS

NGC 5139 (Omega Centauri)
Globular cluster

Boomerang Nebula
Planetary nebula

NGC 3766
Open cluster

NGC 3918 (Blue Planetary Nebula)
Planetary nebula

NGC 5128 (Centaurus A)
Peculiar galaxy and radio source

LUMINOSITIES

Gamma Crucis
148 Suns

Epsilon Crucis
158 Suns

CRUX
THE SOUTHERN CROSS

ALTHOUGH IT IS THE SMALLEST CONSTELLATION OF ALL, CRUX IS ALSO ONE OF THE MOST DISTINCTIVE DUE TO ITS FOUR BRIGHT STARS. CROSSED BY THE MILKY WAY'S STAR-RICH PATH, IT HOSTS ONE OF THE GEMS OF THE SOUTHERN NIGHT SKY: THE JEWEL BOX CLUSTER.

Situated between the legs of Centaurus, Crux is the sky's most compact grouping of four bright stars. Its brilliant stars were known to the Ancient Greeks but only mapped as a separate constellation in the 16th century. Crux first appeared in its modern form on the celestial globe of cartographer Petrus Plancius in 1598. Initially called Crux Australis, the Southern Cross, it is now known simply as Crux. The southern end of its cross-shaped pattern is marked by the constellation's brightest star, Acrux. It, Mimosa, and Gacrux are in the top 25 brightest night-time stars. More distant than Crux's four main stars, at about 600 light-years away, is a wedge-shaped dark patch of sky named the Coalsack. This dark nebula of gas and dust is visible to the naked eye because it blocks out light from the dense Milky Way star fields behind it. Just north and about ten times more distant than the Coalsack is the Jewel Box Cluster (NGC 4755). This appears as a fuzzy star to the naked eye but binoculars reveal individual stars.

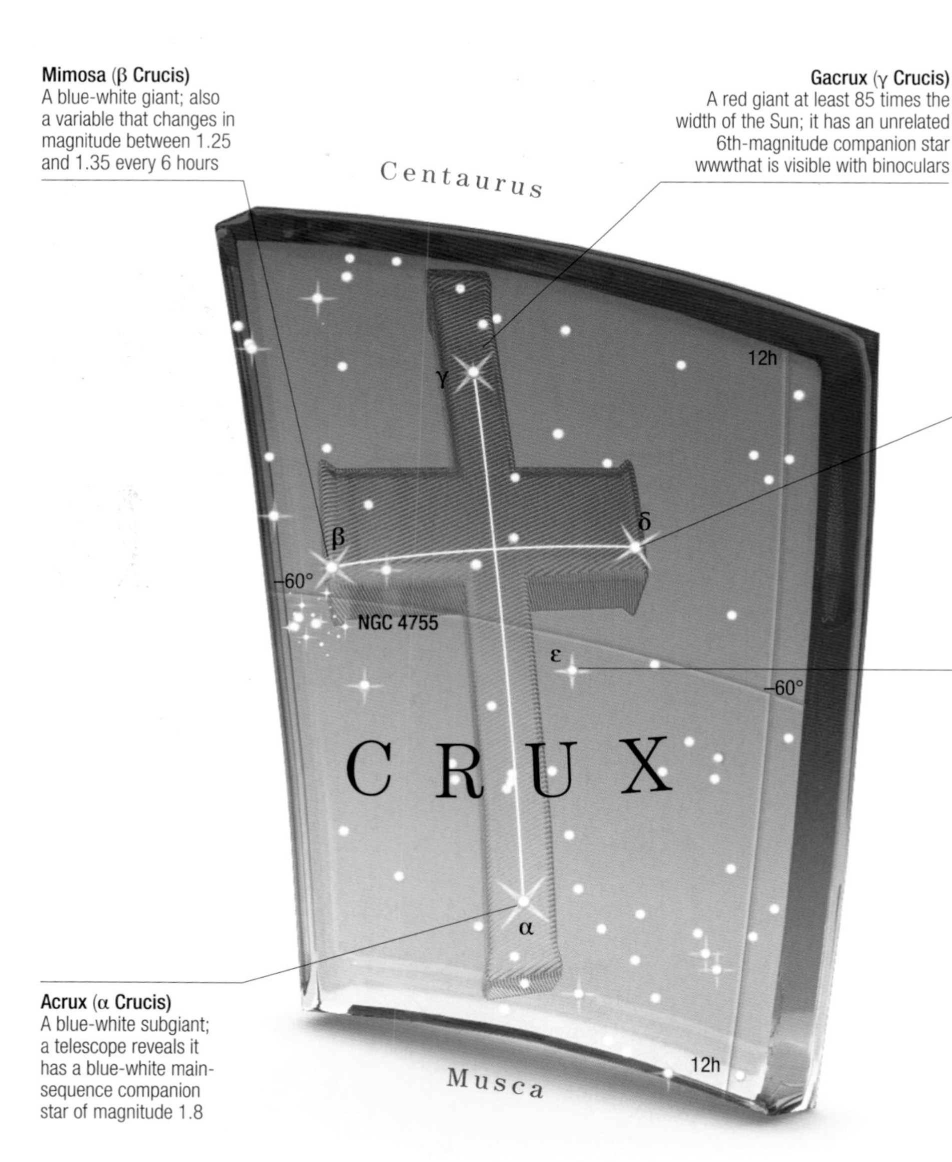

Mimosa (β Crucis)
A blue-white giant; also a variable that changes in magnitude between 1.25 and 1.35 every 6 hours

Gacrux (γ Crucis)
A red giant at least 85 times the width of the Sun; it has an unrelated 6th-magnitude companion star wwwthat is visible with binoculars

Delta (δ Crucis)
A blue-white star that is moving from the main-sequence to the red-giant stage of its life

Epsilon (ε) Crucis
An orange giant with about 1.4 times the Sun's mass and 33 times its width; it lies 230 light-years away and has a magnitude of 3.6

Acrux (α Crucis)
A blue-white subgiant; a telescope reveals it has a blue-white main-sequence companion star of magnitude 1.8

KEY DATA

Size ranking 88

Brightest stars Acrux (α) 0.8, Mimosa (β) 1.25–1.35

Genitive Crucis

Abbreviation Cru

Highest in sky at 10pm April–May

Fully visible 25°N–90°S

CHART 2

MAIN STARS

Acrux Alpha (α) Crucis
Blue-white subgiant; also a double star
0.8 ⟷ 322 light-years

Mimosa Beta (β) Crucis
Blue-white giant; also a variable star
1.25–1.35 ⟷ 278 light-years

Gacrux Gamma (γ) Crucis
Red giant
1.6 ⟷ 89 light-years

Delta (δ) Crucis
Blue-white subgiant
2.8 ⟷ 345 light-years

DEEP-SKY OBJECTS

NGC 4755 (Jewel Box Cluster)
Open cluster

Coalsack Nebula
Dark nebula

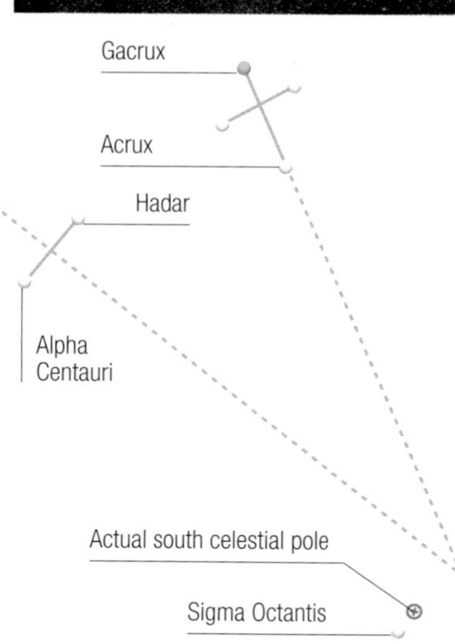

▷ Locating the south celestial pole
Crux has been used for centuries as a pointer to the south celestial pole. Its bright stars and the two brightest in Centaurus (Hadar and Alpha Centauri) are easy to spot, as can be seen in the photograph above. Extend southwards a line connecting Gacrux and Acrux, and an imaginary line bisecting Alpha Centauri and Hadar. The two cross just east of the south pole, the nearest star to which is Sigma Octantis in Octans.

Delta Crucis
750 Suns

Mimosa
2,010 Suns

Acrux
4180 Suns

NGC 4755
Popularly known as the Jewel Box Cluster because of its resemblance to a casket of sparkling jewels, this open star cluster contains several dozen blue-white giants and has a ruby-red supergiant close to its centre. The cluster is about 20 light-years wide and 6,400 light-years from Earth. At about 15 million years old, it is one of the youngest clusters known.

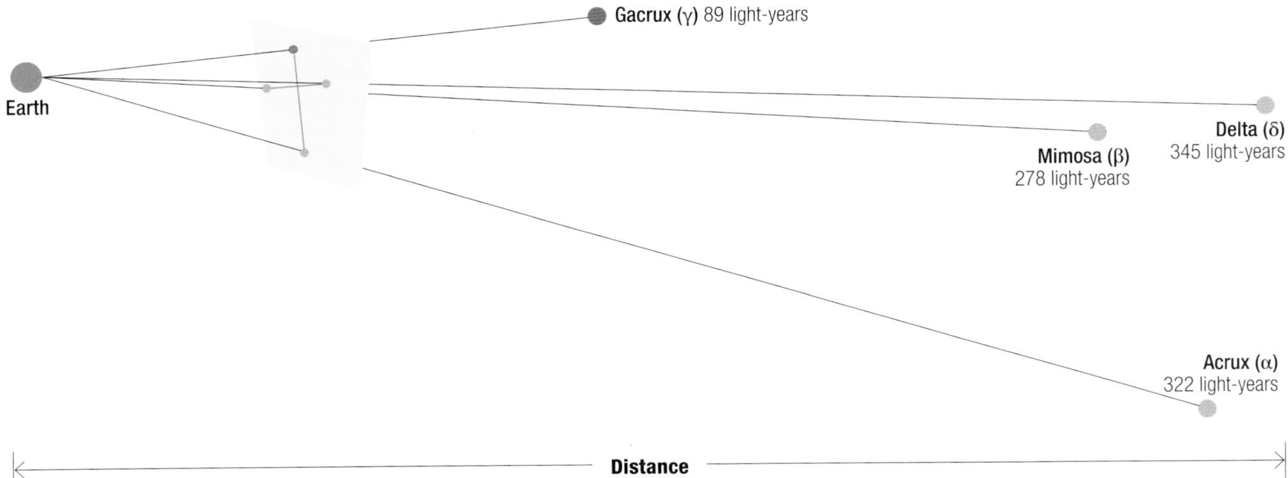

Star distances
hree of Crux's four pattern stars lie at milar distances from Earth: Mimosa at 78 light-years, Acrux at 322 light-years, nd Delta (γ) Crucis at 345 light-years. acrux, marking the northern end of the ross, is the constellation's nearest pattern ar. In fact, at only at 89 light-years away, it also one of the nearest known red giants.

LUPUS
THE WOLF

THIS CONSTELLATION LIES ON THE EDGE OF THE MILKY WAY, BETWEEN SCORPIUS AND CENTAURUS. THE PATTERN OF THE WOLF IS DIFFICULT TO RECOGNIZE, BUT IT CONTAINS STARS OF INTEREST.

Lupus was first outlined as an unspecified wild animal impaled on a pole carried by Centaurus, but is drawn as a separate wolf in modern times. Its two brightest stars, Alpha and Beta mark its hind legs, and the globular cluster NGC 5986 its head. Stargazers might find a wolf easier to picture if Alpha is imagined in its mouth, and Beta in the back of its neck. Small telescopes reveal that stars Kappa and Mu are double stars, and a larger telescope shows that Mu is really a triple star.

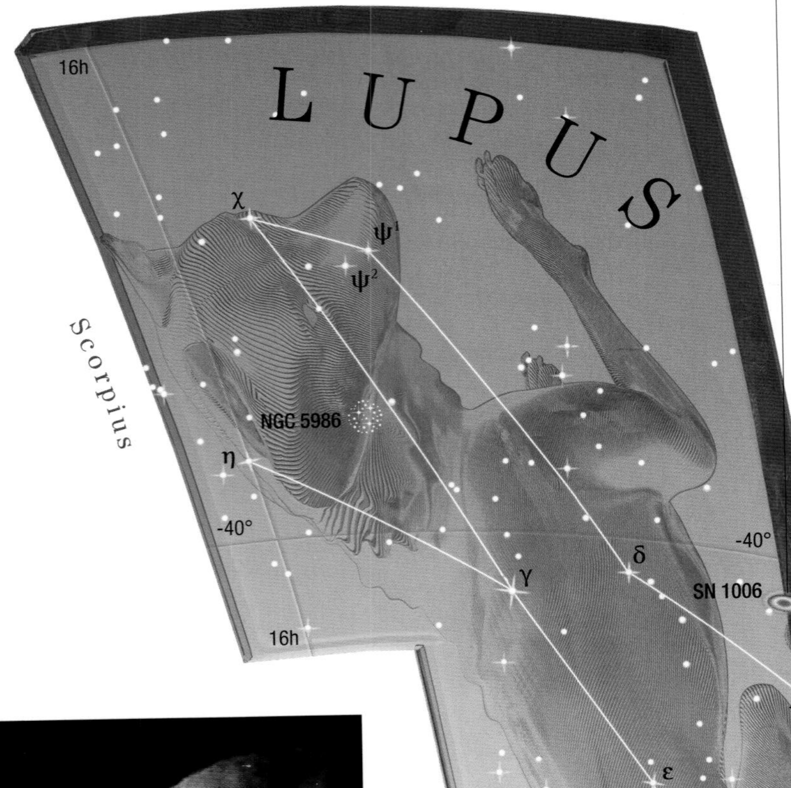

SN 1006
A supernova remnant about 60 light-years across and 7,000 light-years away. It is the remains of the brightest supernova in recorded history

NGC 5882
A planetary nebula with two shells of gas moving out from a central dying star; an elongated shell surrounded by an aspherical shell

Alpha Lupi
At magnitude 2.3, this is the brightest star in Lupus. About ten times the mass of the Sun, its luminosity varies slightly in a seven-hour cycle

△ **The Retina Nebula**
Viewed from Earth, this planetary nebula appears rectangular, but it is ring-shaped and we are viewing from the side. In its centre is a dying star that has pushed off the ring of gas and dust.

▽ **SN 1006**
Ten images, taken over eight days by the Chandra X-ray Space Telescope, are combined in this view of SN 1006. The supernova remnant was created when a white dwarf star exploded and blasted its material into space.

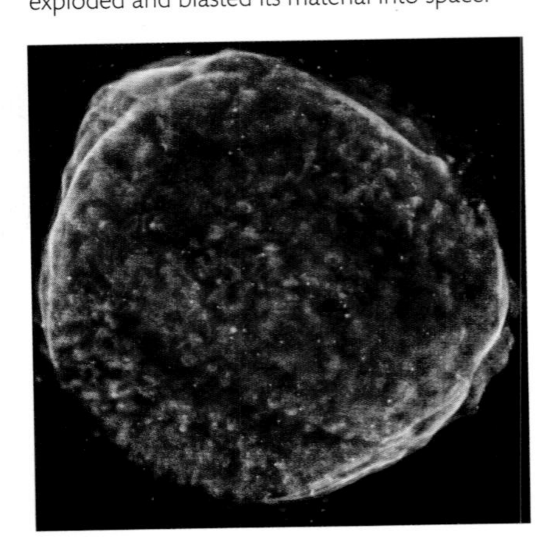

KEY DATA

Size ranking 46
Brightest stars Alpha (α) 2.3, Beta (β) 2.7
Genitive Lupi
Abbreviation Lup
Highest in sky at 10pm May–June
Fully visible 34°N–90°S

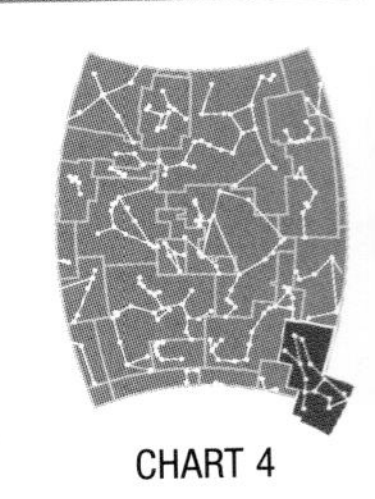
CHART 4

MAIN STARS

Alpha (α) **Lupi**
Blue giant star
2.3 ⟷ 464 light-years

Beta (β) **Lupi**
Blue giant star
2.7 ⟷ 383 light-years

Gamma (γ) **Lupi**
Blue giant star in a binary system
2.8 ⟷ 421 light-years

DEEP-SKY OBJECTS

NGC 5882
Asymmetrically shaped planetary nebula

NGC 5986
Gobular cluster

The Retina Nebula (IC 4406)
Planetary nebula

SN 1006
Supernova remnant

NORMA
THE SET SQUARE

THIS IS A SMALL CONSTELLATION THAT WAS ONLY CREATED IN THE EARLY 1750s, AND LATER REDUCED IN SIZE. IT LIES ON THE PATH OF THE MILKY WAY AND IS RICH IN STAR FIELDS.

When the stars in this region of sky were formed into a constellation by the Frenchman Nicholas Louis de Lacaille it was called Norma et Regula, the square and the ruler. Changes to constellation boundaries saw the stars marking the ruler being re-assigned to neighbouring Scorpius. One result of this is that present-day Norma has no stars designated Alpha or Beta. The set-square pattern is made of a right-angled trio of stars that is difficult to make out against the Milky Way.

△ **ESO 137-001**
False colour highlights two trails of cool gas behind galaxy ESO 137-001. The galaxy is heading towards the centre of the Norma Cluster, the closest massive galaxy cluster to the Milky Way. The trails could have formed when gas was stripped from the galaxy's spiral arms.

KEY DATA

Size ranking 74
Brightest stars Gamma² (γ²) 4.0, Epsilon (ε) 4.5
Genitive Normae
Abbreviation Nor
Highest in sky at 10pm June
Fully visible 29°N–90°S

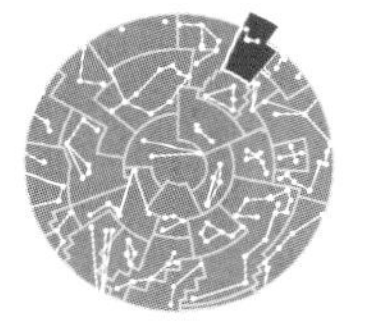

CHART 2

MAIN STARS

Gamma¹ (γ¹) Normae
Yellow supergiant, part of a double star with Gamma²
5.0 ⟷ 1,436 light-years

Gamma² (γ²) Normae
Yellow giant, part of a double star with Gamma¹
4.0 ⟷ 129 light-years

Epsilon (ε) Normae
Double star with components of 5th and 7th magnitude
4.5 ⟷ 400 light-years

Eta (η) Normae
Yellow giant star
4.7 ⟷ 218 light-years

DEEP-SKY OBJECTS

NGC 6067
Open cluster

NGC 6087
Open cluster

NGC 6167
Open cluster

Shapely 1
Planetary nebula, also known as the Fine Ring Nebula

Abell 3627
Cluster of galaxies, also known as the Norma Cluster

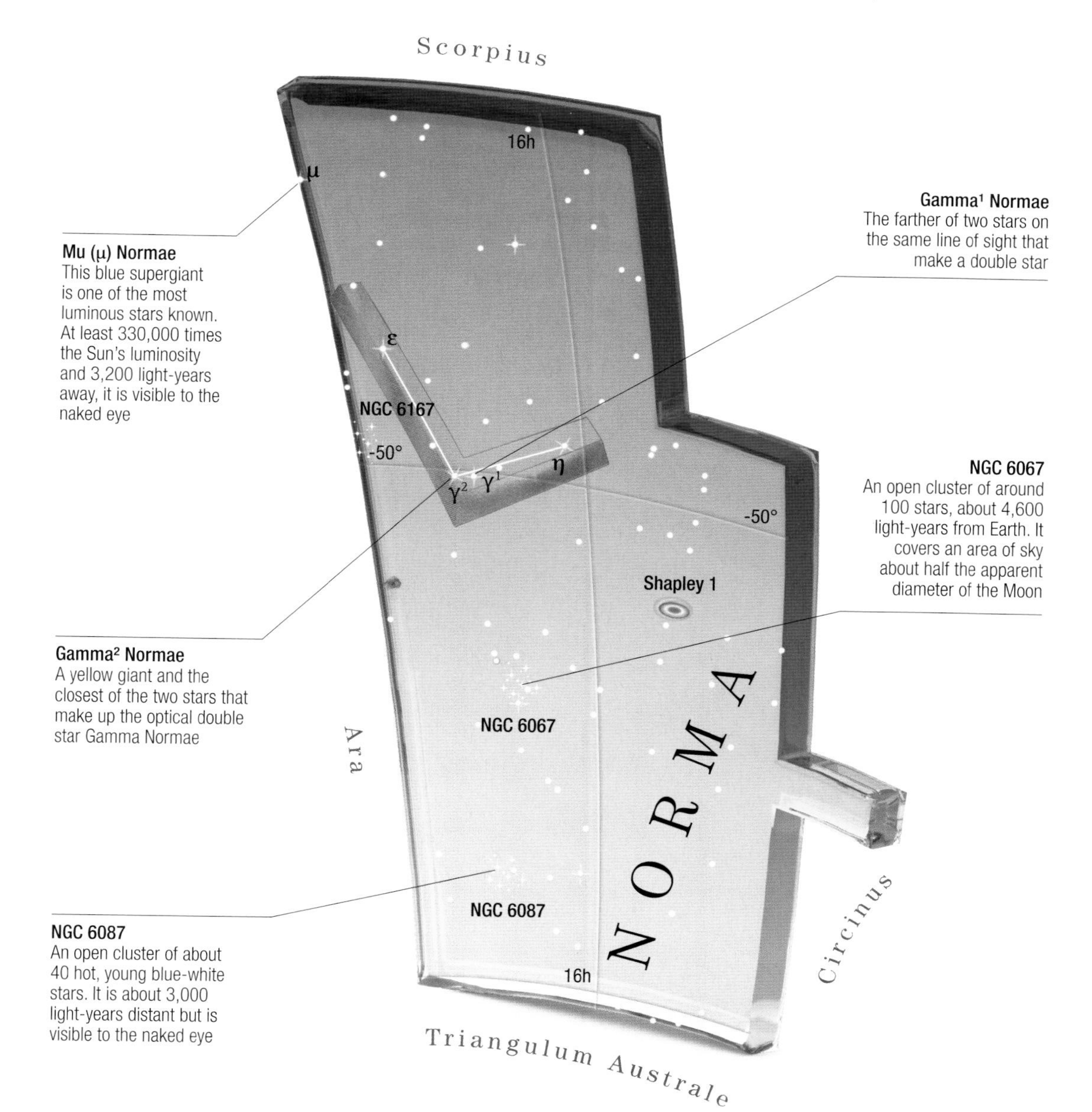

Mu (μ) Normae
This blue supergiant is one of the most luminous stars known. At least 330,000 times the Sun's luminosity and 3,200 light-years away, it is visible to the naked eye

Gamma¹ Normae
The farther of two stars on the same line of sight that make a double star

Gamma² Normae
A yellow giant and the closest of the two stars that make up the optical double star Gamma Normae

NGC 6067
An open cluster of around 100 stars, about 4,600 light-years from Earth. It covers an area of sky about half the apparent diameter of the Moon

NGC 6087
An open cluster of about 40 hot, young blue-white stars. It is about 3,000 light-years distant but is visible to the naked eye

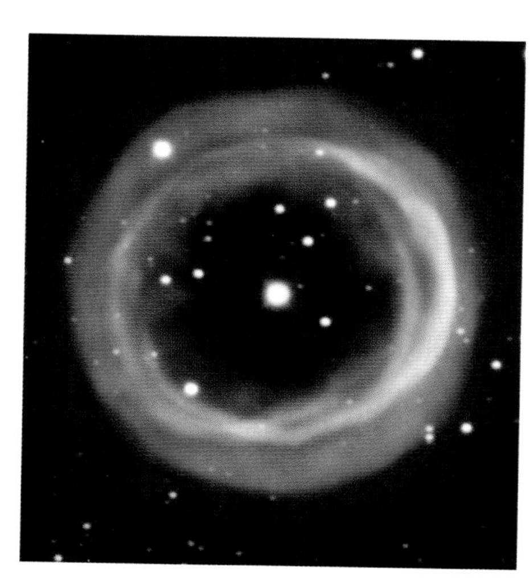

△ **Shapley 1**
Discovered by Harlow Shapley in 1936, this planetary nebula is a ring of gas seen face-on. In its centre is not a single star, but a binary system that cast off the surrounding gas many thousands of years ago. The interaction of the two stars shaped the ejected gas into an almost perfect ring.

ARA
THE ALTAR

SOUTH OF SCORPIUS, ARA LIES WITHIN THE PATH OF THE MILKY WAY. ONE OF THE 48 GREEK CONSTELLATIONS, IT DEPICTS AN ALTAR FROM ANCIENT GREEK MYTHOLOGY.

Ara was the altar where the Greek gods swore allegiance before entering into battle with the Titans for control of the Universe. Eventually, the gods were victorious and the leading god Zeus placed the altar in the sky in gratitude.

The constellation is easy to locate, although its pattern is obscure within the Milky Way's band of stars. Ara's brightest stars, Beta and Alpha, and the star cluster NGC 6193 can be seen with the naked eye. Also noteworthy are globular clusters, such as NGC 6397 and NGC 6362, and Mu Arae, a Sun-like star orbited by at least four planets.

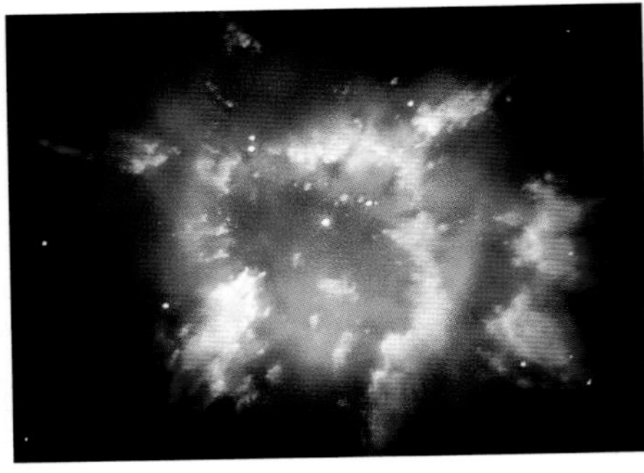

△ **NGC 6326**
Gas is hurtling away from a white dwarf star in the centre of this planetary nebula about 11,000 light-years from Earth. In this image, the red colour indicates hydrogen and the blue is oxygen.

◁ **NGC 6362**
The centre of this globular cluster contains blue stars formed by stellar collisions or the transfer of material between stars, which results in the stars heating up and looking younger than their neighbours.

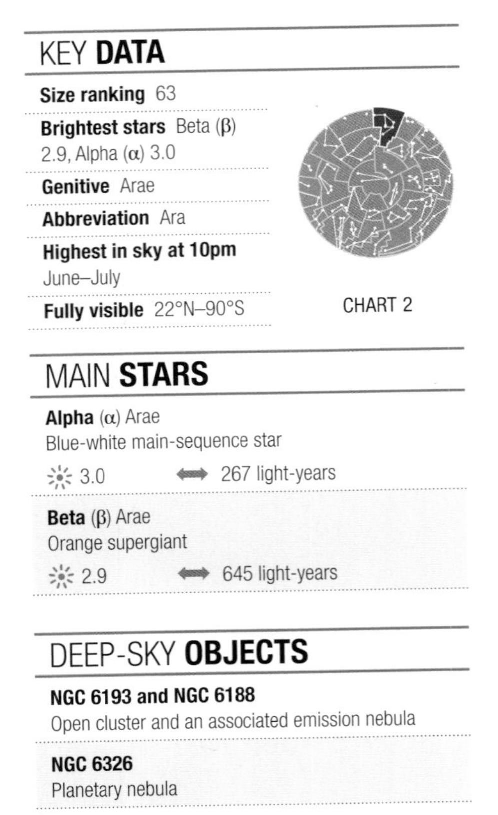

KEY DATA

Size ranking 63
Brightest stars Beta (β) 2.9, Alpha (α) 3.0
Genitive Arae
Abbreviation Ara
Highest in sky at 10pm June–July
Fully visible 22°N–90°S

CHART 2

MAIN STARS

Alpha (α) Arae
Blue-white main-sequence star
3.0 — 267 light-years

Beta (β) Arae
Orange supergiant
2.9 — 645 light-years

DEEP-SKY OBJECTS

NGC 6193 and NGC 6188
Open cluster and an associated emission nebula

NGC 6326
Planetary nebula

NGC 6352
Globular cluster

NGC 6362
Globular cluster

NGC 6397
Globular cluster

Stingray Nebula
Small, young planetary nebula

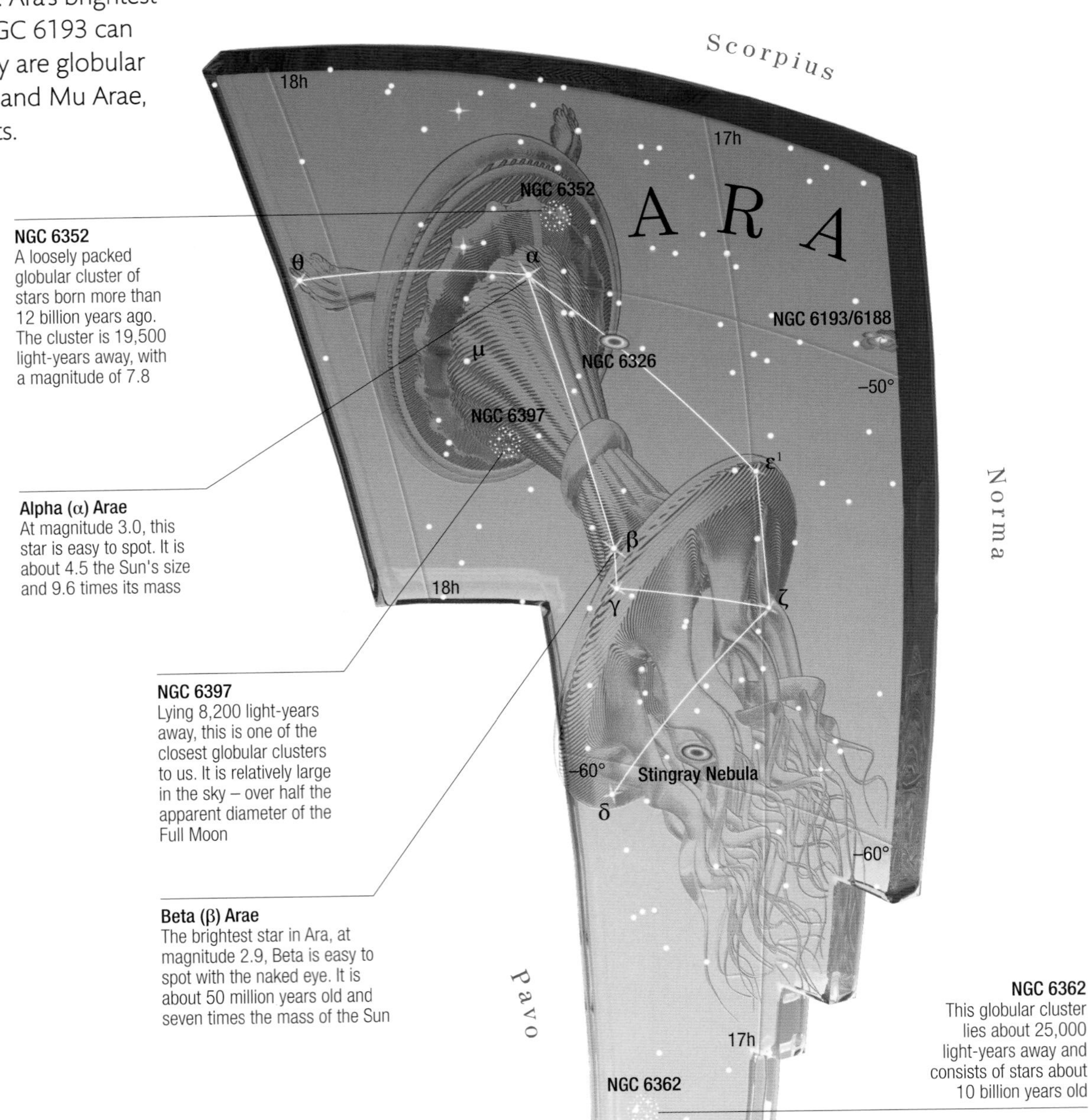

Scorpius
ARA
Norma
Pavo
18h
17h
18h
17h
–50°
–60°
–60°
θ
α
μ
β
γ
ζ
δ
ε1
NGC 6352
NGC 6326
NGC 6397
NGC 6193/6188
Stingray Nebula
NGC 6362
NGC 6352
A loosely packed globular cluster of stars born more than 12 billion years ago. The cluster is 19,500 light-years away, with a magnitude of 7.8
Alpha (α) Arae
At magnitude 3.0, this star is easy to spot. It is about 4.5 the Sun's size and 9.6 times its mass
NGC 6397
Lying 8,200 light-years away, this is one of the closest globular clusters to us. It is relatively large in the sky – over half the apparent diameter of the Full Moon
Beta (β) Arae
The brightest star in Ara, at magnitude 2.9, Beta is easy to spot with the naked eye. It is about 50 million years old and seven times the mass of the Sun
NGC 6362
This globular cluster lies about 25,000 light-years away and consists of stars about 10 billion years old

CORONA AUSTRALIS
THE SOUTHERN CROWN

ONE OF THE SMALLEST CONSTELLATIONS, CORONA AUSTRALIS IS ALSO ONE OF THE ORIGINAL 48 GREEK CONSTELLATIONS, ALTHOUGH IT IS NOT ASSOCIATED WITH ANY PARTICULAR MYTH.

The pattern of the crown of Corona Australis is not golden and jewel-studded like the Northern Crown, Corona Borealis, but a wreath of leaves. Other cultures saw the stars differently. To the ancient Chinese, its stars represented a turtle, while indigenous Australians saw a boomerang or a coolamon (a shallow dish).

None of the constellation's stars is particularly bright but the curved shape they make means the constellation is easy to spot. Its two brightest stars, Alpha and Beta, appear indistinguishable but are very different. Beta is the bigger and more luminous, but at almost four times farther away than Alpha, it shines with the same brightness in our sky. In the north of the constellation a huge region of nebulosity includes NGC 6729, one of the closest star-forming nebulae to us, lying about 400 light-years away.

△ **Coronet Cluster**
This infra-red and X-ray image shows young stars in the Coronet Cluster. Located near NGC 6729 and about 420 light-years away, the cluster is one of the nearest and most active regions of star-birth.

◁ **NGC 6729**
The youngest stars in this nebula are hidden inside dense gas and dust clouds. These young stars are throwing off high-speed jets of material that create shock-waves in the gas and cause it to shine.

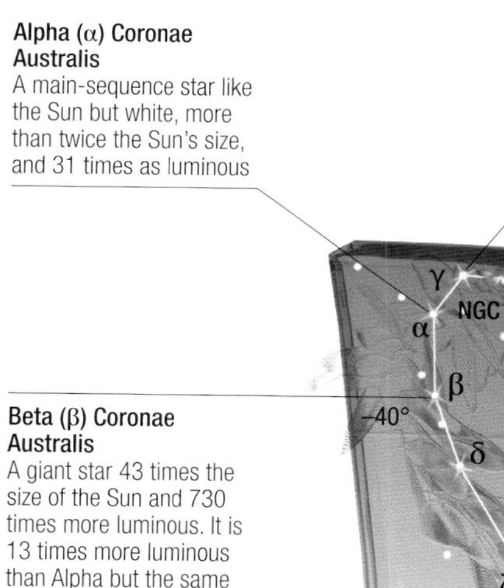

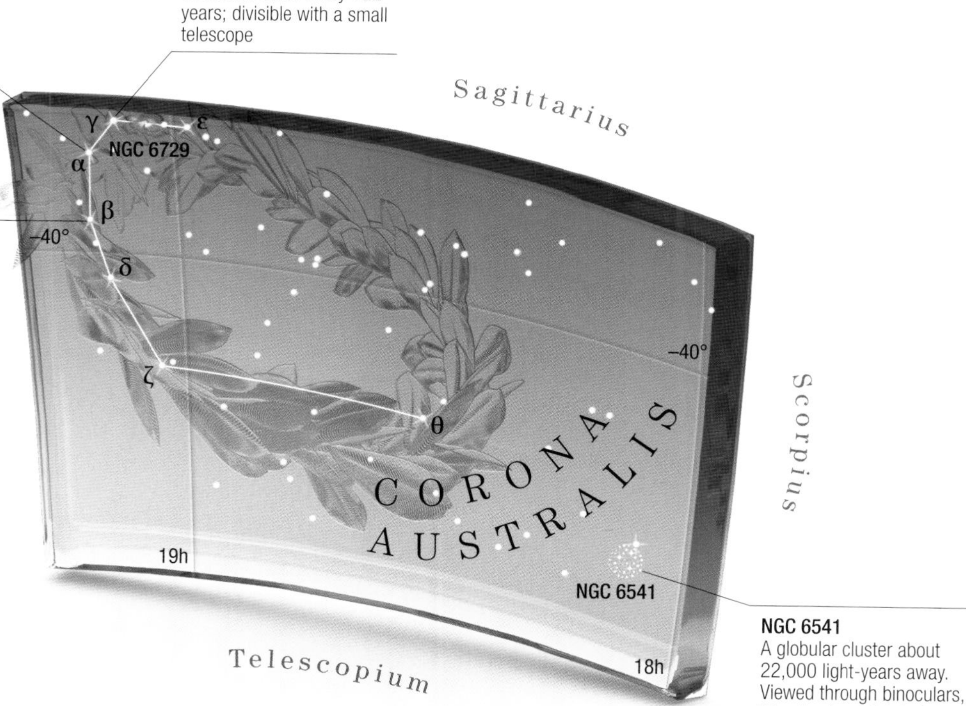
Alpha (α) Coronae Australis
A main-sequence star like the Sun but white, more than twice the Sun's size, and 31 times as luminous
Gamma (γ) Coronae Australis
A binary whose components orbit each other every 122 years; divisible with a small telescope
Beta (β) Coronae Australis
A giant star 43 times the size of the Sun and 730 times more luminous. It is 13 times more luminous than Alpha but the same brightness in the sky due to its greater distance
Sagittarius
γ
ε
NGC 6729
α
β
–40°
δ
ζ
θ
–40°
CORONA AUSTRALIS
Scorpius
19h
NGC 6541
18h
Telescopium
NGC 6541
A globular cluster about 22,000 light-years away. Viewed through binoculars, it is about one third of the apparent diameter of the Full Moon

KEY DATA

Size ranking 80

Brightest stars Alpha (α) 4.1, Beta (β) 4.1

Genitive Coronae Australis

Abbreviation CrA

Highest in sky at 10pm July–August

Fully visible 44°N–90°S

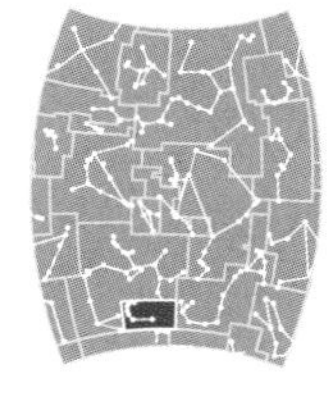

CHART 4

MAIN STARS

Alpha (α) Coronae Australis
White main-sequence star
4.1 ⟷ 125 light-years

Beta (β) Coronae Australis
Yellow giant
4.1 ⟷ 475 light-years

DEEP-SKY OBJECTS

NGC 6541
Globular cluster

NGC 6729
Star-forming nebula

Coronet Cluster
Open cluster

LUMINOSITIES

Omega Sagittarii *8 Suns*

Arkab Posterior *29 Suns*

Alnasl *49 Suns*

Rukbat *70 Suns*

SAGITTARIUS
THE ARCHER

A LARGE ZODIACAL CONSTELLATION, SAGITTARIUS REPRESENTS A MYTHICAL CREATURE CALLED A CENTAUR, PART-MAN, PART-HORSE, HOLDING A BOW AND ARROW. IT LIES IN A RICH AREA OF THE MILKY WAY AND CONTAINS THE CENTRE OF OUR GALAXY.

Sagittarius is most easily identified by the teapot-like shape formed by its main stars'. The Teapot asterism is made up of eight stars. Zeta, Sigma, Tau, and Phi form the handle, Gamma, Delta, and Epsilon form the spout, and Lambda forms the top of the lid. The brightest star in the constellation is Epsilon, not Alpha, as is typically the case in other constellations. In Sagittarius, Alpha is magnitude 4.0, whereas Epsilon is magnitude 1.8.

Sagittarius contains dense Milky Way star fields, because the centre of our Galaxy lies in this direction. The exact centre is marked by the radio source Sagittarius A*, thought to be the site of a supermassive black hole.

Charles Messier catalogued 15 objects in Sagittarius, more than in any of the other constellations. Notable examples include M8 (the Lagoon Nebula), M20 (the Trifid Nebula), and M22, a bright globular cluster.

◁ **Red Spider Nebula** Huge waves sweep through NGC 6537, a spider-like planetary nebula.The waves are caused by the expanding outer layers of the central star compressing and heating the surrounding interstellar gas.

▽ **Star distances** Sagittarius's main pattern stars vary between 78 and about 3,600 light-years from Earth. The nearest, Lambda (λ) Sagittarii, forms the top of the lid of the Teapot asterism. As seen from Earth, Mu (μ) Sagittarii appears relatively close to Lambda but Mu is actually the constellation's most distant pattern star and is more than 3,500 light-years farther from Earth than is Lambda.

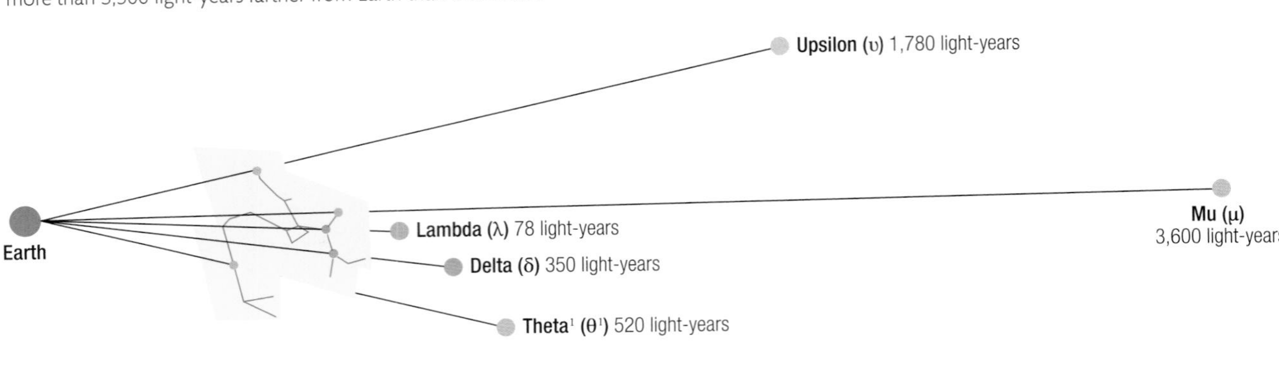

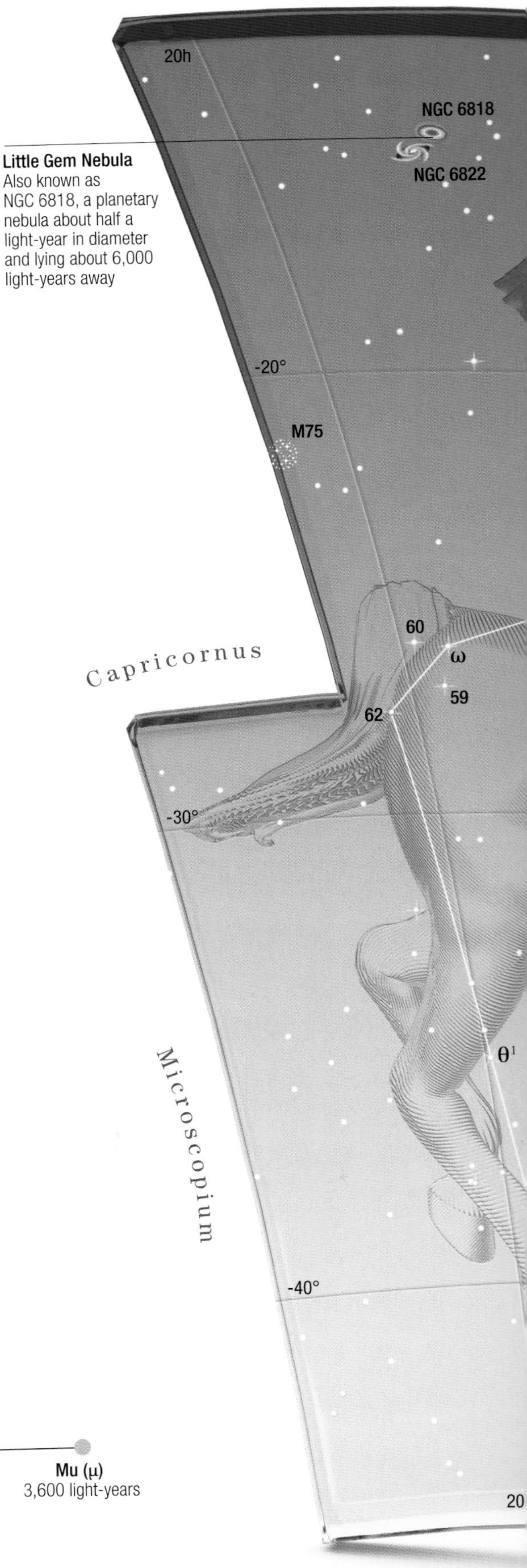

Little Gem Nebula Also known as NGC 6818, a planetary nebula about half a light-year in diameter and lying about 6,000 light-years away

Arkab Prior
210 Suns

Kaus Australis
325 Suns

Nunki
640 Suns

Upsilon Sagittarii
4,050 Suns

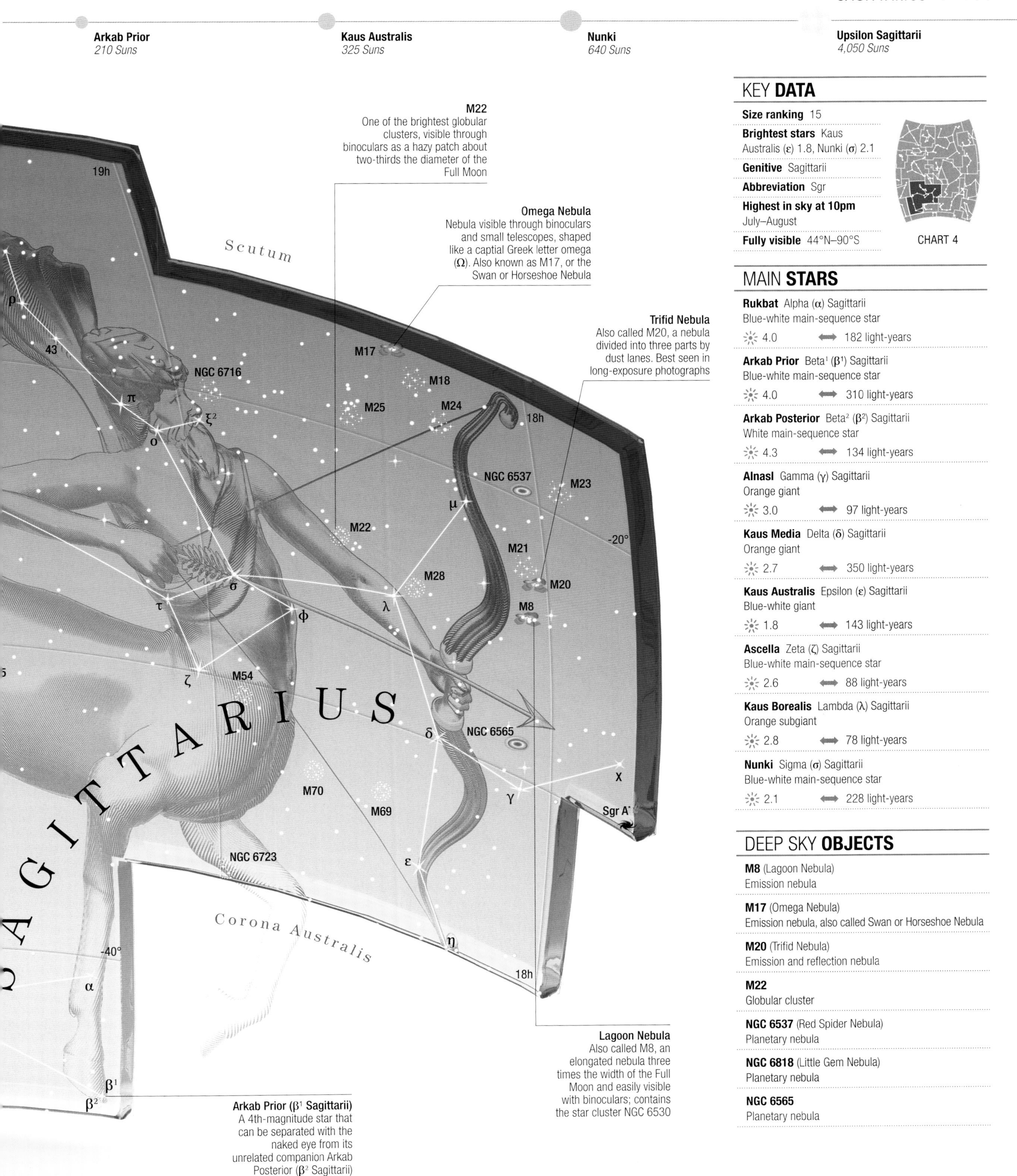

M22
One of the brightest globular clusters, visible through binoculars as a hazy patch about two-thirds the diameter of the Full Moon
Omega Nebula
Nebula visible through binoculars and small telescopes, shaped like a captial Greek letter omega (Ω). Also known as M17, or the Swan or Horseshoe Nebula
Trifid Nebula
Also called M20, a nebula divided into three parts by dust lanes. Best seen in long-exposure photographs
Scutum
19h
43
NGC 6716
M17
M18
M25
M24
18h
NGC 6537
M23
M22
M21
-20°
M28
M20
M8
M54
SAGITTARIUS
NGC 6565
M70
M69
Sgr A*
NGC 6723
Corona Australis
-40°
18h
Lagoon Nebula
Also called M8, an elongated nebula three times the width of the Full Moon and easily visible with binoculars; contains the star cluster NGC 6530
Arkab Prior (β¹ Sagittarii)
A 4th-magnitude star that can be separated with the naked eye from its unrelated companion Arkab Posterior (β² Sagittarii)

KEY DATA

Size ranking 15

Brightest stars Kaus Australis (ε) 1.8, Nunki (σ) 2.1

Genitive Sagittarii

Abbreviation Sgr

Highest in sky at 10pm July–August

Fully visible 44°N–90°S

CHART 4

MAIN STARS

Rukbat Alpha (α) Sagittarii
Blue-white main-sequence star
4.0 ⟷ 182 light-years

Arkab Prior Beta¹ (β¹) Sagittarii
Blue-white main-sequence star
4.0 ⟷ 310 light-years

Arkab Posterior Beta² (β²) Sagittarii
White main-sequence star
4.3 ⟷ 134 light-years

Alnasl Gamma (γ) Sagittarii
Orange giant
3.0 ⟷ 97 light-years

Kaus Media Delta (δ) Sagittarii
Orange giant
2.7 ⟷ 350 light-years

Kaus Australis Epsilon (ε) Sagittarii
Blue-white giant
1.8 ⟷ 143 light-years

Ascella Zeta (ζ) Sagittarii
Blue-white main-sequence star
2.6 ⟷ 88 light-years

Kaus Borealis Lambda (λ) Sagittarii
Orange subgiant
2.8 ⟷ 78 light-years

Nunki Sigma (σ) Sagittarii
Blue-white main-sequence star
2.1 ⟷ 228 light-years

DEEP SKY OBJECTS

M8 (Lagoon Nebula)
Emission nebula

M17 (Omega Nebula)
Emission nebula, also called Swan or Horseshoe Nebula

M20 (Trifid Nebula)
Emission and reflection nebula

M22
Globular cluster

NGC 6537 (Red Spider Nebula)
Planetary nebula

NGC 6818 (Little Gem Nebula)
Planetary nebula

NGC 6565
Planetary nebula

CAPRICORNUS
THE SEA GOAT

THE SMALLEST CONSTELLATION OF THE ZODIAC, CAPRICORNUS DEPICTS A STRANGE CREATURE THAT IS HALF GOAT AND HALF FISH. IT LIES BETWEEN SAGITTARIUS AND AQUARIUS AND CONTAINS SOME INTERESTING STARS.

The Ancient Greeks linked Capricornus with one of their gods, the goat-like Pan, who turned his lower body into that of a fish and hid in a river to escape the monster Typhon. Capricornus lacks bright clusters and nebulae and its galaxies are mostly too faint to be seen with small telescopes but it has stars that can be seen with amateur equipment. Alpha Capricorni is an impressive pair of unrelated stars: a yellow supergiant (Alpha¹, or Algedi Prima) and an orange giant (Alpha², or Algedi Secunda). A small telescope reveals that Alpha¹ is itself a double, and a larger telescope that Alpha² is a triple star.

KEY DATA

Size ranking 40

Brightest stars Deneb Algedi (δ) 2.8, Dabih (β) 3.1

Genitive Capricorni

Abbreviation Cap

Highest in sky at 10pm August–September

Fully visible 62°N–90°S

CHART 4

MAIN STARS

Algedi Secunda Alpha² (α²) Capricorni
Orange giant and triple star
3.6 ⟷ 105 light-years

Dabih Beta (β) Capricorni
Yellow giant and multiple star
3.1 ⟷ 327 light-years

Deneb Algedi Delta (δ) Capricorni
White giant and eclipsing binary star
2.8 ⟷ 37 light-years

DEEP-SKY OBJECTS

M30
Globular cluster

HCG 87 (Hickson Compact Group 87)
Compact group of galaxies

Deneb Algedi (δ Capricorni)
This white giant takes its name from the Arabic for a young goat's tail; a less massive companion star orbits it every 24 hours
Algedi (α Capricorni)
An optical double consisting of the yellow supergiant Algedi Prima (α¹) and the orange giant Algedi Secunda (α²)
Aquila
Aquarius
Sagittarius
Piscis Austrinus
CAPRICORNUS
-10°
-20°
21h
δ
γ
ι
θ
β
α
ρ
ζ
36
ψ
ω
HCG 87
M30

△ **HCG 87 (Hickson Compact Group 87)**
Three of the four galaxies known as HCG 87 are so close that they are affected by their mutual gravity. A faint tidal bridge of stars links the disk-shaped galaxy seen edge-on (bottom centre), to its nearest neighbour, an elliptical galaxy (bottom right). The third, the spiral galaxy at the top of the image, is undergoing intense star formation. The small spiral galaxy near the centre may be far in the distance.

PISCIS AUSTRINUS
THE SOUTHERN FISH

A SMALL, APPROXIMATELY FISH-SHAPED RING OF FAINT STARS, PISCIS AUSTRINUS IS ONE OF THE MOST SOUTHERLY OF THE 48 CONSTELLATIONS DESCRIBED BY THE ASTRONOMERS OF ANCIENT GREECE. IT CAN MOST EASILY BE LOCATED BY ITS BRILLIANT STAR, FOMALHAUT.

Piscis Austrinus is said to be the parent of the two fishes that comprise the less obvious constellation Pisces. It is notable for its bright star Fomalhaut, the 18th-brightest star in the night sky. Fomalhaut's name comes from the Arabic for "fish's mouth", which is where it is located in the constellation. This star is also celebrated as the first known to have a disk of material around it. The disk, which is several times the diameter of our own Solar System, is in the process of forming into planets. One such planet has already been spotted; called Fomalhaut b, it takes about 1,700 years to orbit its parent star. The constellation's other stars are comparatively faint, and has no deep-sky objects of note.

△ **HCG 90 (Hickson Compact Group 90)**
These three galaxies are part of HCG 90, a tight cluster of 16 galaxies about 110 million light-years away. Two are elliptical galaxies; the third is a dusty spiral galaxy that has been distorted by interaction with the closest of the ellipticals. The spiral is being stretched and pulled apart before being engulfed by the other two. Eventually, all three will probably merge to form one super galaxy.

KEY DATA

- **Size ranking** 60
- **Brightest stars** Fomalhaut (α) 1.2, Epsilon (ε) 4.2
- **Genitive** Piscis Austrini
- **Abbreviation** PsA
- **Highest in sky at 10pm** September–October
- **Fully visible** 53°N–90°S

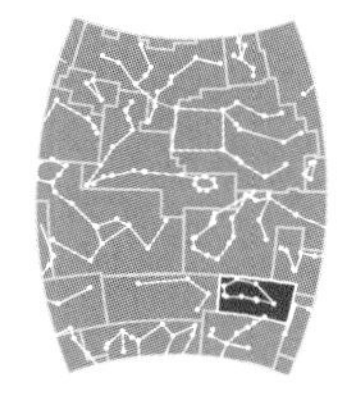

CHART 3

MAIN STARS

Fomalhaut Alpha (α) Piscis Austrini
Blue-white main-sequence star
1.2 — 25 light-years

Epsilon (ε) Piscis Austrini
Blue main-sequence star
4.2 — 744 light-years

DEEP-SKY OBJECTS

Debris disk around Fomalhaut
Ring of planet-forming material

HCG 90 (Hickson Compact Group 90)
Compact group of galaxies

Epsilon (ε) Piscis Austrini
A blue main-sequence star on the fish's back. It is 744 light-years from Earth and is of magnitude 4.2

Aquarius

PISCIS AUSTRINUS

Capricornus

Grus

23h 22h -30° HCG 90

Fomalhaut (α Piscis Austrini)
This brilliant star is only 25 light-years from Earth; it is a blue-white main-sequence star with a debris disk and an orbiting planet

Beta (β) Piscis Austrini
This optical double star is 135 light-years away; its component stars have magnitudes of 4.3 and 7.7

GRUS
THE CRANE

THIS CONSTELLATION WAS INTRODUCED TO THE SKY AT THE END OF THE 16TH CENTURY. ITS DISTINCTIVE FEATURE IS THE LINE OF STARS RUNNING FROM THE CRANE'S BEAK TO ITS TAIL.

Grus is one of several star patterns devised by Dutch navigators Pieter Dirkszoon Keyser and Frederick de Houtman, who made observations of the southern sky during an expedition to the East Indies in 1595. They passed on their observations to the Dutch cartographer Petrus Plancius, who created 12 new constellations based on those observations, all of which are still recognized today. (As well as Grus, these constellations are Apus, Chamaeleon, Dorado, Hydrus, Indus, Musca, Pavo, Phoenix, Triangulum Australe, Tucana, and Volans.)

In Grus, the long line of stars through its neck and body can be extended southwards to the Small Magellanic Cloud in Tucana. Lying within its neck is the naked-eye double Delta Gruis, which consists of two giants: a yellow magnitude 4.0 star, 150 light-years away; and a red magnitude 4.1 star, 420 light-years away. Other notable objects in Grus include the galaxy NGC 7424 and the planetary nebula IC 5148, popularly called the Spare Tyre Nebula.

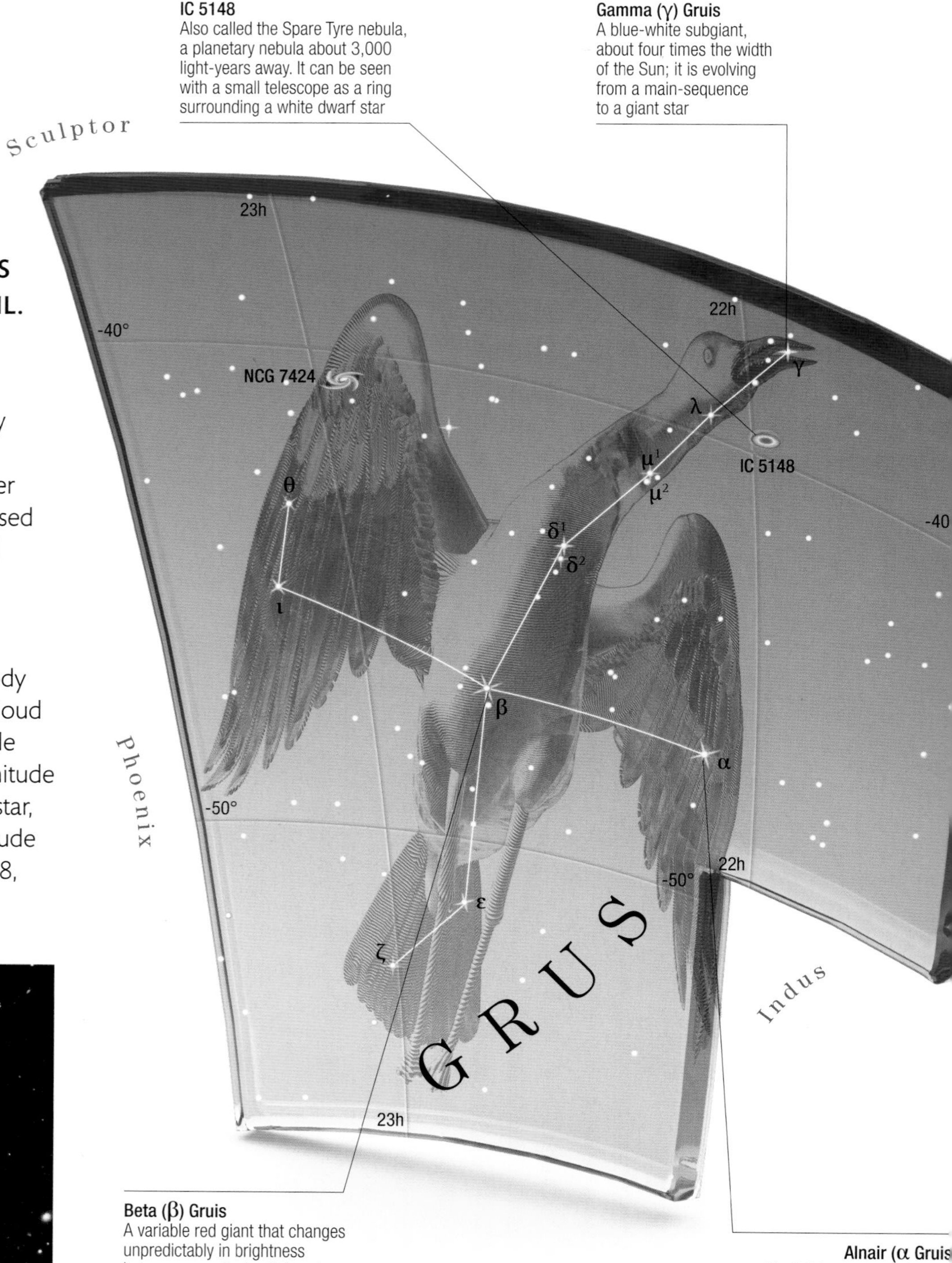

IC 5148
Also called the Spare Tyre nebula, a planetary nebula about 3,000 light-years away. It can be seen with a small telescope as a ring surrounding a white dwarf star

Gamma (γ) Gruis
A blue-white subgiant, about four times the width of the Sun; it is evolving from a main-sequence to a giant star

Beta (β) Gruis
A variable red giant that changes unpredictably in brightness between magnitudes 2.0 and 2.3 as it swells then shrinks

Alnair (α Gruis)
The brightest star in Grus, a magnitude 1.7. A blue-white subgiant, it is about 3.5 times the diamater of the Sun

◁ **NGC 7424**
This galaxy is similar in diameter to the Milky Way – roughly 100,000 light-years – and is about 37 million light-years away. It is classed as an intermediate galaxy, a stage between a spiral and a barred spiral. Its loosely wound arms are dominated by young stars, making them appear blue; the pale orange colour of its central ring-like structure indicates older stars.

The stars that form **Grus** were **originally part of Piscis Austrinus** until the end of the 16th century

MICROSCOPIUM
THE MICROSCOPE

A SMALL AND FAINT SOUTHERN CONSTELLATION ADDED TO THE SKY IN THE MID-18TH CENTURY, MICROSCOPIUM IS A ROUGHLY RECTANGULAR PATTERN OF INDISTINCT STARS.

Microscopium is one of the 14 constellations introduced to the sky by the French astronomer Nicholas Louis de Lacaille. South of Capricornus, it is located between the more prominent constellations of Piscis Austrinus and Sagittarius. Microscopium is an almost featureless constellation with no bright stars and no deep-sky objects except for galaxies too faint for amateur telescopes. Theta is the brightest of several variable stars but its variations are difficult to see, only differing by 0.1 magnitude.

KEY DATA

Size ranking 66

Brightest stars Gamma (γ) 4.7, Epsilon (ε) 4.7

Genitive Microscopii

Abbreviation Mic

Highest in sky at 10pm August–September

Fully visible 62°N–90°S

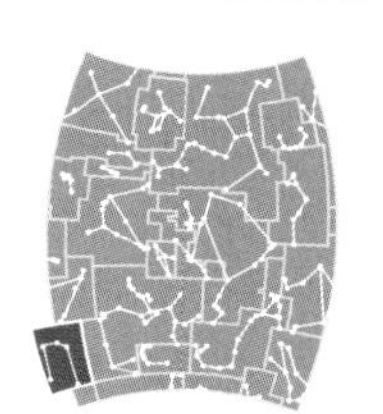

CHART 4

MAIN STARS

Alpha (α) Microscopii
Yellow giant
4.9 ⟷ 380 light-years

Gamma (γ) Microscopii
Yellow giant
4.7 ⟷ 230 light-years

Epsilon (ε) Microscopii
White main-sequence star
4.7 ⟷ 180 light-years

DEEP-SKY OBJECTS

ESO 286-19
Two colliding galaxies

Debris disk around AU Microscopii
Dusty material in orbit around a young star

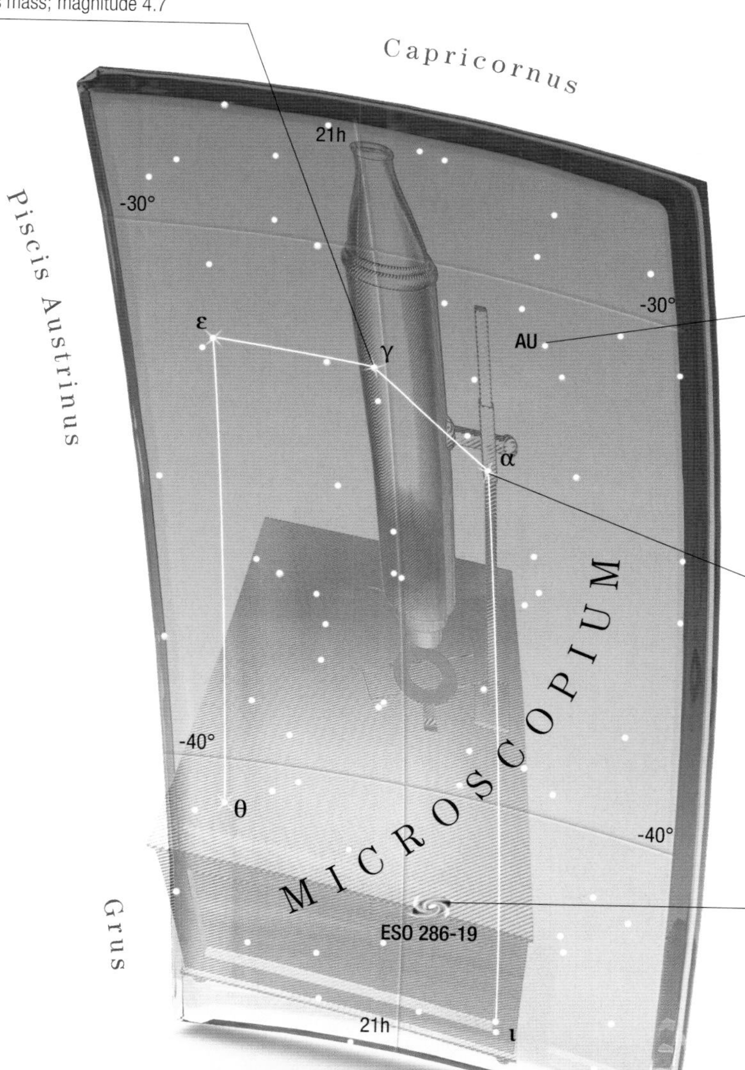

Gamma (γ) Microscopii
Yellow giant about 10 times the size of the Sun and 2.5 times its mass; magnitude 4.7

AU Microscopii
Faint red dwarf, about 32 light-years away and surrounded by a ring of dusty, potentially planet-forming material

Alpha (α) Microscopii
Yellow giant about 16 times the Sun's diameter and 160 times its luminosity. It forms an optical double with a magnitude 10 star

ESO 286-19
An unusual object lying about 600 light-years away and consisting of two previously disk-shaped galaxies in the midst on an ongoing collision

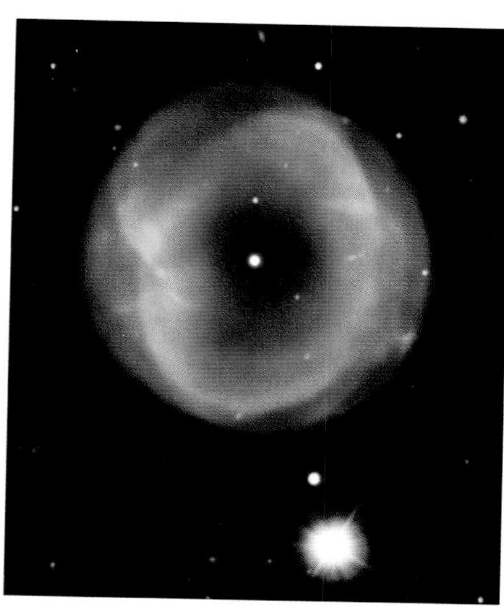

△ **IC 5148**
A ghostly shell of gas cast off by a dying star and resembling a car tyre gives this planetary nebula its popular name of the Spare Tyre Nebula. The gas shell is a couple of light-years across and is speeding away from a white dwarf (the bright white object in the centre of the planetary nebula), which is the remnant of the original star.

KEY DATA

Size ranking 45

Brightest stars Alnair (α) 1.7, Beta (β) 2.0–2.3

Genitive Gruis

Abbreviation Gru

Highest in sky at 10pm September–October

Fully visible 33°N–90°S

CHART 3

MAIN STARS

Alnair Alpha (α) Gruis
Blue-white subgiant
1.7 ⟷ 101 light-years

Beta (β) Gruis
Variable red giant
2.0–2.3 ⟷ 177 light-years

Gamma (γ) Gruis
Blue-white subgiant
3.0 ⟷ 210 light-years

DEEP-SKY OBJECTS

NGC 7424
Intermediate spiral galaxy

IC 5148 (Spare Tyre Nebula)
Planetary nebula

SCULPTOR
THE SCULPTOR

FAINT AND UNREMARKABLE, SCULPTOR IS EASY TO FIND BECAUSE IT LIES DIRECTLY TO THE EAST OF THE BRIGHT STAR FOMALHAUT IN PISCIS AUSTRINUS. IT IS HOME TO SEVERAL INTERESTING GALAXIES.

Sculptor was introduced to the sky by the French astronomer Nicolas Louis de Lacaille in 1754. Originally named Apparatus Sculptoris (the sculptor's studio), it depicts a marble head, mallet, and chisel on a stand. However, the star pattern is more reminiscent of a shepherd's crook. None of the stars are named, and all are 4th magnitude or fainter. It contains the Sculptor Group, a cluster of about a dozen galaxies, one of the nearest to our own Local Group. At the Group's heart is NGC 253, discovered by German-born English astronomer Caroline Herschel in 1783. Close by is globular cluster NGC 288, discovered by her brother William Herschel in 1785.

△ **NGC 253**
The largest and brightest of the Sculptor Group, this spiral galaxy is 11 million light-years away but at magnitude 7.5 appears as a fuzzy oval in binoculars. It is classed as a starburst galaxy due to its high rate of star formation.

▷ **NGC 300**
A spiral galaxy with a poorly defined core and diffuse arms, NGC 300 is close to us at just 6 million light-years away, and probably lies between us and the Sculptor Group.

KEY DATA

Size ranking 36

Brightest stars Alpha (α) 4.3, Beta (β) 4.4

Genitive Sculptoris

Abbreviation Scl

Highest in sky at 10pm October–November

Fully visible 50°N–90°S

CHART 3

MAIN STARS

Alpha (α) Sculptoris
Blue-white giant star
4.3 ⟷ 776 light-years

Beta (β) Sculptoris
Blue-white subgiant
4.4 ⟷ 174 light-years

DEEP-SKY OBJECTS

NGC 55
Irregular galaxy

NGC 253
Spiral galaxy in the Sculptor Group

NGC 288
Globular cluster

NGC 300
Spiral galaxy

NGC 7793
Spiral galaxy in the Sculptor Group

ESO 350-40 (Cartwheel Galaxy)
Combined spiral and ring galaxy

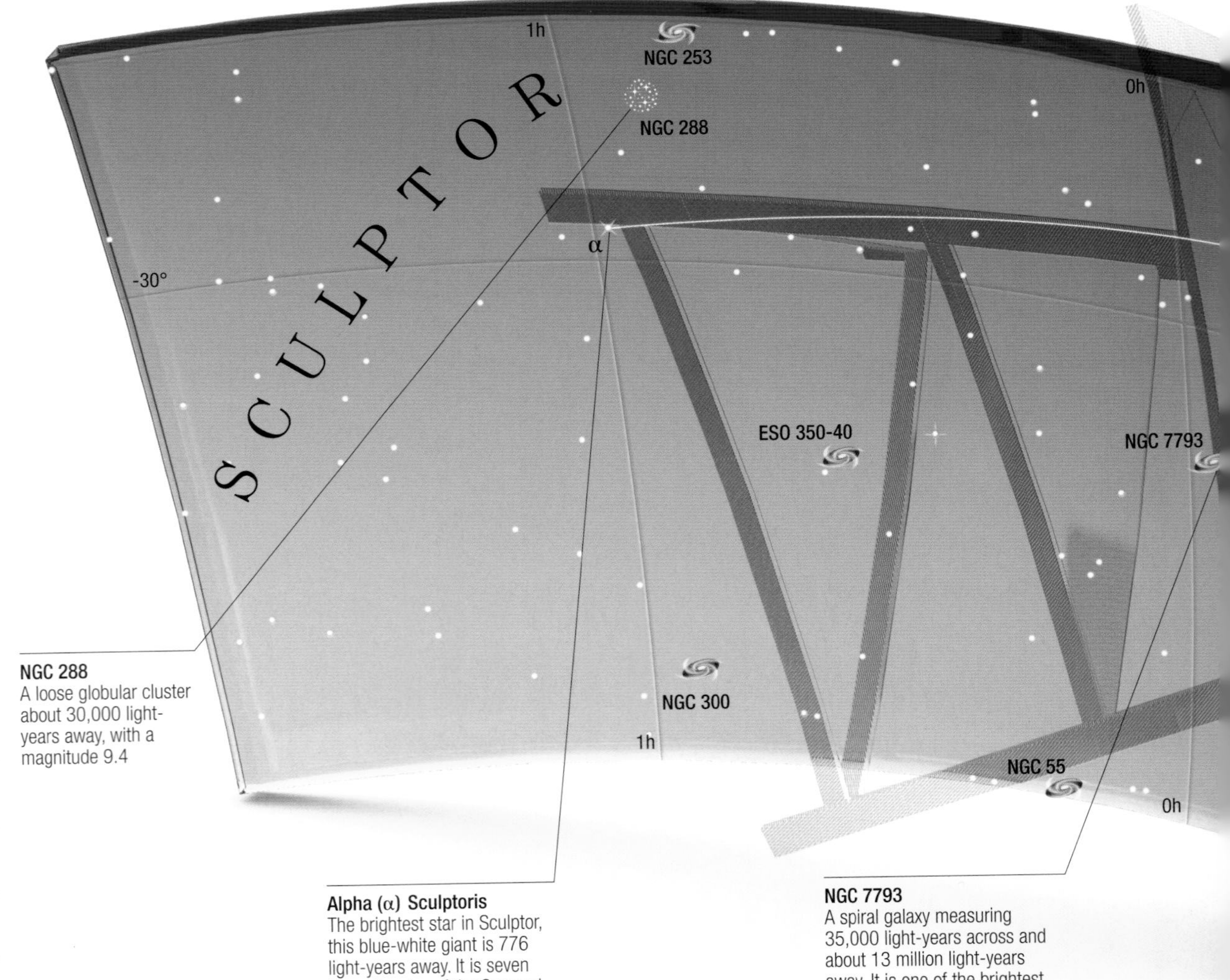

NGC 288
A loose globular cluster about 30,000 light-years away, with a magnitude 9.4

Alpha (α) Sculptoris
The brightest star in Sculptor, this blue-white giant is 776 light-years away. It is seven times the size of the Sun and 1,700 times its luminosity

NGC 7793
A spiral galaxy measuring 35,000 light-years across and about 13 million light-years away. It is one of the brightest members of the Sculptor Group

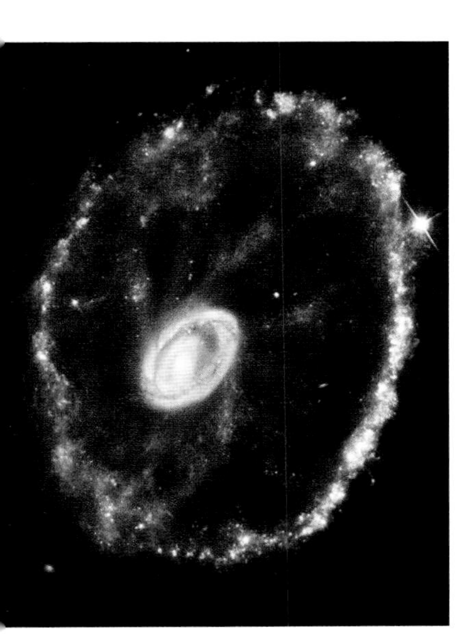

ESO 350-40
›out 500 million light-years away and 150,000 light-ars across, this galaxy's cartwheel shape is the result a huge galactic collision. A smaller galaxy passed ‹ough a larger spiral, producing shock waves that ‹ept up gas and dust. This sparked the birth of ‹lions of stars, seen in blue in the outer ring.

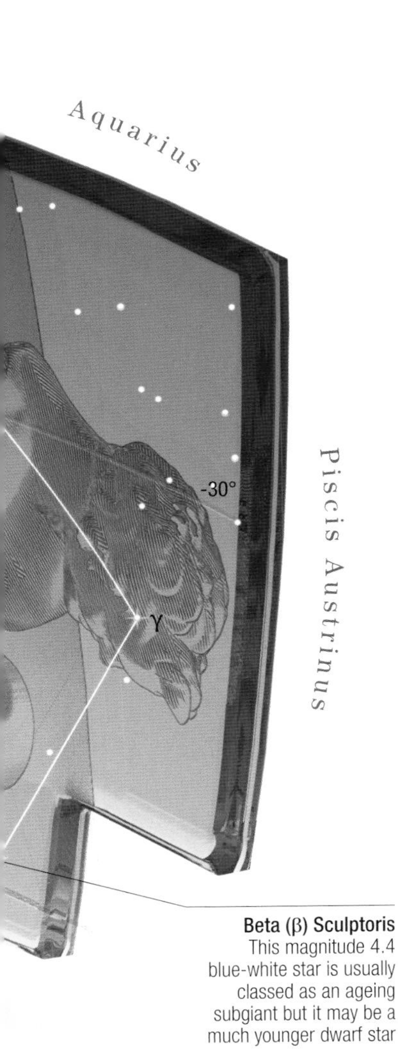

Beta (β) Sculptoris
This magnitude 4.4 blue-white star is usually classed as an ageing subgiant but it may be a much younger dwarf star

CAELUM
THE CHISEL

SMALL AND INSIGNIFICANT, CAELUM IS FORMED FROM TWO FAINT STARS WHICH, LINKED TOGETHER, REPRESENT AN ENGRAVER'S CHISEL.

One of the smallest constellations in the southern sky, Caelum lies between Eridanus and Columba. It was one of the 14 introduced by French astronomer Nicolas Louis de Lacaille in 1754. Caelum's size and position away from the plane of the Milky Way mean that it has few deep-sky objects and so its stars are the main objects of interest. Of them, only Alpha, Beta, and Gamma are brighter than magnitude 5.

CAELUM
Columba
Eridanus
Pictor
Horologium
HE 0450-2958
5h
-30°
-40°
γ
β
α

Gamma (γ) Caeli
An orange giant of magnitude 4.6, this star lies on the constellation's western boundary. A small telescope reveals it is a binary with a companion of magnitude 8.1

Beta (β) Caeli
A white star moving away from the main sequence to become a giant. It lies 93 light-years away and is of magnitude 5.0

Alpha (α) Caeli
This white star is only magnitude 4.5 but it is the brightest in Caelum. Its much dimmer red dwarf companion can be seen with larger telescopes

KEY DATA

Size ranking 81
Brightest stars Alpha (α) 4.5, Beta (β) 5.0
Genitive Caeli
Abbreviation Cae
Highest in sky at 10pm December–January
Fully visible 41°N–90°S

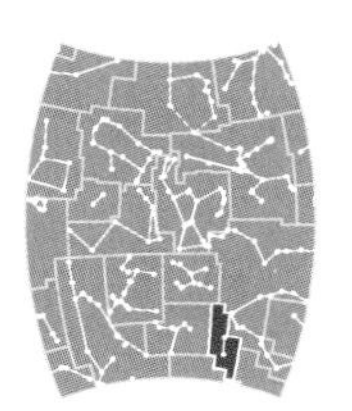

CHART 6

MAIN STARS

Alpha (α) Caeli
White main-sequence star, and a binary star
4.5 ⟷ 66 light-years

Beta (β) Caeli
White subgiant star
5.0 ⟷ 93 light-years

DEEP-SKY OBJECTS

Quasar HE0450-2958
Quasar, also classed as a Seyfert galaxy

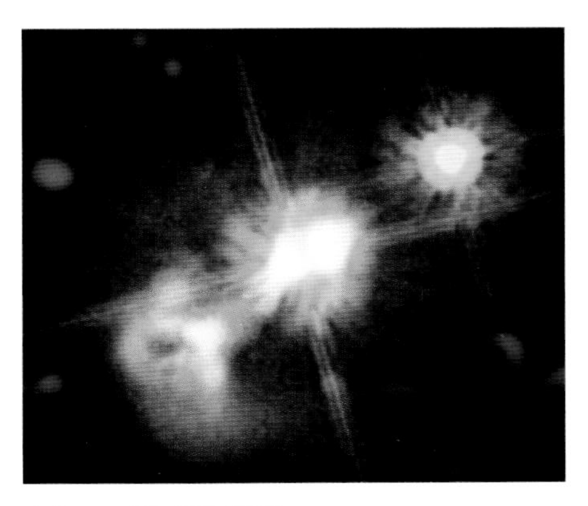

△ **Quasar HE 0450-2958**
Located towards the north of the constellation, quasar HE0450-2958 is unusual in that its host galaxy is too faint to be seen directly because it is swamped by the quasar's light. This composite image of it is made up of an infrared image from ESO's Very Large Telescope and a visible-light image from the Hubble Space Telescope.

FORNAX
THE FURNACE

FOUND SOUTH OF CETUS, THIS CONSTELLATION'S PATTERN IS A WIDE "V" SHAPE LINKING THREE STARS. IT IS KNOWN FOR THE FORNAX CLUSTER OF GALAXIES, AND ONE OF OUR DEEPEST VIEWS INTO THE UNIVERSE.

Originally Fornax Chemica, the chemist's furnace, Fornax is one of 14 constellations devised by Frenchman Nicolas Louis de Lacaille after surveying the southern sky, in 1751-52. It is home to the Fornax Cluster, a rich cluster of galaxies 62 million light-years away. The brighter members of the 58-strong cluster can be seen with amateur equipment. The elliptical galaxy NGC 1316 (also called Fornax A) is the brightest, and also one of the strongest radio sources in the sky. Barred spiral NGC 1365 is the largest spiral galaxy of the cluster. A tiny region in the north of Fornax was specially imaged by the Hubble Space Telescope. Known as the Hubble Ultra Deep Field, the image contains 10,000 galaxies and is one of our deepest views of the Universe.

△ **NGC 1097**
The large Seyfert galaxy NGC 1097 is magnitude 10.3 and one of the sky's brightest barred spirals. It is interacting with the tiny elliptical galaxy NGC 1097A at top right. This is not the first small galaxy to be affected by NGC 1097; it engulfed a dwarf galaxy a few billion years ago.

△ **NGC 1350**
The arms in the inner region of this spiral galaxy form a complete ring, resembling a huge eye in space. The blue tint of the oute arms indicates active star formation. Other galaxies are visible through the outer parts. NGC 1350 is about 85 million light years away and 130,000 light-years across.

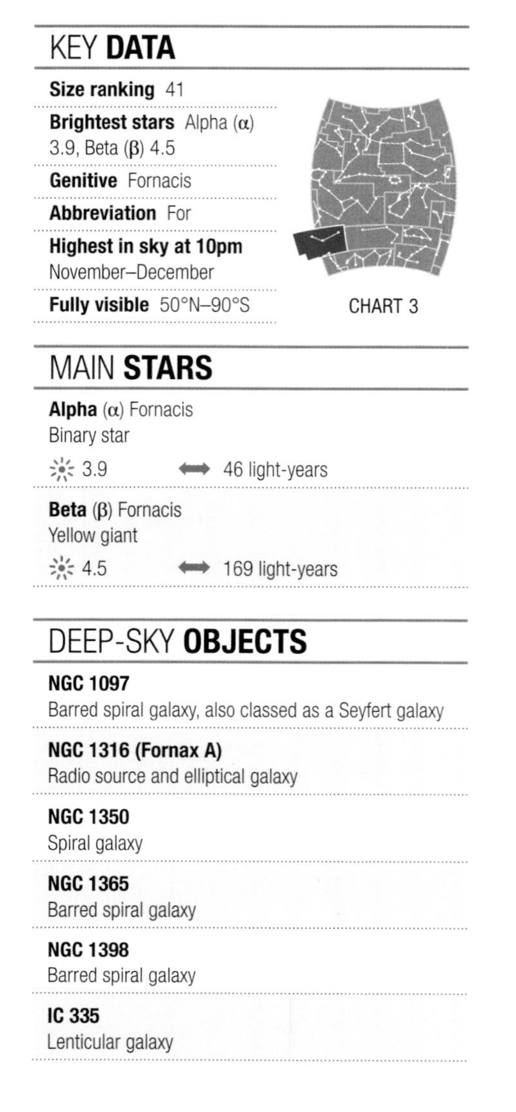

KEY **DATA**

Size ranking 41

Brightest stars Alpha (α) 3.9, Beta (β) 4.5

Genitive Fornacis

Abbreviation For

Highest in sky at 10pm November–December

Fully visible 50°N–90°S

CHART 3

MAIN **STARS**

Alpha (α) Fornacis
Binary star
3.9 ⟷ 46 light-years

Beta (β) Fornacis
Yellow giant
4.5 ⟷ 169 light-years

DEEP-SKY **OBJECTS**

NGC 1097
Barred spiral galaxy, also classed as a Seyfert galaxy

NGC 1316 (Fornax A)
Radio source and elliptical galaxy

NGC 1350
Spiral galaxy

NGC 1365
Barred spiral galaxy

NGC 1398
Barred spiral galaxy

IC 335
Lenticular galaxy

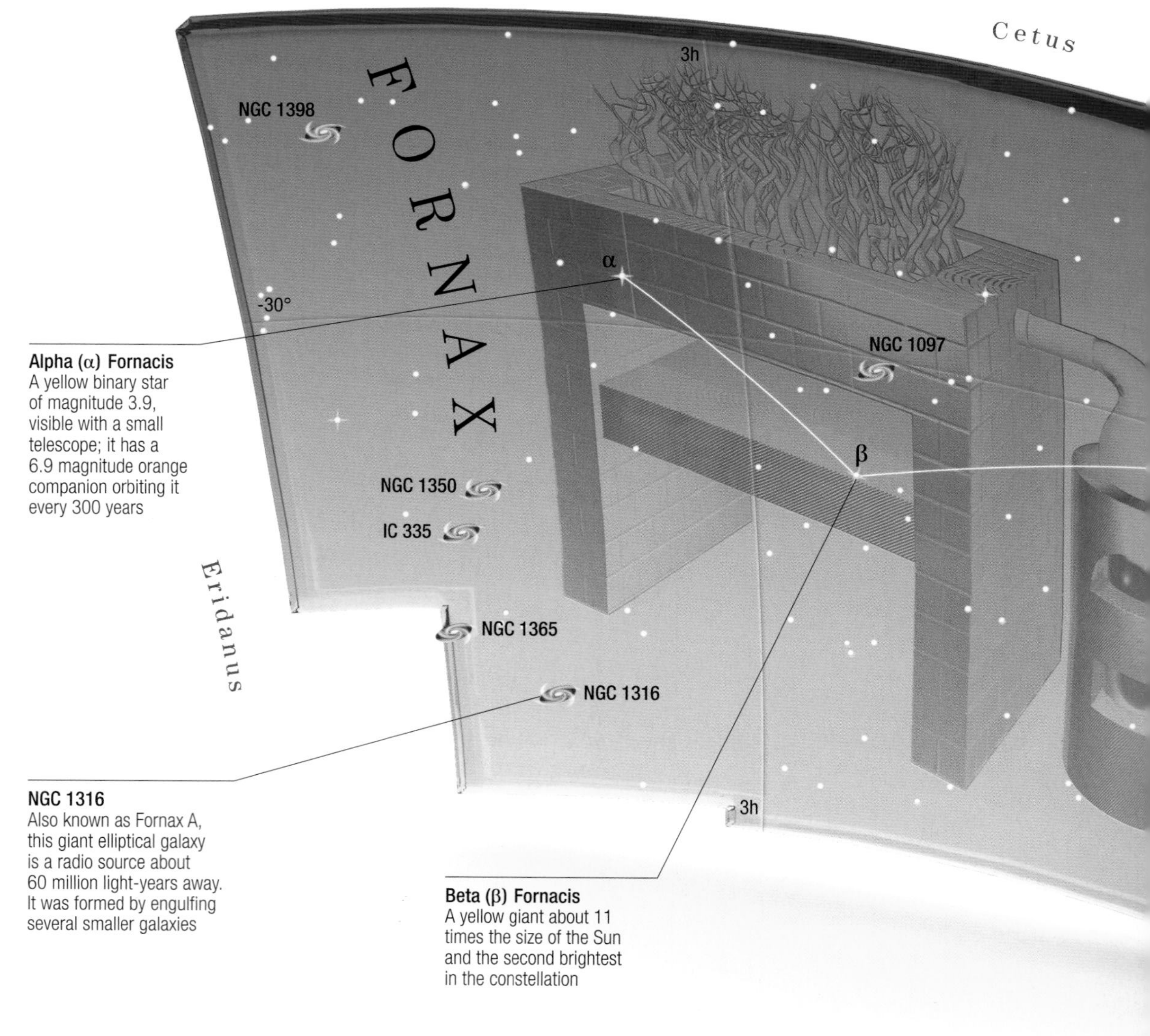

Alpha (α) Fornacis
A yellow binary star of magnitude 3.9, visible with a small telescope; it has a 6.9 magnitude orange companion orbiting it every 300 years

NGC 1316
Also known as Fornax A, this giant elliptical galaxy is a radio source about 60 million light-years away. It was formed by engulfing several smaller galaxies

Beta (β) Fornacis
A yellow giant about 11 times the size of the Sun and the second brightest in the constellation

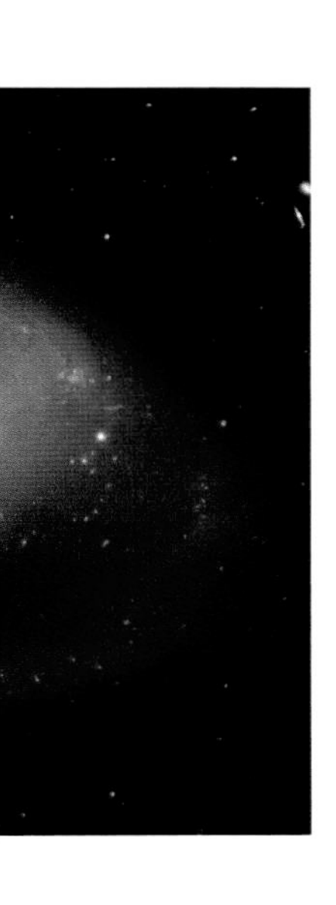

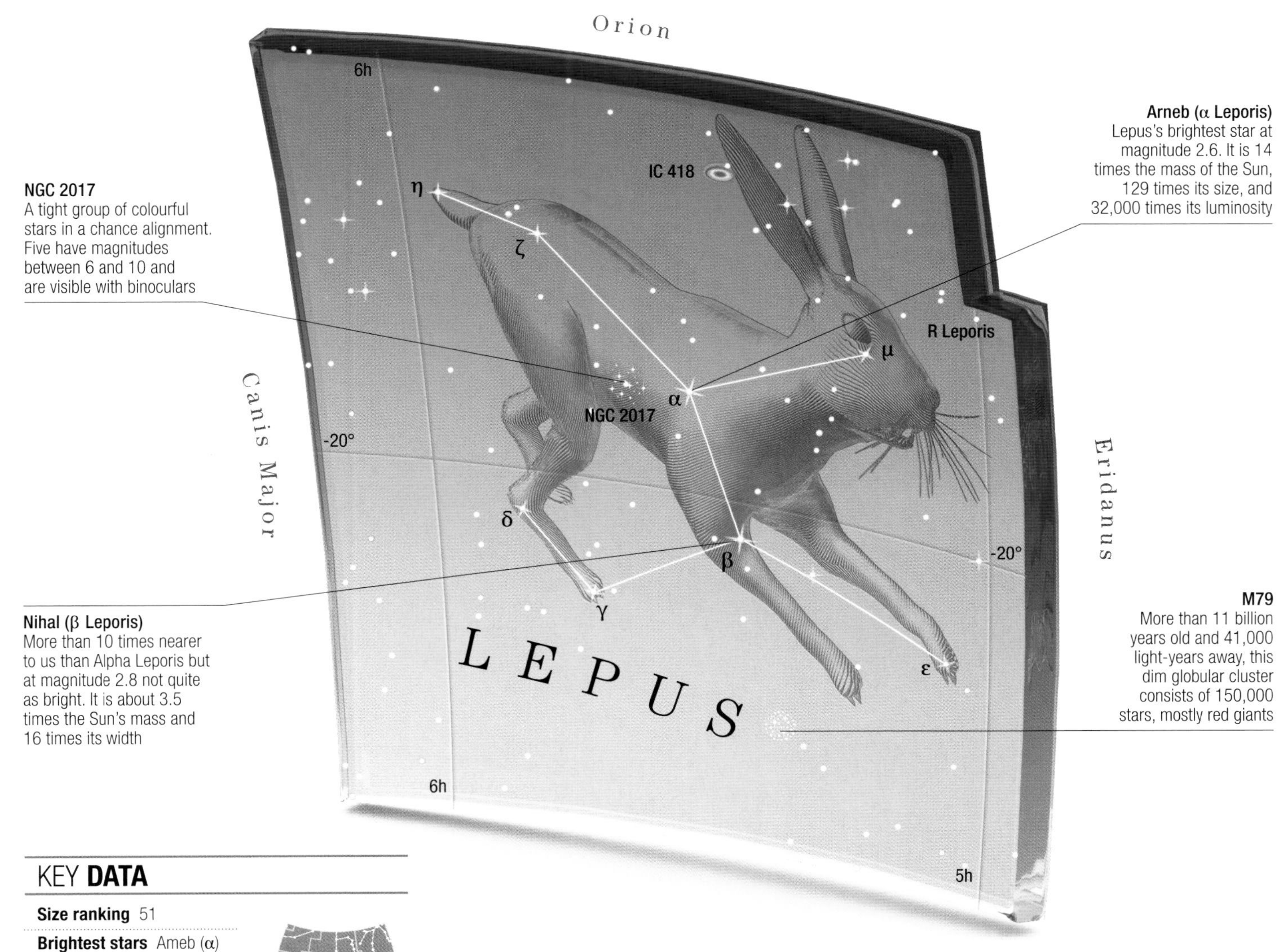
Orion
6h
IC 418
η
ζ
Canis Major
-20°
NGC 2017
α
R Leporis
μ
Eridanus
δ
β
γ
ε
-20°
LEPUS
6h
5h
NGC 2017
A tight group of colourful stars in a chance alignment. Five have magnitudes between 6 and 10 and are visible with binoculars
Arneb (α Leporis)
Lepus's brightest star at magnitude 2.6. It is 14 times the mass of the Sun, 129 times its size, and 32,000 times its luminosity
Nihal (β Leporis)
More than 10 times nearer to us than Alpha Leporis but at magnitude 2.8 not quite as bright. It is about 3.5 times the Sun's mass and 16 times its width
M79
More than 11 billion years old and 41,000 light-years away, this dim globular cluster consists of 150,000 stars, mostly red giants

KEY DATA

Size ranking 51

Brightest stars Ameb (α) 2.6, Nihal (β) 2.8

Genitive Leporis

Abbreviation Lep

Highest in sky at 10pm January

Fully visible 62°N–90°S

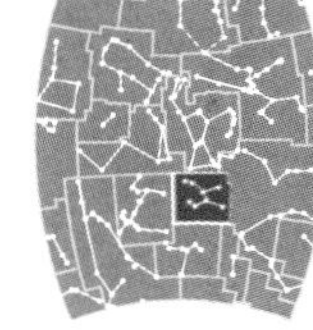

CHART 6

MAIN STARS

Arneb Alpha (α) Leporis
White supergiant
2.6 ⟷ 2,130 light-years

Nihal Beta (β) Leporis
Yellow giant
2.8 ⟷ 160 light-years

Epsilon (ε) Leporis
Orange giant
3.2 ⟷ 213 light-years

DEEP-SKY OBJECTS

M79
Global cluster, also known as NGC 1904

NGC 2017
Multiple star

IC 418 (Spirograph Nebula)
Planetary nebula

LEPUS
THE HARE

LEPUS IS FOUND AS A BOW-TIE SHAPE IMMEDIATELY TO THE SOUTH OF THE EASILY LOCATED CONSTELLATION ORION. ITS STARS, INCLUDING VARIABLES AND MULTIPLES, ARE ITS DISTINCTIVE FEATURE.

Greek mythology relates that this hare was placed in the sky after hares overwhelmed the island of Leros, devastating the land and causing starvation. The constellation was a permanent reminder about the perils of farming too many hares. Lepus is found to the south of Orion, appearing as if in flight from Orion's two hunting dogs Canis Major and Canis Minor. Its brightest star Alpha Leporis, has the name Arneb which comes from the Arabic for "hare". It is one of the most luminous stars visible from Earth. However, because of its distance, it is of only average brightness, with magnitude 2.6. The star R Leporis is a pulsating red giant, a Mira variable changing between magnitude 5.5 and 12 over a 430-day cycle. It is also known as Hind's Crimson Star after English astronomer John Russell Hind who observed it in 1845.

CANIS MAJOR
THE GREATER DOG

AN ANCIENT CONSTELLATION, CANIS MAJOR IS THE LARGER OF ORION'S TWO HUNTING DOGS. IT IS HOST TO SIRIUS, THE BRIGHTEST STAR IN THE ENTIRE NIGHT SKY.

Canis Major is near the heel of its master Orion, and close by (immediately north of Monoceros) is the smaller constellation Canis Minor. Canis Major represents Laelaps, a mythical dog so swift no prey could escape from it. The constellation is dominated by Sirius, which is actually a fairly average star of its type but outshines all other stars in the night sky because it is so close to us. The second-brightest star, Adhara, is a superluminous blue-white giant, which is much more distant but would outshine Sirius if placed next to it. Since the Milky Way crosses it, the constellation contains several notable deep-sky objects, including the clusters M41 and NGC 2362, both of which can be seen with the naked eye.

Sirius is the **brightest star** in the night sky. It is almost **twice as bright as** the next brightest, **Canopus** in the constellation Carina

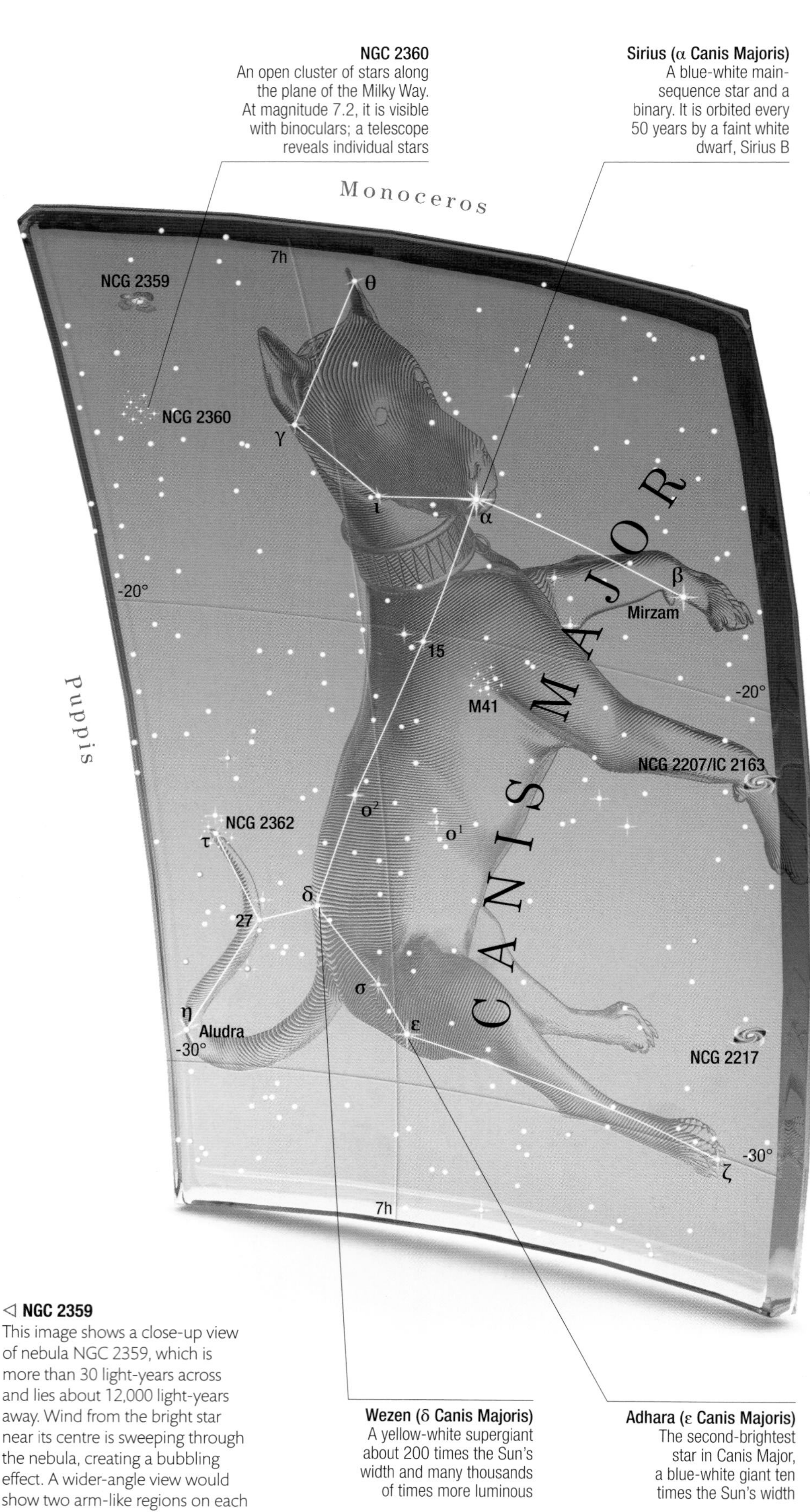

NGC 2360
An open cluster of stars along the plane of the Milky Way. At magnitude 7.2, it is visible with binoculars; a telescope reveals individual stars

Sirius (α Canis Majoris)
A blue-white main-sequence star and a binary. It is orbited every 50 years by a faint white dwarf, Sirius B

Wezen (δ Canis Majoris)
A yellow-white supergiant about 200 times the Sun's width and many thousands of times more luminous

Adhara (ε Canis Majoris)
The second-brightest star in Canis Major, a blue-white giant ten times the Sun's width

◁ **NGC 2359**
This image shows a close-up view of nebula NGC 2359, which is more than 30 light-years across and lies about 12,000 light-years away. Wind from the bright star near its centre is sweeping through the nebula, creating a bubbling effect. A wider-angle view would show two arm-like regions on each side of the nebula, like the wings of a helmet, hence the nebula's popular name of Thor's Helmet.

KEY DATA

Size ranking 43
Brightest stars Sirius (α) -1.5, Adhara (ε) 1.5
Genitive Canis Majoris
Abbreviation CMa
Highest in sky at 10pm January–February
Fully visible 56°N–90°S

CHART 6

MAIN STARS

Sirius Alpha (α) Canis Majoris
Blue-white main-sequence star, also a binary
-1.5 ⟷ 8.6 light-years

Mirzam Beta (β) Canis Majoris
Blue giant
2.0 ⟷ 492 light-years

Wezen Delta (δ) Canis Majoris
Yellow-white supergiant
1.8 ⟷ 1,605 light-years

Adhara Epsilon (ε) Canis Majoris
Blue-white giant
1.5 ⟷ 405 light-years

Aludra Eta (η) Canis Majoris
Blue-white supergiant
2.5 ⟷ 1,985 light-years

DEEP-SKY OBJECTS

M41
Open star cluster

NGC 2207 and IC 2163
Two interacting galaxies

NGC 2217
Barred spiral galaxy

NGC 2359 (Thor's Helmet)
Emission nebula

NGC 2362
Open star cluster centred on Tau (τ) Canis Majoris

△ **NGC 2207 and IC 2163**
These two interacting galaxies have created a huge mask shape in space. The gravitational attraction of the larger, NGC 2207, has distorted IC 2163, flinging out stars and gas into streamers at least 100,000 light-years long. The two will continue to slowly fall closer together, forming one huge galaxy in a few billion years' time.

COLUMBA

THE DOVE

A FAINT CONSTELLATION LYING SOUTH OF LEPUS, COLUMBA WAS FORMED IN THE 16TH CENTURY FROM STARS THAT HAD NOT PREVIOUSLY BEEN ALLOCATED TO ANY OTHER CONSTELLATION.

Invented by the Dutch astronomer Petrus Plancius in 1592, Columba was originally called "Columba Noachi" in reference to the dove Noah sent out from the Ark to find dry land, in the Biblical story of the flood. In the constellation, the dove's body is marked by the star Wezn, and its head end is indicated by the yellow-orange giant Eta Columbae. Its brightest star is Phact, whose name derives from the Arabic for "collared dove". Mu Columbae is a fast-moving 5th-magnitude star that is thought to have been expelled from the Orion Nebula area. Columba's most prominent deep-sky object is globular cluster NGC 1851, visible as a faint patch through binoculars.

KEY DATA

Size ranking 54
Brightest stars Phact (α) 2.7, Wezn (β) 3.1
Genitive Columbae
Abbreviation Col
Highest in sky at 10pm January
Fully visible 46°N–90°S

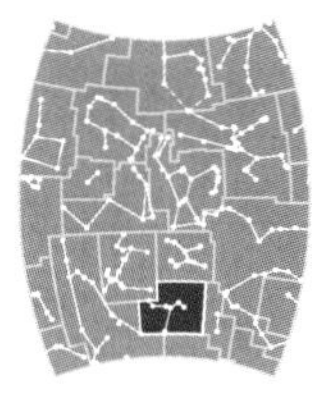

CHART 6

MAIN STARS

Phact Alpha (α) Columbae
Blue-white subgiant
2.7 ⟷ 261 light-years

Wezn Beta (β) Columbae
Yellow giant
3.1 ⟷ 87 light-years

DEEP-SKY OBJECTS

NGC 1792
Spiral galaxy

NGC 1808
Barred spiral galaxy, also a Seyfert galaxy

NGC 1851
Globular cluster

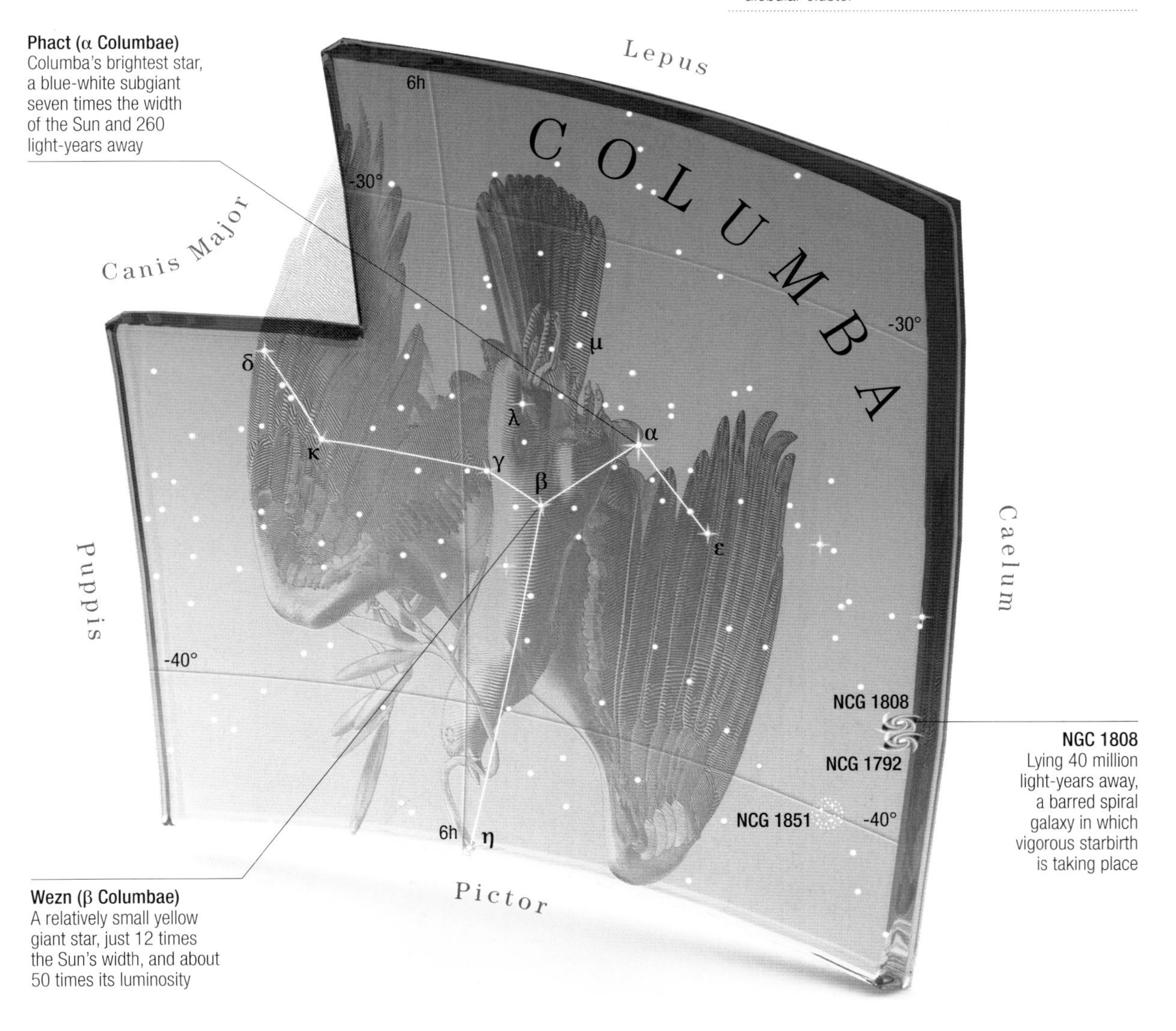
Phact (α Columbae)
Columba's brightest star, a blue-white subgiant seven times the width of the Sun and 260 light-years away
Lepus
6h
COLUMBA
-30°
Canis Major
μ
δ
λ
α
κ
γ
β
ε
Puppis
Caelum
-40°
NCG 1808
NCG 1792
NCG 1851
6h
η
Pictor
Wezn (β Columbae)
A relatively small yellow giant star, just 12 times the Sun's width, and about 50 times its luminosity
NGC 1808
Lying 40 million light-years away, a barred spiral galaxy in which vigorous starbirth is taking place

LUMINOSITIES

Rho Puppis
24 Suns

Tau Puppis
181 Suns

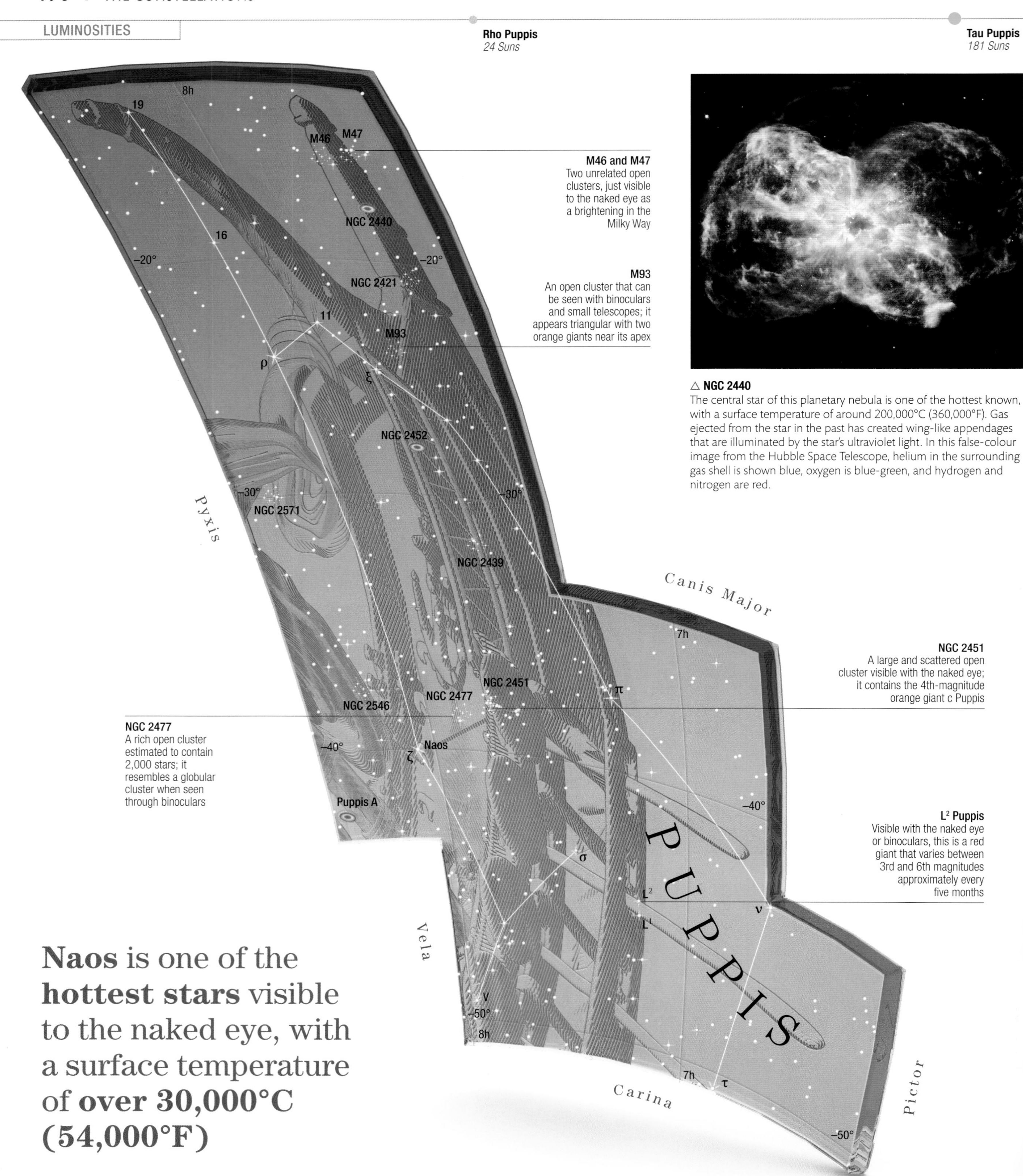

M46 and M47
Two unrelated open clusters, just visible to the naked eye as a brightening in the Milky Way

M93
An open cluster that can be seen with binoculars and small telescopes; it appears triangular with two orange giants near its apex

△ **NGC 2440**
The central star of this planetary nebula is one of the hottest known, with a surface temperature of around 200,000°C (360,000°F). Gas ejected from the star in the past has created wing-like appendages that are illuminated by the star's ultraviolet light. In this false-colour image from the Hubble Space Telescope, helium in the surrounding gas shell is shown blue, oxygen is blue-green, and hydrogen and nitrogen are red.

NGC 2451
A large and scattered open cluster visible with the naked eye; it contains the 4th-magnitude orange giant c Puppis

NGC 2477
A rich open cluster estimated to contain 2,000 stars; it resembles a globular cluster when seen through binoculars

L^2 Puppis
Visible with the naked eye or binoculars, this is a red giant that varies between 3rd and 6th magnitudes approximately every five months

Naos is one of the **hottest stars** visible to the naked eye, with a surface temperature of **over 30,000°C (54,000°F)**

Pi Puppis
4,395 Suns

Naos
12,555 Suns

PUPPIS THE STERN

A MAJOR SOUTHERN CONSTELLATION FOUND NEXT TO CANIS MAJOR, PUPPIS WAS ORIGINALLY DESCRIBED AS PART OF THE MUCH LARGER CONSTELLATION KNOWN TO THE ANCIENT GREEKS AS ARGO NAVIS, THE SHIP. PUPPIS CONTAINS SEVERAL STAR CLUSTERS VISIBLE WITH BINOCULARS AND SMALL TELESCOPES.

For the Ancient Greeks, Puppis represented the stern, or poop, of the legendary ship *Argo* in which Jason and his crew sailed on the quest for the golden fleece.

The early Greek astronomers visualized Argo as a single large constellation, but in the 1750s it was divided into three by the French astronomer Nicholas Louis de Lacaille. The other two parts of the "ship" are the constellations Carina, the hull, and Vela, the sails. Puppis is the largest of the three. However, the brightest stars of Argo are within Carina and Vela, leaving Puppis with only second-magnitude Naos, named for the Greek word for ship, as the brightest member of the constellation.

Two major star clusters in the north of the constellation create a bright patch in the stream of the Milky Way that runs through here. M47 is the closer and larger of the two, about 1,500 light-years away. Next to it lies M46, over three times as distant and hence more difficult to resolve into individual stars. In the far south of the constellation, NGC 2477 is an even richer and brighter cluster.

KEY DATA

Size ranking 20

Brightest stars Naos (ζ) 2.2, Pi (π) 2.7

Genitive Puppis

Abbreviation Pup

Highest in sky at 10pm January–February

Fully visible 39°N–90°S

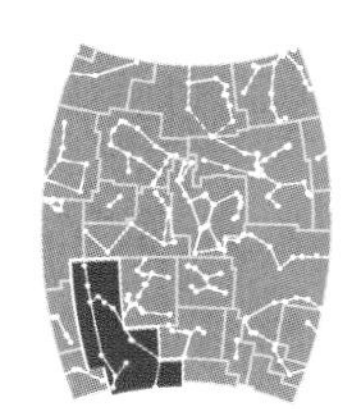

CHART 6

MAIN STARS

Naos Zeta (ζ) Puppis
Blue-white supergiant
2.2 ⟷ 1,080 light-years

Pi (π) Puppis
Orange supergiant
2.7 ⟷ 800 light-years

Rho (ρ) Puppis
White giant
2.8 ⟷ 64 light-years

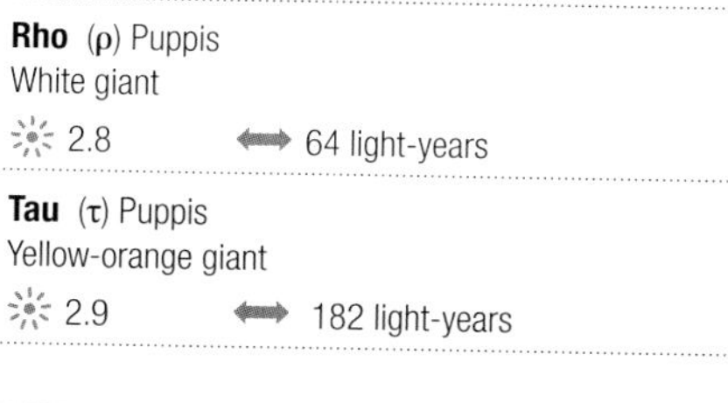

Tau (τ) Puppis
Yellow-orange giant
2.9 ⟷ 182 light-years

DEEP-SKY OBJECTS

M46
Open cluster

M47
Open cluster

M93
Open cluster

NGC 2440
Planetary nebula

NGC 2451
Open cluster

NGC 2452
Planetary nebula

NGC 2477
Open cluster

Puppis A
Supernova remnant

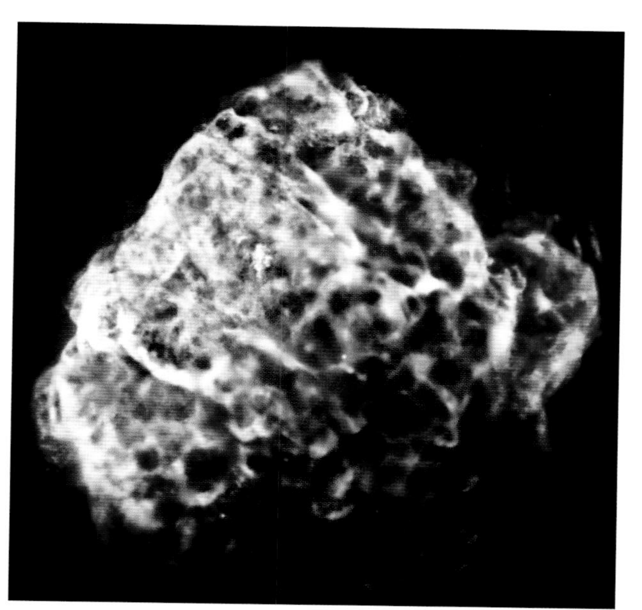

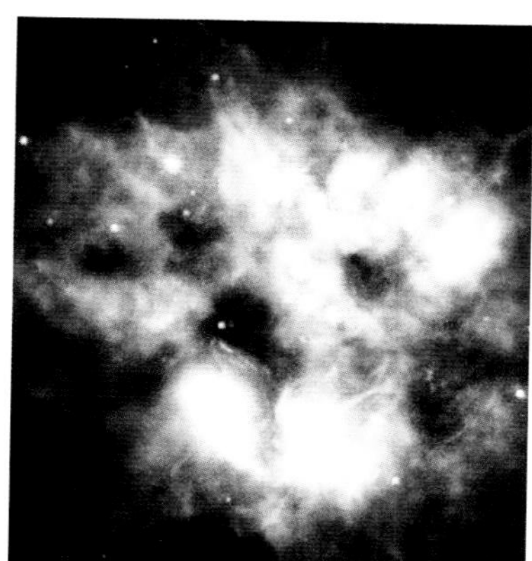

◁ **NGC 2452**
The blue haze in this Hubble Space Telescope image is what remains of the outer layers of a star that have drifted off into space at the end of the star's life, forming a planetary nebula. At the centre of the cloud lies the exposed core of the nebula's progenitor star.

◁ **Puppis A**
Seen here at X-ray wavelengths, Puppis A is the remains of a supernova explosion some 3,700 years ago. It lies about 7,000 light-years away, about eight times farther than the much larger supernova remnant in neighbouring Vela.

▷ **Star distances**
The nearest star to Earth in this constellation's main pattern stars is Rho (ρ) Puppis, which is 64 light-years away. The most distant star is Xi (ξ) Puppis, which is 2,000 light-years from Earth, about 30 times farther away. Naos (ζ Puppis), the constellation's brightest pattern star, is also one of the most distant at 1,080 light-years from Earth.

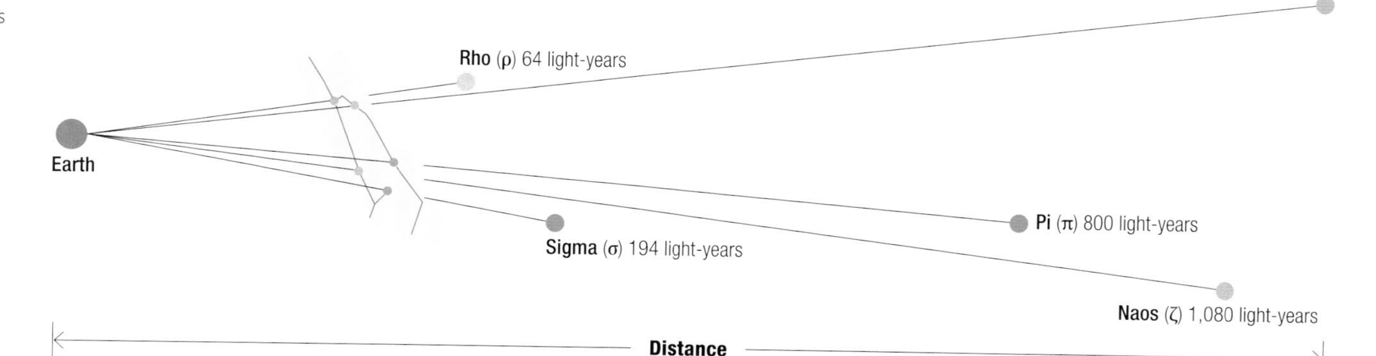

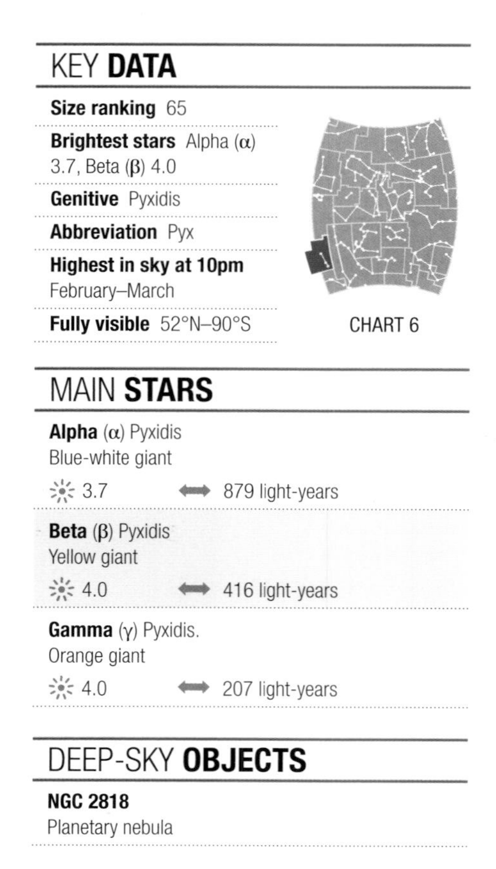

KEY DATA

Size ranking 65

Brightest stars Alpha (α) 3.7, Beta (β) 4.0

Genitive Pyxidis

Abbreviation Pyx

Highest in sky at 10pm February–March

Fully visible 52°N–90°S

CHART 6

MAIN STARS

Alpha (α) Pyxidis
Blue-white giant
3.7 ⟷ 879 light-years

Beta (β) Pyxidis
Yellow giant
4.0 ⟷ 416 light-years

Gamma (γ) Pyxidis.
Orange giant
4.0 ⟷ 207 light-years

DEEP-SKY OBJECTS

NGC 2818
Planetary nebula

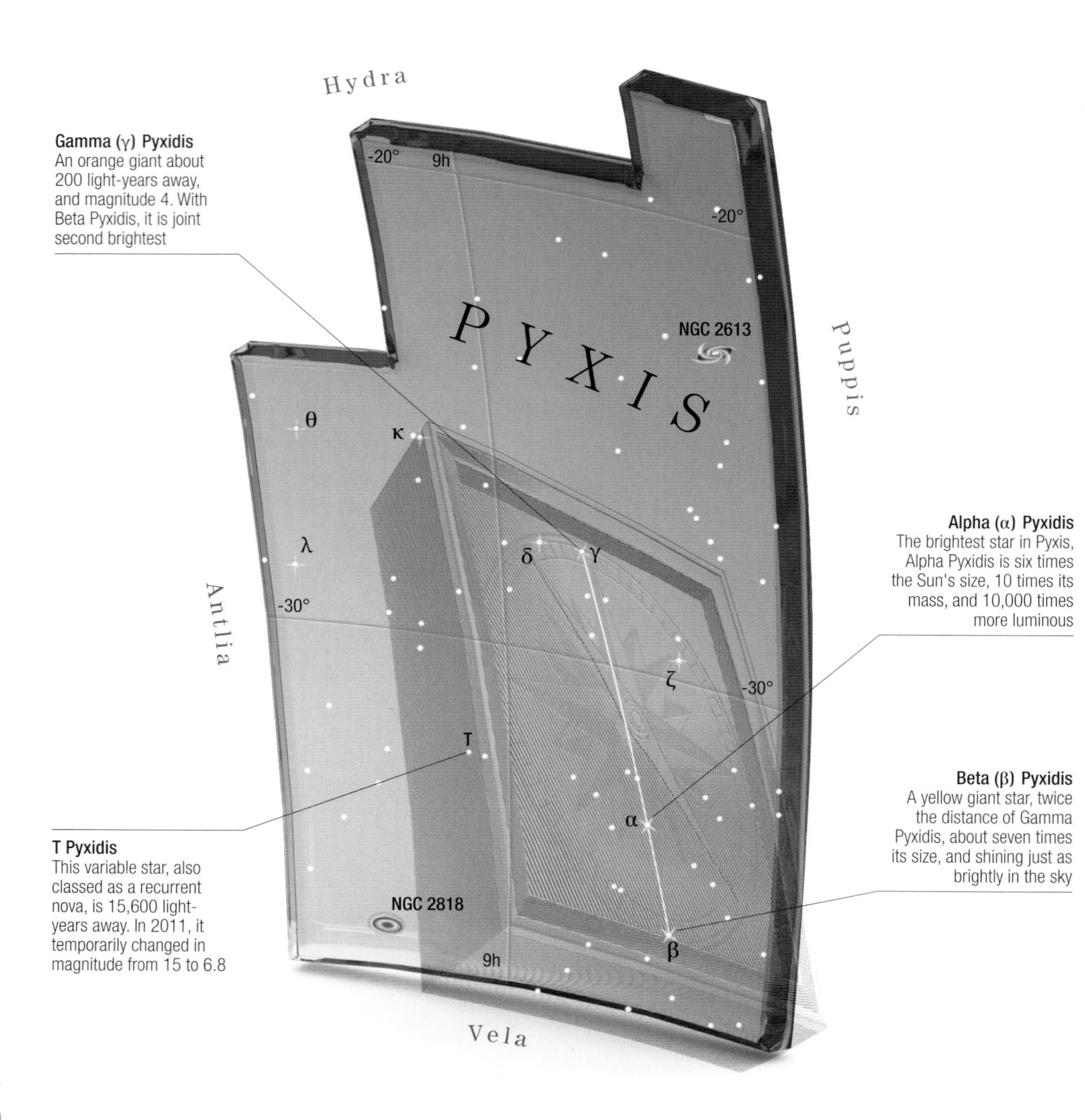

Hydra
Puppis
Antlia
Vela
PYXIS
NGC 2613
NGC 2818
-20°
9h
-30°
θ
κ
λ
δ
γ
ζ
T
α
β
Gamma (γ) Pyxidis
An orange giant about 200 light-years away, and magnitude 4. With Beta Pyxidis, it is joint second brightest
Alpha (α) Pyxidis
The brightest star in Pyxis, Alpha Pyxidis is six times the Sun's size, 10 times its mass, and 10,000 times more luminous
Beta (β) Pyxidis
A yellow giant star, twice the distance of Gamma Pyxidis, about seven times its size, and shining just as brightly in the sky
T Pyxidis
This variable star, also classed as a recurrent nova, is 15,600 light-years away. In 2011, it temporarily changed in magnitude from 15 to 6.8

PYXIS
THE COMPASS

THIS IS A SMALL CONSTELLATION ON THE EDGE OF THE MILKY WAY WHOSE PATTERN IS A ROW OF THREE STARS. PYXIS IS A MAGNETIC COMPASS INTRODUCED TO THE SOUTHERN SKY IN THE 1750s.

Pyxis was devised by the French astronomer Nicholas Louis de Lacaille. After sailing south in 1750 and setting up an observatory at Cape Town, he catalogued the stars of the southern sky and formed some of these into 14 new constellations. The compass is the sort used by seamen and is fittingly next door to Puppis, the ship's stern.

With deep-sky objects such as the barred spiral galaxy NGC 2613 only visible with a large amateur telescope, the constellation's notable objects are its stars. The variable star T Pyxidis consists of a white dwarf pulling material onto its surface from a larger companion. This causes the white dwarf to erupt unpredictably and increase dramatically in brightness. Its last eruption in 2011 was the first since 1966.

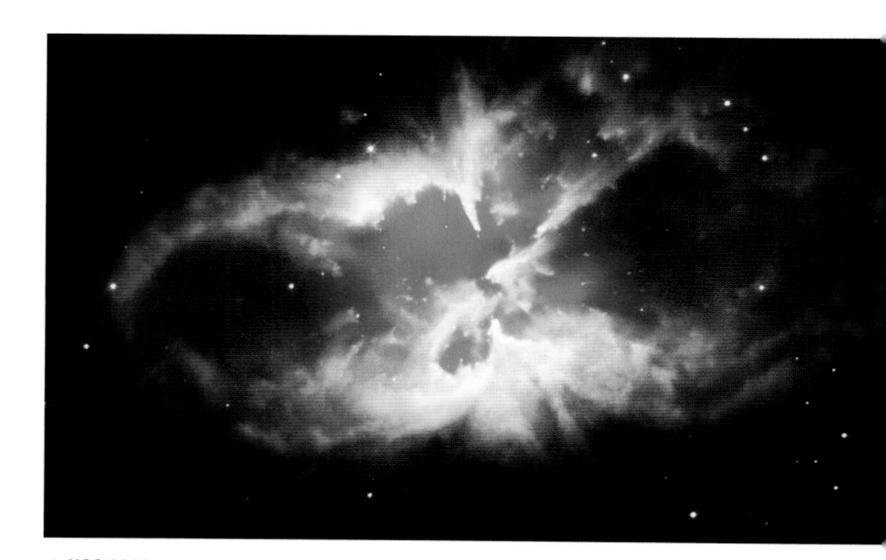

△ **NGC 2818**
More than 10,000 light-years away, this planetary nebula is a star in the process of dying. The outer layers of a once Sun-like star have been pushed off into space. In their centre is a white dwarf, the central remains of the original star. Red represents nitrogen, hydrogen is shown in green, and blue is oxygen.

ANTLIA
THE AIR PUMP

THIS FAINT CONSTELLATION CONTAINS AN INTERESTING CLUSTER OF GALAXIES, BUT ANTLIA IS UNREWARDING FOR THOSE WHO ARE OBSERVING IT WITHOUT USING A LARGE TELESCOPE.

Antlia was introduced by the French astronomer Nicholas Louis de Lacaille. He formed it out of stars seen from his observatory close to Table Mountain, South Africa. On returning to France, he published a star catalogue and a southern-sky map that included Antlia as well as 13 other newly devised constellations. Named Antlia Pneumatica on his map, the constellation represents a vacuum pump.

This inconspicuous grouping of stars has no named stars, bright clusters, or nebulae but includes the Antlia Cluster, the third-nearest cluster of galaxies to us. The two 6th-magnitude stars that are designated Zeta form an optical double and can be seen separately through binoculars.

△ **IC 2560**
The extremely bright nucleus of spiral galaxy IC 2560 can be seen in this Hubble Space Telescope image. It is caused by the ejection of huge amounts of super-hot gas from the region around the galaxy's central black hole. IC 2560 is a member of the Antlia Cluster, a cluster of about 250 galaxies.

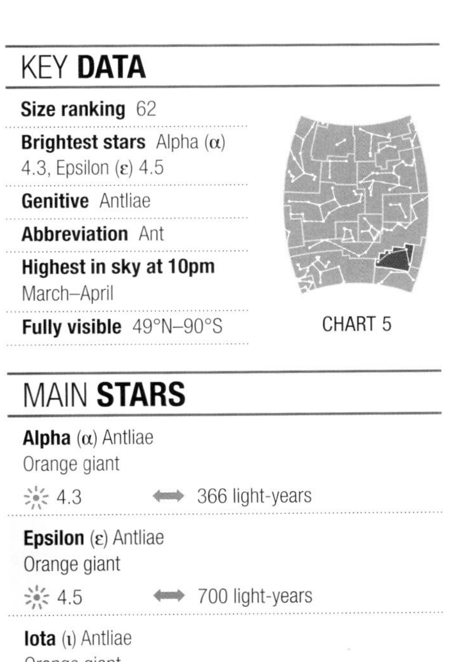

KEY DATA

Size ranking 62

Brightest stars Alpha (α) 4.3, Epsilon (ε) 4.5

Genitive Antliae

Abbreviation Ant

Highest in sky at 10pm March–April

Fully visible 49°N–90°S

CHART 5

MAIN STARS

Alpha (α) Antliae
Orange giant
4.3 ⟷ 366 light-years

Epsilon (ε) Antliae
Orange giant
4.5 ⟷ 700 light-years

Iota (ι) Antliae
Orange giant
4.6 ⟷ 199 light-years

Theta (θ) Antliae
Binary star; a white main-sequence and a yellow giant
4.8 ⟷ 384 light-years

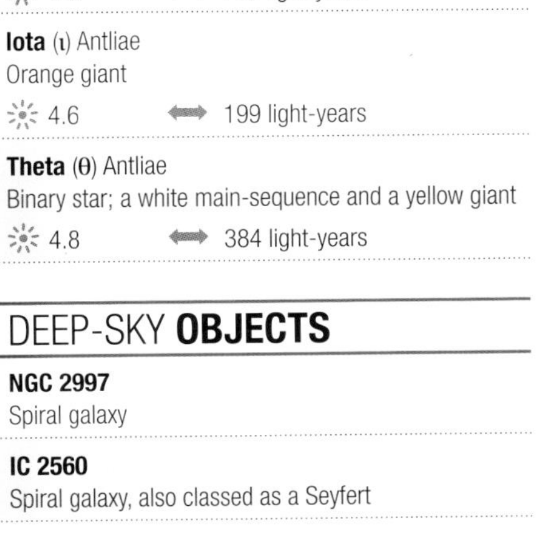

DEEP-SKY OBJECTS

NGC 2997
Spiral galaxy

IC 2560
Spiral galaxy, also classed as a Seyfert

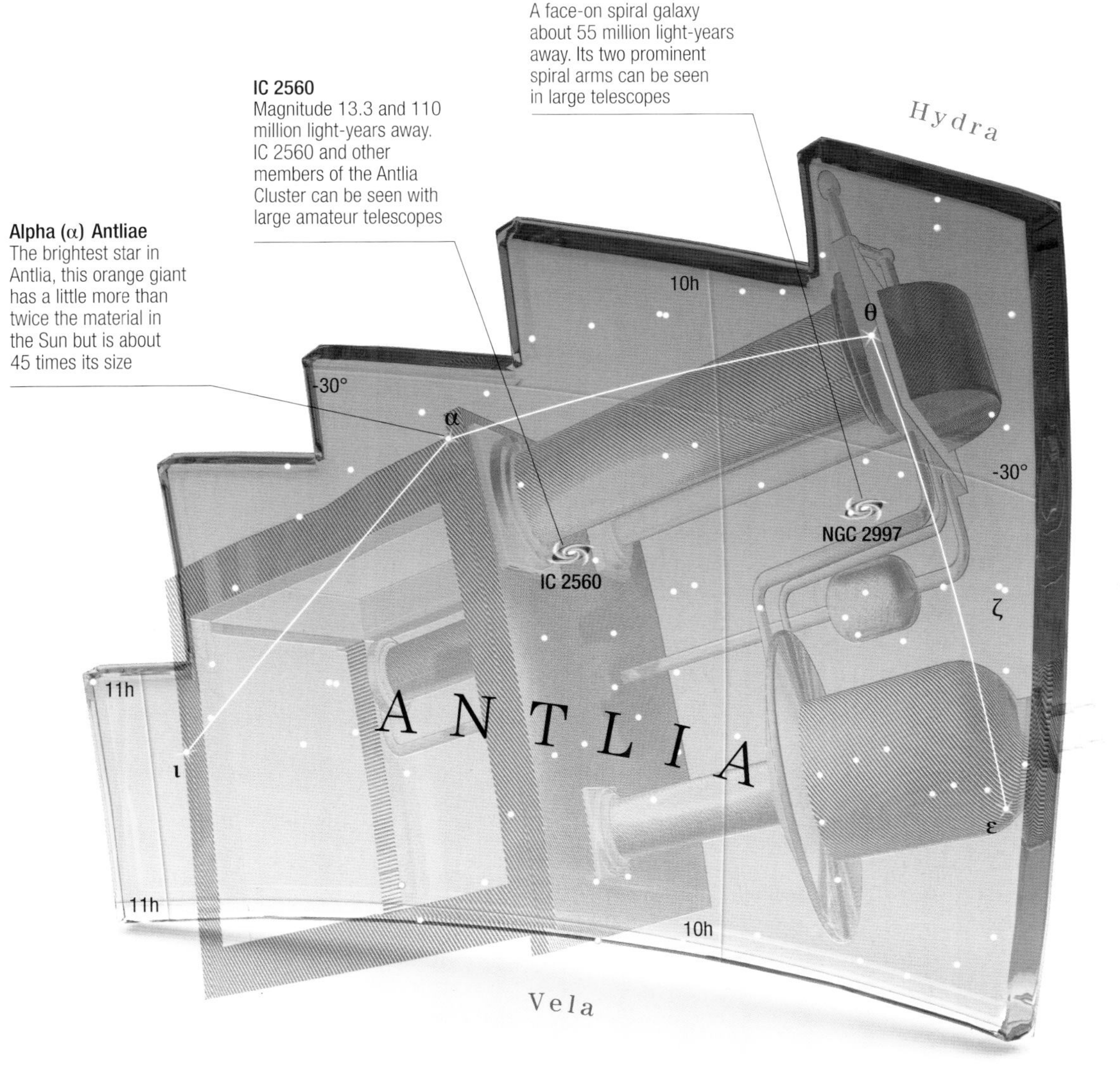

NGC 2997
A face-on spiral galaxy about 55 million light-years away. Its two prominent spiral arms can be seen in large telescopes

IC 2560
Magnitude 13.3 and 110 million light-years away. IC 2560 and other members of the Antlia Cluster can be seen with large amateur telescopes

Alpha (α) Antliae
The brightest star in Antlia, this orange giant has a little more than twice the material in the Sun but is about 45 times its size

LUMINOSITIES

Psi Velorum *11 Suns*

Delta Velorum *90 Suns*

Kappa Velorum *2,760 Suns*

VELA THE SAILS

VELA IS A MAJOR SOUTHERN CONSTELLATION, FORMERLY PART OF THE LARGER ANCIENT GREEK CONSTELLATION OF ARGO NAVIS, THE SHIP. LYING IN A RICH PART OF THE MILKY WAY, VELA CONTAINS THE REMAINS OF A STAR THAT EXPLODED AS A SUPERNOVA ABOUT 11,000 YEARS AGO.

Vela represents the sails of the mythical ship *Argo*, the vessel of Jason and the Argonauts. The Ancient Greeks visualized the ship as a single huge constellation, but the French astronomer Nicolas Louis de Lacaille divided it into three smaller parts in the 1750s. The other two sections are Carina, the hull, and Puppis, the stern. Vela contains several prominent star clusters, including IC 2391, a group of about 50 stars visible to the naked eye. Delta and Kappa Velorum, along with Epsilon and Iota Carinae, form a shape known as the False Cross, which is sometimes mistaken for the true Southern Cross. Vela's most remarkable object is the Vela Supernova remnant. Lying about 800 light-years away, it is among the closest supernova remnants to us. Near its centre lies the fast-spinning Vela pulsar, the remaining core of the star that exploded as a supernova in prehistoric times.

KEY DATA

Size ranking 32

Brightest stars Gamma (γ) 1.8, Delta (δ) 2.0–2.4

Genitive Velorum

Abbreviation Vel

Highest in sky at 10pm February–April

Fully visible 32°N–90°S

CHART 2

MAIN STARS

Gamma (γ) Velorum
Brightest Wolf-Rayet star visible from Earth
1.8 — 1,100 light-years

Delta (δ) Velorum
Eclipsing binary
2.0–2.4 — 80 light-years

Kappa (κ) Velorum
Blue-white subgiant or main-sequence star
2.5 — 570 light-years

Lambda (λ) Velorum
Orange supergiant
2.2 — 545 light-years

DEEP-SKY OBJECTS

IC 2391
Open cluster

NGC 2736 (Pencil Nebula)
Part of the Vela Supernova remnant

NGC 3132 (Eight-Burst Nebula)
Planetary nebula, also called the Southern Ring Nebula

NGC 3228
Open cluster

Vela Supernova remnant
Supernova remnant with central pulsar

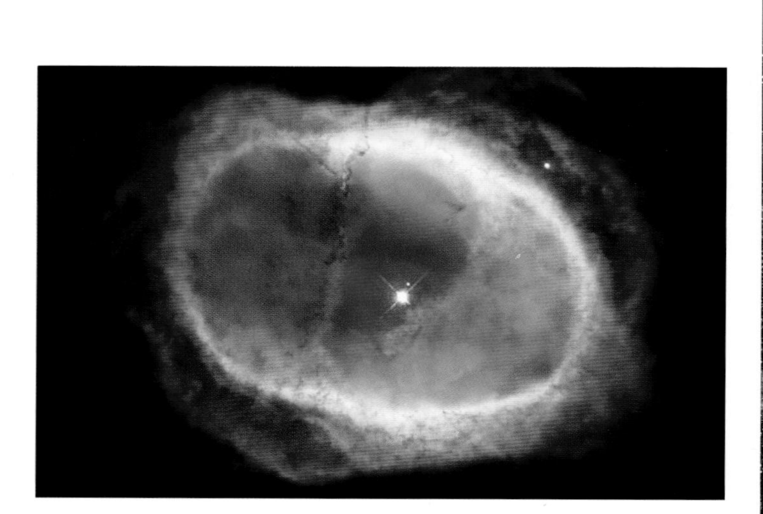

△ **Eight-Burst Nebula**
Also called NGC 3132, this planetary nebula is shaped like overlapping figures-of-eight, hence its popular name. Ultraviolet light from the central star has heated surrounding gas (shown as blue in this image).

▷ **Pencil Nebula**
Part of the Vela Supernova remnant, the Pencil Nebula is a region where the supernova's shock wave slammed into a denser region of interstellar gas, compressing it into the glowing strip seen in this Hubble image.

△ **Vela Supernova remnant**
This wide-field image shows the faint ribbons of gas that are the remains of a star that exploded as a supernova about 11,000 years ago. Lying between the stars Gamma (γ) and Lambda (λ) Velorum, the supernova remnant stretches across a region of sky about the width of 16 Full Moons.

The **Vela pulsar rotates** at more than 11 revolutions per second, **faster than a spinning helicopter rotor**

Lambda Velorum
3,115 Suns

Phi Velorum
8,100 Suns

Gamma Velorum
20,380 Suns

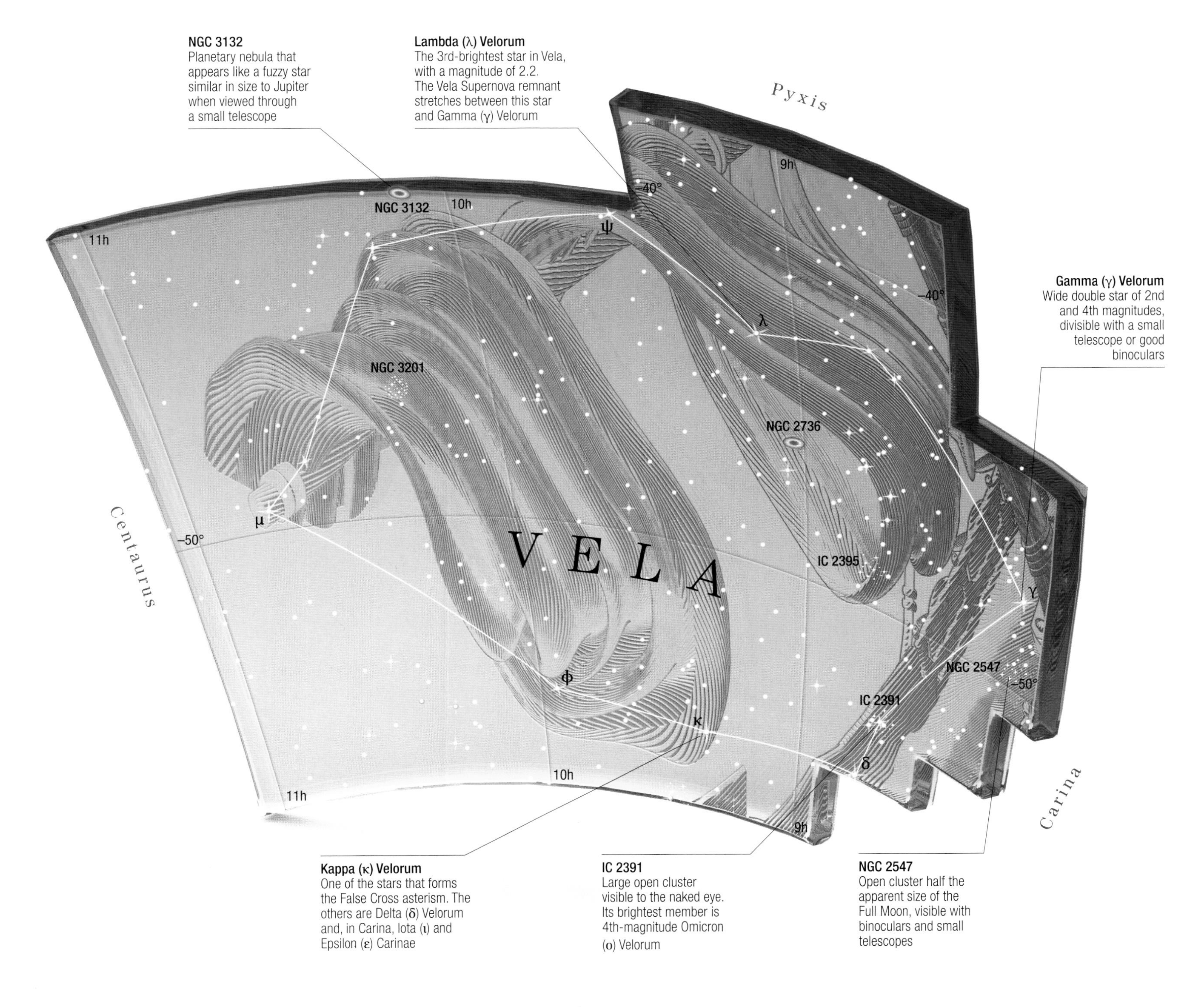

NGC 3132
Planetary nebula that appears like a fuzzy star similar in size to Jupiter when viewed through a small telescope
Lambda (λ) Velorum
The 3rd-brightest star in Vela, with a magnitude of 2.2. The Vela Supernova remnant stretches between this star and Gamma (γ) Velorum
Pyxis
Gamma (γ) Velorum
Wide double star of 2nd and 4th magnitudes, divisible with a small telescope or good binoculars
NGC 3132
10h
11h
9h
-40°
ψ
λ
NGC 3201
NGC 2736
Centaurus
μ
-50°
VELA
IC 2395
γ
NGC 2547
-50°
φ
IC 2391
κ
δ
10h
11h
9h
Carina
Kappa (κ) Velorum
One of the stars that forms the False Cross asterism. The others are Delta (δ) Velorum and, in Carina, Iota (ι) and Epsilon (ε) Carinae
IC 2391
Large open cluster visible to the naked eye. Its brightest member is 4th-magnitude Omicron (ο) Velorum
NGC 2547
Open cluster half the apparent size of the Full Moon, visible with binoculars and small telescopes

▷ **Star distances**
The nearest of Vela's main pattern stars is Psi (ψ) Velorum, which is only 61 light-years away. The farthest is Phi (φ) Velorum, at about 1,590 light-years. Even though Gamma (γ) Velorum is about 1,00 light-years from us, it is the brightest of Vela's pattern stars. It is also the most luminous, emitting a total amount of energy equivalent to more than 20,300 Suns.

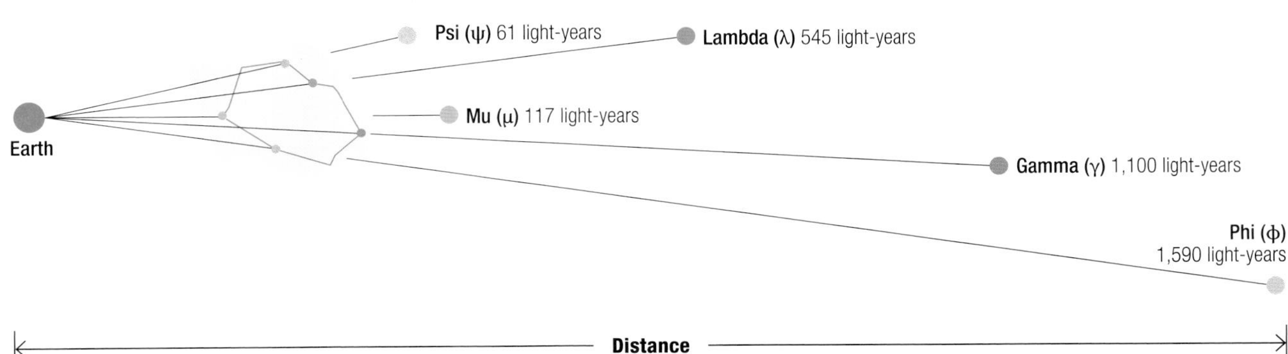

Psi (ψ) 61 light-years
Lambda (λ) 545 light-years
Earth
Mu (μ) 117 light-years
Gamma (γ) 1,100 light-years
Phi (φ) 1,590 light-years
Distance

LUMINOSITIES

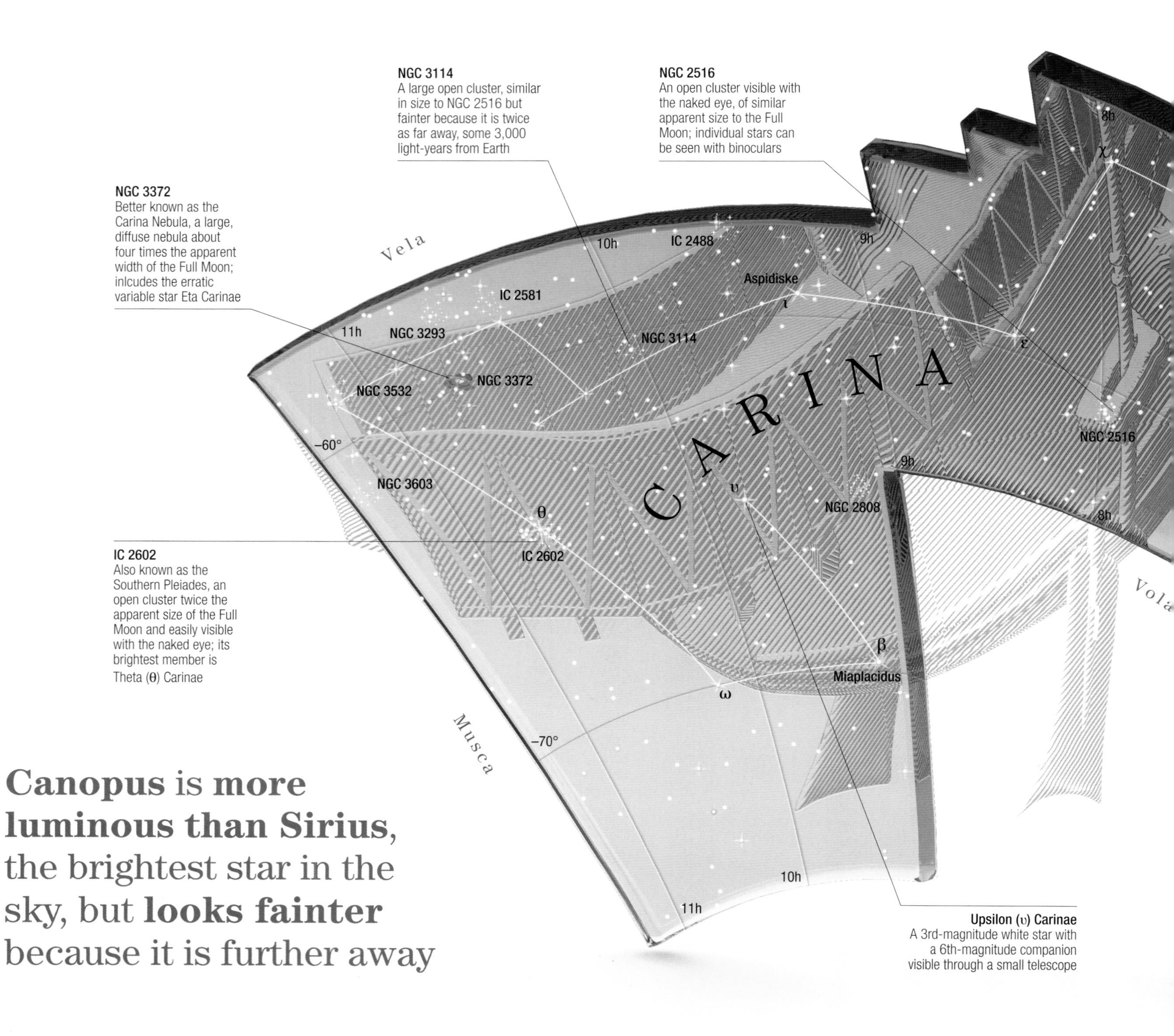

NGC 3114
A large open cluster, similar in size to NGC 2516 but fainter because it is twice as far away, some 3,000 light-years from Earth
NGC 2516
An open cluster visible with the naked eye, of similar apparent size to the Full Moon; individual stars can be seen with binoculars
NGC 3372
Better known as the Carina Nebula, a large, diffuse nebula about four times the apparent width of the Full Moon; inlcudes the erratic variable star Eta Carinae
IC 2602
Also known as the Southern Pleiades, an open cluster twice the apparent size of the Full Moon and easily visible with the naked eye; its brightest member is Theta (θ) Carinae
Upsilon (υ) Carinae
A 3rd-magnitude white star with a 6th-magnitude companion visible through a small telescope
CARINA
Vela
Volans
Musca
IC 2488
IC 2581
Aspidiske
NGC 3293
NGC 3114
NGC 3372
NGC 3532
NGC 3603
IC 2602
NGC 2808
NGC 2516
Miaplacidus
10h
11h
9h
8h
–60°
–70°

Canopus is **more luminous than Sirius**, the brightest star in the sky, but **looks fainter** because it is further away

▷ **Star distances**
Carina's main pattern stars lie between 113 and about 1,400 light-years from Earth. The constellation's two brightest stars, Canopus (α) and Miaplacidus (β), are also the two nearest of the pattern stars. The farthest star, Upsilon (υ) Carinae, is an outlier, lying about twice as far from Earth as the second-farthest pattern star, Iota (ι) Carinae.

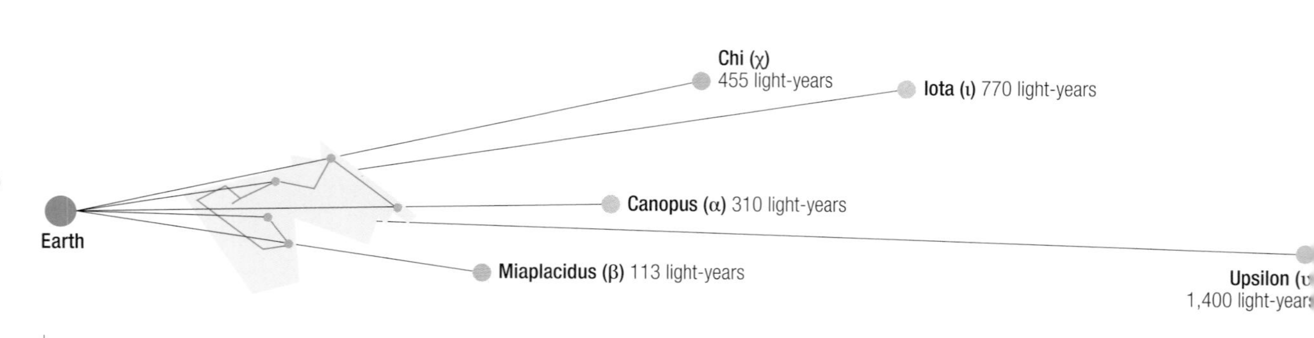

Aspidiske
6,270 Suns

Canopus
13,855 Suns

Eta Carinae
More than 5 million Suns

CARINA
THE KEEL

A PROMINENT SOUTHERN CONSTELLATION, CARINA CONTAINS THE SECOND-BRIGHTEST STAR IN THE SKY, CANOPUS, ALONG WITH RICH MILKY WAY STAR FIELDS.

Carina represents the hull of the mythical ship *Argo*, which the Ancient Greeks visualized as a single large constellation. This constellation was split into three (Carina, Vela, the sails, and Puppis, the stern) by the 18th-century French astronomer Nicolas Louis de Lacaille, and its two brightest stars, Canopus and Miaplacidus, ended up in Carina.

Carina contains one of the most extraordinary stars known, Eta Carinae. Currently just visible to the naked eye, it flared up in 1843 to become temporarily brighter than Canopus. It is thought to be a massive binary, obscured by the debris thrown off in the great eruption. It lies within the Carina Nebula (see pp.204–205), a cloud of gas that is larger and brighter than the Orion Nebula. The stars Epsilon and Iota Carinae, together with Delta and Kappa Velorum, form the so-called False Cross, an asterism that resembles the true Southern Cross.

KEY DATA

Size ranking 34

Brightest stars Canopus (α) -0.7, Miaplacidus (β) 1.7

Genitive Carinae

Abbreviation Car

Highest in sky at 10pm January–April

Fully visible 14°N–90°S

CHART 2

MAIN STARS

Canopus Alpha (α) Carinae
White giant
-0.7 ⟷ 310 light-years

Miaplacidus Beta (β) Carinae
Blue-white giant
1.7 ⟷ 113 light-years

Epsilon (ε) Carinae
Orange giant
2.0 ⟷ 600 light-years

Theta (θ) Carinae
Blue-white main-sequence star
2.8 ⟷ 455 light-years

Aspidiske Iota (ι) Carinae
White supergiant
2.3 ⟷ 770 light-years

Upsilon (υ) Carinae
White supergiant
3.0 ⟷ 1,400 light-years

DEEP-SKY OBJECTS

NGC 2516
Open cluster

NGC 3114
Open cluster

NGC 3372 (Carina Nebula)
Bright diffuse nebula

NGC 3532
Open cluster

IC 2602 (Southern Pleiades)
Open cluster

NGC 3603
his combined visible light and infrared image reveals n enormous cavity in the gas around the massive star uster NGC 3603. The cavity was created by traviolet radiation and stellar winds from young, hot ars in the cluster.

◁ **Eta Carinae and the Keyhole Nebula**
A bright area of glowing gas surrounds the binary star Eta Carinae (centre left of the image), thrown off in its eruption of 1843. Eta Carinae lies within the Carina Nebula, which includes the region called the Keyhole (the elongated darker area just right of Eta Carinae).

DUST CLOUDS IN CARINA

Fantasy-like structures are found throughout the Carina Nebula, a vast molecular cloud of star birth that spans across 300 light-years of space. This detailed false-colour view shows a small part, roughly 15 light-years wide. The fantastic shapes are sculpted out of the cold cloud by stellar winds and ultra-violet radiation emitted by massive stars, as they slowly erode it away. Dark knots of gas and dust are so thick they are opaque, although the cloud is typically less dense than Earth's atmosphere. The dark pillar of cold hydrogen and dust to the right of this image is more than two light-years long, and has so far resisted being worn away. Inside it new stars are taking shape. The image was taken by the Hubble Space Telescope and combines two sets of observations. The first, from 2005, were taken in light emitted by hydrogen atoms. The second, taken in 2010, were in light emitted by oxygen atoms.

MUSCA
THE FLY

A SMALL CONSTELLATION IMMEDIATELY TO THE SOUTH OF CRUX IN THE SOUTHERN SKY, MUSCA'S STARS ARE RELATIVELY BRIGHT BUT CAN GET LOST IN THE BACKGROUND OF THE MILKY WAY.

Musca is best located by first finding the brilliant stars of Crux. The fly is the sky's only insect. Its body is drawn around the constellation's brightest stars. First devised as Apis the bee by Dutch navigators Pieter Dirkszoon Keyser and Frederick de Houtman in the 1590s, it became a fly in the 1750s.

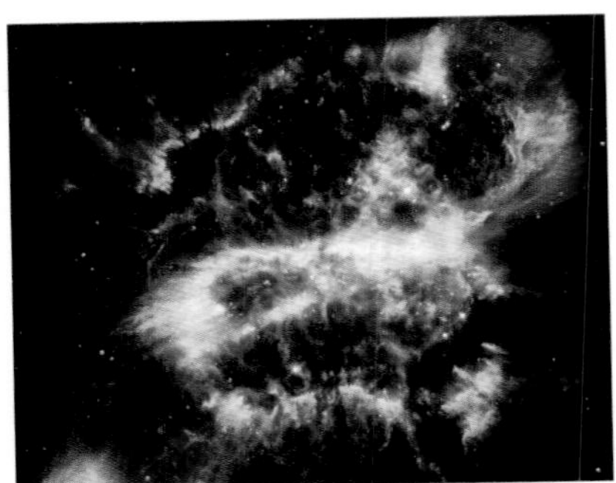

◁ **NGC 5189**
Also sometimes called the Spiral Planetary Nebula, NGC 5189 is a planetary nebula with expelled material rushing away from a dying star, a white dwarf. Unusually, NGC 5189 has two central stars; the white dwarf and a Wolf-Rayet type star. The presence of the two stars explains the complex structure of the surrounding gas.

NGC 4833
A globular cluster 21,500 light-years away with a magnitude of 7.8 and it is seen in binoculars as a hazy ball of light

Alpha (α) Muscae
A blue-white subgiant evolving into a giant star, 315 light-years away. It is a Cepheid variable pulsating every 2 .2 hours

KEY DATA

Size ranking 77

Brightest stars Alpha (α) 2.7, Beta (β) 3.1

Genitive Muscae

Abbreviation Mus

Highest in sky at 10pm April–May

Fully visible 14°N–90°S

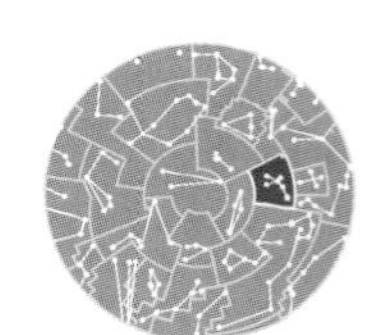

CHART 2

CIRCINUS
THE COMPASSES

ONE OF THE SMALLEST CONSTELLATIONS, CIRCINUS IS SQUEEZED INTO A GAP BETWEEN CENTAURUS AND TRIANGULUM AUSTRALE. IT IS BEST FOUND BY LOCATING THE BRIGHT STAR ALPHA CENTAURI.

Circinus was introduced by Frenchman Nicolas Louis de Lacaille in the 1750s. Drawn around a faint triangle of stars, it represents the compasses used by draughtsmen and navigators. It includes the Circinus Galaxy, which is one of the closest Seyfert galaxies to us. Also noteworthy is RCW 86, the remnant from a supernova explosion witnessed by Chinese astronomers in 185 CE.

◁ **RCW 86**
This colourful band of gas and dust is part of a roughly circular supernova remnant named RCW 86. It is material left over from a time when a white dwarf exploded after siphoning material from a nearby star.

Alpha (α) Circinus
A white main-sequence star, 54 light-years away. Its orange dwarf companion of magnitude 8.6 is seen through a small telescope

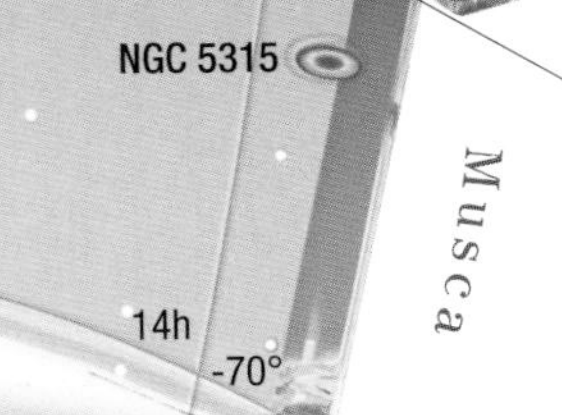

Circinus Galax
A small spiral galax
13 million light-year
away, with an activ
supermassive black ho
in its centr

KEY DATA

Size ranking 85

Brightest stars Alpha (α) 3.2, Beta (β) 4.1

Genitive Circini

Abbreviation Cir

Highest in sky at 10pm May–June

Fully visible 19°N–90°S

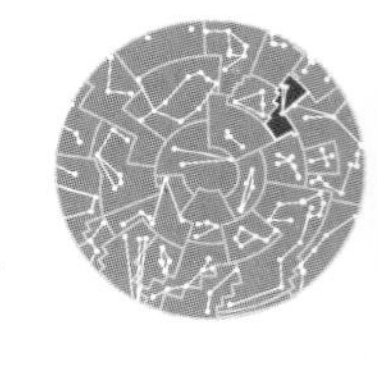

CHART 2

TRIANGULUM AUSTRALE

THE SOUTHERN TRIANGLE

A SMALL CONSTELLATION DEVISED BY LINKING THREE BRIGHT STARS, TRIANGULUM AUSTRALE MAKES A DISTINCTIVE PATTERN AND IS CROSSED BY A STAR-RICH REGION OF THE MILKY WAY.

This bright triangle of stars is easy to spot to the south east of Centaurus. Its not certain who devised the constellation but it was first recorded in Johann Bayer's star atlas, *Uranometria*, in 1603. Although crossed by the Milky Way, Triangulum Australe has little of interest to amateur astronomers, except star cluster NGC 6025.

◁ **ESO 69-6**
Long tails sweep out from each of the galaxies in this interacting pair, together known as ESO 69-6. The tails are gas and stars that have been ripped out of the outer regions of the galaxies. The galaxies are about 650 million light-years from Earth.

KEY DATA

Size ranking 83
Brightest stars Alpha (α) 1.9, Beta (β) 2.8
Genitive Trianguli Australis
Abbreviation TrA
Highest in sky at 10pm June–July
Fully visible 19°N–90°S

CHART 2

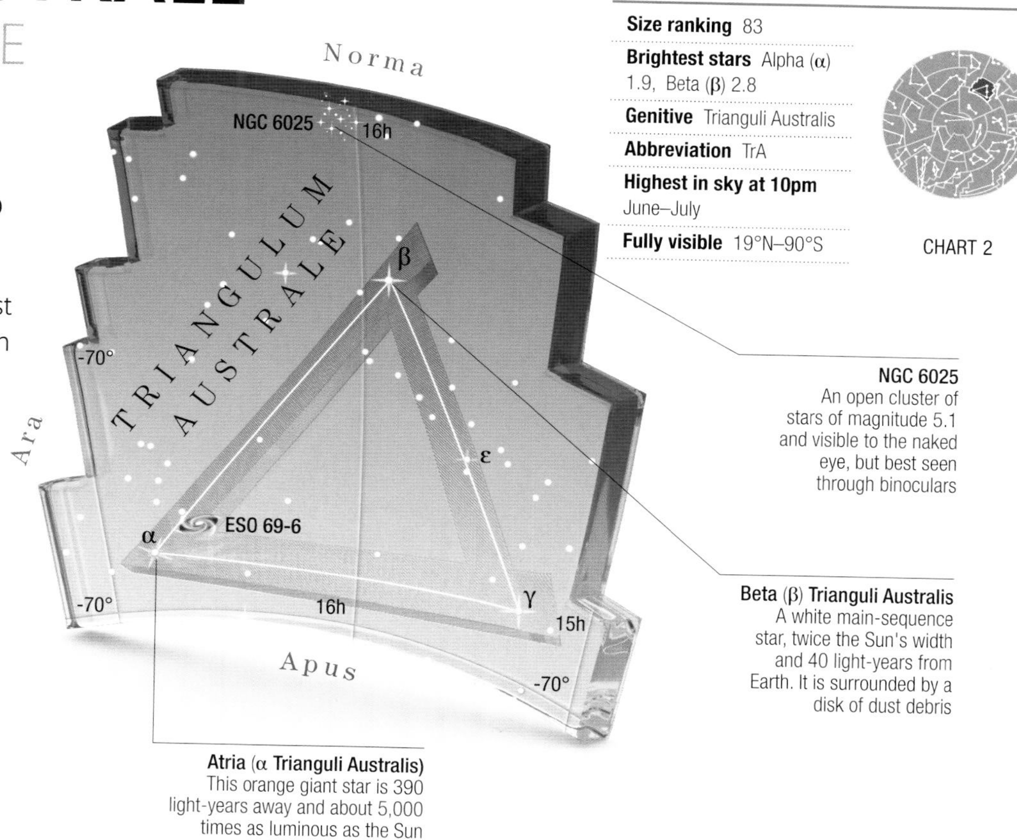

NGC 6025
An open cluster of stars of magnitude 5.1 and visible to the naked eye, but best seen through binoculars

Beta (β) Trianguli Australis
A white main-sequence star, twice the Sun's width and 40 light-years from Earth. It is surrounded by a disk of dust debris

Atria (α Trianguli Australis)
This orange giant star is 390 light-years away and about 5,000 times as luminous as the Sun

TELESCOPIUM

THE TELESCOPE

THIS IS A FAINT AND OBSCURE CONSTELLATION, INTRODUCED TO THE SOUTHERN SKY IN THE 1750s. TELESCOPIUM IS SOUTH OF THE DISTINCTIVE SAGITTARIUS AND CORONA AUSTRALIS.

One of the least recognizable constellations, Telescopium's pattern is drawn around a right angle of linked stars tucked into a corner of the constellation's sky area. It was devised by Frenchman Nicolas Louis de Lacaille, using additional stars from surrounding constellations. These have been returned, leaving Telescopium in its present state.

▷ **NGC 6861**
This is a lenticular galaxy whose disk is tilted to our line of sight. Dark bands within the disk are the result of large clouds of dust particles obscuring the light from more distant stars.

KEY DATA

Size ranking 57
Brightest stars Alpha (α) 3.5, Zeta (ζ) 4.1
Genitive Telescopii
Abbreviation Tel
Highest in sky at 10pm July–August
Fully visible 33°N–90°S

CHART 4

NGC 6861
A lenticular galaxy, magnitude 11.1. It is a member of a group of about a dozen galaxies named the Telescopium Group

Sagittarius
Corona Australis
TELESCOPIUM
20h
19h
ι
δ¹
δ²
α
ε
-50°
NGC 6861
ζ
-50°
ξ
λ
Ara
20h
19h
Pavo

INDUS
THE INDIAN

INTRODUCED TO THE SOUTHERN SKY IN THE 16TH CENTURY, THIS CONSTELLATION REPRESENTS AN INDIAN, ALTHOUGH IT IS UNCLEAR WHETHER THIS REFERS TO A NATIVE OF THE AMERICAS OR ASIA.

Indus is one of the 12 figures invented by Dutch navigators Pieter Dirkszoon Keyser and Frederick de Houtman, who charted the sky of the southern hemisphere during the 1590s. The Indian figure carries a spear and arrows, and is drawn around three stars that form a right angle in the northern part of the constellation. Indus's brightest stars are of 3rd-magnitude and it has no significant star clusters or nebulae. One notable star is Epsilon Indi, a yellow main-sequence, agnitude-4.7 star just 11.2 light-years away, making it one of the closest stars to us.

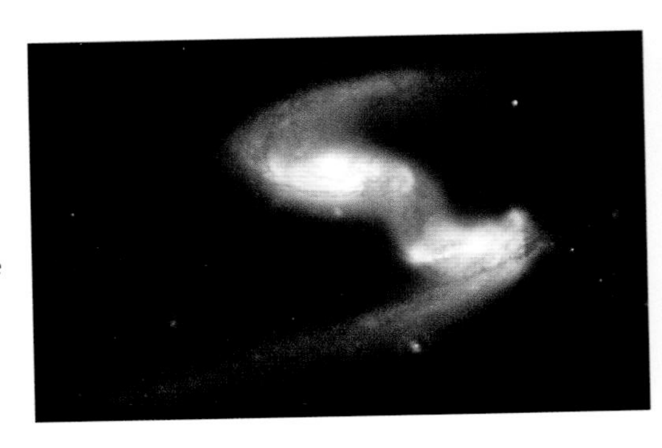

▷ **ESO 77-14**
The once-flat disks of two similar-sized galaxies have become distorted and the pair are connected by material that used to be inside the galaxies. A short, red arm of gas and dust has been pulled out of the top galaxy, the lower galaxy has a longer bluish arm.

Alpha (α) Indi
An orange giant about 12 times the width of the Sun and with 100 times its luminosity. Its two companions can be seen with a medium-sized telescope

Beta (β) Indi
The second-brightest star in this constellation, Beta is an orange giant of magnitude 3.7, and is 600 light-years away

NGC 7090
A spiral galaxy seen edge-on and about 30 million light-years away. It was discovered by British astronomer John Herschel in 1834

ESO 77-14
A pair of galaxies about 550 million light-years away that are distorted by their mutual gravitational pull

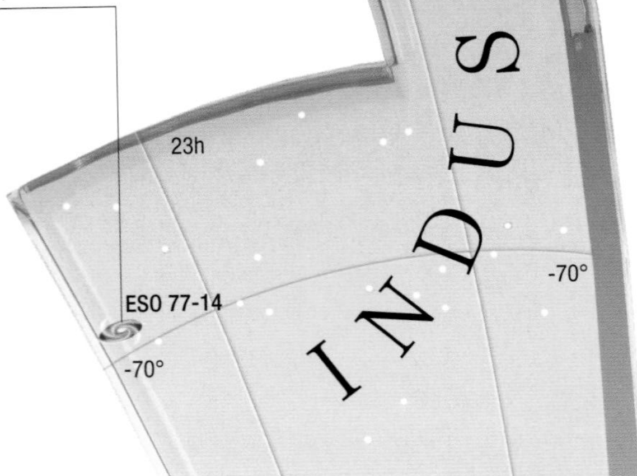

KEY DATA

Size ranking 49

Brightest stars Alpha (α) 3.1, Beta (β) 3.7

Genitive Indi

Abbreviation Ind

Highest in sky at 10pm August–October

Fully visible 15°N–90°S

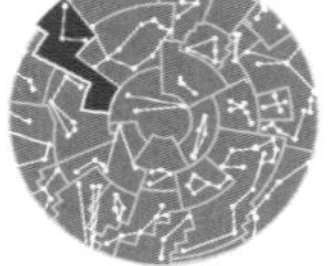

CHART 2

MAIN STARS

Alpha (α) Indi
Orange giant
☼ 3.1 ⟷ 98 light-years

Beta (β) Indi
Orange giant
☼ 3.7 ⟷ 610 light-years

DEEP-SKY OBJECTS

NGC 7049
Lenticular galaxy

NGC 7090
Spiral galaxy

ESO 77-14
A pair of interacting galaxies

△ **NGC 7090**
This side-on view from Earth of the spiral galaxy NGC 7090 shows its disk and bulging central core. The pinkish-red regions indicate clouds of hydrogen gas where star formation is taking place. Dark regions inside the disk are lanes of dust.

PHOENIX
THE PHOENIX

DEPICTING A MYTHICAL BIRD, THE PHOENIX, THIS INDISTINCT CONSTELLATION WAS DEVISED IN THE 16TH CENTURY. IT LIES BETWEEN SCULPTOR TO THE NORTH AND TUCANA TO THE SOUTH.

According to legend, the phoenix lived for hundreds of years, died in flames, and a young phoenix was born from its ashes. The constellation was devised by Dutch navigators Pieter Dirkszoon Keyser and Frederick de Houtman. Ankaa marks the bird's beak at the north end of a rectangle that forms its body, with open wings at either side. Phoenix has interesting double stars and two of the most massive galaxy clusters known: the Phoenix Cluster and the El Gordo Cluster.

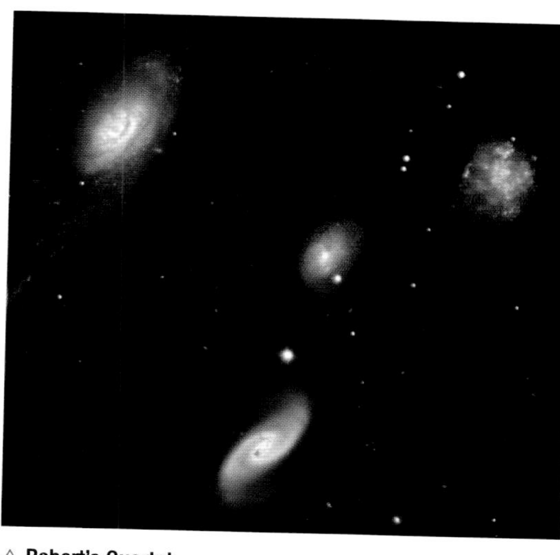

△ **Robert's Quartet**
This is a group of four galaxies about 160 million light-years away that are interacting. They are an irregular galaxy (right) and three spiral galaxies. An arm of the largest spiral (top left) has been distorted and the galaxy has at least 200 areas of intense star formation. There is a diffuse area of material around the central galaxy, and the one below it has two spiral arms.

KEY DATA

Size ranking 37
Brightest stars Alpha (α) 2.4, Beta (β) 3.3
Genitive Phoenicis
Abbreviation Phe
Highest in sky at 10pm October–November
Fully visible 32°N–90°S

CHART 3

MAIN STARS

Ankaa Alpha (α) Phoenicis
Orange giant
2.4 ⟷ 85 light-years

Beta (β) Phoenicis
Yellow giant
3.3 ⟷ 225 light-years

DEEP-SKY OBJECTS

Robert's Quartet
Group of interacting galaxies

El Gordo Cluster
Largest-known galaxy cluster

Phoenix Cluster
Massive galaxy cluster

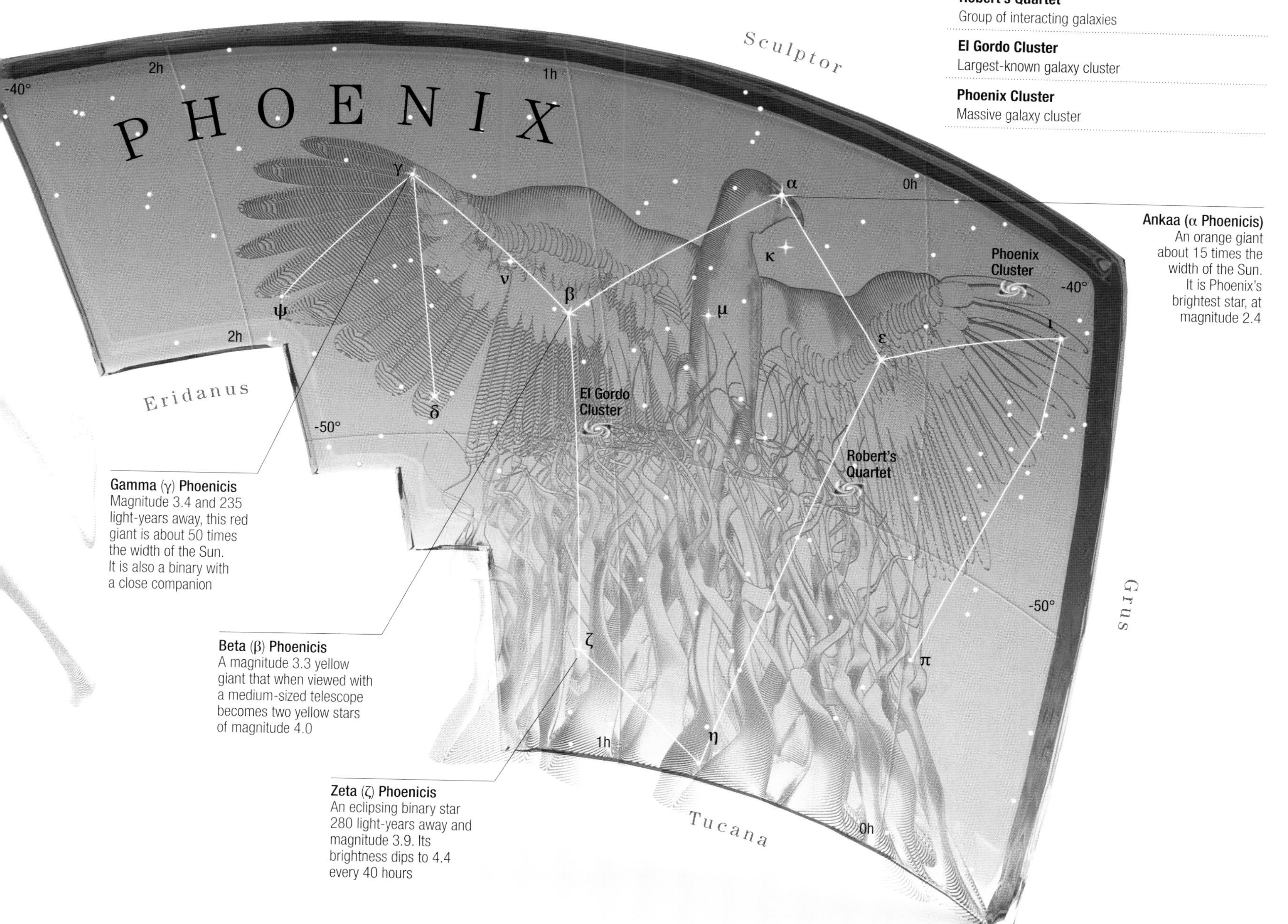

Ankaa (α Phoenicis)
An orange giant about 15 times the width of the Sun. It is Phoenix's brightest star, at magnitude 2.4

Gamma (γ) Phoenicis
Magnitude 3.4 and 235 light-years away, this red giant is about 50 times the width of the Sun. It is also a binary with a close companion

Beta (β) Phoenicis
A magnitude 3.3 yellow giant that when viewed with a medium-sized telescope becomes two yellow stars of magnitude 4.0

Zeta (ζ) Phoenicis
An eclipsing binary star 280 light-years away and magnitude 3.9. Its brightness dips to 4.4 every 40 hours

LUMINOSITIES

Zeta Doradus
2 Suns

Gamma Doradus
7 Suns

DORADO
THE GOLDFISH

A CHAIN OF STARS NEAR THE BRILLIANT STAR CANOPUS IN CARINA, DORADO INCLUDES THE IMPRESSIVE LARGE MAGELLANIC CLOUD, A NEIGHBOURING GALAXY OF THE MILKY WAY.

Although commonly described as a goldfish and sometimes depicted as a swordfish, Dorado represents the dolphinfish, a species found in tropical waters. Its shape is drawn around a faint chain of stars, with the fish swimming towards the south celestial pole. The constellation was introduced to the southern sky in the 1590s by the Dutch navigators Pieter Dirkszoon Keyser and Frederick de Houtman. It has no very bright stars, and none are named. Its most spectacular feature, the Large Magellanic Cloud (LMC), is visible to the naked eye but binoculars reveal its numerous star clusters and nebulous patches in more detail. The LMC is named after Ferdinand Magellan, the Portuguese explorer who recorded it in the early 1520s. The Tarantula Nebula is part of the LMC, and the Supernova 1987A was seen in the outskirts of this nebula in 1987.

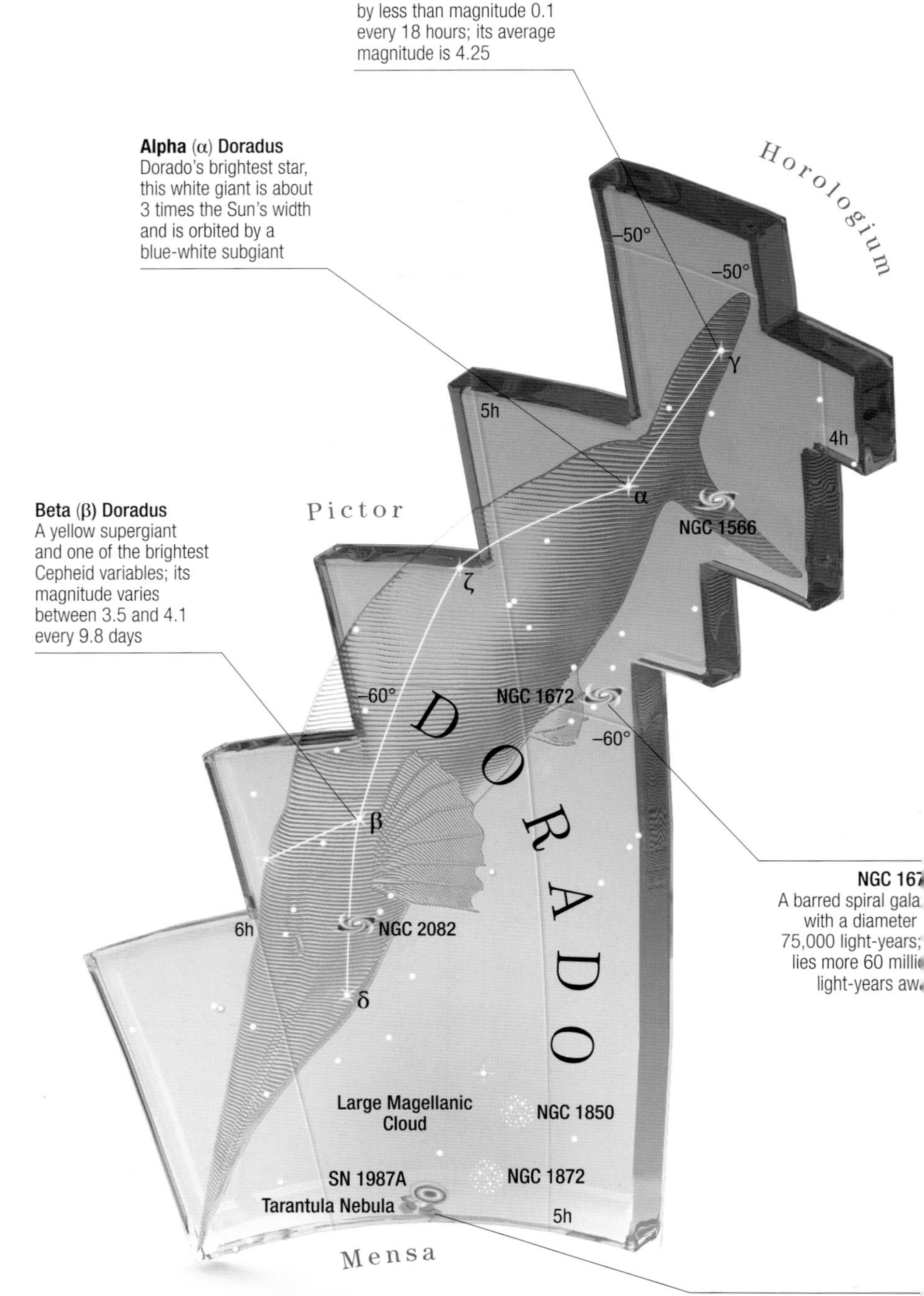
Gamma (γ) Doradus
A pulsating variable star whose brightness varies by less than magnitude 0.1 every 18 hours; its average magnitude is 4.25
Alpha (α) Doradus
Dorado's brightest star, this white giant is about 3 times the Sun's width and is orbited by a blue-white subgiant
Beta (β) Doradus
A yellow supergiant and one of the brightest Cepheid variables; its magnitude varies between 3.5 and 4.1 every 9.8 days
Horologium
Pictor
DORADO
Mensa
−50°
−50°
−60°
−60°
5h
4h
6h
5h
γ
α
ζ
β
δ
NGC 1566
NGC 1672
NGC 2082
Large Magellanic Cloud
NGC 1850
NGC 1872
SN 1987A
Tarantula Nebula
NGC 167
A barred spiral gala
with a diameter
75,000 light-years;
lies more 60 milli
light-years aw

Tarantula Nebu
Also known as 30 Doradu
a massive star-formi
region about 800 light-yea
in diameter; it is visible
the naked eye as a fuzzy st

The **Tarantula Nebula**, named for its spider-like shape, is the only nebula **outside the Milky Way** visible with the **naked eye**

▷ **Star distances**
Dorado's closest pattern star to Earth and relatively nearby at just 38 light-years is Zeta (ζ) Doradus, a white main-sequence star. The most distant pattern star is the yellow supergiant Beta (β) Doradus, which is more than 26 times farther away, at 1,005 light-years from Earth.

Earth

Gamma (γ) 67 light-years
Alpha (α) 170 light-years
Zeta (ζ) 38 light-years
Delta (δ) 150 light-years
Beta (
1,005 light-yea

Distance

Delta Doradus
34 Suns

Alpha Doradus
110 Suns

Beta Doradus
2,600 Suns

KEY **DATA**

Size ranking 72

Brightest stars Alpha (α) 3.3, Beta (β) 3.8

Genitive Doradus

Abbreviation Dor

Highest in sky at 10pm December–January

Fully visible 20°N–90°S

CHART 2

MAIN **STARS**

Alpha (α) Doradus
White giant and binary star
3.3 ⟷ 170 light-years

Beta (β) Doradus
Yellow supergiant and Cepheid variable star
3.5–4.1 ⟷ 1,005 light years

DEEP-SKY **OBJECTS**

NGC 1566
Spiral galaxy; also a Seyfert galaxy

NGC 1672
Barred spiral galaxy; also a Seyfert galaxy

NGC 1850
Compact star cluster in the Large Magellanic Cloud

NGC 1929
Star cluster in the Large Magellanic Cloud

NGC 2080 (Ghost Head Nebula)
Star-forming region in the Large Magellanic Cloud

NGC 2082
Barred spiral galaxy

Large Magellanic Cloud
Disrupted barred spiral galaxy

Tarantula Nebula (30 Doradus)
Star-forming region in the Large Magellanic Cloud

Supernova 1987A
Supernova in the Large Magellanic Cloud

△ **Tarantula Nebula**
This huge region of star clusters, glowing gas, and dark dust is one of the largest star-forming nebulae known. A bright star (centre left) appears to shine in a clearing. This is actually a cluster of stars that is emitting most of the energy that makes the nebula so clearly visible.

◁ **NGC 1929**
Massive stars in the star cluster NGC 1929 expel matter at high speed and explode as supernovae. The supernova shock-waves and winds carve out huge cavities called superbubbles (the blue areas) in the surrounding gas of the N44 nebula.

▷ **The Large Magellanic Cloud**
This satellite galaxy of the Milky Way is about 180,000 light-years away. Once thought to be an irregular galaxy, it is now believed to be a disrupted barred spiral. The red patch (centre right) is the Tarantula Nebula.

PICTOR

THE PAINTER'S EASEL

CONTAINING ONLY FAINT STARS, PICTOR WAS INTRODUCED IN THE 1750S. IT IS FOUND BETWEEN THE STAR CANOPUS IN CARINA AND THE LARGE MAGELLANIC CLOUD IN DORADO.

Pictor is said to represent an artist's easel although its star pattern bears little resemblance to one. It is one of 14 constellations introduced by French astronomer Nicolas Louis de Lacaille after observing the southern stars in the 1750s. A generally unremarkable constellation, it nevertheless has some interesting stars. Close-up views of Beta Pictoris reveal it is surrounded by a disk of planet-making material that extends more than 1,000 times the distance from the Earth to the Sun. A planet, named Beta Pictoris b, has already been identified in the inner disk. It has the mass of about nine Jupiters and is almost as close to its parent star as Saturn is to the Sun. The red dwarf Kapteyn's Star is the second-fastest moving star in the sky (after Barnard's Star in Ophiuchus).

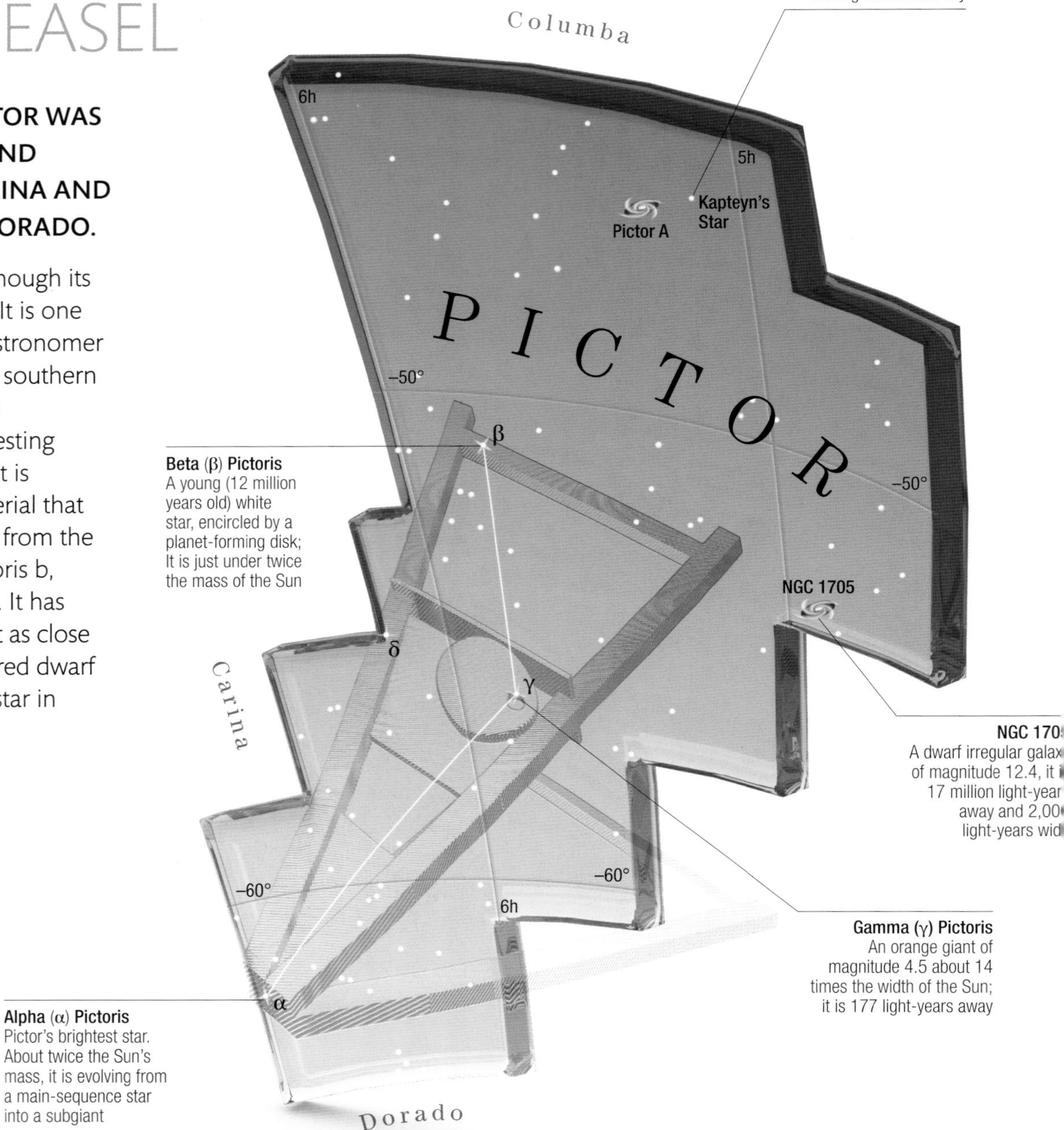

Kapteyn's Star
A red dwarf of magnitude 8.9, 13 light-years away and one of the fastest-moving stars in the sky

Beta (β) Pictoris
A young (12 million years old) white star, encircled by a planet-forming disk; It is just under twice the mass of the Sun

NGC 170
A dwarf irregular galax of magnitude 12.4, it 17 million light-year away and 2,00 light-years wid

Gamma (γ) Pictoris
An orange giant of magnitude 4.5 about 14 times the width of the Sun; it is 177 light-years away

Alpha (α) Pictoris
Pictor's brightest star. About twice the Sun's mass, it is evolving from a main-sequence star into a subgiant

KEY DATA

Size ranking 59

Brightest stars Alpha (α) 3.3, Beta (β) 3.9

Genitive Pictoris

Abbreviation Pic

Highest in sky at 10pm December–February

Fully visible 23°N–90°S

CHART 2

MAIN STARS

Alpha Alpha (α) Pictoris
White main-sequence star
3.3 · 97 light-years

Beta Beta (β) Pictoris
White main-sequence star
3.9 · 63 light-years

Gamma (γ) Pictoris
Orange giant
4.5 · 177 light-years

DEEP-SKY OBJECTS

NGC 1705
Dwarf irregular galaxy; also a starburst galaxy

Pictor A
Radio galaxy; also a Seyfert galaxy

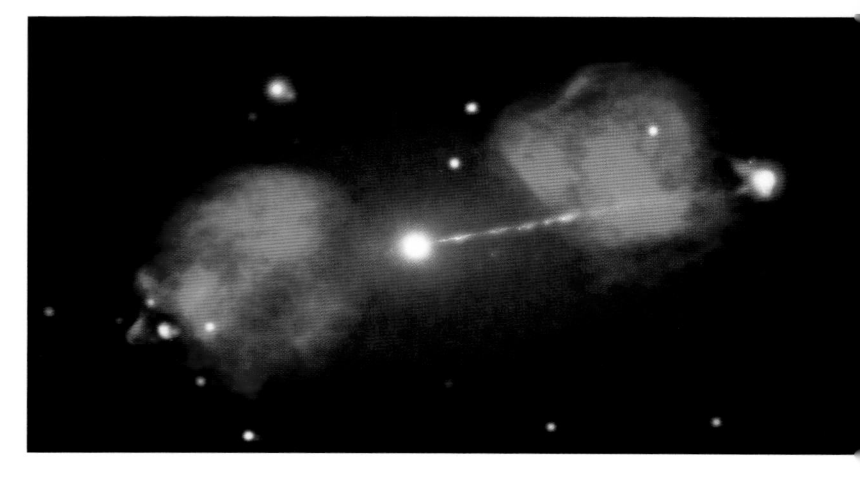

▷ **Pictor A**
The bright centre of this double-lobed radio galaxy is host to a supermassive black hole. As material swirls around the black hole, energy is released as an enormous beam of particles 300,000 light-years in length.

RETICULUM

THE NET

A FAINT DIAMOND SHAPE OF STARS IN THE SOUTHERN SKY, RETICULUM IS LOCATED NORTHWEST OF THE LARGE MAGELLANIC CLOUD IN DORADO.

The stars in this region were first grouped together in 1621 as the constellation Rhombus, the diamond, by German astronomer Isaac Habrecht. It was given its current description in the 1750s by French astronomer Nicolas Louise de Lacaille. The name refers to the reticule, the grid-like crosshairs in a telescope's eyepiece used to measure the positions of the stars. One of the smallest constellations, its principal attractions are the double star Zeta Reticuli and some faint galaxies, including NGC 1313, a starburst galaxy where unusually large numbers of hot young stars are forming, and NGC 1559, a spiral galaxy 50 million light-years away.

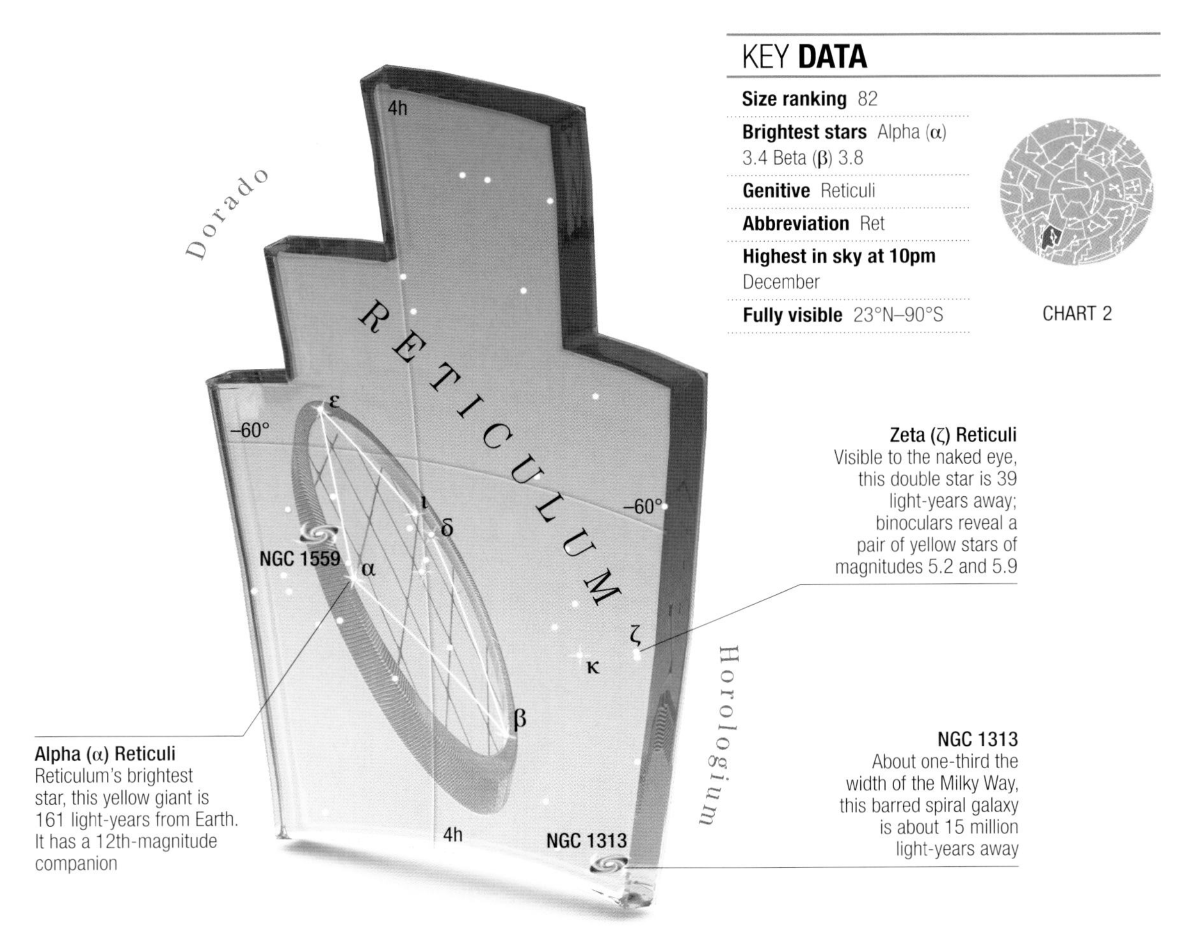

Alpha (α) Reticuli Reticulum's brightest star, this yellow giant is 161 light-years from Earth. It has a 12th-magnitude companion

Zeta (ζ) Reticuli Visible to the naked eye, this double star is 39 light-years away; binoculars reveal a pair of yellow stars of magnitudes 5.2 and 5.9

NGC 1313 About one-third the width of the Milky Way, this barred spiral galaxy is about 15 million light-years away

KEY DATA

Size ranking 82

Brightest stars Alpha (α) 3.4 Beta (β) 3.8

Genitive Reticuli

Abbreviation Ret

Highest in sky at 10pm December

Fully visible 23°N–90°S

CHART 2

VOLANS

THE FLYING FISH

THIS INDISTINCT CONSTELLATION IS SITUATED BETWEEN THE BRIGHT STARS OF CARINA AND THE SOUTH CELESTIAL POLE.

Volans was described by the Dutch explorers Peter Dirkszoon Keyser and Frederick de Houtman in the 1590s. Its most interesting features are its double stars, such as Gamma Volantis, visible with a small telescope, and its galaxies, which can be seen only with large telescopes. Like a huge figure "S", NGC 2442 has an arm coming from each end of a bar. Nicknamed the "Meathook", its distorted shape is the result of a near-miss with a smaller galaxy. Formerly a spiral, AM 0644-741 is a now a ring galaxy as a result of a collision with another galaxy.

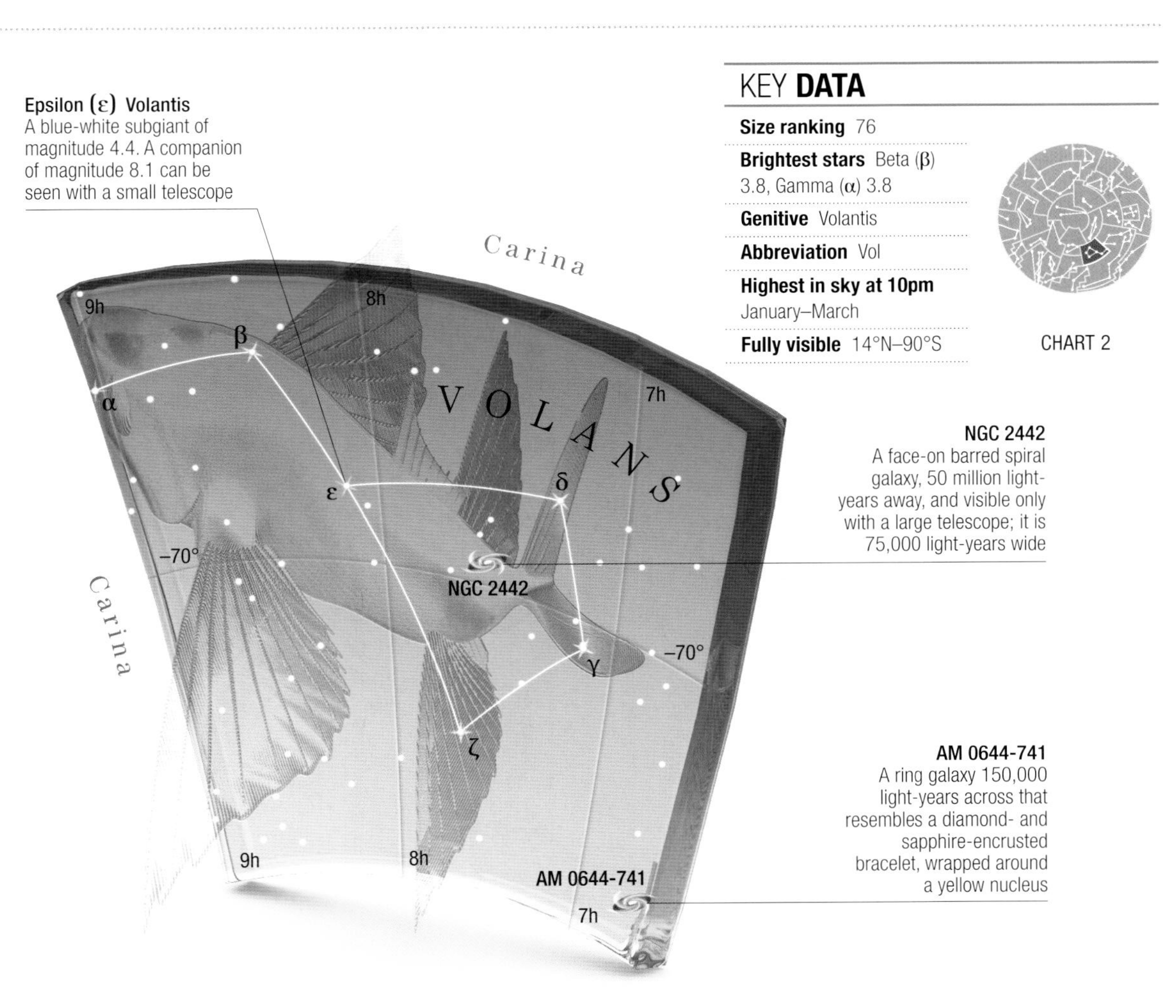

Epsilon (ε) Volantis A blue-white subgiant of magnitude 4.4. A companion of magnitude 8.1 can be seen with a small telescope

NGC 2442 A face-on barred spiral galaxy, 50 million light-years away, and visible only with a large telescope; it is 75,000 light-years wide

AM 0644-741 A ring galaxy 150,000 light-years across that resembles a diamond- and sapphire-encrusted bracelet, wrapped around a yellow nucleus

KEY DATA

Size ranking 76

Brightest stars Beta (β) 3.8, Gamma (α) 3.8

Genitive Volantis

Abbreviation Vol

Highest in sky at 10pm January–March

Fully visible 14°N–90°S

CHART 2

CHAMAELEON

THE CHAMELEON

A SMALL AND INSIGNIFICANT CONSTELLATION NEAR THE SOUTH CELESTIAL POLE, CHAMAELEON WAS ONE OF THE CONSTELLATIONS INTRODUCED IN THE 1590S BY PETRUS PLANCIUS.

Chamaeleon's faint diamond pattern of stars lies between the star fields of Carina and the South Celestial Pole in Octans. None of its stars is bright and there are no associated legends. Eta Chamaeleontis is the brightest star in an open star cluster, and Delta Chamaeleontis is a pair of unrelated stars. The Chamaeleon I Cloud is a star-forming nebula some 500 light-years away.

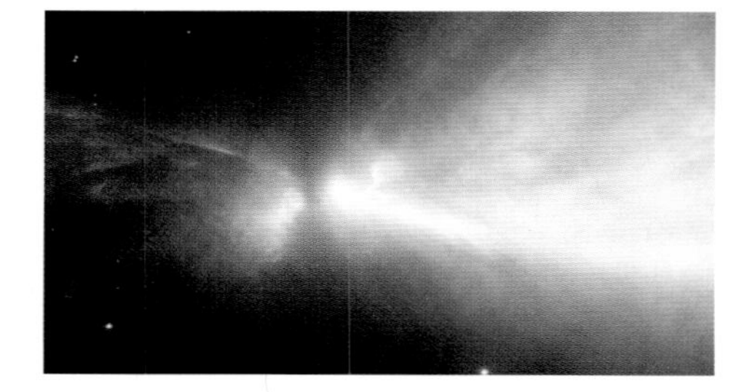

▷ **The Chamaeleon I Cloud** A star is pictured in the process of forming within the Chamaeleon I Cloud. Gas jetting out from its poles collides with surrounding gas and lights up the region.

KEY DATA

Size ranking 79

Brightest stars Alpha (α) 4.1, Gamma (γ) 4.1

Genitive Chamaeleontis

Abbreviation Cha

Highest in sky at 10pm February–May

Fully visible 7°N–90°S

CHART 2

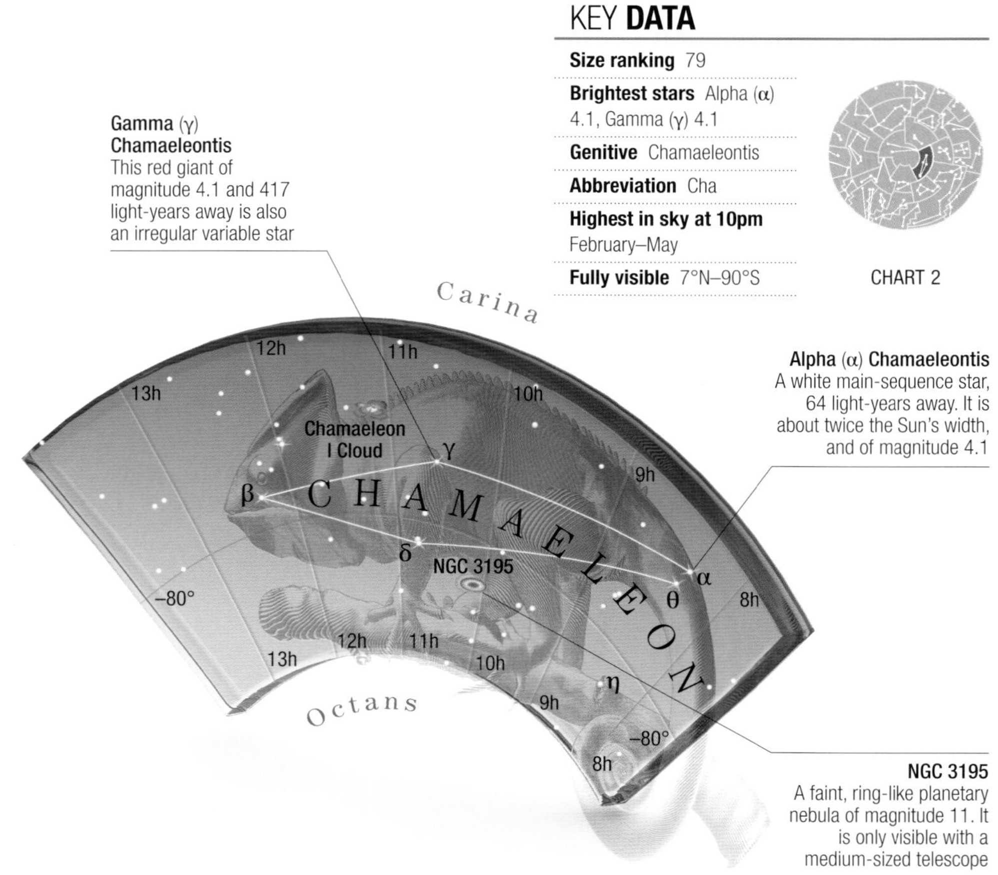

Gamma (γ) Chamaeleontis This red giant of magnitude 4.1 and 417 light-years away is also an irregular variable star

Alpha (α) Chamaeleontis A white main-sequence star, 64 light-years away. It is about twice the Sun's width, and of magnitude 4.1

NGC 3195 A faint, ring-like planetary nebula of magnitude 11. It is only visible with a medium-sized telescope

APUS

THE BIRD OF PARADISE

ONE OF THE 12 FAR SOUTHERN CONSTELLATIONS INTRODUCED AT THE END OF THE 16TH CENTURY, THIS TROPICAL BIRD IS DRAWN AROUND A CHAIN OF FOUR INDISTINCT STARS.

Apus lies south of an obvious triangle of stars, Triangulum Australe, and occupies an almost featureless area near the South Celestial Pole. It was devised by Dutch navigators Pieter Dirkszoon Keyser and Frederick de Houtman after seeing the exotic bird of paradise during their explorations of New Guinea in the 1590s. Sharp eyes or binoculars show that the constellation's most interesting star, Delta Apodis, is a double – a wide pair of unrelated red giants of magnitudes 4.7 and 5.3, 310 light-years away. Other notable features are Theta Apodis, a red giant varying between magnitudes 6.4 and 8.0 about every four months, and the globular clusters IC 4499 and NGC 6101.

KEY DATA

Size ranking 67

Brightest stars Alpha (α) 3.8, Gamma (γ) 3.9

Genitive Apodis

Abbreviation Aps

Highest in sky at 10pm May–July

Fully visible 7°N–90°S

CHART 2

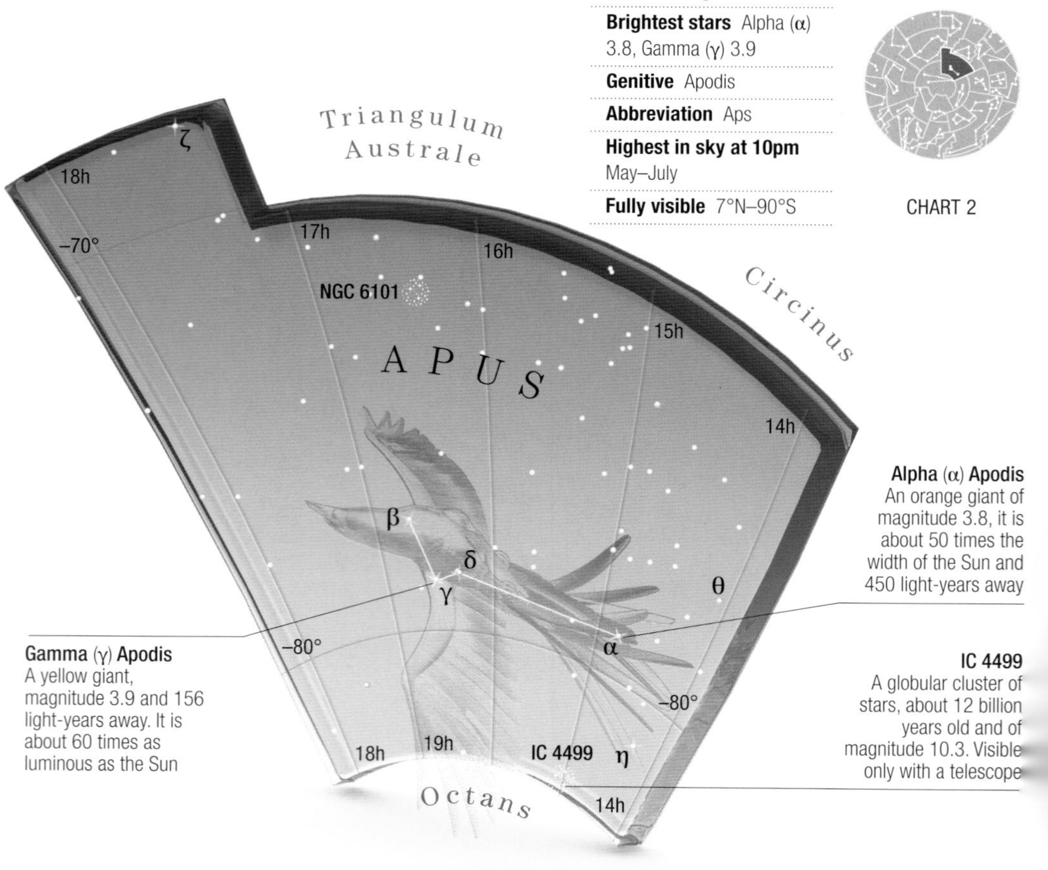

Gamma (γ) Apodis A yellow giant, magnitude 3.9 and 156 light-years away. It is about 60 times as luminous as the Sun

Alpha (α) Apodis An orange giant of magnitude 3.8, it is about 50 times the width of the Sun and 450 light-years away

IC 4499 A globular cluster of stars, about 12 billion years old and of magnitude 10.3. Visible only with a telescope

TUCANA

THE TOUCAN

THIS CONSTELLATION DEPICTING THE TOUCAN, A LARGE-BEAKED TROPICAL BIRD, WAS INTRODUCED TO THE FAR SOUTHERN SKY IN THE LATE 16TH CENTURY. IT IS FAINT AND HAS AN INDISTINCT PATTERN, BUT IT IS HOST TO SOME SIGNIFICANT CELESTIAL OBJECTS.

Located south of Phoenix and Grus, west of Hydrus, and southwest of the bright star Achernar in Eridanus, Tucana is one of 12 constellations devised by Dutch navigators Pieter Dirkszoon Keyser and Frederick de Houtman. It was first depicted on a globe by fellow Dutchman Petrus Plancius in 1598. None of the constellation's stars is named and there are no legends associated with it. However, Tucana is notable for two important features: the Small Magellanic Cloud (SMC) and 47 Tucanae. The SMC is the smaller of the Milky Way's two major satellite galaxies (the other is the Large Magellanic Cloud, in Dorado and Mensa). A compact globular cluster, 47 Tucanae (also known as NGC 104) contains several million stars and is the second-brightest globular cluster visible from Earth in the night sky.

Alpha (α) Tucanae
Tucana's brightest star, this orange giant marks the end of the bird's beak. It is about 37 times the Sun's width and 424 its luminosity

Small Magellanic Cloud
This irregular-shaped galaxy is visible to the naked eye as a hazy patch but binoculars reveal rich star fields and star-forming nebula

47 Tucanae
Also called NGC 104, a globular cluster 120 light-years across. It is 16,700 light-years away and can be seen as a fuzzy star with the naked eye

KEY DATA

Size ranking 48

Brightest stars Alpha (α) 2.8, Gamma (γ) 4.0

Genitive Tucanae

Abbreviation Tuc

Highest in sky at 10pm September–November

Fully visible 14°N–90°S

CHART 2

MAIN STARS

Alpha (α) Tucanae
Orange giant
2.9 — 200 light-years

DEEP-SKY OBJECTS

47 Tucanae
Globular cluster, also known as NGC 104

NGC 121
Globular cluster in the Small Magellanic Cloud

NGC 346
Star cluster and nebula in the Small Magellanic Cloud

NGC 362
Globular cluster

NGC 406
Spiral galaxy

Small Magellanic Cloud (NGC 292)
Irregular galaxy in orbit around the Milky Way

N81
Star-forming nebula in the Small Magellanic Cloud

▽ **NGC 346**
Located within the Small Magellanic Cloud, this star-forming region contains more than 2,500 infant stars. A cluster of dozens of hot, blue stars lies at its heart. Energy from these stars is sculpting the surrounding nebula. Other newborn stars, which have not yet started the nuclear fusion process that will make them shine, are within the nebula.

PAVO
THE PEACOCK

REPRESENTING THE EYE-CATCHING BIRD FROM INDIA, PAVO WAS FIRST DEPICTED ON A CELESTIAL GLOBE IN 1598. THE CONSTELLATION IS FOUND ON THE EDGE OF THE MILKY WAY, SOUTH OF TELESCOPIUM AND BETWEEN ARA AND INDUS.

Pavo is one of the 12 constellations devised by the Dutch navigators Pieter Dirkszoon Keyser and Frederick de Houtman, who voyaged in the southern hemisphere in the late 16th century and catalogued its stars. It is in a fairly featureless area of the sky but easily spotted because of its brightest star, Alpha Pavonis. Named Peacock in the late 1930s, it marks the bird's head. Pavo's flamboyant open tail feathers are drawn around a rectangle of stars. In the middle of this is Kappa Pavonis, a yellow supergiant, and one of the brightest Cepheid variables in the sky. Visible to the naked eye, this star varies in brightness every 9.1 days between magnitudes 3.9 and 4.8 as it expands and contracts. Pavo also contains an impressively bright globular cluster, NGC 6752, also visible to the naked eye.

KEY DATA

Size ranking 44

Brightest stars Alpha (α) 1.9, Beta (β) 3.4

Genitive Pavonis

Abbreviation Pav

Highest in sky at 10pm July–September

Fully visible 15°S–90°S

CHART 2

MAIN STARS

Peacock Alpha (α) Pavonis
Blue-white giant
1.9 ⟷ 179 light-years

DEEP-SKY OBJECTS

NGC 6744
Barred spiral galaxy

NGC 6752
Globular cluster

NGC 6782
Barred spiral galaxy

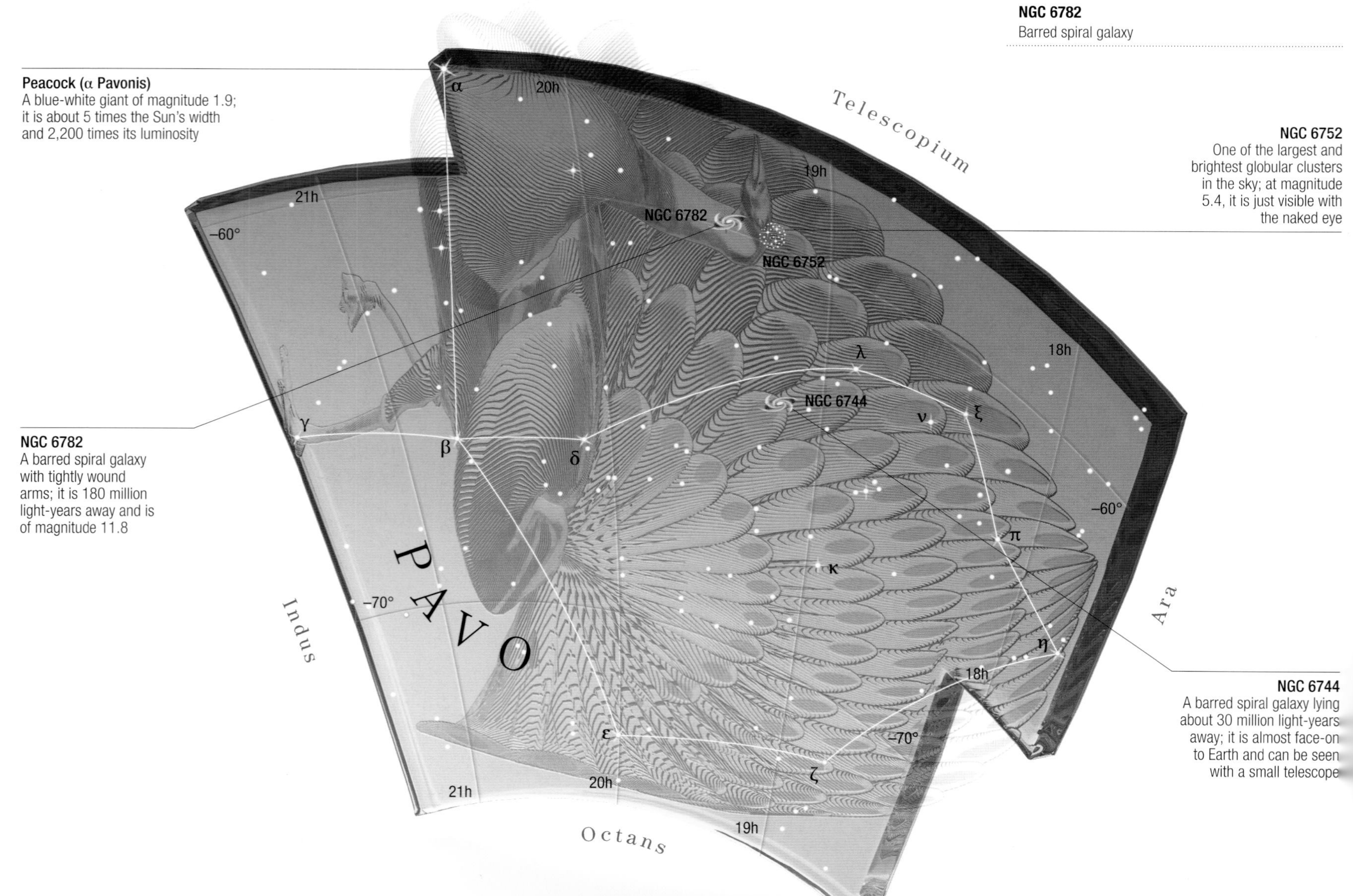

Peacock (α Pavonis)
A blue-white giant of magnitude 1.9; it is about 5 times the Sun's width and 2,200 times its luminosity

NGC 6752
One of the largest and brightest globular clusters in the sky; at magnitude 5.4, it is just visible with the naked eye

NGC 6782
A barred spiral galaxy with tightly wound arms; it is 180 million light-years away and is of magnitude 11.8

NGC 6744
A barred spiral galaxy lying about 30 million light-years away; it is almost face-on to Earth and can be seen with a small telescope

HYDRUS

THE LITTLE WATER SNAKE

HYDRUS FORMS A ZIGZAG IN THE SKY SOUTH OF THE BRILLIANT STAR ACHERNAR IN NEIGHBOURING ERIDANUS. IT IS SOMETIMES CONFUSED WITH HYDRA, THE WATER SNAKE, BUT THE LATTER IS MUCH BIGGER AND LIES FARTHER NORTH.

Hydrus is situated between the Large Magellanic Cloud in Dorado and the Small Magellanic Cloud in Tucana, with the star Achernar to the north. It is one of the 12 constellations introduced by the Dutch navigators Pieter Dirkszoon Keyser and Frederick de Houtman in the 16th century. The snake pattern of the constellation is indistinct, and its bright stars are more easily thought of as a triangle, with Alpha, Beta, and Gamma Hydri at the corners. North of Gamma is VW Hydri, a nova that explodes roughly once a month, and is easy to see with a small telescope. An outlying region of the Small Magellanic Cloud lies just within Hydrus.

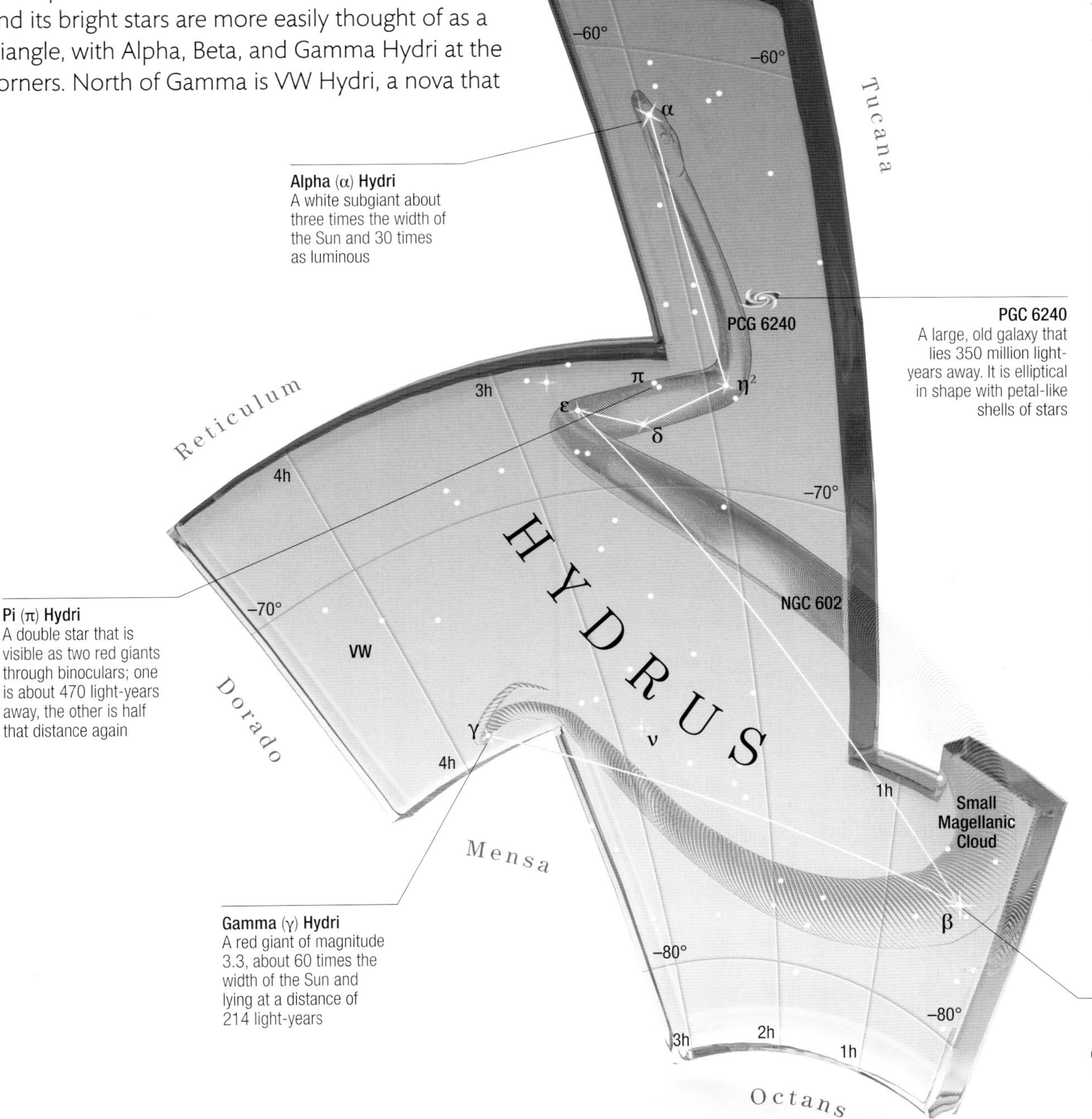

Alpha (α) Hydri
A white subgiant about three times the width of the Sun and 30 times as luminous

PGC 6240
A large, old galaxy that lies 350 million light-years away. It is elliptical in shape with petal-like shells of stars

Pi (π) Hydri
A double star that is visible as two red giants through binoculars; one is about 470 light-years away, the other is half that distance again

Gamma (γ) Hydri
A red giant of magnitude 3.3, about 60 times the width of the Sun and lying at a distance of 214 light-years

Beta (β) Hydri
A yellow subgiant lying only 24 light-years away; it is about the same mass as the Sun but is older and slightly more evolved

KEY DATA

Size ranking 61

Brightest stars Beta (β) 2.8, Alpha (α) 2.8

Genitive Hydri

Abbreviation Hyi

Highest in sky at 10pm October–Decembeer

Fully visible 8°N–90°S

CHART 2

MAIN STARS

Alpha (α) Hydri
White subgiant
2.8 ⟷ 72 light-years

Beta (β) Hydri
Yellow subgiant
2.8 ⟷ 24 light-years

DEEP-SKY OBJECTS

PGC 6240 (White Rose Galaxy)
Elliptical galaxy

NGC 602
Cluster of young stars

△ **NGC 602**
This star cluster lies at the heart of a huge star-forming nebula known as N90. Star formation started in the centre of the cluster NGC 602 then moved outwards. Radiation from the brilliant, blue, newly formed stars is continuing to sculpt the inner edges of the nebula, and the youngest stars are still forming along the nebula's long ridges of dust.

HOROLOGIUM

THE PENDULUM CLOCK

A FAINT AND UNREMARKABLE SOUTHERN-SKY CONSTELLATION, HOROLOGIUM HOSTS A DISTANT GLOBULAR CLUSTER BUT CONTAINS NO BRIGHT STARS.

Horologium occupies a region of sky that contains a sparse collection of stars, none of which is brighter than magnitude 3.9. It is one of the 14 constellations introduced by French astronomer Nicolas Louis de Lacaille in the 1750s. A horologium was the type of clock used for precision timekeeping in astronomical observatories of the time. The centre of the clock's face is marked by the star Alpha Horologii and the pendulum "swings" between Beta and Lambda, but the picture can be reversed so that Alpha marks the bottom of the pendulum. Horologium's deep-sky objects include Arp–Madore 1, the most distant globular cluster orbiting the Milky Way.

KEY DATA

Size ranking 58

Brightest stars Alpha (α) 3.9, R Horologii 4.7

Genitive Horologii

Abbreviation Hor

Highest in sky at 10pm October–December

Fully visible 8°N–90°S

CHART 2

MAIN STARS

Alpha (α) Horologii
Orange giant
3.9 115 light-years

DEEP-SKY OBJECTS

NGC 1261
Globular cluster

NGC 1512
Barred spiral galaxy

Arp–Madore 1 (AM1)
Globular cluster

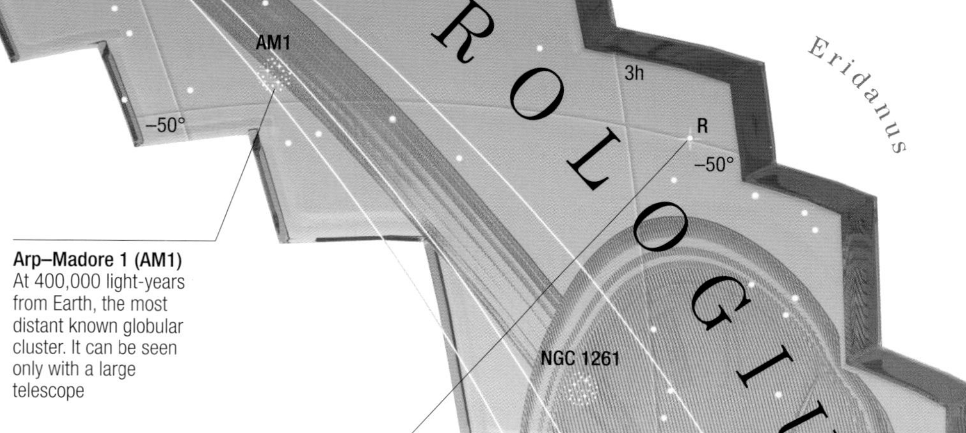

Alpha (α) Horologii
An orange giant about 11 times the width of the Sun. It is the constellation's brightest star and is 115 light-years from Earth

NGC 1512
Visible with a small telescope, this barred spiral galaxy is about 38 million light-years away and 70,000 light-years across

Arp–Madore 1 (AM1)
At 400,000 light-years from Earth, the most distant known globular cluster. It can be seen only with a large telescope

R Horologii
A red giant variable star that ranges between 5th and 14th magnitudes about every 13 months. It is 685 light-years from Earth

TW Horologii
A semi-regular red giant about 1,000 light-years away. Its brightness varies as it alternately expands and contracts

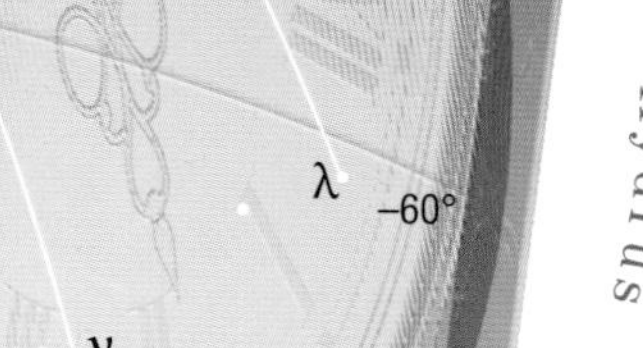

◁ **NGC 1261**
This compact globular cluster of old stars lies about 50,000 light-years away. The combined light from the stars gives the cluster a magnitude of 8.6, making it a good object for observing through binoculars or a small telescope.

△ **NGC 1512**
The bright centre of this barred spiral galaxy is dominated by a region of star formation and infant star clusters 2,400 light-years across. The birth of stars is fuelled by gas funnelled into the heart of the galaxy. Blue stars and red star-forming clouds of glowing hydrogen outline the spiral arms on the galaxy's outer edge.

MENSA

THE TABLE MOUNTAIN

THE FAINTEST CONSTELLATION OF ALL, MENSA IS NEAR THE SOUTH CELESTIAL POLE. THE LARGE MAGELLANIC CLOUD IS ON ITS BORDER WITH DORADO.

Small and faint, Mensa was devised by French astronomer Nicolas Louis de Lacaille. He named it "Mons Mensae" after Table Mountain, near Cape Town, South Africa, close to where he observed the southern skies in the 1750s. Mensa is the only one of the 14 constellations he defined that is not a scientific or artistic tool. Its most notable feature is the part of the Large Magellanic Cloud that Mensa includes.

▷ **Quasar PKS 0637-752**
This high-luminosity quasar is 6 billion light-years away and can only be studied using space-based telescopes. It radiates with the power of 10 trillion suns from a region smaller than our Solar System. Its energy source is a supermassive black hole at its heart.

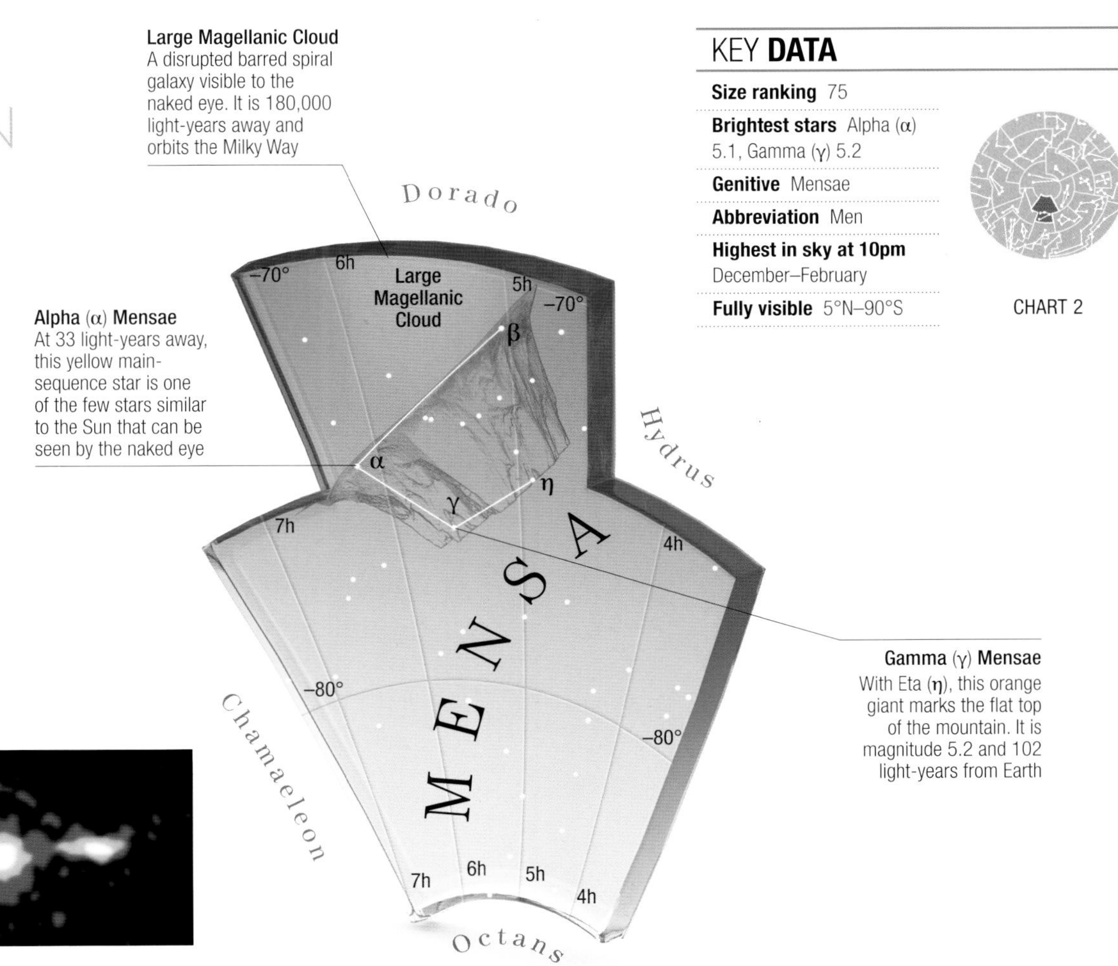

Large Magellanic Cloud
A disrupted barred spiral galaxy visible to the naked eye. It is 180,000 light-years away and orbits the Milky Way
Alpha (α) Mensae
At 33 light-years away, this yellow main-sequence star is one of the few stars similar to the Sun that can be seen by the naked eye
Gamma (γ) Mensae
With Eta (η), this orange giant marks the flat top of the mountain. It is magnitude 5.2 and 102 light-years from Earth
Dorado
Hydrus
Chamaeleon
Octans
MENSA
Large Magellanic Cloud
–70°
–80°
7h
6h
5h
4h
α
β
γ
η

KEY DATA

Size ranking 75
Brightest stars Alpha (α) 5.1, Gamma (γ) 5.2
Genitive Mensae
Abbreviation Men
Highest in sky at 10pm December–February
Fully visible 5°N–90°S

CHART 2

OCTANS

THE OCTANT

LOCATED IN A BARREN AREA OF SKY, OCTANS ENCOMPASSES THE SOUTH CELESTIAL POLE AND CONTAINS FEW SIGNIFICANT CELESTIAL OBJECTS.

Devised in the 1750s by the French astronomer Nicolas Louis de Lacaille, this constellation was named for the then recently invented navigational instrument and forerunner to the better-known sextant, the octant. Its most notable feature is Sigma Octantis, the nearest naked-eye star to the south celestial pole. Gamma Octantis is a chain of three unrelated stars, two yellow and one orange giant, usually able to be distinguished by the naked eye.

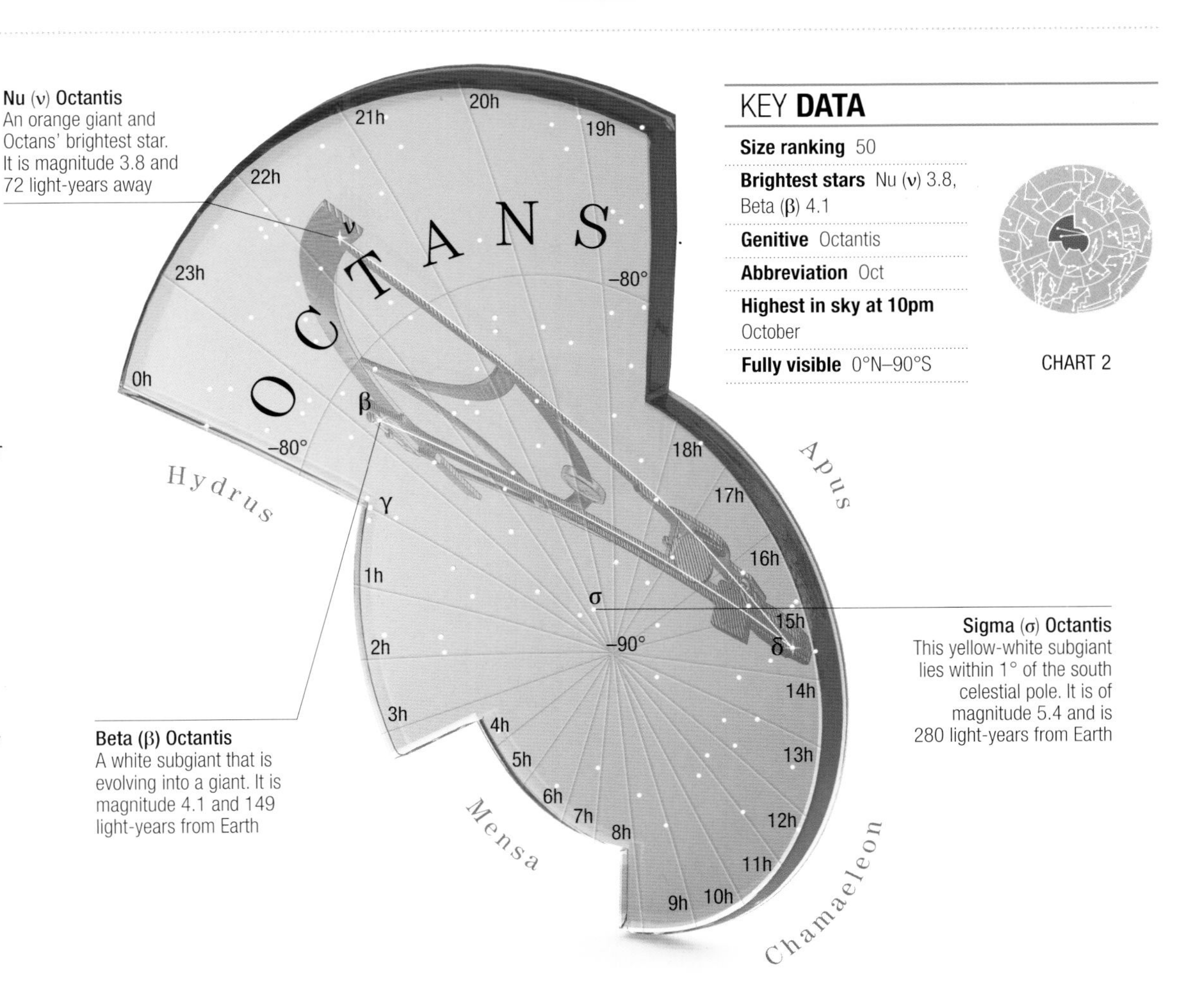

KEY DATA

Size ranking 50
Brightest stars Nu (ν) 3.8, Beta (β) 4.1
Genitive Octantis
Abbreviation Oct
Highest in sky at 10pm October
Fully visible 0°N–90°S

CHART 2

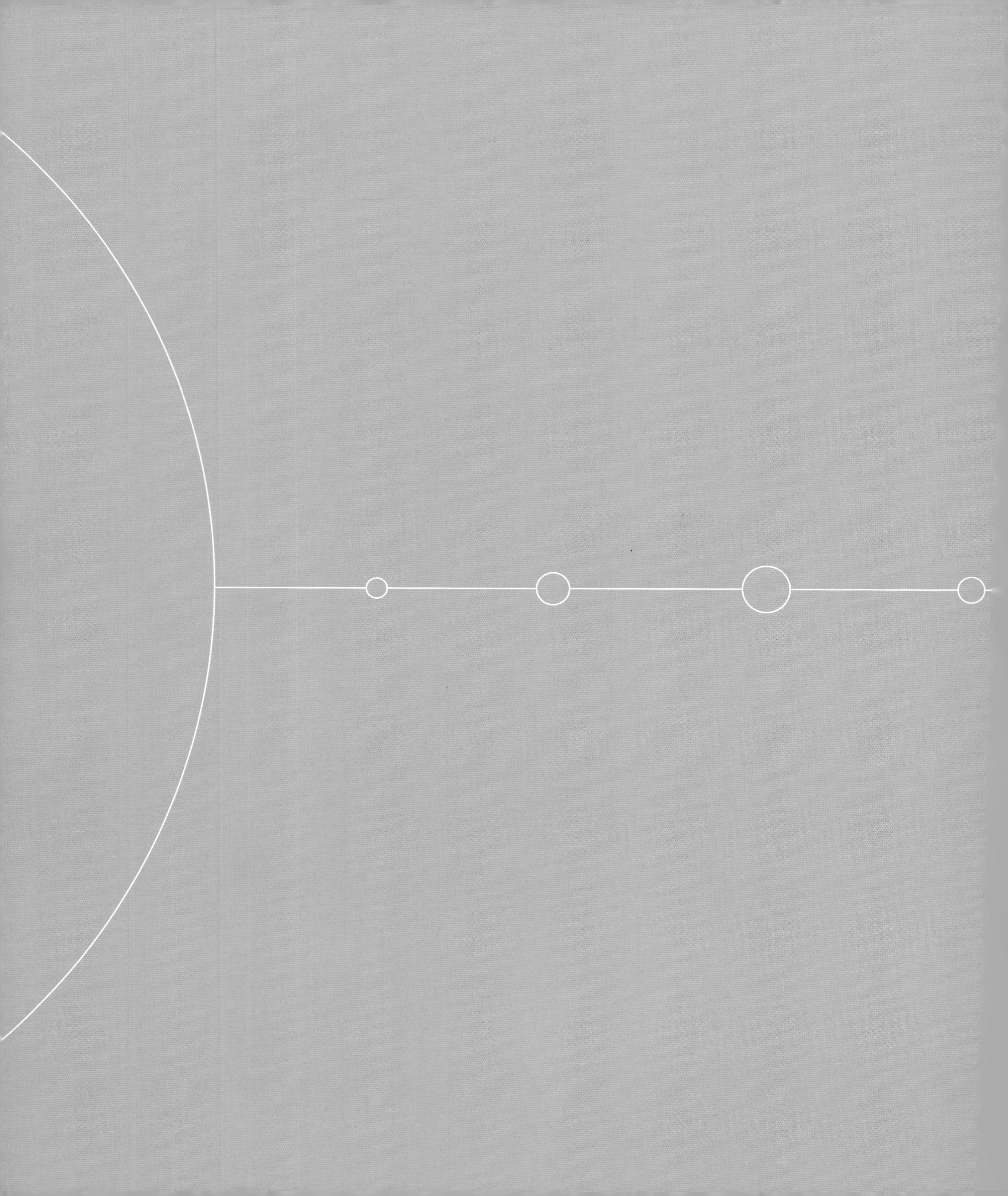

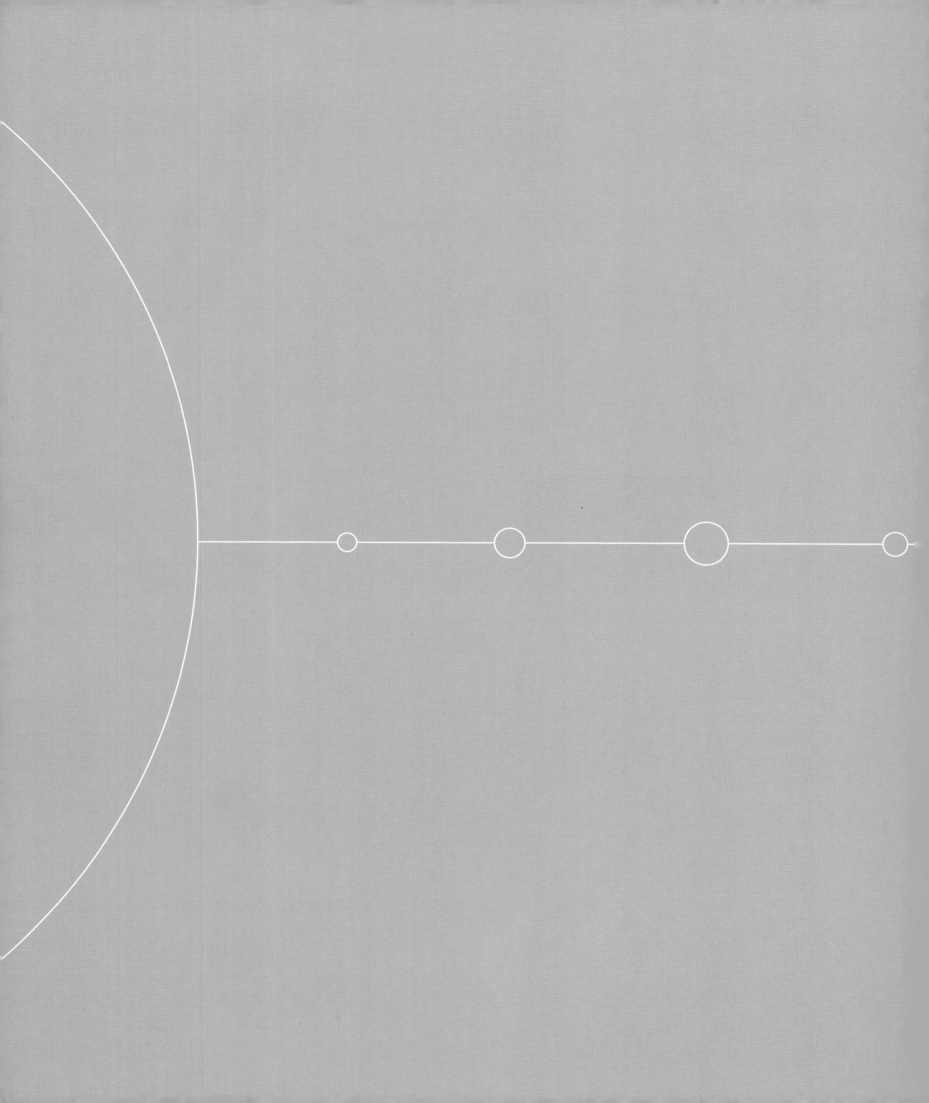

THE
SOLAR SYSTEM

Of the estimated 200 billion stars that make up the Milky Way, our home galaxy, only one is vital to our existence: a fairly unremarkable main-sequence star that we call the Sun. As is the case for most other stars in the galaxy, the Sun did not form in isolation. Its mighty gravitational pull keeps hold of a family of bodies that formed with it, the largest of which we know as the planets. Five of these other worlds

AROUND THE **SUN**

have been known since ancient times, from their wanderings amongst the stars in our night skies. Through the use of telescopes and space-based observatories, two more planets, hundreds of moons, and more than a million smaller objects, including comets and asteroids left over from the Solar System's formation, have also been found. The planets orbit the Sun in a disk, with the paths of smaller bodies generally becoming more scattered farther away from the Sun. The innermost planets – Mercury, Venus, Earth, and Mars – are small, solid globes made up primarily of rock and metals. In contrast, the outer worlds – Jupiter, Saturn, Uranus, and Neptune – are giants formed of gas and liquid, each accompanied by a multitude of their own natural satellites. Despite the extensive knowledge we have of our neighbours, we may not yet fully appreciate the scale of our planetary system. Indeed, large bodies may lie yet undiscovered in the dark extremities of our Solar System.

◁ **Surface of the Sun**
At the high temperatures close to our star, almost all the gas present is split into charged particles – a plasma – through which weaves a tangled jumble of magnetic fields. The complex churning of the Sun's braided surface overlies the primary source of heat and light in our Solar System: the nuclear furnace that is buried in the star's interior.

THE **SOLAR SYSTEM**

THE SOLAR SYSTEM FORMED FROM A SLOWLY ROTATING NEBULA OF GAS AND DUST AROUND 4.6 BILLION YEARS AGO. THE PLANETS AND COUNTLESS MINOR BODIES THAT ORBIT THE SUN WERE FORMED FROM ACCUMULATIONS OF GAS AND DUST.

The distances from the Sun at which planets formed largely control their compositions, with the ices of water, methane, and other molecules becoming more dominant in the planets far from our star's warmth. The Solar System is much larger than just the orbits of the planet, continuing far beyond Neptune. Beyond the eighth planet lies a scattered disk of icy worlds, some of which are hundreds of kilometres across. Further than that is thought to lie a spherical cloud of small icy objects, known as the Oort Cloud.

▽ The orbits of the planets
All the planets follow stable paths around the Sun. These paths are almost circular and are close to lying in a flat disk. The speed of the planets' motion depends on their distances from the Sun, so Mercury travels far more rapidly than Neptune. Comets, trans-Neptunian objects, and many asteroids follow more elliptical orbits, during which their distance from the Sun varies considerably. Some long-period comets follow extremely elongated paths that take them from very close to the Sun to vast distances away over periods of thousands of years.

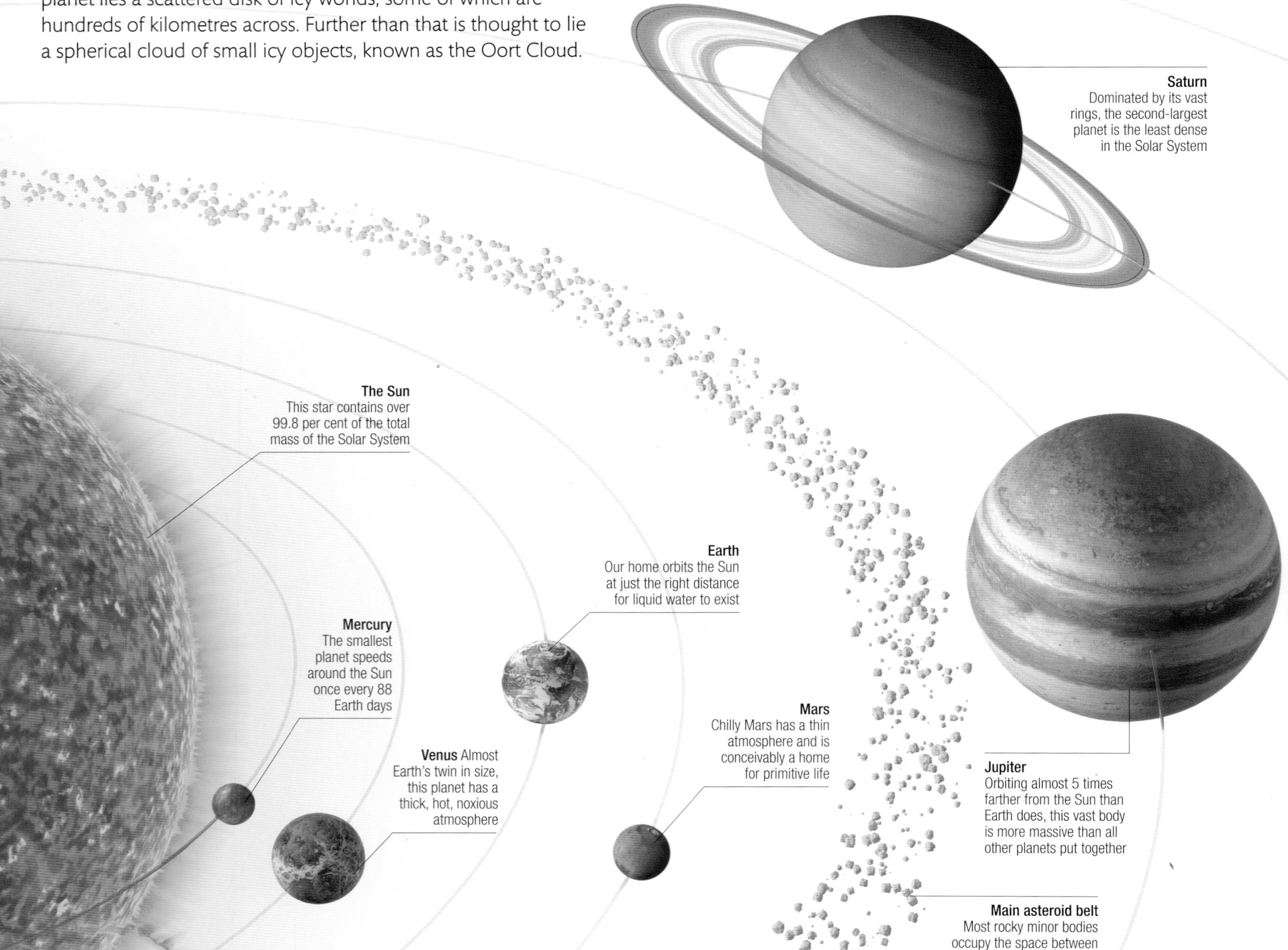

Saturn
Dominated by its vast rings, the second-largest planet is the least dense in the Solar System
The Sun
This star contains over 99.8 per cent of the total mass of the Solar System
Earth
Our home orbits the Sun at just the right distance for liquid water to exist
Mercury
The smallest planet speeds around the Sun once every 88 Earth days
Mars
Chilly Mars has a thin atmosphere and is conceivably a home for primitive life
Venus Almost Earth's twin in size, this planet has a thick, hot, noxious atmosphere
Jupiter
Orbiting almost 5 times farther from the Sun than Earth does, this vast body is more massive than all other planets put together
Main asteroid belt
Most rocky minor bodies occupy the space between Mars and Jupiter

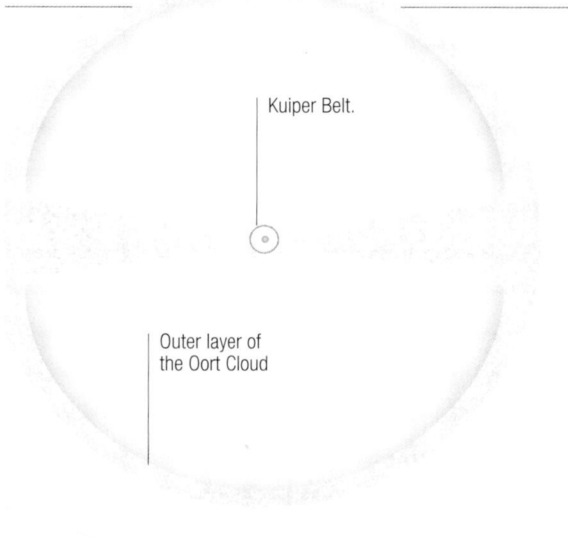

◁ **The Oort Cloud**
Billions of the minor planets were thrown out of the Solar System by the gravity of the newly formed planets. These minor planet now form this vast cloud of comets that may stretch one quarter of the way to the nearest star.

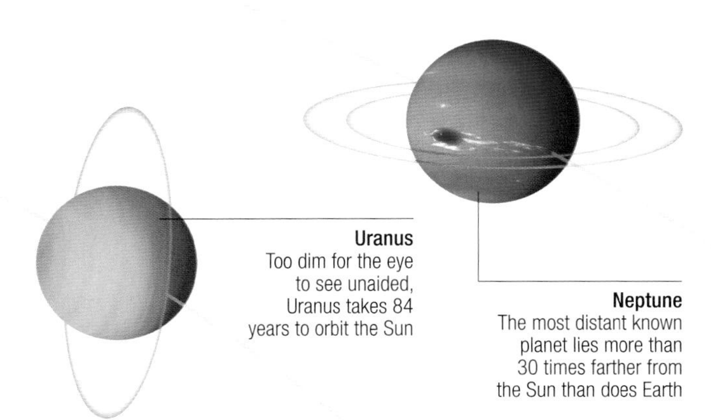

Uranus
Too dim for the eye to see unaided, Uranus takes 84 years to orbit the Sun

Neptune
The most distant known planet lies more than 30 times farther from the Sun than does Earth

Rocky planets

The planets nearest to the Sun – Mercury, Venus, Earth, and Mars – contain far more rock and metal than gas. Each one has a surface shaped by volcanoes, where hot, molten material from the interior has broken through the solid crust. Apart from Mercury, these planets possess a significant atmosphere that has protected its surface, to varying degrees, from impacts. These atmospheres may have largely originated in asteroid and comet impacts, with the latter possibly also delivering large quantities of water to the planets. Only Earth and Mars have moons. Mars's natural satellites, Phobos and Deimos, are believed to be captured asteroids.

Gas planets

The four outer worlds – Jupiter, Saturn, Uranus, and Neptune – are bloated giants largely formed of gas surrounding a dense core, each accompanied by a large retinue of moons. When young, these planets grew large enough to draw in gas from the surrounding nebula. The motions of their churning atmospheres are driven by internal heat as well as energy from sunlight. Given their cold environments far from the Sun, these planets' many moons are predominantly worlds with water ice crusts, some heated inside due to the effects of tides. Some of these satellites have atmospheres, and others have active volcanoes.

Minor bodies

In the early Solar System, dust and ice grains first formed small bodies, and these then accumulated to form the planets. Billions of the small bodies did not, however, become parts of planets, and they remain as minor bodies today. These are the asteroids and comets, and studying their make-up can tell us much about conditions in the early Solar System. The smallest of these – some are mere dust grains – enter our atmosphere, appearing as meteors. Many larger minor bodies, however, present Earth with a threat should a collision occur. Their rare impacts can potentially lead to global disruption, as is thought to have occurred 65 million years ago, leading to the extinction of the dinosaurs.

The term **planet** originates from the ancient Greek term for **wandering star**

△ **Asteroids**
Most of these bodies, such as 951 Gaspra (shown here), are too small to form spheres. Despite being mostly rocky, several are now known to possess water ice under their surface.

△ **Dwarf planets and Trans-Neptunian objects**
Numerous icy bodies orbit beyond Neptune, in a flat disk known as the Kuiper Belt. Several of these are large enough to be classed as dwarf planets, including Pluto (shown here).

△ **Comets**
Comets, such as 67P/Churyumov-Gerasimenko (shown here), are small, icy bodies. As they approach the Sun, their ice turns into gas, releasing dust to create tails millions of kilometres long.

THE SUN

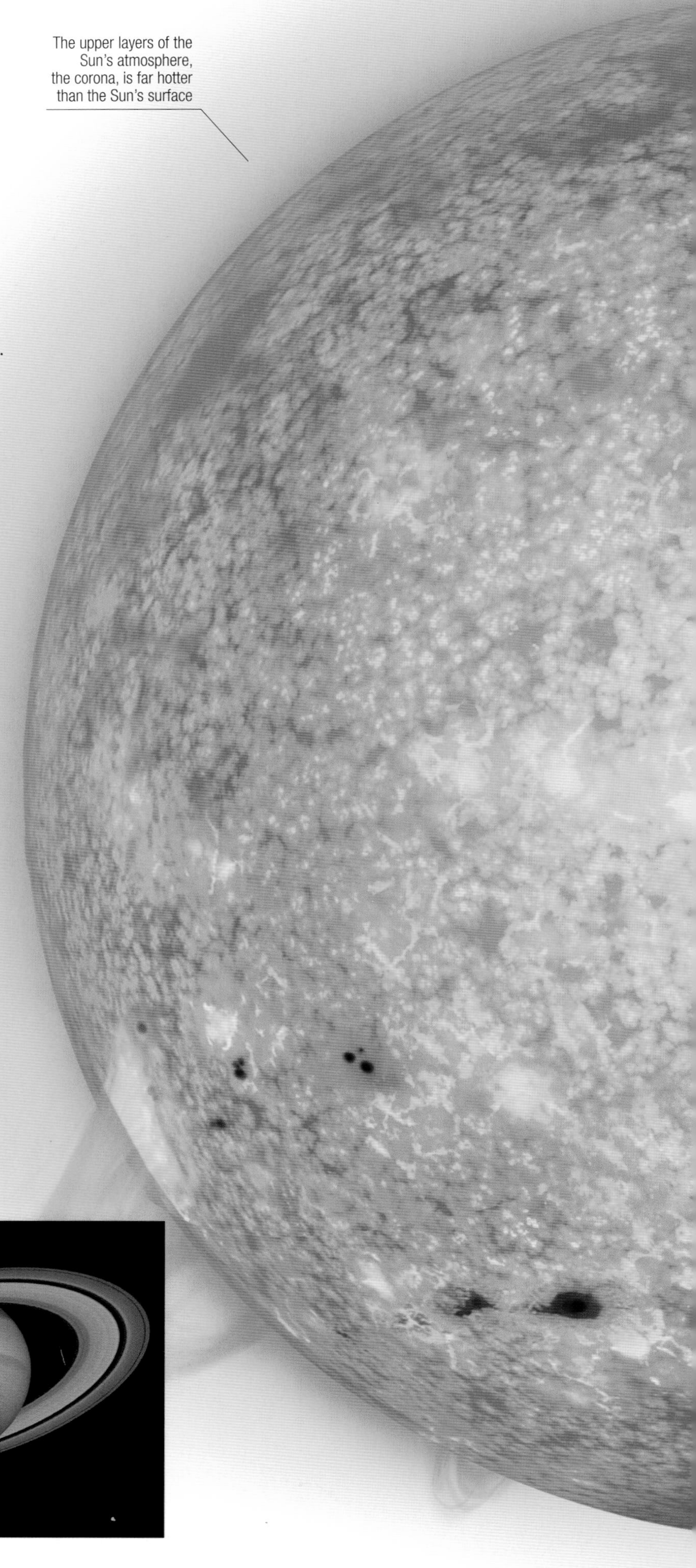
The upper layers of the Sun's atmosphere, the corona, is far hotter than the Sun's surface

ESSENTIAL TO LIFE ON EARTH BY PROVIDING US WITH WARMTH AND LIGHT, THE SUN MAY SEEM SPECIAL, BUT IT IS JUST A TYPICAL STAR. AS IT IS SO CLOSE TO US, IT HAS BEEN STUDIED IN GREAT DETAIL.

Measuring 1.39 million km (863,000 miles) across and rotating every 24.5 days, the Sun is a nuclear fusion furnace in which atoms are crushed together in its core, releasing vast quantities of heat and light. The surface that we see, at 6,000°C (10,800°F), is only part of a complex, seething jumble of magnetic field and charged particles. Almost everything we can see on the Sun is in a state of matter called a plasma, where gas has separated into negatively charged electrons and positively charged ions (atoms or molecules that have lost electrons). The Sun has an activity cycle of around 11 years, during which the numbers of sunspots, flares, and eruptions rise and fall significantly.

The solar wind

A flow of charged particles, called the solar wind, flows continuously from the Sun into space at hundreds of kilometres per second. This wind carves out an enormous bubble in space called the heliosphere; Earth and all the other planets orbit within it, shielded from interstellar space. Like weather on the Earth, the solar wind is variable. It can be gusty, and its effects on planets and comets can change very abruptly.

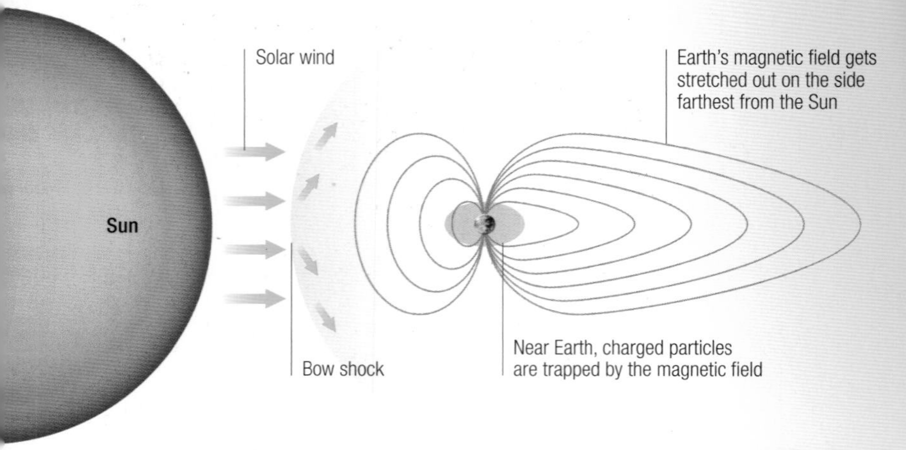
Solar wind
Earth's magnetic field gets stretched out on the side farthest from the Sun
Sun
Bow shock
Near Earth, charged particles are trapped by the magnetic field

△ **Earth magnetosphere**
Our planet is protected from the solar wind by its magnetic field. This magnetic bubble, called the magnetosphere, allows the solar wind to usually only affect Earth near the poles, where the northern and southern lights – the aurorae – are seen.

▷ **Saturn's aurora**
Like Earth and most other planets, Saturn has a magnetosphere. The interaction between the solar wind and the planet's magnetic field generates aurora near its poles, as occurs on our planet. Here, ultraviolet light shows the southern lights near the south pole.

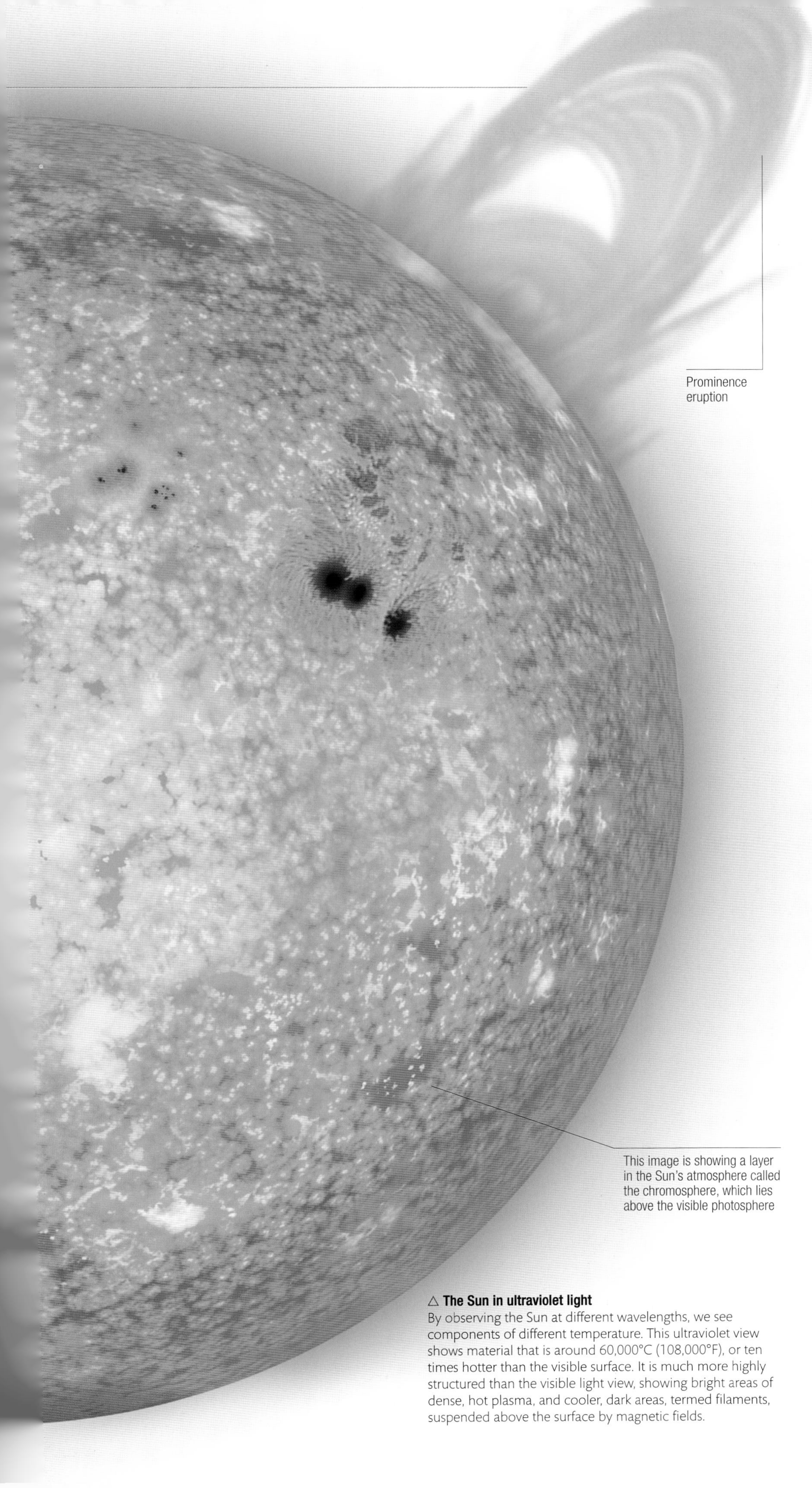

△ **The Sun in ultraviolet light**
By observing the Sun at different wavelengths, we see components of different temperature. This ultraviolet view shows material that is around 60,000°C (108,000°F), or ten times hotter than the visible surface. It is much more highly structured than the visible light view, showing bright areas of dense, hot plasma, and cooler, dark areas, termed filaments, suspended above the surface by magnetic fields.

Surface features

Almost all the features seen on the Sun are controlled by its magnetic field. Large-scale dark patches in hot plasma indicate coronal holes, which are the source of much of the solar wind. The magnetic field at these holes escapes into space with the wind. Bright patches indicate tightly bunched, twisted magnetic fields trapping hot plasma, and are termed active regions. They usually overlie sunspots. When an active region erupts, it becomes a solar flare, which is bright at all wavelengths for minutes or longer.

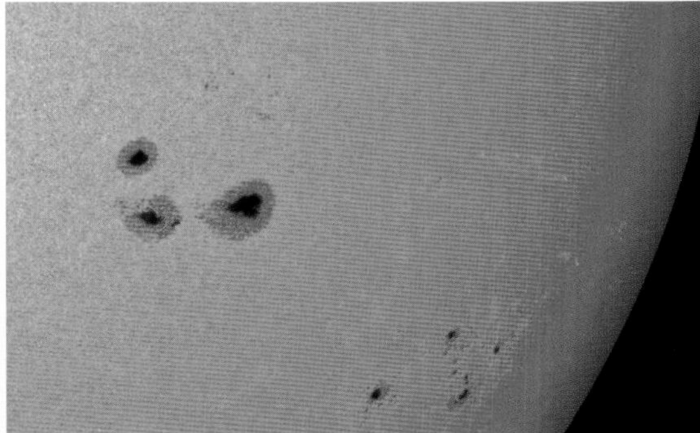

△ **Sunspots**
These dark regions are cooler than their surroundings, but are still extremely hot. They indicate where bundles of magnetic fields from the Sun's interior have broken through the surface.

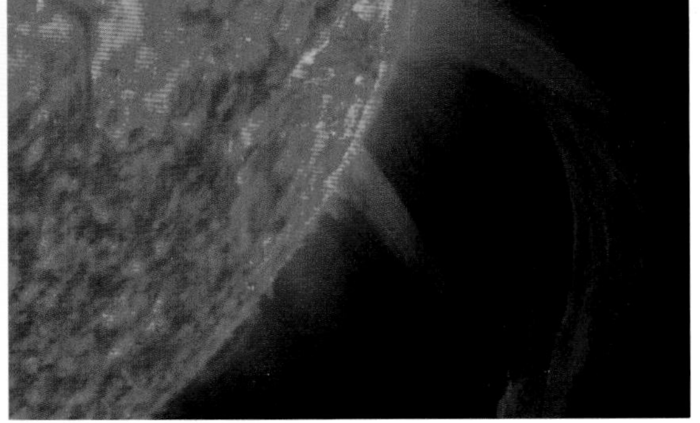

△ **Prominence**
A lifted, plume-like region of cooler, denser plasma that emerges from the Sun's visible surface (the photosphere) is called a prominence.

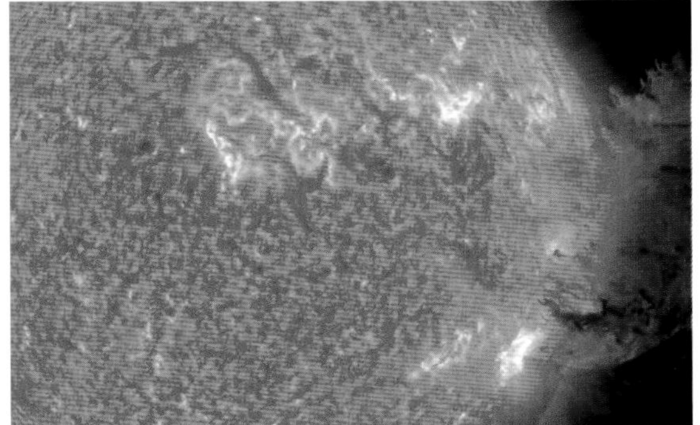

△ **Coronal mass ejections**
These are eruptions of plasma that are sent out into space. They occur with a range of sizes and speeds. Some can have dramatic effects on Earth's magnetosphere.

THE INNER PLANETS

THE INNER PLANETS, INCLUDING OUR OWN PLANET, ARE ALL ROCKY BODIES. BASKING IN THE WARMTH RELATIVELY CLOSE TO THE SUN, ALL THESE FOUR BODIES HAVE SOLID, ROCKY CRUSTS, BUT ARE DIVERSE WORLDS.

The sizes of the planets' atmospheres are controlled by their gravitational pull. Mercury is too small to sustain any significant atmosphere. Mars's air is gradually being lost, and was previously be much more extensive. Venus and Earth can retain thick atmospheres.

Venus

Despite almost being Earth's twin in terms of size, with a diameter of 12,104 km (7,521 miles), Venus has evolved along a very different path to our planet. The planet rotates very slowly every 243 Earth days and is covered in outflows from many volcanoes. Its extremely thick atmosphere gives it a surface pressure 90 times higher than Earth's, at a temperature of around 460°C (860°F).

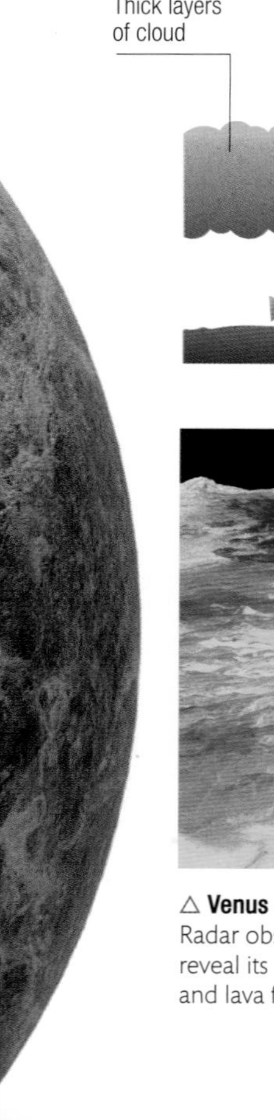

The filtering of sunlight by Venus's thick clouds makes its mostly grey surface appear orange

Mercury

With a diameter of 4,879 km (3,032 miles), Mercury is the smallest of the terrestrial planets. It is, however, very dense. It has a huge iron core that generates a planet-wide magnetic field, similar to that of Earth. The planet has an extremely thin atmosphere that barely differs from a vacuum, much of which is kicked up by particles from the Sun striking the surface. It is a world of temperature extremes, extending from -170 to 420°C (-274 to 788°F).

Mercury's surface is similar to the Moon in appearance: covered in countless impact craters, but also some smooth areas formed from lava flows.

◁ **Mercury crater**
Fresh impact craters like this one show that Mercury's soil and rocks are varied in composition. This strike by a small asteroid has thrown up bright material from just under the surface.

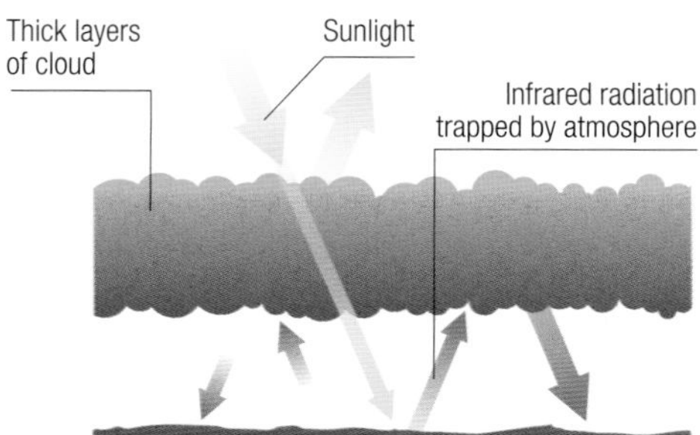

◁ **The greenhouse effect**
The extremely high surface temperature on Venus is due to its thick atmosphere and clouds. The gases surrounding the planet allow in some sunlight, which heats the planet's surface. The warmed ground glows at infrared wavelengths. The atmosphere prevents this heat from escaping into space.

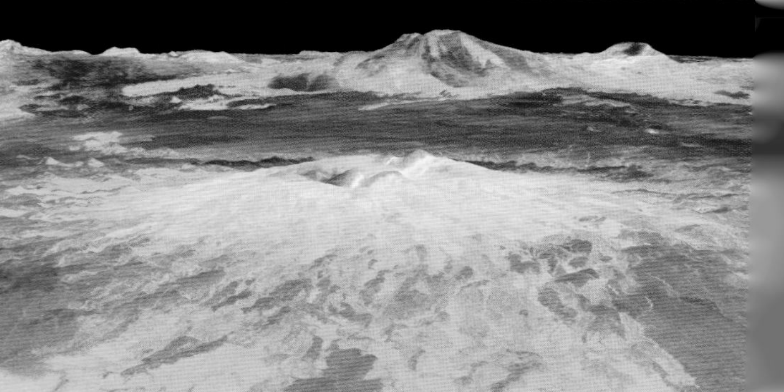

△ **Venus surface**
Radar observations that can penetrate Venus's thick clouds reveal its surface to be almost entirely covered by volcanoes and lava flows; there are very few impact craters.

Earth

Our home is the largest of the inner planets, measuring 12,756 km (7,926 miles) across. Earth's surface has few visible impact craters—their remains have mostly been erased by the effects of air and water. The vast seas hide a unique feature: the ocean floor is being gradually replenished by fresh rock from long chains of underwater volcanoes. The sea floor moves very gradually like a giant conveyor belt, sinking back into the interior under the continents.

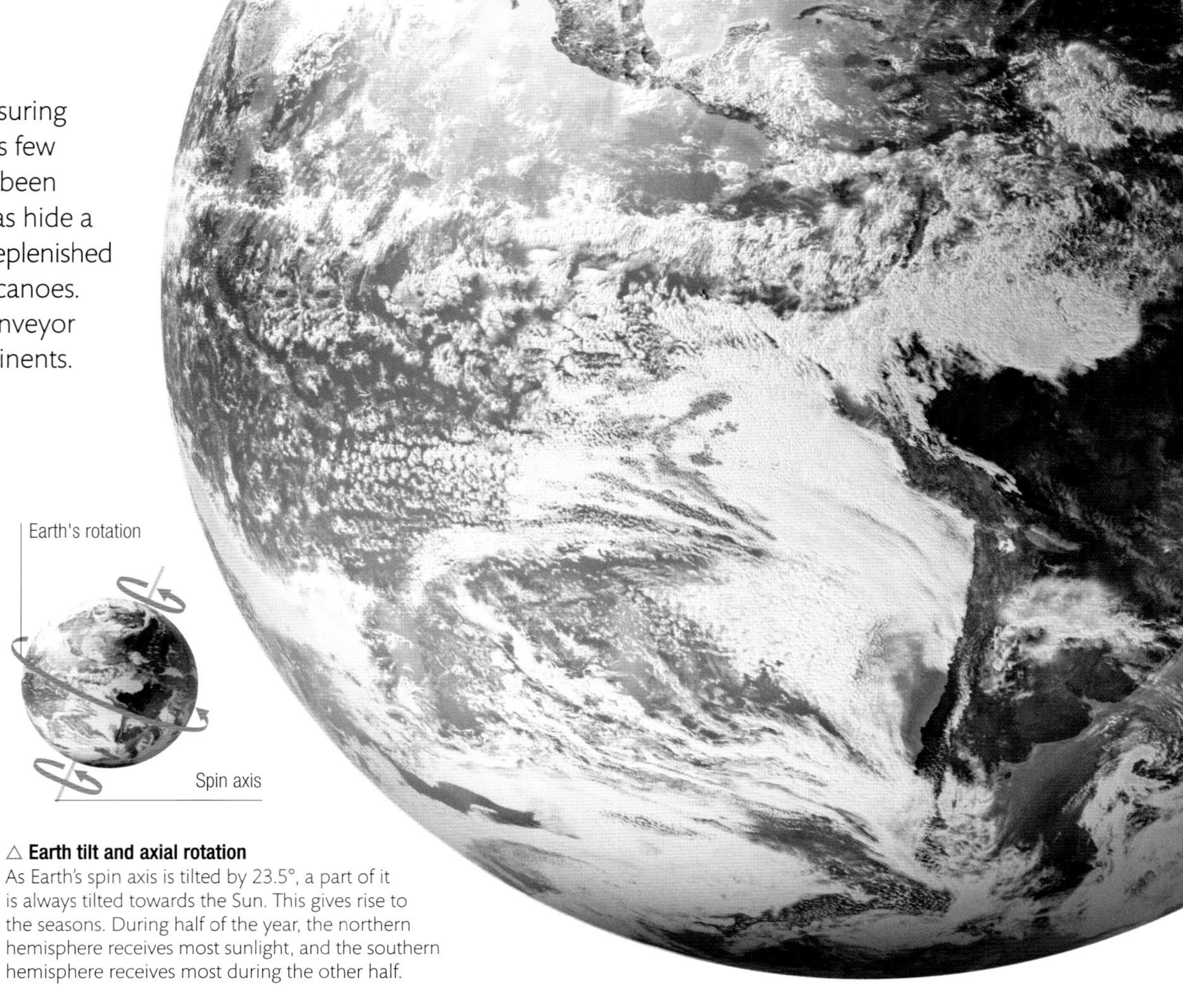

△ **Water and life**
Earth's distance from the Sun and the thickness of its atmosphere allow water to exist as a liquid, solid, and gas. The presence of liquid water is thought to have been essential for life to begin here. Why Earth has quite so much water is currently unknown.

△ **Earth tilt and axial rotation**
As Earth's spin axis is tilted by 23.5°, a part of it is always tilted towards the Sun. This gives rise to the seasons. During half of the year, the northern hemisphere receives most sunlight, and the southern hemisphere receives most during the other half.

Most the surface rocks on Mars have been oxidized, like the rust seen on some metals, giving them an orange colour.

Mars

Mars has a surface almost equal in area to that of Earth's continents. The so-called Red Plant, at 6,792 km- (4,220 miles-) wide and most similar to Earth at its surface, is one of the few other places in our Solar System where life could have arisen, and may even exist today (see p.82). The planet's air is now too thin to support liquid water for long, even when the temperature creeps above 0°C (32°F). There is, however, plenty of evidence that the air was once thicker, and conditions for Martian life were much better billions of years ago.

△ **Mars surface**
Few worlds have terrains as diverse as Mars, from its heavily-cratered ancient lands to smooth plains and plunging canyons. Its highest volcano, Olympus Mons, reaches 25 km (15 miles) above the planet's rocky surface, while the huge Valles Marineris canyon plunges 7 km (4 miles) below the surrounding plains.

△ **Evidence of water**
Mars displays features such as channels and canyons that clearly indicate that water flowed on its surface in the ancient past. There are also signs of brief water flows today, as seen on the slopes of this crater. Sunlight is thought to melt water ice buried under the surface, which flows before evaporating in the thin air.

THE OUTER PLANETS

UNLIKE THE ROCKY COMPOSITION OF THE INNER PLANETS, THE FOUR GIANT OUTER PLANETS ARE LARGELY MADE OF GAS. THESE MASSIVE BODIES HAVE DOZENS OF MOONS, EACH FORMING A MINIATURE PLANETARY SYSTEM.

All four outer planets are much larger than Earth. Despite their size, they all have days shorter than Earth's. This rapid rotation leads to their atmosphere splitting up into bands. Visits by spacecraft have shown that they all possess a magnetic field, and aurora occur in their atmospheres. The four bodies all have ring systems, but Saturn has the most extensive by far.

Jupiter's bands alternate between light-coloured regions of rising air and darker regions of falling air

Saturn

Measuring over 120,536 km (75,898 miles) across, Saturn is the second-largest planet. It has a banded atmosphere similar to Jupiter's, but its cloud structures are more muted. The planet is surrounded by an extensive ring system, probably the remains of a destroyed moon. Its largest moon, Titan, has a thick atmosphere with a surface pressure higher than that at Earth's surface.

△ **Saturn's rings**
Saturn's rings are composed of chunks of almost pure water ice, with a tiny amount of dust. One of the largest rings, called the E ring, is still fed by ice grains erupting from the moon Enceladus.

Jupiter

Measuring 142,984 km (88,846 miles) across, ten Earths could fit side-by-side across Jupiter's equator. The largest planet's great mass affects the orbits of many bodies, and numerous comets' orbits have been altered by its presence. This vast planet has an extremely strong magnetic field. High energy particles are trapped by this field, making it a dangerous place for human exploration.

◁ **Giant Red Spot**
This spectacular, churning storm has existed in Jupiter's atmosphere for at least 300 years. Typically measuring 30,000 km (18,600 miles) across, Earth would fit inside it two or three times. It varies in shape and darkness, and its red colour is probably due to chemicals being drawn up from deeper in the atmosphere.

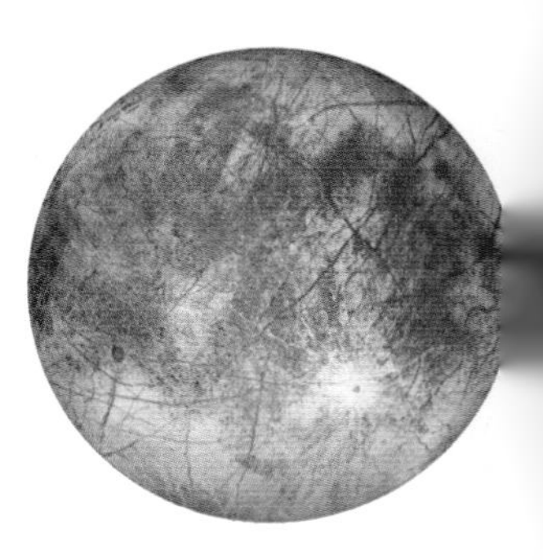

▷ **Europa**
This large moon of Jupiter is a world of ice. Europa's cracked surface covers a global ocean of water heated by the effects of tides. This combination of water and heat means it's one of the few places where life could have arisen.

The rings are split into regions of varying density, shepherded by the gravitational effects of Saturn's moons

◁ **Enceladus**
This moon, at 504 km (313 miles) wide, is the most reflective body in the Solar System. Its bright water-ice surface covers a liquid ocean, which might be a haven for life. Icy grains and water vapour erupt into space from ice volcanoes near the moon's south pole.

Uranus

The first planet to be discovered through the use of a telescope, Uranus is a bizarre world. Measuring at 51,118 km (31,763 mile) wide, it spins on its side, resulting in extreme seasons. The atmosphere is very bland around mid-winter, but bursts into activity when the Sun heats the equator. The planet's thin rings were first discovered in 1977, when they briefly blocked the light from a distant star.

Neptune

This planet's presence and rough location were predicted due to its effect on other planets. Neptune is 49,775 km (30,500 miles) wide and has a much more active atmosphere than Uranus, possessing huge storms. It has a large moon, Triton, whose backward orbit indicates that it was captured by Neptune in the distant past. Neptune's rings are uneven, consisting of concentrated arcs of material trapped inside tenuous rings.

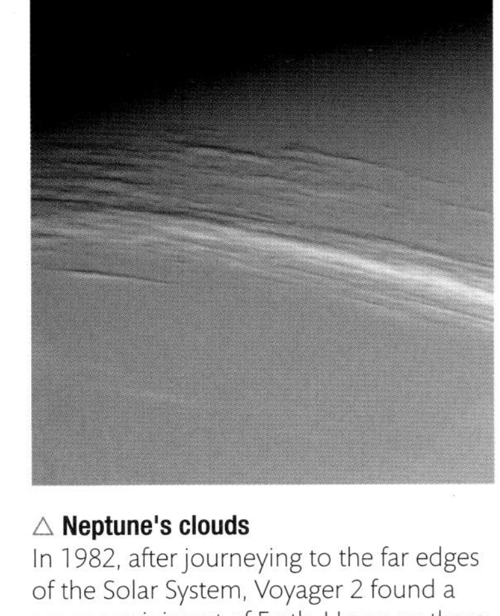

△ **Neptune's clouds**
In 1982, after journeying to the far edges of the Solar System, Voyager 2 found a scene reminiscent of Earth. However, these white clouds in Neptune's atmosphere are thought to be frozen methane.

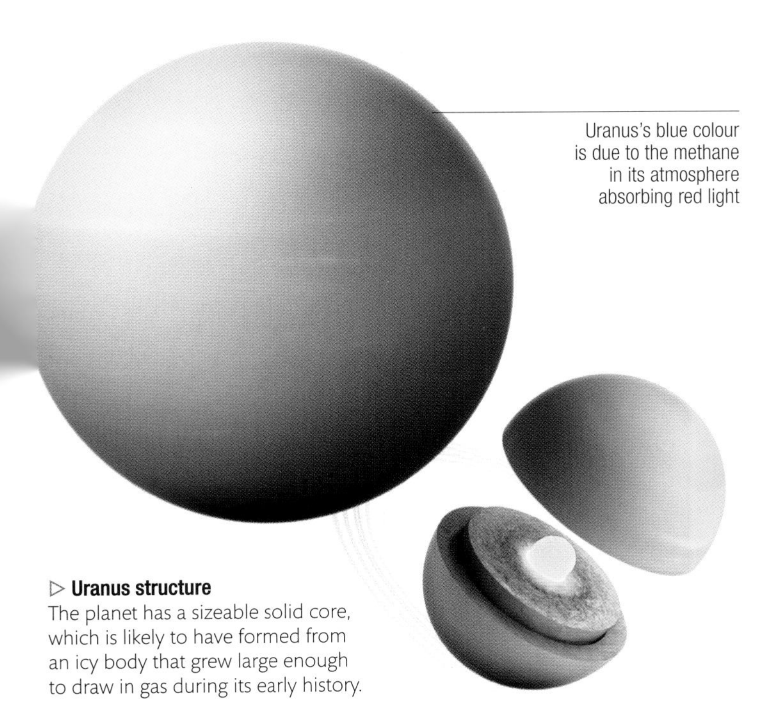

Uranus's blue colour is due to the methane in its atmosphere absorbing red light

▷ **Uranus structure**
The planet has a sizeable solid core, which is likely to have formed from an icy body that grew large enough to draw in gas during its early history.

As for all the giant planets, Neptune's atmosphere is primarily hydrogen and helium. Like Uranus, it also contains methane

Neptune's weather activity and fast winds indicate that there's a source of heat in the planet's interior

THE MOON

EARTH IS THE ONLY PLANET IN THE SOLAR SYSTEM WITH A MOON COMPARABLE IN SIZE TO ITSELF, BUT OUR NATURAL SATELLITE IS A COMPLETELY DRY, AIRLESS ENVIRONMENT.

Formation and structure

The Moon was almost certainly formed when a large, Mars-sized body struck a young Earth, throwing debris into space that eventually merged to form the body we see today. Its orbit has gradually widened and lengthened, and today, the Moon orbits Earth every 27.3 days. The Moon's interior was once warm enough for many volcanic eruptions, but such activity has now stopped.

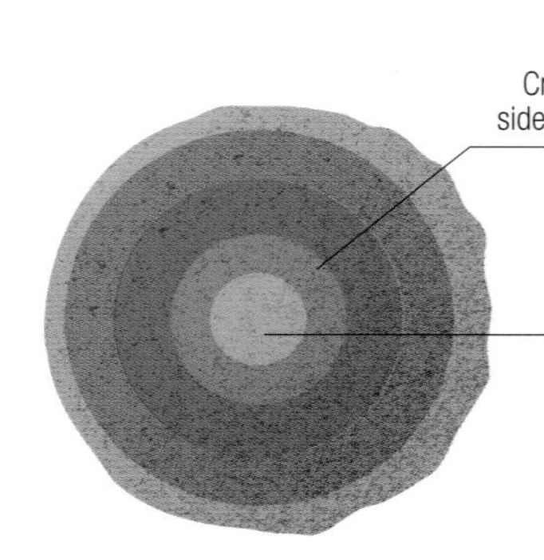
Crust is thicker to the side farthest from Earth
Core offset towards Earth

△ **Offset structure**
Tidal forces created by Earth's gravity early in the Moon's history distorted the Moon's symmetry.

Surface features

The Moon's surface has been pounded by countless impacts for billions of years. Much of the ground is covered by craters of all sizes. Looking by eye from Earth, the obvious features on the Moon are its dark patches. These, the seas, or maria, are where very fluid lava flooded large impact basins. These large eruptions ended just over a billion years ago.

◁ **Hayn impact crater**
This typical impact crater was formed when an asteroid or comet hit the Moon. It has a flat floor and central peaks, where the Moon's surface rebounded.

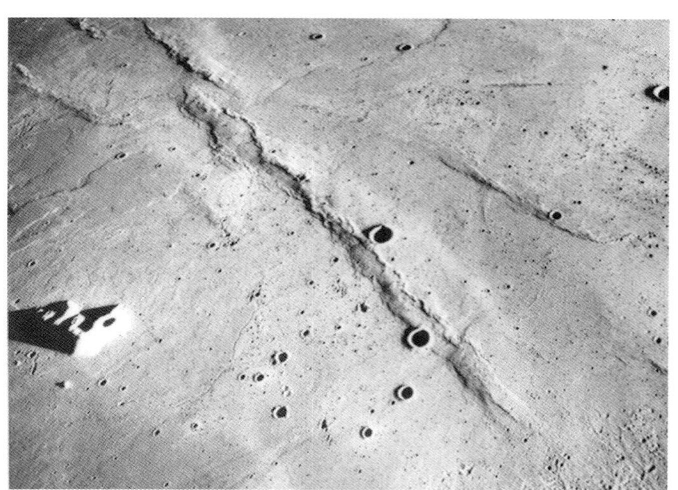

◁ **Lava plain**
The maria are vast expanses of lava that have covered several large areas of the Moon. They are not, however, perfectly smooth. Wrinkle ridges show where the lava cooled and shrank.

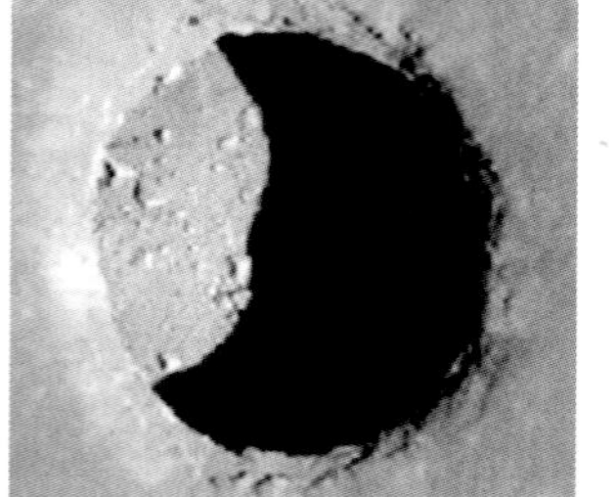

◁ **Sink hole**
This pit, less than 100 metres (330 ft) across, is where a sub-surface channel that once carried lava has collapsed. Similar features are seen in volcanic areas on Earth.

Mare Imbrium (Sea of Rains) is one of the largest of the lunar maria

◁ **Near side of the Moon**
It takes the Moon the same amount of time to spin on its own axis as it takes for it to orbit Earth. Known as synchronous rotation, it means that the side of the Moon shown here is always facing Earth.

Earth's satellite

The Moon is responsible for the tides in Earth's seas: it pulls water towards it, thereby raising a bulge on one side of the planet. Another bulge forms on the opposite side of the planet, where the Moon's gravity is weakest. The Moon's presence may also have stabilized Earth's spin axis, which has helped life develop here.

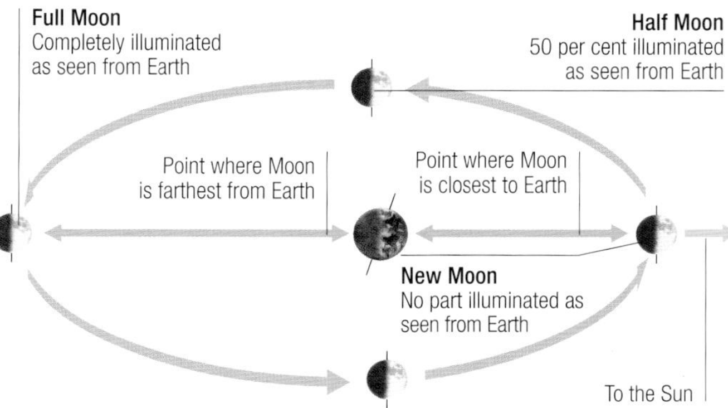

△ **Phases of the Moon**
The Moon shows phases during its orbit around the Earth, as the amount of sunlit ground seen from Earth varies. The orbit isn't circular since the distance between Earth and the Moon varies from 362,600 to 405,400 km (225,300 to 251,900 miles).

△ **Astronauts on the Moon**
Twelve astronauts walked on the Moon between 1969 and 1972, during six Apollo missions. Their work, and the study of the rocks and soil that they returned to Earth, transformed lunar science.

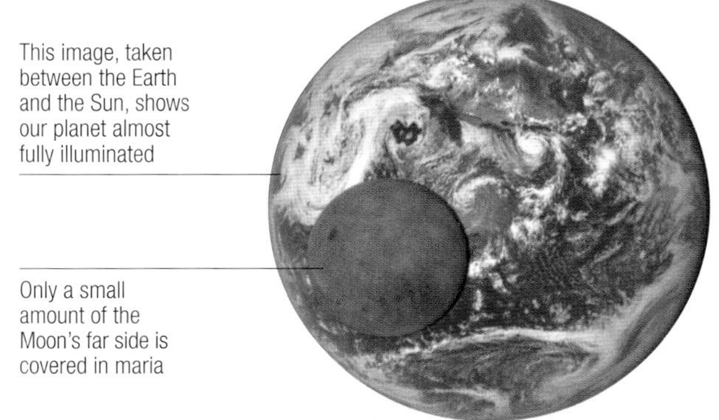

This image, taken between the Earth and the Sun, shows our planet almost fully illuminated
Only a small amount of the Moon's far side is covered in maria

△ **Earth and Moon still**
The differences between the surfaces of our planet and the Moon are obvious in this image. The greyish lunar surface reflects only 12 per cent of the light falling on it, whilst the colourful Earth reflects around 30 per cent. The Moon's far side, never seen from Earth, is visible here.

REFERENCE SECTION

STARS AND STAR GROUPS

Bright stars

The stars seen at night have different brightnesses. Sirius, the brightest, appears about 1,000 times brighter than the faintest seen with the naked eye. The Ancient Greek astronomer, Hipparchus, grouped the stars into "importance" categories, 1 being the brightest and 6 the faintest. Today we call these importances "apparent magnitudes". Each unit of apparent magnitude is 2.51 times fainter than the one before. A star's absolute magnitude is the apparent magnitude it would have if at a standard distance of 32.6 light-years.

BRIGHTEST STARS

Name	Constellation	Apparent magnitude	Absolute magnitude	Distance from Earth (in light-years)	Type
Sirius	Canis Major	-1.47	1.42	8.6	Blue-white main-sequence
Canopus	Carina	-0.72	-5.53	310	White giant
Rigel Kentaurus	Centaurus	-0.28	4.07	4.4	Yellow main-sequence
Arcturus	Boötes	-0.10	-0.31	37	Orange giant
Vega	Lyra	0.03 (variable)	0.58	25	Blue-white main-sequence
Capella	Auriga	0.08	-0.48	43	Yellow giant
Rigel	Orion	0.13 (variable)	-7.92	860	Blue supergiant
Procyon	Canis Minor	0.40	2.68	11	White main-sequence
Achernar	Eridanus	0.50	-2.77	144	Blue-white main-sequence
Betelgeuse	Orion	0.45 (variable)	-5.14	498	Red supergiant
Hadar	Centaurus	0.61 (variable)	-5.23	390	Blue-white giant
Altair	Aquila	0.76 (variable)	2.20	17	White main-sequence
Acrux	Crux	0.77	-4.19	322	Blue-white subgiant
Aldebaran	Taurus	0.87	-0.63	67	Red giant
Spica	Virgo	0.98 (variable)	-3.55	250	Blue-white giant
Antares	Scorpius	0.90 (variable)	-5.28	550	Orange giant
Pollux	Gemini	1.16	1.09	34	Orange giant
Fomalhaut	Piscis Austrinus	1.16	1.73	25.0	Blue-white main-sequence
Mimosa	Crux	1.25 (variable)	-3.92	278	Blue-white giant
Deneb	Cygnus	1.25	-8.38	1,400	Blue-white supergiant

Stellar giants

The vast majority of stars are too far away to have their radii measured directly. Size is usually estimated using a physical law that links radius, energy output, and surface temperature. As many of the biggest stars pulsate, the resulting size accuracy is only about 10 per cent. Theoretically, it has been estimated that giant stars become unstable if bigger than about 1,500 times the size of the Sun. The orbits of Earth and Jupiter are 215 and 1,120 the radius of the Sun, so all the stars in this list are bigger than Jupiter's orbit

LARGEST KNOWN STARS (BY RADIUS)

Name	Estimated radius (1=Radius of the Sun)	Type
UY Scuti	1,700	Red supergiant
NML Cygni	1,640	Red hypergiant
WOH G64	1,540	Red hypergiant
RW Cephei	1,535	Orange hypergiant
Westerlund 1-26	1,530	Red supergiant
V354 Cephei	1,520	Red supergiant
VX Sagittarii	1,520	Red hypergiant
VY Canis Majoris	1,420	Red hypergiant
KY Cygni	1,420	Red hypergiant
AH Scorpii	1,410	Red supergiant

Nearby stars and star groups

Over 90 per cent of the Sun's neighbours are main-sequence stars, and 50 per cent are in binary or triple groups. Typically, the average spacing is about seven light-years. It would take spacecraft such Voyager 1 about 100,000 years to travel this far.

Proxima Centauri will remain the closest star to the Sun for the next 25,000 years, after which Alpha Centauri takes over. This list will slowly change as the Sun travels around its Galactic orbit every 225,000,000 years.

CLOSEST STARS AND GROUPS

Name	Group	Component stars	Apparent magnitude	Absolute magnitude	Distance from Earth (in light-years)	Type
Sun	Single		-26.78	4.82	0.000016	Yellow main-sequence
Alpha Centauri	Triple	Proxima Alpha Centauri A Alpha Centauri B	11.09 0.01 1.34	15.53 4.38 5.71	4.2 4.4 4.4	Red main-sequence Yellow main-sequence Orange main-sequence
Barnard's Star	Single		9.53	13.22	5.9	Red main-sequence
Wolf 359	Single		13.44	16.55	7.8	Red main-sequence
Lalande 21185	Single		7.47	10.44	8.3	Red main-sequence
Sirius	Double	Alpha Canis Majoris A Alpha Canis Majoris B	-1.43 8.44	1.47 11.34	8.6 8.6	Blue-white main-sequence White dwarf
Luyten 726-8	Double	BL Ceti UV Ceti	12.54 12.99	15.40 15.85	8.7 8.7	Red main-sequence Red main-sequence
Ross 154	Single		10.43	13.07	9.7	Red main-sequence
Ross 248	Single		12.29	14.79	10.3	Red main-sequence
Epsilon Eridani	Single		3.73	6.19	10.5	Orange main-sequence
Lacaille 9352	Single		7.34	9.75	10.7	Red main-sequence
Ross 128	Single		11.13	13.51	10.9	Red main-sequence
EZ Aquarii	Triple	EZ Aquarii A EZ Aquarii B EZ Aquarii C	13.33 13.27 14.03	15.64 15.58 16.34	11.3 11.3 11.3	Red main-sequence Red main-sequence Red main-sequence
Procyon	Double	Alpha Canis Minoris A Alpha Canis Minoris B	2.66 12.98	2.66 12.98	11.4 11.4	White main-sequence White dwarf
61 Cygni	Double	61 Cygni A 61 Cygni B	7.49 8.31	7.49 8.31	11.4 11.4	Orange main-sequence Orange main-sequence

CONSTELLATIONS

Patterns in the sky

The sky is divided into 88 areas, most of which contain a recognizable pattern of stars. These constellations help astronomers name stars, describe the positions of planets and comets, and generally find their way around. The naming of celestial regions started around 4,000 years ago. Around 150 CE, Ptolemy listed the 48 constellations that could be seen from the Mediterranean region. In the 1590s Dutch explorers increased this list when they travelled across the equator to the southern oceans. More additions were made by astronomers in the 17th Century.

THE CONSTELLATIONS (RANKED BY AREA)

Rank	Name	Abbreviation	Named by
1	Hydra	Hya	Ptolemy
2	Virgo	Vir	Ptolemy
3	Ursa Major	UMa	Ptolemy
4	Cetus	Cet	Ptolemy
5	Hercules	Her	Ptolemy
6	Eridanus	Eri	Ptolemy
7	Pegasus	Peg	Ptolemy
8	Draco	Dra	Ptolemy
9	Centaurus	Cen	Ptolemy
10	Aquarius	Aqr	Ptolemy
11	Ophiuchus	Oph	Ptolemy
12	Leo	Leo	Babylonian origin
13	Boötes	Boo	Ptolemy
14	Pisces	Psc	Ptolemy
15	Sagittarius	Sgr	Ptolemy
16	Cygnus	Cyg	Ptolemy
17	Taurus	Tau	Babylonian origin
18	Camelopardalis	Cam	Peter Plancius
19	Andromeda	And	Ptolemy
20	Puppis	Pup	Nicholas de Lacaille
21	Auriga	Aur	Ptolemy
22	Aquila	Aqi	Ptolemy
23	Serpens	Ser	Ptolemy
24	Perseus	Per	Ptolemy
25	Cassiopeia	Cas	Ptolemy
26	Orion	Ori	Ptolemy
27	Cepheus	Cep	Ptolemy
28	Lynx	Lyn	Johannes Hevelius
28	Libra	Lib	Ptolemy
30	Gemini	Gem	Ptolemy
31	Cancer	Cnc	Ptolemy
32	Vela	Vel	Nicholas de Lacaille
33	Scorpius	Sco	Babylonian origin
34	Carina	Car	Nicholas de Lacaille
35	Monoceros	Mon	Petrus Plancius
36	Sculptor	Scl	Nicholas de Lacaille
37	Phoenix	Phe	Keyser/De Houtman
38	Canes Venatici	CVn	Johannes Hevelius
39	Aries	Ari	Ptolemy
40	Capricornus	Cap	Babylonian origin
41	Fornax	For	Nicholas de Lacaille
42	Coma Berenices	Com	Gerardus Mercator
43	Canis Major	CMA	Ptolemy
44	Pavo	Pav	Keyser/De Houtman
45	Grus	Gru	Keyser/De Houtman
46	Lupus	Lup	Ptolemy
47	Sextans	Sex	Johannes Hevelius
48	Tucana	Tuc	Keyser/De Houtman
49	Indus	Ind	Keyser/De Houtman
50	Octans	Oct	Nicholas de Lacaille
51	Lepus	Lep	Ptolemy
52	Lyra	Lyr	Ptolemy
53	Crater	Crt	Ptolemy
54	Columba	Col	Pertus Plancius
55	Vulpecula	Vul	Johannes Hevelius
56	Ursa Minor	UMi	Ptolemy
57	Telescopium	Tel	Nicholas de Lacaille
58	Horologium	Hor	Nicholas de Lacaille
59	Pictor	Pic	Nicholas de Lacaille
60	Piscis Austrinus	PsA	Ptolemy
61	Hydrus	Hyi	Keyser/De Houtman
62	Antlia	Ant	Nicholas de Lacaille
63	Ara	Ara	Ptolemy
64	Leo Minor	LMi	Johannes Hevelius
65	Pyxis	Pyx	Nicholas de Lacaille
66	Microscopium	Mic	Nicholas de Lacaille
67	Apus	Aps	Keyser/De Houtman
68	Lacerta	Lac	Johannes Hevelius
69	Delphinus	Del	Ptolemy
70	Corvus	Crv	Ptolemy
71	Canis Minor	CMi	Ptolemy
72	Dorado	Dor	Keyser/De Houtman
73	Corona Borealis	CrB	Ptolemy
74	Norma	Nor	Nicholas de Lacaille
75	Mensa	Men	Nicholas de Lacaille
76	Volans	Vol	Keyser/De Houtman
77	Musca	Mus	Keyser/De Houtman
78	Triangulum	Tri	Ptolemy
79	Chamaeleon	Cha	Keyser/De Houtman
80	Corona Australis	Cra	Ptolemy
81	Caelum	Cae	Nicholas de Lacaille
82	Reticulum	Ret	Nicholas de Lacaille
83	Triangulum Australe	TrA	Keyser/De Houtman
84	Scutum	Sct	Johannes Hevelius
85	Circinus	Cir	Nicholas de Lacaille
86	Sagitta	Sge	Ptolemy
87	Equuleus	Equ	Ptolemy
88	Crux	Cru	João Faras

MILKY WAY AND OTHER GALAXIES

The Local Group

The Local Group of galaxies is a gravitationally bound cluster of over 54 galaxies, mainly dwarfs, and is about 10 million light-years across. It is dominated by three giant galaxies; the Milky Way, Andromeda, and Triangulum. Each of these has a swarm of orbiting smaller satellite galaxies. The Local Group was first recognized in 1936 by the American astronomer Edwin Hubble. The membership of some of the outliers (such as Antlia Dwarf, Sextans A, and NGC 3109) is debatable. Other, as yet undiscovered, members could be hidden behind the giant galaxies.

THE LOCAL GROUP OF GALAXIES

Name	Type	Distance from Solar System light-years	Diameter
Milky Way	Barred spiral	0	100,000
Sagittarius Dwarf	Dwarf elliptical	78,000	20,000
Ursa Major II	Dwarf elliptical	100,000	1,000
Large Magellanic Cloud	Disrupted barred spiral	165,000	25,000
Small Magellanic Cloud	Irregular	195,000	15,000
Boötes Dwarf	Dwarf elliptical	197,000	2,000
Ursa Minor Dwarf	Dwarf elliptical	215,000	2,000
Sculptor Dwarf	Dwarf elliptical	258,000	3,000
Draco Dwarf	Dwarf elliptical	267,000	2,000
Sextans Dwarf	Dwarf elliptical	280,000	3,000
Ursa Major I	Dwarf elliptical	325,000	3,000
Carina Dwarf	Dwarf elliptical	329,000	2,000
Fornax Dwarf	Dwarf elliptical	450,000	5,000
Leo II	Dwarf elliptical	669,000	3,000
Leo I	Dwarf elliptical	815,000	3,000
Phoenix Dwarf	Irregular	1,450,000	2,000
NGC 6822	Irregular	1,520,000	8,000
NGC 185	Dwarf elliptical	2,010,000	8,000
Andromeda II	Dwarf elliptical	2,165,000	3,000
Leo A	Irregular	2,250,000	4,000
IC 1613	Irregular	2,365,000	10,000
NGC 147	Dwarf elliptical	2,370,000	10,000
Andromeda III	Dwarf elliptical	2,450,000	3,000
Cetus Dwarf	Dwarf elliptical	2,485,000	3,000
Andromeda I	Dwarf elliptical	2,520,000	2,000
LGS 3	Irregular	2,520,000	2,000
Andromeda Galaxy (M31)	Barred spiral	2,560,000	140,000
M32	Dwarf elliptical	2,625,000	8,000
M110	Dwarf elliptical	2,960,000	15,000
IC 10	Irregular	2,960,000	8,000
Triangulum Galaxy (M33)	Spiral	2,735,000	55,000
Tucana Dwarf	Dwarf elliptical	2,870,000	2,000
Pegasus Dwarf	Irregular	3,000,000	6,000
WLM	Irregular	3,020,000	10,000
Aquarius Dwarf	Irregular	3,345,000	3,000
SAGDIG	Irregular	3,460,000	3,000
Antlia Dwarf	Dwarf elliptical	4,030,000	3,000
NCG 3109	Irregular	4,075,000	25,000
Sextans A	Irregular	4,350,000	10,000
Sextans B	Irregular	4,385,000	8,000

Galaxy clusters and groups

Galaxies in the Universe are not distributed at random. They are in gravitationally bound groups containing tens to thousands of individuals. Dominating our region is The Great Attractor (the main component of which is the Norma Cluster). This is so massive it affects the normal expansion of the Universe discovered by Edwin Hubble. Clusters accumulate together to form superclusters. Cluster diameters are between 6 and 30 million light-years.

Galaxy cluster MACS J0416.1–2403 in Eridanus

GALAXY CLUSTERS AND GROUPS

Name	Distance millions of light-years	Recessional velocity km per second (miles per second)
Local Group	0	
M81 Group	11	334 (207)
Centaurus Group	12	299 (186)
Sculptor Group	12.7	292 (181)
Canes Venatici I Group	13	483 (300)
Canes Venatici II Group	26	703 (387)
M51 Group	31	555 (345)
Leo Triplet	35	662 (386)
Leo I Group	38	680 (423)
Draco Group	40	704 (437)
Ursa Major Group	55	1,016 (631)
Virgo Cluster	59	1,139 (708)

MESSIER OBJECTS

Deep-sky catalogue

The French astronomer Charles Messier (1730–1817) produced a catalogue of nebulae and star clusters easily visible in small telescopes. His designations (for example M31 for Andromeda Galaxy) are still much in use today. Messier was a comet hunter (he discovered 13) and did not want to confuse transient comets with similar-looking permanent bodies. He used a 10-cm refractor telescope in Paris, so objects south of -35.7° declination were not included. His catalogue, started in 1760, finally listed 110 objects.

MESSIER CATALOGUE

Messier number	Constellation	Common name	Object type
M1	Taurus	Crab Nebula	Supernova remnant
M2	Aquarius		Globular cluster
M3	Canes Venatici		Globular cluster
M4	Scorpius		Globular cluster
M5	Serpens (Caput)		Globular cluster
M6	Scorpius	Butterfly Cluster	Open cluster
M7	Scorpius	Ptolemy Cluster	Open cluster
M8	Sagittarius	Lagoon Nebula	Emission nebula
M9	Ophiuchus		Globular cluster
M10	Ophiuchus		Globular cluster
M11	Scutum	Wild Duck Cluster	Open cluster
M12	Ophiuchus		Globular cluster
M13	Hercules		Globular cluster
M14	Ophiuchus		Globular cluster
M15	Pegasus		Globular cluster
M16	Serpens (Cauda)	Eagle Nebula	Open cluster emission nebula
M17	Sagittarius	Omega/Swan Nebula	Emission nebula
M18	Sagittarius		Open cluster
M19	Ophiuchus		Globular cluster
M20	Sagittarius	Trifid Nebula	Emission/reflection dark nebula
M21	Sagittarius		Open cluster
M22	Sagittarius		Globular cluster
M23	Sagittarius		Open cluster
M24	Sagittarius	Sagittarius Star Cloud	Starfield in Milky Way
M25	Sagittarius		Open cluster
M26	Scutum		Open cluster
M27	Vulpecula	Dumbbell Nebula	Planetary nebula
M28	Sagittarius		Globular cluster
M29	Cygnus		Open cluster
M30	Capricornus		Globular cluster
M31	Andromeda	Andromeda Galaxy	Spiral galaxy
M32	Andromeda		Dwarf elliptical galaxy
M33	Triangulum	Triangulum Galaxy	Spiral galaxy
M34	Perseus		Open cluster
M35	Gemini		Open cluster
M36	Auriga		Open cluster
M37	Auriga		Open cluster
M38	Auriga		Open cluster
M39	Cygnus		Open cluster
M40	Ursa Major	Winnecke 4	Double star
M41	Canis Major		Open cluster
M42	Orion	Orion Nebula	Emission/reflection nebula
M43	Orion	De Mairan's Nebula	Emission/reflection nebula
M44	Cancer	Beehive Cluster/Praesepe	Open cluster
M45	Taurus	Pleiades/Seven Sisters	Open cluster
M46	Puppis		Open cluster
M47	Puppis		Open cluster
M48	Hydra		Open cluster
M49	Virgo		Elliptical galaxy
M50	Monoceros		Open cluster
M51	Canes Venatici	Whirlpool Galaxy	Spiral galaxy
M52	Cassiopeia		Open cluster
M53	Coma Berenices		Globular cluster
M54	Sagittarius		Globular cluster
M55	Sagittarius		Globular cluster
M56	Lyra		Globular cluster
M57	Lyra	Ring Nebula	Planetary nebula
M58	Virgo		Barred spiral galaxy
M59	Virgo		Elliptical galaxy
M60	Virgo		Elliptical galaxy

MESSIER CATALOGUE CONTINUED

Messier number	Constellation	Common name	Object type
M61	Virgo		Spiral galaxy
M62	Ophiuchus		Globular cluster
M63	Canes Venatici	Sunflower Galaxy	Spiral galaxy
M64	Coma Berenices	Black Eye Galaxy	Spiral galaxy
M65	Leo		Spiral galaxy
M66	Leo		Spiral galaxy
M67	Cancer		Open cluster
M68	Hydra		Globular cluster
M69	Sagittarius		Globular cluster
M70	Sagittarius		Globular cluster
M71	Sagitta		Globular cluster
M72	Aquarius		Globular cluster
M73	Aquarius		Asterism
M74	Pisces		Spiral galaxy
M75	Sagittarius		Globular cluster
M76	Perseus	Little Dumbbell Nebula	Planetary nebula
M77	Cetus		Barred spiral galaxy
M78	Orion		Reflection nebula
M79	Lepus		Globular cluster
M80	Scorpius		Globular cluster
M81	Ursa Major	Bode's Galaxy	Spiral galaxy
M82	Ursa Major	Cigar Nebula	Spiral galaxy
M83	Hydra	Southern Pinwheel Galaxy	Barred spiral galaxy
M84	Virgo		Elliptical or lenticular galaxy
M85	Coma Berenices		Lenticular galaxy
M86	Virgo		Lenticular galaxy
M87	Virgo	Virgo A	Elliptical galaxy
M88	Coma Berenices		Spiral galaxy
M89	Virgo		Elliptical galaxy
M90	Virgo		Spiral galaxy
M91	Coma Berenices		Barred spiral galaxy
M92	Hercules		Globular cluster
M93	Puppis		Open cluster
M94	Canes Venatici		Spiral galaxy
M95	Leo		Barred spiral galaxy
M96	Leo		Spiral galaxy
M97	Ursa Major	Owl Nebula	Planetary nebula
M98	Coma Berenices		Spiral galaxy
M99	Coma Berenices		Spiral galaxy
M100	Coma Berenices		Spiral galaxy
M101	Ursa Major	Pinwheel Galaxy	Spiral galaxy
M102		Identification unknown	Possibly lenticular galaxy NGC 5866 in Virgo
M103	Cassiopeia		Open cluster
M104	Virgo	Sombrero Galaxy	Spiral galaxy
M105	Leo		Elliptical galaxy
M106	Canes Venatici		Spiral galaxy
M107	Ophiuchus		Globular cluster
M108	Ursa Major		Barred spiral galaxy
M109	Ursa Major		Barred spiral galaxy
M110	Andromeda		Dwarf elliptical galaxy

Star-forming gas clouds the Lagoon Nebula (M8)

GLOSSARY

A

Absolute magnitude
A measure of the true brightness of a star or other astronomical object, defined as the apparent magnitude it would have at a distance of 10 parsecs (32.6 light-years). See also *apparent magnitude, luminosity*

Accretion
(1) The colliding and sticking together of small astronomical bodies to make larger ones. (2) The process whereby a body grows in mass by accumulating matter from its surroundings.

Active galaxy
A galaxy that emits an exceptional amount of electromagnetic radiation over a wide range of wavelengths. The radiation comes from a central "active galactic nucelus" and is thought to be powered by the accretion of gas onto a supermassive black hole. There are several named types, although the apparent differences between them may be because we are viewing them at different angles as seen from Earth. See also *blazar, quasar, Seyfert galaxy*

Apparent magnitude
A measure of the brightness of a star or other astronomical object as seen from Earth, which depends on its closeness as well as how luminous it really is. The brighter the object, the smaller the numerical value of its apparent magnitude. See also *absolute magnitude, luminosity*

Asterism
A pattern of bright stars in the night sky that is usually only part of a constellation. For example, the Plough or Big Dipper is an asterism within the constellation Ursa Major. See also *constellation*

Asteroid
A small, irregularly shaped Solar System object of rock or metal less than 1,000 km (600 miles) in diameter. Most, but not all, asteroids are found in the Asteroid Belt between Mars and Jupiter.

Astronomical unit (AU)
A unit of distance based on the average distance between Earth and the Sun. It is approximately 150 million km (93 million miles).

Atmosphere
A gaseous surround to a planet or a low-density region of plasma around a star.

Atom
A building block of ordinary matter. It consists of a central heavy nucleus of protons (positively charged, with a different number for each chemical element) and neutral neutrons, surrounded by orbiting electrons (negatively charged, and equal in number to the protons). See also *electron, ion*

Aurora
A display of glowing light in the sky that is most common in polar regions. It is caused by high-energy particles from the Sun being deflected by Earth's magnetic field and colliding with atoms in Earth's atmosphere.

Axis
The imaginary line about which a body rotates.

B

Background radiation
See *cosmic microwave background radiation*

Barred spiral galaxy
A galaxy with spiral arms that originate from the ends of an elongated, bar-shaped central region. See also *spiral galaxy*

Big Bang
The event in which the Universe was born. According to Big Bang theory, the Universe originated around 13.8 billion years ago in an extremely hot dense state and has been expanding ever since.

Binary star
A pair of stars that orbit each other around a common centre of mass. See also *centre of mass*

Black dwarf star
A former white dwarf star, which has cooled so much that it emits no detectable light. The Universe is not yet old enough for black dwarfs to have formed. See also *brown dwarf star, white dwarf star*

Black hole
A region of space into which so much matter has collapsed that nothing, not even light, can escape its gravitational pull. The supermassive black holes found in the centres of galaxies can be up to billions of times the mass of the Sun.

Blazar
The most luminous and variable type of active galaxy in terms of its radiation. See also *active galaxy*

Blueshift
The opposite of redshift: the shifting of electromagnetic radiation to a higher frequency when it radiates from an object moving towards the observer. See also *redshift*

Bok globule
A type of compact dark nebula believed to be the precursor of a protostar. See also *protostar*

Brown dwarf
A body that forms out of a contracting cloud of gas, like a star, but because it contains too little mass never becomes hot enough to sustain nuclear fusion.

C

Celestial equator
An imaginary circle on the celestial sphere that is a projection of the Earth's own equator onto it. See also *celestial sphere*

Celestial poles
The points in the sky directly above Earth's north and south poles. The celestial sphere appears to rotate around an axis joining the celestial poles.

Celestial sphere
An imaginary sphere that surrounds the Earth. As the Earth rotates from west to east, the sphere appears to rotate from east to west. In order to define the position of stars and other celestial bodies, it is convenient to think of them as being attached to the inside surface of this sphere.

Centre of mass
The point around which two or more bodies revolve, as for example when two stars revolve around each other. If one of the bodies has more mass than the other, the centre of mass lies towards the larger one.

Cepheid variable
A type of variable star whose luminosity alters in a regular rhythm. Cepheids vary in brightness as they physically expand and contract. The more luminous the Cepheid, the longer its period of variation. See also *luminosity, variable star*

Chromosphere
The relatively thin layer of the Sun's atmosphere that lies between the photosphere and the corona. See also *corona, photosphere*

Comet
A small body composed mainly of dust-laden ice that orbits the Sun. When a comet enters the inner Solar System some of its material evaporates, often forming a long "tail" of gas and dust.

Constellation
(1) A named pattern of stars in the night sky, used for convenience as a way of describing the position of astronomical objects seen from Earth. (2) One of the 88 regions into which the celestial sphere is divided for reference purposes, based on these traditional constellations.

Corona
The outermost part of the atmosphere of the Sun or another star, stretching thousands of miles into space. It has a very high temperature but a low density.

Cosmic microwave background radiation (CMBR)
The radiation left over from the Big Bang, appearing from all directions of the sky.

D

Dark energy
A little-understood phenomenon that appears to account for about 70 per cent of the total "mass plus energy" in the Universe. It is thought necessary to explain why the expansion of the Universe is currently accelerating.

Dark matter
A mysterious kind of matter that seems to interact only via gravity and not by emitting or absorbing electromagnetic radiation, in contrast to ordinary matter made of atoms. Scientists think it exists in large quantities in the Universe because without it, galaxies should fly apart as they rotate.

Declination
The equivalent on the celestial sphere of latitude on Earth. The declination of a star is its angular distance north or south of the celestial equator. See also *right ascension*

Diffuse nebula
A nebula lacking sharp outer boundaries and without obvious internal features. See also *nebula*

Double star
Two stars that appear close together in the sky. If they actually orbit each other, the system is called a binary. An optical double consists of stars that

look close together only because they lie in the same line of sight from Earth. See also *binary star*

Dwarf planet
A rounded orbiting body similar to a planet but not massive enough to have cleared its orbital path of other objects. Pluto is an example. See also *planet*

E

Eclipsing binary
A binary star system in which each star passes alternately in front of the other, cutting off all or part of its light and causing a periodic variation in the system's overall brightness.

Ecliptic
(1) The track along which the Sun appears to travel around the celestial sphere, relative to the background stars, in the course of a year. (2) The plane of Earth's orbit around the Sun (which determines the position of the ecliptic in sense 1). See also *celestial sphere, zodiac*

Electromagnetic radiation
Radiation that transmits energy throughout the Universe as waves of fluctuating electric and magnetic fields which all travel at "the speed of light". See also *electromagnetic spectrum*

Electromagnetic spectrum
The whole range of electromagnetic radiation, from radio waves (which have the lowest frequencies and the longest wavelengths) through microwaves, infrared radiation, visible light, ultraviolet radiation, and X-rays, to gamma rays (with highest frequencies and energies, and shortest wavelengths).

Electron
A subatomic particle with a negative electric charge, found in all atoms. Electrons are much lighter than the protons and neutrons that make up atomic nuclei. See also *atom*

Elliptical galaxy
A galaxy that is elliptical or round in shape. Elliptical galaxies typically contain older stars and show little evidence of current star creation.

ESA
Short for European Space Agency, an organization supported by most European countries, with headquarters in Paris.

Exoplanet
See *extrasolar planet*

Extrasolar planet (exoplanet)
A planet orbiting a star other than the Sun.

Extremophile
Any life form that thrives under extreme conditions, such as high pressure, very high or low temperatures, or unusual chemical environments.

F

Fusion (nuclear fusion)
The process whereby atomic nuclei join to form heavier atomic nuclei at high temperatures. Stars are powered by fusion reactions that take place in their cores and release large amounts of energy.

G

Galactic plane
The plane of the flat disk of a galaxy, especially the Milky Way, where most of its stars are found.

Galaxy
A huge aggregation of star systems, gases, dust, and dark matter, held together by gravity and distinct from other surrounding galaxies. Galaxies can hold from millions up to trillions of stars. The Milky Way Galaxy is often referred to as "The Galaxy" with a capital letter. See also *active galaxy, elliptical galaxy, irregular galaxy, lenticular galaxy, spiral galaxy*

Galaxy cluster
An aggregation of 50 to 1,000 galaxies held together by gravity.

Galaxy supercluster
A cluster of galaxy clusters. A supercluster may contain 10,000 or more galaxies, spread through a volume of space with a diameter of up to about 200 million light-years.

Gamma radiation
Electromagnetic radiation of extremely short wavelengths and high frequencies and energy. See also *electromagnetic radiation, electromagnetic spectrum*

Gas giant
A large planet composed mainly of hydrogen and helium. Jupiter and Saturn are examples in our Solar System. See also *rocky planet*

Globular cluster
A near-spherical cluster of between 10,000 and more than 1 million stars. Globular clusters consist of very old stars and are located mainly in the spherical halo regions around galaxies.

Gravitationally bound
Phrase applied to any astronomical system that is kept together by gravitational attraction between its parts. The Solar System and the Milky Way are examples.

Gravity
An attractive force between all objects that have mass or energy, experienced on Earth as weight. The force of gravity keeps planets in orbit around the Sun and stars in orbit around the Galaxy.

H

Hertzsprung-Russell (HR) diagram
A diagram where stars are plotted according to their luminosity and surface temperature/colour. See also *luminosity, main-sequence star*

Hot Jupiter
An extrasolar planet similar to Jupiter in size and composition, but orbiting much closer to its star and therefore hotter. See also *extrasolar planet*

Hubble constant
A mathematical constant that relates a galaxy's distance to the speed that it is receding from our own galaxy. It represents an estimate of the expansion rate of the Universe.

Hypergiant star
A star of exceptionally high mass, larger than a supergiant. Hypergiants may be 100 times or more the mass of the Sun, but are short-lived, burning themselves out quickly.

I

Infrared radiation
Electromagnetic radiation with wavelengths longer than visible light but shorter than microwaves, experienced as heat radiation in everyday life. See also *electromagnetic radiation, electromagnetic spectrum*

Interferometry
A technique involving measuring the overlap between electromagnetic waves from a distant source, used to achieve sharper images of an object. Information from arrays of telescopes or radio telescopes many kilometres apart can be combined, resulting in images approximating those from an imaginary huge telescope the size of the array.

Ion
An atom which has lost or gained one or more electrons and therefore has an overall positive or negative charge. The process of this happening is called ionization. See also *electron*

Irregular galaxy
A galaxy that lacks a well defined structure or symmetry.

K

Kuiper Belt
A region of the Solar System beyond Neptune containing bodies of icy and rocky composition. See also *Oort cloud*

L

Lenticular galaxy
A galaxy shaped like a convex lens. It has a central bulge that merges into a flattened disk, but has no spiral arms.

Light-year
A unit of measurement, defined as the distance that light travels through a vacuum in one year. 1 light-year is equal to 9,460 billion km (5,878 billion miles).

Local Group
A small cluster of over 50 galaxies that includes our own galaxy, the Milky Way. The group also contains two other large spiral galaxies, including the well known Andromeda galaxy, although most of its members are small elliptical or irregular galaxies. See also *galaxy cluster*

Look-back distance
The distance that light has travelled from a distant object to reach us today. It is farther than the original distance to the object (since the Universe has expanded while the light was travelling) but less than its present distance (since the object is now farther away than when it sent the light).

Luminosity
The total amount of energy emitted in one second by a source of radation, such as the Sun or a star. See also absolute magnitude.

M

Magnetic field
The region around a magnetized body within which magnetic forces affect the motion of electrically charged particles.

Magnitude
A measure of the brightness of a star or other astronomical object. See also *absolute magnitude, apparent magnitude*

Main-sequence star
Any star that falls within the main diagonal band on a Hertzsprung-Russell diagram, in which luminosity is plotted against temperature. Main-sequence stars are converting hydrogen in their cores into helium, and may stay in the same position on the sequence for billions of years, the exact position depending mainly on the star's original mass. The Sun is a main-sequence star. See also *Hertzsprung-Russell diagram.*

Messier catalogue
A catalogue of nebulae (including some objects now known to be galaxies) compiled, and published in 1781, by French astronomer Charles Messier and his assistant Pierre Méchain. Objects in this catalogue are assigned an individual number, preceded by "M". For example, the Andromeda galaxy is M31. See also *New General Catalogue*

Meteor
The short-lived streak of light or "shooting star" seen when a small Solar System body burns up on entering Earth's atmosphere. If the body survives to reach the ground it is termed a meteorite.

Meteorite
A small solid Solar System body that has survived passing through the atmosphere of Earth or another planet and has reached the ground.

Microwave radiation
Electromagnetic radiation with wavelengths shorter than radio waves but longer than infrared and visible light. See also *electromagnetic radiation, electromagnetic spectrum*

Milky Way
(1) Originally, the luminous band across the night sky that represents the combined light of vast numbers of stars and nebulae in the disk of our home galaxy. (2) Now used as a name for the galaxy itself.

Mira variable
A class of giant variable stars whose brightness varies over a period of around 100 to 500 days. See also *variable star*

Moon
A natural satellite orbiting a planet. Spelled with a capital, "the Moon" refers to the Earth's moon.

Multiple star
A system consisting of three or more stars bound together by gravity and orbiting around one another. See also *binary star*

N

NASA
Short for National Aeronautics and Space Administration, the main space agency of the United States.

Nebula
A cloud of gas and dust in interstellar space. Some nebulae are sites of star formation, while others are produced at the end of a star's life. See also *planetary nebula*

Neutrino
A particle of exceedingly low mass and zero electrical charge that travels close to the speed of light and rarely interacts with other matter.

Neutron star
An extremely dense compact star made of tightly packed neutrons (neutral subatomic particles). Neutron stars are formed by supernova explosions not massive enough to create a black hole. See also *pulsar*

New General Catalogue (NGC)
A catalogue of star clusters and nebulae (including objects now known to be galaxies) compiled by J.L.E. Dreyer in 1888. Objects are assigned an individual number, preceded by "NGC". With amendments, this system is still in use today. See also *Messier catalogue*

Nova
A star that suddenly brightens, then fades back to its original brightness over a period of weeks or months. The brightening happens when a fusion reaction is triggered on the surface of a white dwarf star by gas flowing from another star. See also *fusion, supernova*

Nuclear fusion
See *fusion*

Observable Universe
That part of the Universe from which light has had time, since the Big Bang, to reach Earth.

Oort cloud
A spherically distributed collection of trillions of icy bodies such as cometary nuclei that is believed to surround the Solar System and extend more than 1 light-year distant from the Sun.

Open cluster
A relatively spread-out cluster of stars that all formed at the same time. See also *globular cluster*

Optical double
See *double star*

Orbit
The path of an astronomical body when it is revolving around another under the influence of gravity.

P

Particle
In astronomical contexts this usually means a subatomic particle, such as a proton or neutron, or more exotic particles of similarly tiny size.

Photosphere
The layer of the Sun or other star from which most of the light is emitted and which forms its visible surface. See also *chromosphere, corona*

Planet
A large body orbiting a star. A planet is sufficiently massive for its gravity to have formed it into a round shape and also for it to have cleared its orbital path of other objects. See also *dwarf planet*

Planetary nebula
A glowing shell of gas ejected by a star of similar mass to the Sun when coming to the end of its life. The term was first used by William Herschel for circular nebulae that looked similar to a planet.

Plasma
A mixture of electrons and positive ions that behaves like a gas, but conducts electricity and is affected by magnetic fields. The Sun and other stars are made of hot plasma. See also *ion*

Precession
A slow cyclic change in the direction of Earth's axis (i.e. the direction in which the north and south poles "point"), which takes 25,800 years to complete. A similar kind of movement is seen in a spinning top. The term is also applied to other astronomical cycles, such as the slow change in position of the farthest point of a planet's orbit.

Prominence
A huge eruption of glowing plasma into the Sun's corona, often in a looping shape. See also *corona, plasma*

Protoplanet
A precursor of a planet, which develops through the gradual aggregation of smaller bodies in the protoplanetary disk that forms around many new stars. Planets are thought to form by collision of protoplanets.

Protostar
A star in the early stages of formation, before hydrogen fusion has begun.

Pulsar
A rapidly rotating neutron star that is sending out powerful jets of radiation from its magnetic poles. Pulses are detected if the jets happen to sweep by in Earth's directions as the neutron star spins. See also *neutron star*.

Pulsating variable
See variable star.

Quasar
A compact but extremely powerful source of radiation, now believed to be a type of highly luminous active galaxy. Most quasars are at extreme distances from our own galaxy and we are observing them as they were early in the history of the Universe. See also *active galaxy*

Radio telescope
An instrument that is designed to detect radio waves from astronomical sources. The most familiar type is a concave dish that collects radio waves and focuses them onto a detector.

Red dwarf star
A cool, red, low-luminosity star. Red dwarfs are common in the Universe and are very long-lived.

Red giant star
A greatly expanded reddish star with a low surface temperature that forms at the end of the life of Sun-like stars. It is "giant" in its size and luminosity, rather than in mass. See also *supergiant*

Redshift
The shifting to a lower frequency of electromagnetic radiation when it comes from an object moving away from an observer. It can be compared to the siren on an emergency vehicle that sounds a lower note when speeding away.

Reflector
A telescope in which the light is collected and focused by a curved mirror. See also *refractor*.

Refractor
A telescope in which the light is collected and focused by a lens. See also reflector.

Relativity
Two theories developed in the early 20th century by Albert Einstein. The special theory of relativity describes how the relative motion of observers affects their measurements of mass, length, and time. One consequence is that mass and energy are equivalent. The general theory of relativity treats gravity as a distortion of spacetime. See also *spacetime*

Right ascension
The equivalent on the celestial sphere of longitude on Earth. The right ascension of a star is its angular distance east of a point in the sky called the first point of Aries. It is expressed in hours, minutes, and seconds, 1 hour being the equivalent of 15 degrees. See also *declination*

Rocky planet
A planet composed mainly of rock.The four rocky planets in the Solar System are Mercury, Venus, Earth, and Mars. See also *gas giant*

Satellite
A natural satellite is an astronomical body that orbits a planet, otherwise known as a moon. An artificial satellite is an object deliberately put in orbit around Earth or another planet.

Seyfert galaxy
A spiral galaxy with an exceptionally bright central region. Seyfert galaxies are believed to be similar to quasars, although less powerful and found closer to our own galaxy. See also *active galaxy*

Singularity
A point of infinite density into which matter has been compressed by gravity, and a point at which the known laws of physics break down. Theory implies that a singularity exists at the centre of a black hole. See also *black hole*

Solar flare
A violent release of huge amounts of energy from a localized region on the surface of the Sun.

Solar System
The Sun together with the eight planets, smaller bodies (dwarf planets, moons, asteroids, comets, trans-Neptunian objects), dust, and gas that orbit the Sun.

Solar wind
A constant stream of fast-moving particles that escapes from the Sun and flows outwards through the Solar System.

Spacetime
The combination of the three dimensions of space (length, breadth, height) and the single time dimension. See also *relativity*

Spiral galaxy
A galaxy that consists of a central concentration of stars surrounded by a flattened disk of stars, gas, and dust, within which the major visible features are clumped together into spiral arms. See also *barred spiral galaxy*

Star
A huge sphere of glowing plasma that generates energy by means of nuclear reactions at its centre. See also *fusion, plasma*

Star cluster
A group of stars bound together by gravity. See also *globular cluster, open cluster*

Subgiant star
A star that is significantly more luminous than a main-sequence star of the same surface temperature and colour.

Sunspot
A region of intense magnetic activity in the Sun's photosphere. Sunspots appear dark in images because their temperature is lower than the rest of the photosphere. See also *photosphere*

Super-Earth
An extra-solar planet whose mass is greater than Earth's but less than planets such as Uranus and Neptune. See also *extra-solar planet.*

Supergiant star
An exceptionally luminous star with a very large diameter.

Supernova
A violent explosion of a massive star, during which it expels most of its matter, and its brightness increases hugely for a short time. A different kind of supernova happens when a white dwarf explodes after attracting material from a neighbouring star.

Trans-Neptunian object
A Solar System body orbiting the Sun beyond the orbit of Neptune.

Ultraviolet radiation
Electromagnetic radiation with wavelengths shorter than visible light but longer than X-rays. See also *electromagnetic radiation*

Universe
The totality of matter, energy, and space that came into being as a result of the Big Bang.

Variable star
A star that varies in brightness. A pulsating variable star physically expands and contracts in a regular rhythm, varying in brightness as it does so. An eruptive variable star brightens and fades abruptly. See also *Cepheid variable, Mira variable, eclipsing binary*

Wavelength
The distance between two successive crests in a wave motion.

White dwarf star
A small, but very hot and dense, glowing body that remains after a star of similar mass to our Sun dies and sheds its outer layers into space.

Wolf-Rayet star
A massive, very hot star from which gas is escaping at an exceptionally rapid rate.

X-ray
Electromagnetic radiation with wavelengths shorter than ultraviolet radiation but longer than gamma rays. See also *electromagnetic radiation*

Zenith
The point in the sky directly above an observer.

Zodiac
An imaginary band around the celestial sphere, through which the Sun, Moon, and planets appear to travel. It represents the plane of the Solar System as seen from Earth. See also *ecliptic*

INDEX

Bold page numbers refer to mainentries.

1,2,3

1 Lacertae 129
3C 273 136, 137
5 Lyncis 108
6 Trianguli 128
8 Monocerotis see Epsilon Monocerotis
12 Lyncis 108
13 Monocerotis 171
17 Sextantis 174
18 Sextantis 174
19 Lyncis 108
21 Monocerotis 170
24 Sextantis 48
36 Ophiuchi 144, 145
38 Lyncis 108
39 Draconis 104
40 Eridani 160
40 Leonis 134, 135
46 Leonis Minoris 132
47 Tucanae 45, 215
61 Cygni A and B 23
67 Ophiuchi 144
67P/Churyumov-Gerasimenko 225
70 Ophiuchi 145
95 Herculis 118, 119
100 Herculis 118, 119
104 Aquarii 152
110 Herculis 118
951 Gaspra 225

A

Abell 39 119
Abell 383 68–9
Abell 2065 117
Abell 2744 19
Abell 3627 181
absolute magnitude scale 22, 238
Acamar (Theta Eridani) 160, 161
accretion disks 41, 64
  superheated 60
Achernar (Alpha Eridani) 160, 161, 217
Acrux (Alpha Crucis) 178, 179
active galactic nucleus (AGN) 60, 114, 115, 129
active galaxies **60–1**
Acubens (Alpha Cancri) 168, 169
ADaptive Optics Near Infrared System (ADONIS) 43
Adhafera (Zeta Leonis) 134, 135
Adhara (Epsilon Canis Majoris) 194, 195
AE Aurigae 133
aerial telescopes 78
Aesculapius (legendary healer) 144
AGN see active galactic nucleus
al-Sufi 89
Albireo (Beta Cygni) 40, 41, 124, 125
Alcor 110
Alcyone (Eta Tauri) 157
Aldebaran (Alpha Tauri) 20, 25, 156, 157
Alderamin (Alpha Cephei) 103
Alfirk (Beta Cephei) 103
algae, photosynthesis 83
Algedi (Alpha¹ Capricorni) 40, 186
Algedi Secunda (Alpha² Capricorni) 186
Algenib (Gamma Pegasi) 150
Algieba (Gamma Leonis) 134, 135
Algol (Beta Persei) 43, 130, 131
aliens, search for intelligent 83
Alioth (Epsilon Ursae Majoris) 110, 111
Alkaid (Eta Ursae Majoris) 110, 111
Alkalurops (Mu Boötis) 117
Alkes (Alpha Crateris) 175
ALMA see Atacama Large Millimeter/sub-millimeter Array
Almagest (Ptolemy) 89
Alnair (Alpha Gruis) 188, 189
Alnasi (Gamma Sagittarii) 184, 185
Alnilam (Epsilon Orionis) 162, 163
Alnitak (Zeta Orionis) 162, 163
Alpha Antliae 199
Alpha Apodis 214
Alpha Arae 182
Alpha Caeli 191
Alpha Camelopardalis 109
Alpha Centauri see Rigil Kentaurus
Alpha Centauri B 12, 23
Alpha Chamaeleontis 214
Alpha Circinus 206
Alpha Coronae Australis 183
Alpha Doradus 210, 211
Alpha Fornacis 192
Alpha Horologii 218
Alpha Hydri 217
Alpha Indi 208
Alpha Lacertae 129
Alpha Lupi 180
Alpha Lyncis 108
Alpha Mensae 219
Alpha Microscopii 189
Alpha Monocerotis 171
Alpha Muscae 206
Alpha Persei Cluster 130
Alpha Pictoris 212
Alpha Pyxidis 198
Alpha Reticuli 213
Alpha Sculptoris 190
Alpha Scuti 147
Alpha Sextantis 174
Alpha Trianguli 128
Alpha Tucanae 215
Alpha Vulpeculae 148
Alphard (Alpha Hydrae) 172, 173
Alphekka (Alpha Coronae Borealis) 117
Alpheratz (Alpha Andromedae) 126, 127
Alrescha (Alpha Piscium) 154, 155
Alshain (Beta Aquilae) 146
Altair (Alpha Aquilae) 121, 146
Altarf (Beta Cancri) 168, 169
Aludra (Eta Canis Majoris) 195
Alya (Theta Serpentis) 142, 143
AM 1 see Arp-Madore 1
AM 0644-741 213
Anatres, magnitude and luminosity 22
Andromeda **126–7**, 150
Andromeda, Princess 102, 126, 130
Andromeda Galaxy (M31) 13, 50, 63, 65, 126, 127, 241
Ankaa (Alpha Phoenicis) 209
Antares (Alpha Scorpii) 140, 141
Antennae Galaxies 174
Antila Dwarf Galaxy 241
Antlia **199**
Antlia Cluster 199
Aphrodite (Greek goddess) 154
Apollo (Greek god) 174, 175
Apollo missions 233
Apparatus Sculptoris see Sculptor
apparent magnitude scale 22, 238
Apus 188, **214**
Aquarius **152–3**
Aquila **146–7**
Ara **182**
Aratus 88
Arcturus (Alpha Boötis) 100, 116, 117
Argo (mythological ship) 196, 200, 202
Argo Navis (obsolete constellation) 196, 200, 202
Ariadne, Princess 117
Aries 92, **157**
  first point of 90, 91, 93, 157
Aristarchus of Samos 17
Aristotle 16
Arkab Posterior (Beta 2 Sagittarii) 185
Arkab Prior (Beta 1 Sagittarii) 185
Arkel, Hanny van 132
Arneb (Alpha Leporis) 193
Arp 147 159
Arp 220 143
Arp 256 158, 159
Arp 273 65
Arp-Madore 1 218
Ascella (Zeta Sagittarii) 184, 185
Aspidiske (Iota Carinae) 202, 203
Assellus Australis (Delta Cancri) 168, 169
Assellus Borealis (Gamma Cancri) 168, 169
asterisms 202
asteroid belt 224
asteroids 30, 61, 166, 223, 224, 225
astronauts, on the Moon 233
astronomers
  charting the heavens 88–9
  study of the Universe 16–17
Atacama Desert 79
Atacama Large Millimeter/sub-millimeter Array (ALMA) (Chajnantor plateau, Chile) 56–7, 86
Athena (Gamma Geminorum) 166
Atlas Coelestis (Flamsteed) 88
atlases, star 88, 89
atmosphere
  Earth 76, 78, 79, 83, 225
  exoplanets 83
  gas giant moons 225
  hot Jupiters 47
  inner (rocky) planets 225, 228, 229
  Mars 224, 225
  Neptune 231
  outer (gas) planets 225
  oxygen in 83
  Sun 226
  Uranus 231
  Venus 224, 225, 228
atoms
  formation of 14
  nuclei 14, 16
  primeval atom 16
Atoms for Peace Galaxy (NGC 7252) 153
Atria (Alpha Trianguli Australis) 207
AU Microscopii 189
Auriga **132–3**
aurorae
  Earth 226
  Saturn 226
autumn equinox 93
axis of rotation
  black holes 38
  Earth 90, 229, 233
  neutron stars 36, 37
  star formation 30

B

B Lac objects 129
Babylonians 88, 156
Baily, Francis 132
Barnard's Star 23, 144
barred spiral galaxies 19, **46**, 47, **51**
  Milky Way 54, 56, 66
Bayer, Johann 88, 207
Beehive Cluster (M44) 168, 169
Bell Labs (New Jersey) 16
Bellatrix (Gamma Orionis) 25, 162, 163
Bellerophon (mythological hero) 150
Berenices II, Queen of Egypt 138
Berlin Observatory 89
Beta Arae 182
Beta Caeli 191
Beta Camelopardalis 109
Beta Canum Venaticorum 112, 113
Beta Comae Berenices 138
Beta Coronae Australis 183
Beta Doradus 210, 211
Beta Fornacis 192
Beta Gruis 188, 189
Beta Horologii 218
Beta Hydri 217
Beta Indi 208
Beta Lacertae 129
Beta Leonis Minoris 132
Beta Lupi 180
Beta Lyrae 120
Beta Monocerotis 170, 171
Beta Octantis 219
Beta Phoenicis 209
Beta Pictoris 212
Beta Pictoris b 212
Beta Piscis Austrini 187
Beta Piscium 155
Beta Pyxidis 198
Beta Sculptoris 190, 191
Beta Scuti 147
Beta Serpentis 142, 143
Beta Tauri 132
Beta Trianguli 128
Beta Trianguli Australis 207
Betelgeuse (Alpha Orionis) 20, 24, **162**, 163, 169
Big Bang 11, 12, **14–15**
  and cosmic expansion 70, 75
  and creation of the Universe 72
  distribution of matter 66
  first use of term 16
  inflationary Big Bang theory 17
Big Chill 75

Big Crunch 75
Big Dipper 102, 110
Big Rip 75
binary star systems 40, 42, **43**
  supernovas 34
  X-ray 114
biosignature 83
birth, of stars **30–1**
BL Lacertae 129
black dwarfs 28, 29, 33
Black Eye Galaxy (M64) 138
black holes 11, 18, 28, 29, 35, **38–9**, 125, 137
  in galaxies 50, 114, 199
  in Milky Way 54
  supermassive 38–9, 59, 60, 61, 116, 174, 184, 212
blazars 60, 129
Blaze Star (T Coronae Borealis) 117
blue dwarfs 28
blue giants 21, 25
blue hypergiants 24
Blue Planetary Nebula (NGC 3918) 177
blue stars 20, 24, 25, 44, 133
blueshift 70
BM Scorpii 140
Bode, Johann Elert 89
Bok gobules 30
The Book of the Fixed Stars (al-Sufi) 89
Boomerang Nebula 176, 177
Boötes 112, **116–17**
Brahe, Tycho 88, 89
bright emission nebulae 163
brightness
  dips in 42, 43, 46
  and size 25
  stars 20, 21, 22
Brocchi's Cluster 148
brown dwarfs 18, 25
Bruno, Giordano 17
"bubble Universes" 73
Bug Nebula (NGC 6302) 141
Bullet Cluster 75
Butterfly Cluster (M6) 140, 141
Butterfly Nebula see Bug Nebula

C

Cacciatore, Niccolò 149
Caelum **191**
calendars 87
Camelopardalis **109**
Cancer **168–9**
Canes Venatici **112–13**, 116
Canis Major 94, 162, **194–5**
Canis Minor 162, **169**
Canopus (Alpha Carinae) 177, 194, 202, 203
Capella (Alpha Aurigae) 132, 133
Caph (Beta Cassiopeiae) 106, 107
Capricornus 40, **186**
carbon 29, 33
  and life 82
Carina 196, 200, **202–5**
Carina Nebula (NGC 3372) 18–19, 59, 202, 203
  dust clouds 204–5
Cartwheel Galaxy (ESO 350-40) 190, 191
Cassini space probe 83
Cassiopeia **106–7**
Cassiopeia, Queen 102, 106, 126
Cassiopeia A 107
Castor (Alpha Geminorum) 166, 167
cataclysmic variables 42
catalogues, star 88, 89
Cat's Eye Nebula (NGC 6543) 104, 105
Cebalrai (Beta Ophiuchi) 144
Celeris 151
celestial coordinates 91
celestial equator 91, 93, 144, 154, 174
  charts centred on 96, 98–101
celestial meridian 91
celestial objects **18–19**
celestial sphere **90–1**
  mapping the sky **94–5**
  and zodiac 92
Centaurus **176–7**, 178
Centaurus A (NGC 5128) 176, 177
Centre for Nuclear Research (CERN) 14
Cepheid variables 42, 102, 146, 148, 167, 216
Cepheus **102–3**
Cepheus, King of Ethiopia 102, 106, 126
Cepheus OB2 association 59
Cetus 41, **158–9**
Cetus (sea monster) 126, 130, 158
Chamaeleon 188, **214**
The Chamaeleon I Cloud 214
Chandra X-ray Observatory 37, 80
  images from 80, 115
Charles I, King of England 112
charts, star 89
Chelae Scorpionis 139
Chertan (Theta Leonis) 135
Chi Carinae 203
Chi Eridani 160
Chinese, ancient 89, 109, 183
Chiron (mythological centaur) 177
chromosphere 27, 227
Cl 0024+17 74
Cigar Galaxy (M82) 110, 111
Circinus **206**
Circinus Galaxy 206
the Circlet 154
circumpolar stars 96, 97
civilizations, number of 83
clocks 87
clouds
  Neptune 231
  Venus 228
clusters of galaxies see galaxy cluster
CMBR see Cosmic Microwave Background Radiation
Coalsack Nebula 178
the Coathanger 148
COBE satellite 17
cold gas giants 46
colliding galaxies **62–5**, 149, 174, 189
Collinder 399 148
collisions
  galaxies **64–5**, 66, 67
  galaxy clusters 75
  star 45
colour
  and size 24, 25
  spectral classification 20–1
  and temperature 20
Columba **195**
Coma Berenices 51, 68, 136, **138**
Coma Cluster 66, 138
coma (head) (comets) 18
comets 18, 30, 223, 224, 225, 230
  and transfer of life 82
Cone Nebula 170
constellations
  Andromeda **126–7**, 150
  Antlia **199**
  Apus 188, **214**
  Aquarius **152–3**
  Aquila **146–7**
  Ara **182**
  Aries 92, **157**
  Auriga **132–3**
  Boötes 112, **116–17**
  boundaries 94, 95
  Caelum **191**
  Camelopardalis **109**
  Cancer **168–9**
  Canes Venatici **112–13**, 116
  Canis Major 94, 162, **194–5**
  Canis Minor 162, **169**
  Capricornus 40, **186**
  Carina 196, 200, **202–5**
  Cassiopeia **106–7**
  Centaurus **176–7**, 178
  Cepheus **102–3**
  Cetus 41, **158–9**
  Chamaeleon 188, **214**
  changing shape of 106
  charting the heavens 88–9
  Circinus **206**
  Columba **195**
  Coma Berenices 68, 136, **138**
  Corona Australis **183**
  Corona Borealis **117**, 183
  Corvus **174**, 175
  Crater **175**
  Crux **178–9**, 200
  Cygnus 40, 41, **124–5**
  Delphinus **149**
  Dorado 188, **210–11**
  Draco **104–5**, 118
  Equuleus **151**
  Eridanus **160–1**
  Fornax **192–3**
  Gemini 92, **166–7**
  Grus **188–9**
  Hercules 88, **118–19**
  Horologium **218**
  Hydra 95, **172–3**, 175, 217
  Hydrus 188, **217**
  Indus 188, **208**
  Lacerta **129**
  Leo 51, **134–5**
  Leo Minor 88, **132**
  Lepus **193**
  Libra 136, **139**
  locators 97
  Lupus **180**
  Lynx **108**
  Lyra **120–1**
  mapping the sky **94–5**
  Mensa **219**
  Microscopium **189**
Constellations (continued)
  Monoceros **170–1**
  Musca 188, **206**
  Norma **181**
  number of 87, 94
  Octans **219**
  Ophiuchus 92, 142, **144–5**
  Pavo 188, **216**
  Pegasus 51, 68, 89, **150–1**
  Perseus 106, **130–1**
  Phoenix 188, **209**
  Pictor **212**
  Pisces 51, **154–5**, 157
  Piscis Austrinus 152, **187**, 188, 189
  Puppis **196–7**, 200
  Pyxis **198**
  Reticulum **213**
  Sagitta **149**
  Sagittarius **184–5**, 189
  Scorpius 92, **140–1**, 162, 181
  Sculptor **190–1**
  Scutum **147**
  Serpens 30, **142–3**
  Sextans **174**
  sky charts 96–101
  Taurus 88, 89, **156–7**
  Telescopium **207**
  Triangulum 13, **128**
  Triangulum Australe 188, **207**
  Tucana 188, **215**
  Ursa Major **110–11**, 112
  Ursa Minor **102**, 110
  Vela 196, **200–1**
  Virgo 43, 50, 92, **136–7**
  Volans 188, **213**
  Vulpecula 125, **148**
  zodiacal 92
convection 26
convective zone 27
coordinates, celestial 91
Copernicus, Nicolaus 17
Cor Caroli (Charles's Heart) (Alpha Canum Venatricorum) 112, 113
core
  dying stars 32
  end of fusion in 34
  following supernova explosions 29
  Mercury 228
  stars 20, 26, 27, 28
  supergiant 34
  Uranus 231
  Whirlpool Galaxy 114, 115
corona
  stars 27
  Sun 226
Corona Australis **183**
Corona Borealis **117**, 183
coronal mass ejections 227
Coronet Cluster 183
CoRoT spacecraft 48
Corvus **174**, 175
cosmic expansion 70–1
  and dark energy 75
cosmic golden egg 16
Cosmic Microwave Background Radiation (CMBR) 16, 72
  variations in 17

cosmic rays 18
cosmological constants 75
cosmology **16–17**
Cosmos **12–13**
Crab Nebula (M1) 37, 59, 157
Crater **175**
craters
 Mars 229
 Mercury 228
 the Moon 232
 Venus 228
crust
 gas planets 225
 Moon 232
 rocky planets 225, 228
Crux **178–9**, 200
Cygnus 40, 41, **124–5**
Cygnus A 125
Cygnus Loop Nebula 125
Cygnus Rift 59, 125
Cygnus X-1 124, 125

D

Dabih (Beta Capricorni) 186
dark energy 17, **75**
dark matter 11, 19, **75**
 mapping 74
 in Milky Way 58
dark nebula 178
De revolutionibus orbium coelestium (Copernicus) 17
death of stars 29
 and planetary nebulae 32–3
debris discs, Formalhaut 187
declination (Dec) 91
Deimos 225
Delphinus **149**
Delta Andromedae 127
Delta Apodis 214
Delta Boötis 117
Delta Cephei 102, 103
Delta Ceti 158
Delta Chamaeleontis 214
Delta Corvi 174
Delta Crateris 175
Delta Crucis 178, 179
Delta Cygni 125
Delta Doradus 210
Delta Equulei 151
Delta Gruis 188
Delta Herculis 118, 119
Delta Hydrae 173
Delta Librae 139
Delta Lyrae 120, 121
Delta Monocerotis 170, 171
Delta Ophiuchi 144
Delta Persei 130
Delta Serpentis 143
Delta Velorum 200, 202
Delta Virginis 137
Deltaa Scuti 147
Demeter (Greek goddess) 136
Deneb (Alpha Cygni) 121, 124, 125, 146
Deneb Algedi (Delta Capricorni) 186
Deneb Kaitos (Beta Ceti) 159
Denebola (Beta Leonis) 135
density
 black holes 38
 earlier Universe 72
 Ring Nebula 120
 Saturn's rings 131
 white dwarfs 33
density waves 54, 55
Di Cha system 41
Diadem (Alpha Comae Berenices) 138
diameters 25
digital cameras 76
Dike 136
dinosaurs, mass extinction of 225
Dionysus (Greek god) 117
distance
 exoplanets 48–9
 stars 20, **23**
Dog Star see Sirius
Doppler effect 70, 71
Doppler spectroscopy 46
Dorado 188, **210–11**
double binaries 40, 41
Double Cluster 130
Double Double see Epsilon Lyrae
double star systems 30
Draco **104–5**, 118
Drake, Frank 83
Drake Equation 83
Dschubba (Delta Scorpii) 141
Dubhe (Alpha Ursae Majoris) 110, 111
Dumbell Nebula (M27) 148
Dürer, Albrecht 89
dust
 in galactic disk 58
 and magnetism 58
 molecular clouds 28, 29, 30
 in star and planet formation 58
dust clouds
 Carina Nebula 204–5
 Cygnus Rift 59
 NGC 6729 183
 Whirlpool Galaxy 115
dust lanes 115
dust rings 114, 115
dust torus 60
dwarf galaxies 66, 120, 192
dwarf irregular galaxies 51, 212
dwarf novae 149
dwarf planets 225
dwarf stars 25
DX Cancri 23

E

E ring (Saturn) 230
Eagle Nebula (M16) 30, 31, 142, 143
Earth 12, 223, 224, 225, 228, **229**
 aurorae 226
 gravity 232
 light year distances from 72–3
 magnetic field 226, 227
 mapping the sky from 94–5
 the Moon 232–3
 tides 233
Earth sized (Terran) planets 48–9
earth-based telescopes 80
eclipses, stellar 43
eclipsing binaries 25, 43, 120, 132, 156
eclipsing and ellipsoidal variables 43
the ecliptic 90, 92, 93
ecliptic plane 90
Egg Nebula 125
Eight-Burst Nebula (NGC 3132) 200, 201
Einstein 16–17, 75
 spacetime 73
El Gordo Cluster 68, 69, 209
electromagnetic radiation, neutron stars 36
electromagnetic spectrum 80, 81
electromagnetic waves 80
electrons 226
elements
 made in supernovas 35
 in stars 28, 29
Elephant's Trunk Nebula 102
elliptical galaxies 19, **46**, 47, **50**, 72
 formation of 64, 66
 in galaxy clusters 66
elliptical orbits 224
Elnath (Beta Tauri) 156, 157
Eltanin see Etamin
Enceladus 230, **231**
 seas of 83
energy
 active galaxies 60
 Big Bang 14
 and life 82
 stars 20, 26
Enif (Epsilon Pegasi) 150, 151
Epsilon Antliae 199
Epsilon Aurigae 132, 133
Epsilon Carinae 200, 202, 203
Epsilon Cassiopeiae 106, 107
Epsilon Centauri 177
Epsilon Crucis 178
Epsilon Cygni 125
Epsilon Eridani 23, 160, 161
Epsilon Herculis 118
Epsilon Hydrae 172
Epsilon Indi 208
Epsilon Indi System 23
Epsilon Leonis 135
Epsilon Leporis 193
Epsilon Lyrae 120, 121
Epsilon Microscopii 189
Epsilon Monocerotis 170, 171
Epsilon Normae 181
Epsilon Persei 130
Epsilon Piscis Austrini 187
Epsilon Piscium 154
Epsilon Scorpii 141
Epsilon Volantis 213
equator, celestial 91, 93
equilibrium, stars in 27
Equuleus **151**
ergosphere 38
Eridanus **160–1**
Eros (Greek god) 154
Eros-MP J0032-4405 25
Errai (Gamma Cephei) 103
Eskimo Nebula (NGC 2392) 166, 167
ESO 69-6 207
ESO 77-14 208
ESO 137-001 181
ESO 286-19 189
ESO 350-40 see Cartwheel Galaxy
ESO 381-12 50
ESO 510-913 172
Eta Aquarid meteor shower 152
Eta Aquarii 152
Eta Aquilae 146
Eta Aurigae 133
Eta Carinae 59, 202
Eta Cassiopeiae 106, 107
Eta Centauri 177
Eta Chamaeleontis 214
Eta Draconis 105
Eta Geminorum 167
Eta Herculis 118, 119
Eta Hydrae 173
Eta Leonis 135
Eta Lyrae 121
Eta Normae 181
Eta Piscium 154, 155
Eta Serpentis 143
Etamin (Gamma Draconis) 104, 105
Europa 230
Europa (Greek mythology) 156
European Extremely Large Telescope (Chile) 76, 79
European Paranal Observatory (Chile) 164
European Southern Observatory (Chile) 43, 79, 134
Euxodus 88
event horizon 38
exoplanets **46–9**, 158
 detecting 46–7
 life on 82, 83
 multiplanetary systems 48–9
 properties of 47
extrasolar planetary systems **46–7**
"extremophiles" 82
extrinsic variables 42, 43
Eye Nebula see Ghost of Jupiter
EZ Aquarii 23

F

False Cross 200, 201, 202
Fermi telescope 80
filaments 71, 227
Firework Nebula (GK Persei) 42, 130
First point of Aries 90, 91, 93, 154, 157
first star generation 44
Flaming Star Nebula (IC 405) 133
Flamsteed, John 88
flares, solar 226, 227
flaring variables 42
Fleming 1 33
Formalhaut (Alpha Piscis Austrini) 46, 187, 190
Fornax **192–3**
Fornax Cluster of Galaxies 51, 192
FS Comae Berenices 138

G

GaBany, Jay 112
Gacrux (Gamma Crucis) 178, 179
Gaia spacecraft 89
Gaia telescope 81
galactic disk 54, 58
galactic plane 23, 58
galaxies 19, **50–1**, 241
 active **60–1**
 coding 50
 colliding **64–5**, 66, 67, 108, 189, 213
 evolving 15, 64
 and the expanding Universe 16, 70–1
 first 11, 14
 formation of 11
 light from 13
 mergers 64, 65, 66, 155, 158
 number of 50
 receding 71
 recessional velocities 17
 types of 19, **46–7**, **50–1**
galaxy clusters 12, 13, 15, 19, 64, **66–9**, 241
 collissions between 75
 space between 70
Galaxy Redshift Survey 71
galaxy superclusters 71
Galex 80
Galilei, Galileo 78

Gama Ceti 159
Gamma Apodis 214
Gamma Caeli 191
Gamma Cassiopeiae 106, 107
Gamma Centauri 177
Gamma Chamaeleontis 214
Gamma Comae Berenices 138
Gamma Coronae Australis 183
Gamma Coronae Borealis 117
Gamma Crateris 175
Gamma Delphini 149
Gamma Doradus 210
Gamma Equulei 151
Gamma Gruis 188, 189
Gamma Herculis 119
Gamma Hydrae 172
Gamma Hydri 217
Gamma Librae 139
Gamma Lupi 180
Gamma Lyrae 120
Gamma Mensae 219
Gamma Microscopii 189
Gamma Monocerotis 171
Gamma Normae (1 and 2) 181
Gamma Octantis 219
Gamma Persei 130
Gamma Phoenicis 209
Gamma Pictoris 212
Gamma Piscium 155
Gamma Pyxidis 198
gamma rays 36, 60, 80
Gamma Sagittae 149
Gamma Serpentis 143
Gamma Trianguli 128
Gamma Velorum 200, 201
Gamma Volantis 213
Gamow, George 16
Ganymede (mythological hero) 146, 152
Garnet Star (Mu Cephei) 22, 102
gas
  clouds of star forming 64
  colliding gas clouds 63
  ejected by dying stars 32, 33
  in galactic disk 58
  in galaxy clusters 67
  molecular clouds 28, 29, 30
gas giants 46, 223, 225, **230–1**
Gemini 92, **166–7**
Geminid meteor shower 166
Gendler, Robert 112
General Theory of Relativity 16, 17, 73
geocentrism 16
Ghost Head Nebula (NGC 2080) 211
Ghost of Jupiter (NGC 3242) 172, 173
giant elliptical galaxies 46, 47, 68
Giant Magellan Telescope (Chile) 76
giant stars 24
Gienah (Epsilon Cygni) 125
Gienah Corvi 174
GK Persei 42, 130
Gliese 667c 48
Gliese 676 49
Gliese 876 48
Gliese 1061 23
globular clusters 18, **44–5**
  evolution of 65
  Milky Way 54, 58, 59
gold 35
Golden Fleece 157, 166, 196
Goldilocks zone see habitable zone
Gomeisa (Beta Canis Minoris) 169
Goodricke, John 102
Graffias (Beta Scorpii) 140, 141
Gran Telescopio Canarias 76
Grasshopper (UGC 4881) 108
gravitational fields 73
gravitational lensing 66, 74
gravitational microlensing 46
gravitational pull
  black holes 38
  inner planets 228
gravitational waves 17
gravity 14, 16
  black holes 38
  dark matter and dark energy 75
  and distortion of spacetime 73
  Earth 232
  galaxies 50, 63
  galaxy clusters 66
  Moon 233
  multiple stars 40
  neutron stars 36, 37
  star clusters 18, 44
  and star formation 30, 64
  stars 27
  Sun 223
Great Attractor 12, 241
Great Red Spot (Jupiter) 230
Greeks, ancient 16, 17
  charting the heavens 87, 88–9
  mythology 87, 88, 94
  see also constellations by name
greenhouse effect 228
Groombridge 34 A and B 23
Grus **188–9**
Gum 29 11
Guth, Alan 17

H

habitable zone 47, 48, 49, **82**
Hadar (Beta Centauri) 177, 178
  magnitude and luminosity 22
Hale Reflector (California) 76
Halley, Edmond 88
Halley's comet 152
halo, Milky Way 54, 58, 59
Hamal (Alpha Arietis) 157
Hanny's Voorwerp 132
Hayn Impact Crater (Moon) 232
HCG see Hickson Compact Group
HD 10180 49
HD 40307 49
HD 98800 system 41
HD 215497 49
HE 1450-2958 191
Heart Nebula 106
heavens, charting the **88–9**
heliocentrism 17
Helios (Greek Sun-god) 160
heliosphere 226
helium 14, 28, 29, 32, 33, 197, 231
  nuclei 26
  Ring Nebula 122
Helix Nebula (NGC 7293) 152, 153
Herchel Space Telescope 164, 165
Hercules 88, **118–19**
Hercules (mythical hero) 104, 118, 134, 168, 172
Hercules A galaxy 118
Hercules Cluster 119
Herschel, Caroline 190
Herschel, William 79, 103, 175, 190
Herschel far-infrared telescope 81
Hertzsprung, Ejnar 21
Hertzsprung-Russell diagram 21
Hevelius, Johannes 88, 108, 112, 129, 132, 148, 149, 174
Hickson Compact Group 87 (HCG 87) 186
Hickson Compact Group 90 (HCG 90) 187
high-mass stars
  and formation of supernovas 34
  inside 26
  life of 28–9
high-velocity stars, Milky Way 54
Hind, John Russell 193
Hind's Crimson Star 193
Hind's Variable Nebula (NGC 1555) 157
Hipparchus 89, 238
Hipparcos satellite 89
Hiranyagarbha 16
Hoag's Object 142, 143
Hooker Telescope 79
Horologium **218**
hot Jupiters 46–7
hot Neptunes 46
Houtman, Frederick de 88, 188, 206, 208, 209, 210, 213, 214, 215, 216, 217
Hoyle, Fred 16
Hubble, Edwin 16, 241
  and classification of galaxies 50, 51
Hubble Space Telescope 76, **78**, 81
  images from 10, 15, 22, 41, 42, 46, 47, 110, 112, 115, 120, 122, 125, 132, 134, 141, 142, 144, 149, 152, 160, 171, 176, 191, 197, 199, 200, 204–5
Hubble's Law 71
the Hyades 156, 157
Hydra 95, **172–3**, 175, 217
Hydra (mythical monster) 168, 172
hydrogen 14, 28, 29, 32, 42, 162, 197, 231
  nuclei 26
  Ring Nebula 123
Hydrus 188, **217**
hypergiants 24, 25

I

IAU see International Astronomical Union
IC 335 192
IC 405 see Flaming Star Nebula
IC 418 see Spirograph Nebula
IC 1396 102, 103
IC 1805 106
IC 2006 46
IC 2163 195
IC 2391 200, 201
IC 2497 132
IC 2560 199
IC 2602 see Southern Pleiades
IC 3568 109
IC 4406 see Retina Nebula
IC 4499 214
IC 4539 119
IC 4665 144, 145
IC 4756 142, 143
IC 5148 see Spare Tyre Nebula
ice 224
IDCS J1426 67
Indus 188, **208**
infrared radiation 30, 39, 78
infrared telescopes 80
infrared waves 81
inner planets **228–9**
interacting binaries **41**
interacting galaxies
  Arp 273 65
  ESO 77-14 208
  ESO 96-6 207
  NGC 2207 and IC 2163 195
  Robert's Quartet 209
interferometry 76, 78
intergalactic space 54, 64
intermediate galaxies 188
International Astronomical Union (IAU) 87, 88, 89, 94
interstellar medium 29, 30
intrinsically variable stars 42
ions 226
Iota Antliae 199
Iota Cancri 168
Iota Carinae 200
Iota Ceti 158
Iota Piscium 154
Iota Scorpii 141
iron 28, 34, 35
irregular galaxies 19, 44, **47**, 50, **51**, 64, 215
  in galaxy clusters 66
Izar (Epsilon Boötis) 116, 117

J

James Webb Space Telescope 76, 79, 80, 81
Jansky, Karl 79
Jason and the Argonauts 157, 196, 200
Jewel Box Cluster (NGC 4755) 178
John III Sobiesci, King of Poland 147
Jupiter 223, 224, 225, **230**
Jupiter-sized (Jovian) planets 48–9

K

Kant, Immanuel 17
Kappa Cygni 125
Kappa Draconis 105
Kappa Lupi 180
Kappa Lyrae 121
Kappa Pavonis 216
Kappa Persei 131
Kappa Serpentis 143
Kappa Tauri 156
Kappa Ursae Majoris 110
Kappa Velorum 200, 201, 202
Kapteyn's star 212
Kaus Australis (Epsilon Sagittarii) 184, 185
Kaus Borealis (Lambda Sagittarii) 184, 185
Kaus Media (Delta Sagittarii) 184, 185
Keck Telescope (Hawaii) 76
Kemble's Cascade 109
Kepler, Johannes 17
Kepler Space Telescope 47, 48, 81
Kepler-37 system 48
Kepler-47 system 49
Kepler-62 system 47, 48
Kepler-69 system 48
Kepler-90 system 48–9
Kepler-186 system 48
Keyser, Pieter Dirkszoon 88, 188, 206, 208, 209, 210, 213, 214, 215, 216, 217
the Keystone (Hercules) 118
Kitalpha (Alpha Equulei) 151
Kitt Peak National Observatory (USA) 123
Kochab (Beta Ursae Minoris) 102
Kornephoros (Beta Herculis) 118, 119

L

L Puppis 197
La Superba (Y Canum Venaticorum) 112, 113
Lacaille 9352 23

Lacaille, Nicolas Louis de 88, 181, 189, 190, 191, 192, 196, 198, 199, 200, 202, 206, 207, 212, 218, 219
Lacerta **129**
Laelaos (mythical dog) 194
Lagoon Nebula (M8) 184, 185
Lagrangian points 1 and 2 80
Lalande 21185 23
Lambda Arietis 157
Lambda Draconis 105
Lambda Eridani 160
Lambda Geminorum 167
Lambda Horologii 218
Lambda Leonis 135
Lambda Tauri 156
Lambda Velorum 200, 201
Laniakea 12
Large Hadron Collider 14
Large Magellanic Cloud 34, 51, 63, 210, 211, 217, 219
latitude 90
lava plains (Moon) 232
lead 35
Leda, Queen of Sparta 125, 166
legends 87, 88, 94
  see also constellations by name
Lemaître, Georges 16
lens technology 76, 78
lensing 66
lenticular gallaxies **46**, 47, **50**, 64, 66
Leo 51, **134–5**
Leo Minor 88, 132
Leonid meteor storm (1833) 134
Lepus **193**
Leros, island of 193
Lesath (Upsilon Scorpii) 141
Leviathan of Parsonstown 79
Libra 136, **139**
  first point of 93
life 11
  on Earth 229
  on Enceladus 231
  on Europa 230
  habitable zone 47, 48, 49
  on Mars 229
  requirements for **82**
  search for **82–3**
  signature of **83**
light
  bending 37, 38
  and measuring distance 12, 13, 23
  rays 36, 38, 39, 60
  speed of 12, 13, 23, 71, 72
  studying optical 76
  visible 80
light curves
  eclipsing binaries 43
  pulsating variables 42
"light echo" 61
light-years 23
Lippershey, Hans 78
Little Dipper 102
Little Dumbell 130
Little Gem Nebula (NGC 6818) 184, 185
Little Ghost Nebula (NGC 6369) 144
lives, stars' 28–9
LMC see Large Magellanic Cloud
local Bubble 59
Local Group 12, 13, 66, 67, 190, 241
  collision **62–3**
longest day
  northern hemisphere 101
  southern hemisphere 99
longitude 90
low-mass stars
  inside 27
  life of 28–9
luminosity
  scale 97
  visual 21, 22
luminous matter 75
Lupus **180**
Luyten 726-8 A and B 23
Luyten's Star 23
Lynx **108**
Lynx Arc 108
Lyra **120–1**

M

M1 see Crab Nebula
M2 153
M3 112
M4 141
M5 143
M6 see Butterfly Cluster
M7 44, 140, 141
M8 see Lagoon Nebula
M10 144, 145
M11 see Wild Duck Cluster
M12 144, 145
M13 118, 119
M15 150, 151
M16 30, 31, 142, 143
M17 see Omega Nebula
M20 see Trifid Nebula
M22 184, 185
M26 147
M27 see Dumbell Nebula
M30 186
M31 see Andromeda Galaxy
M32 50
M33 see Triangulum Galaxy
M34 130, 131
M35 166, 167
M36 132, 133
M37 132, 133
M38 132, 133
M39 124
M41 194, 195
M42 see Orion Nebula
M44 see Beehive Cluster
M45 see the Pleiades
M46 196
M47 196, 197
M49 137
M50 170, 171
M51 see Whirlpool galaxy
M52 106, 107
M53 138
M56 120, 121
M57 see Ring Nebula
M58 137
M59 137
M60 50, 137
M61 137
M63 112, 113
M64 see Black Eye Galaxy
M65 134, 135
M66 134, 135
M67 169
M68 172
M71 149
M72 153
M73 153
M74 51, 154, 155
M76 130
M77 (NGC 1068) 158, 159
M78 163
M79 193
M80 141
M81 110, 111
M82 see Cigar Galaxy
M83 see Southern Pinwheel
M84 137
M85 138
M86 137
M87 68, 136, 137
M88 138
M89 50
M90 137
M91 51, 138
M92 119
M93 196, 197
M94 112
M95 51, 134, 135
M96 134, 135
M97 see Owl Nebula
M99 138
M100 138
M101 see Pinwheel Galaxy
M103 106, 107
M104 see Sombrero Galaxy
M106 112
M110 50
MACS J0416.1-2403 241
Magellan, Ferdinand 210
magnetic fields
  Earth 226
  Jupiter 230
  Mercury 228
  Milky Way 58
  neutron stars 36
  outer planets 230
  Sun 223, 226, 227
magnetosphere 226, 227
magnitudes
  apparent and absolute 22
  origin of system of 89
main-sequence stars 21, 25
  formation of 30
  lives of 28–9
mapping
  dark matter 74
  from redshift 71
  the sky **94–5**
  sky charts **96–101**
  space telescopes 80
maria (Moon) 232, 233
Markab (Alpha Pegasi) 150
Mars 19, 223, 224, 225, 228, **229**
  methane on 83
mass
  loss of 26
  multiple stars 40
  stars 27
  Sun 224
mass warp spacetime 16
massive main sequence stars 28
Matar (Eta Pegasi) 150
matter 14
  Big Bang and 66, 72
  distribution of 66
  in galaxies 50
the "Meathook" see NGC 2442
Mebsuta (Epsilon Geminorum) 166, 167
medium-mass stars, lives of 28–9
Medusa the Gorgon (mythological monster) 130
Megrez (Delta Ursae Majoris) 22, 110, 111
Melotte 111 (Coma Star Cluster) 138
Menkalinan (Beta Aurigae) 133
Menkar (Alpha Ceti) 158, 159
Mensa **219**
Merak (Beta Ursae Majoris) 110, 111
Mercury 223, 224, 225, **228**
meridian, celestial 91
Mesartim (Gamma Arietis) 157
Messier, Charles 130, 184, 242
Messier objects 242–3
Messier star clusters 133
metabolism 83
metallicity 29, 54
meteorites, and transfer of life 82
meteors 225
  meteor showers 130, 134, 152, 166
methane 224, 231
  on Mars 83
Miaplacidus (Beta Carinae) 202, 203
microorganisms 83
Microscopium **189**
microwaves 72
"Mikomeda" 63
Milky Way 11, 12, **54–9**
  as active galaxy 61
  and Andromeda Galaxy 63, 127
  Aquila 146
  Ara 182
  as barred spiral galaxy 54, 56, 66
  Canis Major 194
  in the celestial sphere 95
  central bulge 54, 58, 59
  cross-section 54
  Crux 178
  dark matter 58
  data 241
  from above **58–9**
  galaxies beyond 17, 50
  Lacerta 129
  Large Magellanic Cloud 211
  and Local Group 13, 63, 66
  Lupus 180
  magnetic field 58
  mass 58
  Monoceros 170
  multiple star systems 40
  Norma 181
  number of stars in 20, 25, 58, 223
  Pavo 216
  planets in 58
  planets orbiting stars in 46
  pulsars 37
  Puppis 196
  Pyxis 198
  rotation curves 58
  Sagittarius 184
  Scorpius 141
  Scutum **147**
  search for life in 82
  Small Magellanic Cloud 215
  spiral arms 54, 58–9
  star clusters 44
  starbirth 30
  Triangulum Australe 207
  Vela 200
  Vulpecula 148
Mimas 19
Mimosa (Beta Crucis) 178, 179
Minkowski 2-9 see Twin Jet Nebula
minor bodies 225
Mintaka (Delta Orionis) 162, 163

Mira (Omicron Ceti) 158, 159
Mira system 41
Mira variables 193
Mirach (Beta Andromedae) 127
Mirphak (Alpha Persei) 130, 131
mirror segments 78
mirrored reflectors 76, 78, 79, 80
Mirzam (Beta Canis Majoris) 195
Mizar (Zeta Ursae Majoris) 110, 111
Modified Big Chill 75
molecular clouds 30
Monoceros **170–1**
monster clusters 68
MOO J1142+1527 68
the Moon 12, **232–3**
moons 19, 223, 225
  Earth 225
  Jupiter 225, 230
  Mars 225
  Neptune 225, 231
  outer (gas) planets 225, 230–1
  possible life on 82
  Saturn 225, 230–1
  Uranus 225
Mount Wilson Observatory 79
mountaintop telescopes 77, 78
Mu Centauri 177
Mu Cephei see Garnet Star
Mu Columbae 195
Mu Cygni 125
Mu Herculis 118
Mu Lupi 180
Mu Normae 181
Mu Sagittarii 184
Mu Ursae Majoris 110
Mu Velorum 201
Multi Mirror Telescope (Arizona) 76
multiplanetary systems 47, **48–9**
multiple star systems 30, **40–1**, 162, 168, 193
multiverse 73
Murphrid (Eta Boötis) 117
Musca 188, **206**
mythology 87, 88, 94
  see also constellations by name

N

N44 Nebula 211
N81 215
N90 217
Naos (Zeta Puppis) 20, 196
navigation 87, 88
nebulae 18–19
  star-forming 11, 30, 31, 59
Necklace Nebula 149
Needle Galaxy (NGC 4565) 138
Nekkar (Beta Boötis) 116, 117
Neptune 12, 223, 225, **231**
Neptune-sized (Neptunian) planets 48–9
neutrinos 35
neutron stars 24, 25, 28, 29, 35, **36–7**
neutrons 14, 16, 35, 36
Newton, Isaac 76, 79
NGC 55 190
NGC 104 215
NGC 121 215
NGC 201 159
NGC 246 159
NGC 247 159
NGC 253 190
NGC 288 190
NGC 300 190
NGC 346 215
NGC 362 215
NGC 406 215
NGC 457 106, 107
NGC 520 155
NGC 602 217
NGC 604 128
NGC 663 107
NGC 695 157
NGC 752 127
NGC 784 128
NGC 799 159
NGC 800 159
NGC 869 130
NGC 884 130
NGC 891 126, 127
NGC 908 19
NGC 925 128
NGC 1068 159
NGC 1097 192
NGC 1097A 192
NGC 1261 218
NGC 1291 160, 161
NGC 1300 46, 47, 160, 161
NGC 1309 160, 161
NGC 1313 213
NGC 1316 (Fornax A) 192
NGC 1350 192
NGC 1365 192
NGC 1376 161
NGC 1398 192
NGC 1427A 51
NGC 1499 130
NGC 1502 109
NGC 1512 218
NGC 1514 157
NGC 1528 131
NGC 1535 161
NGC 1555 see Hind's Variable Nebula
NGC 1664 133
NGC 1672 210
NGC 1705 212
NGC 1792 195
NGC 1808 195
NGC 1850 211
NGC 1851 195
NGC 1904 193
NGC 1929 211
NGC 2017 193
NGC 2070 see Tarantula Nebula
NGC 2080 see Ghost Head Nebula
NGC 2082 211
NGC 2207 195
NGC 2217 195
NGC 2232 170
NGC 2237 see Rosette Nebula
NGC 2244 170, 171
NGC 2264 170, 171
NGC 2281 133
NGC 2359 see Thor's Helmet
NGC 2360 194
NGC 2362 194, 195
NGC 2392 see Eskimo Nebula
NGC 2403 109
NGC 2419 108
NGC 2440 196, 197
NGC 2451 196, 197
NGC 2452 196
NGC 2477 196, 197
NGC 2516 203
NGC 2547 201
NGC 2613 198
NGC 2736 see Pencil Nebula
NGC 2787 46, 47
NGC 2818 198
NGC 2903 135
NGC 2997 199
NGC 3109 241
NGC 3114 202, 203
NGC 3115 see Spindle Galaxy
NGC 3132 see Eight-Burst Nebula
NGC 3195 214
NGC 3228 200
NGC 3242 see Ghost of Jupiter
NGC 3372 see Carina Nebula
NGC 3511 175
NGC 3532 203
NGC 3603 202
NGC 3628 135
NGC 3766 177
NGC 3808 134, 135
NGC 3808A 134
NGC 3887 175
NGC 3918 177
NGC 3981 175
NGC 3982 110, 111
NGC 4038 see Antennae Galaxies
NGC 4039 see Antennae Galaxies
NGC 4214 47
NGC 4244 112
NGC 4254 138
NGC 4258 38–9
NGC 4321 138
NGC 4382 138
NGC 4449 112
NGC 4501 138
NGC 4548 138
NGC 4565 see Needle Galaxy
NGC 4631 112
NGC 4647 50
NGC 4755 see Jewel Box Cluster
NGC 4826 138
NGC 4833 206
NGC 4951 51
NGC 5024 138
NGC 5128 176, 177
NGC 5139 176, 177
NGC 5189 33, 206
NGC 5195 112, 114
NGC 5248 117
NGC 5460 176
NGC 5466 117
NGC 5548 116, 117
NGC 5676 117
NGC 5752 117
NGC 5754 117
NGC 5882 180
NGC 5897 139
NGC 5986 180
NGC 6025 207
NGC 6067 181
NGC 6087 181
NGC 6101 214
NGC 6167 181
NGC 6188 182
NGC 6193 182
NGC 6210 119
NGC 6231 140
NGC 6302 see Bug Nebula
NGC 6326 182
NGC 6352 182
NGC 6362 182
NGC 6369 see Little Ghost Nebula
NGC 6397 182
NGC 6503 105
NGC 6530 185
NGC 6537 see Red Spider Nebula
NGC 6541 183
NGC 6543 see Cat's Eye Nebula
NGC 6565 185
NGC 6621 105
NGC 6622 105
NGC 6633 144
NGC 6709 146, 147
NGC 6729 183
NGC 6744 216
NGC 6745 120, 121
NGC 6751 32, 146, 147
NGC 6752 216
NGC 6782 216
NGC 6786 105
NGC 6818 see Little Gem Nebula
NGC 6861 207
NGC 6934 149
NGC 7009 see Saturn Nebula
NGC 7023 103
NGC 7049 208
NGC 7090 208
NGC 7217 51
NGC 7243 129
NGC 7252 see Atoms for Peace Galaxy
NGC 7293 see Helix Nebula
NGC 7331 150, 151
NGC 7354 103
NGC 7424 188, 189
NGC 7479 51
NGC 7635 107
NGC 7662 126, 127
NGC 7714 155
NGC 7793 190
night sky
  December to February 101
  June and December 92
  June to August 99
  mapping **94–101**
  March to May 100
  September to November 98
Nihal (Beta Leporis) 193
nitrogen 197
  Ring Nebula 123
non-rotating black holes 38
Norma **181**
Norma Arm 59
Norma Cluster 241
North America Nebula 125
North celestial pole 90, 91, 102
  chart centred on 96
north polar sky 96
Northern Cross see Cygnus
Northern Crown see Corona Borealis
northern hemisphere
  apparent star movement 91
  longest day 101
  shortest day 99
  solstices 92, 93
northern lights 226
Nova Persei 130
novae 34, 41, 42, 217
Nu Andromedae 126
Nu Coronae Borealis 117
Nu Draconis 104
Nu Octantis 219
Nu Virginis 136
nuclear explosions 42
nuclear fusion 18, 25, 26, 30, 32, 33, 34, 226
nuclei
  active galaxies 60, 61, 129, 199
  atomic 14, 16, 26
Nunki (Sigma Sagittarii) 184, 185
Nusakan (Beta Coronae Borealis) 117

O

observable Universe 13, **72–3**
observatories
  ground-based 76–7
  orbiting 78
  solar 80
  space-based 76, 80
Octans **219**
OGLE-2012-BLG-0026L 48
Olbers' paradox 16, 17
Olympus Mons (Mars) 229
Omega Aquarii 152
Omega Centauri 44, 59, 177
Omega Centauri (NGC 5139) 176, 177
Omega Draconis 105
Omega Nebula (M17) 185
Omicron Andromedae 127
Omicron Centauri 177
Omicron Eridani 160
Omicron Serpentis 143
Oort Cloud 224, **225**
open clusters 18, **44**
Ophiuchus 92, 142, **144–5**
optical doubles 40, 199
optical telescopes, space-based 81
orange giants 25, 132
orange stars 20
orbits
  chaotic 64, 65
  density waves 54, 55
  elliptical 224
  the Moon 233
  multiple stars 40
  perfectly ordered 55
  Solar System planets 90, 223, 224
  space telescopes 80
Orion 94, 160, **162–3**, 194
Orion (mythical hero) 141, 162
Orion Nebula (M42) 30, 133, 162, 163, **164–5**, 202
Orion Spur 59
Orion's Belt 162, 163
Orion's Sword 164
Orpheus 120
Outer Arm (Milky Way) 58
outer planets **230–1**
Owl Nebula (M97) 110, 111
oxygen 29, 33, 197
  and life 83
  Ring Nebula 122, 123

P

Pan (Greek god) 186
parallax method 23
Parsons, William 79
particles 14, 16
  collisions 30
  subatomic 16, 18, 26, 35, 36, 75
Pavo 188, **216**
Peacock (Alpha Pavonis) 216
Pegasus 51, 68, 89, **150–1**
Pelican Nebula 29
Pencil Nebula (NGC 2736) 200
Penzias, Arno 16
Perseid meteor shower 130
Perseus 106, **130–1**
Perseus A (NGC 1275) 130
Perseus Arm 58, 59
Perseus (mythological hero) 106, 126, 130, 158
Persian astronomers 89
PGC 6240 see White Rose Galaxy
Phact (Alpha Columbae) 195
Phad (Gamma Ursae Majoris) 110, 111
Phaenomena (Euxodus) 88
Phaethon (asteroid) 166
Phaethon (Greek mythology) 160
Pherkad (Gamma Ursae Minoris) 102
Phi Andromedae 126
Phi Sagittarii 184
Phi Velorum 201
philosophers, study of the Universe 16–17
Phobos 225
Phoenix 188, **209**
Phoenix Cluster 209
photometers 76
photons 26, 32
photosphere 20, 26, 27, 227
photosynthesis 83
Pi Andromedae 127
Pi Aquarii 152
Pi Herculis 118, 119
Pi Hydrae 173
Pi Hydri 217
Pi Puppis 196
pictogram messages 83
Pictor **212**
Pictor A 212
"pillars of creation" 142
Pinwheel Galaxy (M101) 46, 110, 111
Pioneer 10 probe 83
Pioneer 11 probe 83
Pipe Nebula 144, 145
Pisces 51, **154–5**, 157
Piscis Austrinus 152, **187**, 188, 189
Pistol star 24
PKS 0637-752 219
Plancius, Petrus 88, 109, 171, 178, 188, 195, 214, 215
Planck satellite 58, 72
planetary nebulae 28, 29, **32–3**
  evolution of 122
  Ring Nebula 122–3
planetary systems
  extrasolar **46–7**
  life-supporting worlds in 83
  multiplanetary systems 47, **48–9**
planets 11, 19
  discovery of 223
  extrasolar planetary systems **46–7**
  formation of 30, 187, 224, 225
  inner planets **228–9**
  multiplanetary systems 47, **48–9**
  outer planets **230–1**
  transfer of life between 82
plants, photosynthesis 83
plasma 222, 226, 227
the Pleiades (M45) 156, 157
The Plough 110
Polaris (Alpha Ursae Minoris) 96, **102**, 103, 110
  magnitude and luminosity 22
poles
  black holes 38
  celestial sphere 90, 91
  ejection of material from 30, 38
  protostars 30
pollution 83
Pollux (Beta Geminorum) 25, 166, 167
Porrima (Gamma Virginis) 137
Poseidon (Greek sea god) 106, 149
Praesepe see Beehive Cluster
pressure, in stars 26, 27
pressure waves 30
primeval atom 16
"Primordial soup" 82
Procyon A (Alpha Canis Minoris) 20, 21, 23, 169
Procyon B 21, 23, 169
prominences 20, 227
protons 14, 16
  proton-proton chain reaction 26
protoplanetary disks 30
protostars 28, 29
  formation of 30
Proxima Centauri 12, 25, 177, 239
  distance 23
  magnitude and luminosity 22
Psi Draconis 104
Psi Piscium 154
Psi Velorum 201
PSR 1257 + 12 48
Ptolemy 88, 89
Pulsar 3C58 37
pulsars 11, 37, 46, 48, 49
pulsating variables 42, 210
pulsation, dying stars 32
Puppis **196–7**, 200, 202
Puppis A 196
Pyxis **198**

Q

Quadrans 116
Quadrantid meteor shower 116
quadruple star systems 40, 41, 120
quasars 60, 61, 129, 136, 137, 175
  HE 1450-2958 191
  PKS 0637-752 219
quintessences 75

R

R Coronae Borealis 117
R Horologii 218
R Hydrae 172, 173
R Leonis 135
R Leporis 193
R Lyrae 121
R Scuti 147
R Serpentis 143
R Trianguli 128
radiant energy 26
radiation 16, 26
  active galaxies 60, 61
  detecting 76
  electromagnetic 36
  from first stars 72
  and life 82
  map 72
  pulses 37
radiative zone 26
radii
  exoplanets 48
  largest known stars 239
radio antennae 77
radio astronomy 79
radio galaxies 60, 212
radio lobes 60
radio signals 79
  as signs of life 83
radio telescopes 77, 81
radio waves 36, 60, 79, 81
Rasalgethi (Alpha Herculis) 118, 119
Rasalhague (Alpha Ophiuchi) 144
Rastaban (Beta Draconis) 104, 105
RCW 86 206
recycling, stellar **29**
red dwarfs 12, 21, 25, 28, 29, 46
  flare stars 158
  inside 27
red giants 21, 24–5, 28, 41, 46
  and planetary nebulae 32, 33
  pulsating 193
red hypergiants 24
red nova 42
Red Rectangle 171
Red Spider Nebula (NGC 6537) 184, 185
red stars 20, 24, 25, 44, 54
red supergiants 21, 24, 28, 103, 119, 162
redshift 70
  mapping from 71
reflection nebulae 163
reflector telescopes 76, 79
refractor telescopes 76
Regulus (Alpha Leonis) 134, 135
relativity 16–17, 73
Reticulum **213**
Retina Nebula (IC 4406) 180
Rho Cassiopeiae 107
Rho Herculis 118
Rho Leonis 135
Rho Persei 130
Rho Puppis 196
Rigel (Beta Orionis) 20, 162, 163
  magnitude and luminosity 22
Rigel A 24
right ascension (RA) 91, 157
Rigil Kentaurus (Alpha Centauri) 12, **176**, 177, 178, 206, 239
  distance 23
  magnitude and luminosity 22
Rigveda 16
ring galaxies 190, 191, 213
Ring Nebula (M57) 120, 121, **122–3**
rings
  outer planets 230–1
  planetary nebulae 32
  Saturn 230–1
Robert's Quartet 209
Robur Carolinum (Charles's Oak) 88
rocky planets 223, 225, **228–9**
Rosette Nebula (NGC 2237) 170, 171
Ross 128 34
Ross 154 23
Ross 238 34
Rosse, Lord 112
Rotanev (Beta Delphini) 149
rotating black holes 38
rotating ellipsoidal binaries 43
rotation curves, Milky Way 58
RR Lyrae 121
RS Canum Venaticorum 112
RS Puppis 42
Ruchbah (Delta Cassiopeiae) 107
Rukbat (Alpha Sagittarii) 184, 185
Russell, Henry Norris 21
RXJ 1131 175

S

S Monocerotis 170
Sadachbia (Gamma Aquarii) 152, 153
Sadalmelik (Alpha Aquarii) 153
Sadalsuud (Beta Aquarii) 152, 153
Sadr (Gamma Cygni) 124, 125
Sagitta **149**
Sagittarius **184–5**, 189

Sagittarius A* 59, 184
Sagittarius Arm 59
satellite galaxies 12, 46
Saturn 223, 224, 225, **230–1**
  aurora 226
  magnetic field 226
  moons 19, 230, 231
  rings 230–1
Saturn Nebula (NGC 7009) 152, 153
Scheat (Beta Pegasi) 150
Sco X-1 140
Scorpius 92, **140–1**, 162, 181
Scorpius-Centaurus Association 141
Sculptor **190–1**
Sculptor Group 190
Scutum **147**
Scutum Star Cloud 147
Scutum-Centaurus Arm 59
SDSS J1531+3414 117
Search for Extraterrestrial Intelligence (SETI) 83
seas
  Earth 229
  Moon 232
seasons
  Earth 229
  Uranus 231
  and zodiac 92
second star generation 44
Seginus (Gamma Boötis) 117
segmented-mirror telescopes 77, 78
Serpens 30, **142–3**, 144
Serpens Caput 142
Serpens Cauda 142
SETI see Search for Extraterrestrial Intelligence
Sextans **174**
Sextans A 241
Seyfert galaxies 60, 158, 159, 192, 206
Seyfert's Sextet 142, 143
Shapely 1 181
Shapley, Harlow 181
Shaula (Lambda Scorpii) 141
Shedir (Alpha Casiopeiae) 106, 107
Sheliak (Beta Lyrae) 121
Sheratan (Beta Arietis) 157
shock waves 37
shortest day
  northern hemisphere 99
  southern hemisphere 101
The Sickle 134, 135
Sigma Coronae Borealis 117
Sigma Leonis 135
Sigma Octantis 178, 219
Sigma Orionis 163
Sigma Puppis 196
Sigma Scorpii 141
Sigma Tauri 156
singularities 38
sink holes (Moon) 232
Sirius (Alpha Canis Majoris) (Sirius A) 20, 169, 177, **194**, 195, 203
  distance 23
  magnitude and luminosity 22
Sirius B 23, 25, 194
size
  exoplanets 48–9
  stars 20, **24–5**
Skat (Delta Aquarii) 153
sky see night sky
Small Magellanic Cloud (SMC) 12, 51, 63, 188, 215, 217
SN 1006 180
SN 1572 107
solar flares 226, 227
solar observatories 80
Solar System 12, 223, **224–5**, 226–33
  age of 15
  and celestial sphere 90
  formation of 15, 224
  and Milky Way 56, 59
  and zodiac 92
solar wind 226, 227
Sombrero Galaxy (M104) 136, 137
South celestial pole 91, 214, 219
  chart centred on 97
  locating 178
south polar sky 96
Southern Cross see Crux
Southern Crown see Corona Australis
Southern Fish see Piscis Austrinus
southern hemisphere
  apparent star movement 91
  longest day 99
  shortest day 101
southern lights 226
Southern Pinwheel (M83) 172–3
Southern Pleiades (IC 2602) 202, 203
southern sky 88
space, cataloguing stars from 89
space telescopes **80–1**
space-based astronomy 76, 78–9
spacetime 16, 17, 72, **73**, 75
  warping 38
Spare Tyre Nebula (IC 5148) 188, 189
Special Theory of Relativity 17, 73
spectral classification **20–1**
spectrometers 76
spectroscopy 78
speed of light 12, 13, 23, 71, 72
Spektr-R orbiting radio telescope 81
spheres, Aristotle's theory of 16
Spica (Alpha Virginis) 43, 136, 137, 174
Spindle Galaxy (NGC 3115) 174
spiral galaxies 13, 15, 19, **46–7**, **50–1**, 60
  cluster distribution 44
  in galaxy clusters 66
  merged 64
Spiral Planetary Nebula (NGC 5189) 206
Spirograph Nebula (IC 418) 193
Spitzer Space Telescope 122, 160, 164
star clusters 10, 18, **44–5**
star remnants 18, 28, 36
star-forming areas
  Chamaelion I Cloud 214
  The Lynx Arc 108
  N90 217
  NGC 346 215
  Orion Nebula 164–5
  Southern Pinwheel 172–3
  star-forming nebulae 11, 30, 31, 59
starbirth **30–1**
stars 18
  apparent movement 91
  brightest 238
  brightness and distance **22–3**
  celestial coordinates 91
  closest 239
  collisions 63
  in equilibrium 27
  first 14
  formation of **30–1**, 50, 63, 64
  inside a star **26–7**
  largest known 238
  lives of **28–9**
  newborn 18
  planetary systems 46–9
stars (continued)
  size **24–5**
  spectral classification 20–1
  what is a star? 20–1
stellar black holes 28, 29, 38
stellar recycling **29**
stellar speeds 58
Stephan's Quintet 68–9, 151
Stingray Nebula 182
storms
  Jupiter 230
  Neptune 231
storytellers 87
Struve 1694 109
Struve 2398 A and B 23
Sualocin (Alpha Delphini) 149
subatomic particles 16, 18, 26, 35, 36, 75
Sulafat (Gamma Lyrae) 121
summer solstice 92
Summer Triangle of stars 124, 146
the Sun 12, 25, 223, **226–7**
  and celestial sphere 90
  magnitude and luminosity 22
  mass 224
  spectral classification 20
  surface 222–3
  and the zodiac 92–3
Sun-like stars
  inside 26
  nuclear fusion 26
Sunflower Galaxy (M63) 113
sunspots 226, 227
Super-Earth (Superterran) planets 48
super-galaxies 127, 187
superclusters 12, 19, 65, **66–7**
supergiants 24, 25, 27, 34, 38
supermassive black holes 38–9, 59, 60, 61, 116, 174, 184, 212
Supernova 1987A 210, 211
supernova remnants 29, 157, 180, 200, 206
supernovae 18, **34–5**
  explosions 28, 29, 30, 34, 35, 36, 37, 38, 106, 200
  type 1a 34
  types of 34
surface
  inner planets 228, 229
  Moon 232, 233
  neutron stars 36
  rocky 47
  stars 26
  Sun 222–3, 226, 227

T

T Coronae Borealis 117
T Pyxidis 198
T Tauri 156
T Vulpeculae 148
tachocline 27
Tadpole Galaxy (UGC 10214) 104, 105
tails, comets 18
Tarantula Nebula (NGC 2070) 210, 211
Tarazed (Gamma Aquilae) 146
tardigrades 82
Tau Aquarii 152
Tau Boötis 116
Tau Ceti 23, 158, 159
Tau Eridani 160
Tau Puppis 196
Tau Sagittarii 184
Taurus 88, 89, 156–7
Teapot asterism 184
technosignatures 83
telescopes
  earth-based 80
  history of **78–9**
  infrared 80
  mirrored 80
  observing the skies **76–7**
  space **80–1**
  telescope arrays 78
Telescopium **207**
temperature
  and existence of water 47
  inner planets 228, 229
  neutron stars 36
  stars 20, 21
  white dwarfs 33
Theta Andromidae 126
Theta Antliae 199
Theta Apodis 214
Theta Aurigae 133
Theta Carinae 203
Theta Centauri 177
Theta Draconis 105
Theta Herculis 118
Theta Sagittarii 184
Theta Scorpii 141
Theta Tauri 156
Theta Virginis 136
Thor's Helmet (NGC 2359) 194, 195
Thuban (Alpha Draconis) 104, 105
tidal forces 30, 83
tides
  Earth 233
  gas planet moons 225
tilt, Earth's angle of 90
Titan 230
Titans (Greek mythology) 182
trans-Neptunian objects 224, 225
transit method 46
transition zone 27
the Trapezium (Theta Orionis) 162, 163, 164, 165
Triangulum 13, **128**
Triangulum Australe 188, **207**
Triangulum Galaxy (M33) 63, 128, 241
Trifid Nebula (M20) 184, 185
triple star systems 108
Triton 231
"true" binaries 40, 41
Tucana 188, **215**
TW Horologii 218
Twin Jet Nebula (Minkowski 2-9) 144
TX Piscium 154, 155
Tycho's Star 106
Tyndareus, King of Sparta 166
Typhon 186

U

UGC 4881 108
UGC 10214 see Tadpole Galaxy
Uhuru 78
ultraviolet light 197, 200
ultraviolet radiation 33, 60, 78, 80
Universe **12–13**
  composition of 75
  dark energy and expansion of 75
  expansion of 15, 16, **70–1**, 72, 79
  fates of 75
  formation of 11, **14–15**
  more than one 73
  nature of **16–17**
  observable 13, **72–3**

Universe (continued)
  search for life **82–3**
  shape of 73
  size and structure of **72–3**
  young 72
unstable stars 32
Unukalhai (Alpha Serpentis) 142, 143
Upsilon Andromedae 126, 127
Upsilon Carinae 202, 203
Upsilon Sagittarii 184
Uranographia (Bode) 89
Uranometria (Bayer) 88, 207
Uranus 223, 225, **231**
  discovery of 79
Ursa Major **110–11**, 112, 116
Ursa Major Cluster 13
Ursa Major Moving Group 110
Ursa Minor **102**, 110, 116
UV Ceti 158

V

V-2 rockets 79
V434 Cephei 59
V838 Monocerotis 42
Valles Marineris (Mars) 229
variable stars 42–3
Vega (Alpha Lyrae) 120, 121, 146
  magnitude and luminosity 22
Veil Nebula 125
Vela 196, **200–1**, 202
Vela pulsar 200
Vela Supernova remnant 200, 201
Venator, Nicolaus 149
Venus 223, 224, 225, **228**
vernal equinox 154, 157
Very Large Array 78
Very Large Telescope (Chile) 76, 79, 191
Vindemiatrix (Epsilon Virginis) 136, 137
Virgo 43, 50, 92, **136–7**
Virgo A 137
Virgo Cluster 67, **68**, 136, 137, 138
Virgo Supercluster 12, 13, 67
visible light 80, 81
VISTA infrared telescope 164
voids 71
Volans 188, **213**
volcanoes
  Earth 229
  Enceladus 231
  gas giant moons 225
  Mars 229
  the Moon 232
  rocky planets 225
  Venus 228
Vopel, Caspar 138
Voyager 2 231
Vulpecula 125, **148**
VW Hydri 217
VY Canis Majoris 24

W

W Cephei 103
Wasat (Delta Geminorum) 166
water
  Earth 224, 229
  Enceladus 231
  Europa 230
  and life 82
  Mars 229
  temperatures and surface 47
water ice 224, 225
Water Jar asterim 152
waves
  detection by space telescopes 80–1
  length and frequency of 70
weather, hot Jupiters 47
Westerhout 31 59
Westerlund 2 10
Wezen (Delta Canis Majoris) 194, 195
Wezn (Beta Columbae) 195
Whirlpool galaxy (M51) 17, 112, 113, **114–15**
white dwarfs 21, 25, 28, 29, **33**, 41, 42
  and supernovas 34
White Rose Galaxy (PGC 6240) 217
white stars 20
white supergiants 21, 132, 133
Wild Duck Cluster (M11) 147
Wilkinson Microwave Anisotropy Probe (WMAP) 72 86–7, 94, 240
Wilson, Robert 16
wind, solar 226, 227
winter solstice 92
Winter Triangle 169
Wolf 359 23
Wolf-Rayet type stars 206
WZ Sagittae 149

X

X-ray binary star systems 114
X-ray gas 67
X-rays 26, 36, 37, 38, 39, 41, 68, 69, 79, 80, 140
Xi Draconis 104
Xi Ophiuchi 144
Xi Persei 131
Xi Puppis 196
Xi Ursae Majoris 110, 111
XMM-Newton Space Telescope 164, 165

Y

yellow dwarfs 25
yellow stars 20, 54
Yerkes Observatory (Wisconsin) 76
YZ Ceti 23

Z

Zavijava (Beta Virginis) 136, 139
Zeta Andromidae 126
Zeta Antliae 199
Zeta Aquarii 152, 153
Zeta Aquilae 146, 147
Zeta Aurigae 133
Zeta Cancri 168
Zeta Centauri 177
Zeta Coronae Borealis 117
Zeta Cygni 125
Zeta Doradus 210
Zeta Geminorum 167
Zeta Herculis 118, 119
Zeta Lyrae 120, 121
Zeta Monocerotis 171
Zeta Ophiuchi 144
Zeta Persei 130, 131
Zeta Phoenicis 209
Zeta Puppis 20
Zeta Reticuli 213
Zeta Scorpii 140, 141
Zeta Tauri 156, 157
Zeus (Greek god) 102, 125, 146, 152, 156, 166, 182
Zodiac **92–3**
  Aquarius **152–3**
  Aries 92, **157**
  Cancer **168–9**
  Capricornus 40, **186**
  Gemini **166–7**
  Leo **134–5**
  Libra **139**
  Ophiuchus 92, 142, **144–5**
  Pisces 51, **154–5,** 157
  Sagittarius **184–5**
  Scorpius **140–1**
  seasonal view 92
  Taurus 88, 89, **156–7**
  Virgo **136–7**
Zosma (Delta Leonis) 135
Zubenelgenubi (Alpha Librae) 139
Zubeneschamali (Beta Librae) 139
ZW II 96 149

ACKNOWLEDGMENTS

The publisher would like to thank the following people for their assistance in the preparation of this book:

Peter Frances for intial editorial work; Shahid Mahmood and Charlotte Johnson for design; Constance Novis for proofreading; Helen Peters for the index. Special thanks also to Adam Block (http://adamblockphotos.com) for his help with images.

The publisher would like to thank the following for their kind permission to reproduce their photographs:

(**Key:** a-above; b-below/bottom; c-centre; f-far; l-left; r-right; t-top)

4–5 NASA: ESA
6–7 NASA: ESA, N. Smith (University of California, Berkeley), and The Hubble Heritage Team (STScI / AURA)
10 NASA: ESA, the Hubble Heritage Team (STScI / AURA), A. Nota (ESA / STScI), and the Westerlund 2 Science Team
12 Science Photo Library: Mark Garlick (tr)
14 © CERN : Mona Schweizer (br)
15 Carnegie Mellon University and NASA: ESA / S. Beckwith (STScI) and the HUDF Team (bl)
16 Alamy Stock Photo: Keystone Pictures USA (cr). **Corbis:** Bettmann (cl); Roger Ressmeyer (bc). **Exotic India:** (tc)
17 Corbis: Stefano Bianchetti (tc). **From Nichol 1846 plate VI:** (c). **NASA:** (bl); C. Henze (br). **Thinkstock:** Photos.com (tl)
18 Professor Justin R. Crepp: (c). **ESA:** Hubble & NASA (bl). **ESO:** B. Tafreshi (twanight.org) (cl); TRAPPIST / E. Jehin (tr). **NASA:** The Hubble Heritage Team (AURA / STScI) (bc)
18–19 NOAO / AURA / NSF: N. Smith (b)
19 ESO: (tr). **NASA:** ESA and J. Lotz, M. Mountain, A. Koekemoer, and the HFF Team (STScI) (br); The Hubble Heritage Team (AURA / STScI) / J. Bell (Cornell University), and M. Wolff (Space Science Institute, Boulder) (tl); JPL / Space Science Institute (tc)
22 ESA: Hubble & NASA (br). **ESO:** John Colosimo (tr)
25 SOHO (ESA & NASA): (cr)
29 Alamy Stock Photo: Stocktrek Images, Inc. (tr)
31 NOAO / AURA / NSF: T.A. Rector NRAO / AUI / NSF and NOAO / AURA / NSF) and B.A. Wolpa (NOAO / AURA / NSF)
32 NASA: The Hubble Heritage Team (AURA / STScI) (tr)
33 ESO: H. Boffin (cr). **NASA:** ESA and The Hubble Heritage Team (STScI / AURA) (tl)
35 Corbis: Ikon Images / Oliver Burston (b)
37 NASA: CXC / SAO (tr, tl)
38–39 NASA: X-ray: NASA / CXC / Caltech / P.Ogle et al; Optical: NASA / STScI; IR: NASA / JPL-Caltech; Radio: NSF / NRAO / VLA
41 John Chumack www.galacticimages.com: (bl). **ESA:** Hubble & NASA (br). **NASA:** CXC / SAO / M. Karovska et al. (cra); JPL-Caltech / UCLA (c)
42 NASA: ESA, and the Hubble Heritage Team (STScI / AURA) - Hubble / Europe Collaboration (tr); STScI (bc); ESA and The Hubble Heritage Team (STScI / AURA) (br)
44 ESO: (br). **NASA: ESA** / A. Feild (STScI) (cr)
45 ESO: M.-R. Cioni / VISTA Magellanic Cloud survey
46 NASA: ESA, and P. Kalas (University of California, Berkeley) (cl, bl)
48 ESA: CNES / D. Ducros (tl). **NASA:** (bl)
50 NASA: ESA, and the Hubble Heritage Team (STScI / AURA) - ESA / Hubble Collaboration (tr); ESA, P. Goudfrooij (STScI) (crb); ESA (clb); ESA, Digitized Sky Survey 2 (cb)
51 Adam Block: Pat Balfour / NOAO / AURA / NSF (bl); Mount Lemmon SkyCenter / University of Arizona (adamblockphotos.com) (cl, c, cr, bc, br). **NASA:** ESA and The Hubble Heritage Team (STScI / AURA) (tr)
52 Corbis: Science Faction / Tony Hallas
53 ESA: Hubble & NASA / Judy Schmidt and J. Blakeslee (Dominion Astrophysical Observatory) (tr). **NASA:** and The Hubble Heritage Team (STScI / AURA) (cr); ESA and The Hubble Heritage Team (STScI / AURA) (tc); ESA, and the Hubble Heritage (STScI / AURA)-ESA / Hubble Collaboration (cb)
54–55 NASA: JPL-Caltech / ESA / CXC / STScI
56–57 ESO: A. Duro
58–59 NASA: JPL-Caltech
58 ESA: and the Planck Collaboration (bl)
60–61 ESA: NASA, the AVO project and Paolo Padovani
61 NASA: CXC / Caltech / M.Muno et al. (br)
62–63 NASA: ESA, Z. Levay and R. van der Marel (STScI), T. Hallas, and A. Mellinger
64 ESA: P. Jonsson (Harvard-Smithsonian Center for Astrophysics, USA), G. Novak (Princeton University, USA), and T.J. Cox (Carnegie Observatories, Pasadena, Calif., USA) (right top to bottom)
65 NASA: ESA and The Hubble Heritage Team (STScI / AURA)
66 NASA: ESA, J. Rigby (NASA Goddard Space Flight Center), K. Sharon (Kavli Institute for Cosmological Physics, University of Chicago), and M. Gladders and E. Wuyts (University of Chicago) (bl); JPL-Caltech / L. Jenkins (GSFC) (cl)
67 NASA: ESA, and M. Brodwin (University of Missouri)
68 Rogelio Bernal Andreo, www.deepskycolors.com: (t)
69 NASA: ESA, C. McCully (Rutgers University), A. Koekemoer (STScI), M. Postman (STScI), A. Riess (STScI / JHU), S. Perlmutter (UC Berkeley, LBNL), J. Nordin (NBNL, UC Berkeley), and D. Rubin (Florida State University) (tr); ESA, J. Jee (Univ. of California, Davis), J. Hughes (Rutgers Univ.), F. Menanteau (Rutgers Univ. & Univ. of Illinois, Urbana-Champaign), C. Sifon (Leiden Obs.), R. Mandelbum (Carnegie Mellon Univ.), L. Barrientos (Univ. Catolica de Chile), and K. Ng (Univ. of California, Davis) (br); ESA and the Hubble SM4 ERO Team (cl); JPL-Caltech / Gemini / CARMA (cr)
71 The 2dFGRS Team: (crb)
72 ESA: and the Planck Collaboration (br). **NASA:** WMAP Science Team (bl)
73 Science Photo Library: Mark Garlick (br)
74 NASA: ESA, M.J. Jee and H. Ford (Johns Hopkins University)
75 NASA: CXC / CfA / M.Markevitch et al.; Optical: NASA / STScI; Magellan / U.Arizona / D.Clowe et al.; Lensing Map: NASA / STScI; ESO WFI; Magellan / U.Arizona / D. Clowe et al (tr)
76 Barnaby Norris: (bl)
77 ESO: L. Calçada (t). **NRAO:** AUI and NRAO (b)
78 123RF.com: Chris Hill (tc). **Dorling Kindersley:** Andy Crawford (bc). **NRAO:** AUI and NRAO / AUI Photographer: Bob Tetro www.photojourneysabroad.com (cl). **Wikipedia:** Fig. AA from *Machinae coelestis*, 1673, by Johannes Hevelius (1611–1687). Typ 620.73.451, Houghton Library, Harvard University (tr)
79 Corbis: Dennis di Cicco (cr). **Dorling Kindersley:** Dave King / Courtesy of The Science Museum, London (tl). **ESO:** L. Calçada (bc). **NASA:** Northrop Grumman (br); US Army (cl). **Wikipedia:** (tr)
80 NASA: JPL-Caltech / UCLA (bl)
82 NASA: ESA / Giotto Project (tr). **Science Photo Library:** Steve Gschmeissner (b)
83 NASA
86 ESO: B. Tafreshi (twanight.org)
88 123RF.com: perseomedusa (tr). **akg-images:** Serge Rabatti / Domingie (tc). **Alamy Stock Photo:** Pictorial Press Ltd (cl). **courtesy of Barry Lawrence Ruderman Antique Maps - www.RareMaps.com:** (c). **University of Cambridge, Institute of Astronomy Library:** (br)
89 Alamy Stock Photo: The Art Archive / Gianni Dagli Orti (cl). **Corbis:** Heritage Images (cr). **courtesy of Barry Lawrence Ruderman Antique Maps - www.RareMaps.com. Eon Images:** (tl). **ESA:** D. Ducros (br). **Science Photo Library:** British Library (tr). **Wikipedia:** National Gallery of Art (c)
102 NOAO / AURA / NSF: WIYN / T.A. Rector / University of Alaska Anchorage (br)
103 ESA: Hubble & NASA (tr)
105 NASA: ESA, HEIC, and The Hubble Heritage Team (STScI / AURA) (c); H. Ford (JHU), G. Illingworth (UCSC / LO), M.Clampin (STScI), G. Hartig (STScI), the ACS Science Team, and ESA (tc)
106 NASA: X-ray: NASA / CXC / SAO; Optical: NASA / STScI; Infrared: NASA / JPL-Caltech / Steward / O.Krause et al. (clb)
108 ESA: NASA and Robert A.E. Fosbury (European Space Agency / Space Telescope-European Coordinating Facility, Germany) (tc)
110 NASA: ESA and The Hubble Heritage Team (STScI / AURA) (cl, cb)
112 NASA: ESA, A. Aloisi (STScI / ESA), and The Hubble Heritage (STScI / AURA)-ESA / Hubble Collaboration (br); STScI / R. Gendler (bl)
114 NASA: ESA, S. Beckwith (STScI), and The Hubble Heritage Team (STScI / AURA) (t)

115 NASA: CXC / UMd. / A.Wilson et al. (tc); H. Ford (JHU / STScI), the Faint Object Spectrograph IDT, and NASA (c); ESA, M. Regan and B. Whitmore (STScI), and R. Chandar (University of Toledo) (b); CXC / Wesleyan Univ. / R.Kilgard, et al; Optical: NASA / STScI (tr)
116 ESA: Hubble & NASA (bl)
118 Adam Block: Mount Lemmon SkyCenter / University of Arizona (adamblockphotos.com) (tr). **NASA:** ESA, S. Baum and C. O'Dea (RIT), R. Perley and W. Cotton (NRAO / AUI / NSF), and the Hubble Heritage Team (STScI / AURA) (clb)
120 Adam Block: Jim Rada / NOAO / AURA/NSF (br). **NASA:** and The Hubble Heritage Team (STScI/AURA) (bc)
122 NASA: ESA, C.R. O'Dell (Vanderbilt University), and D. Thompson (Large Binocular Telescope Observatory) (cb); The Hubble Heritage Team (AURA / STScI) (tl); JPL-Caltech / J. Hora (Harvard-Smithsonian CfA) (bl). **NOAO / AURA / NSF:** C.F.Claver / WIYN / NOAO / NSF (c); Bill Schoening / NOAO / AURA / NSF (tc)
123 Science Photo Library: Robert Gendler
125 NASA: The Hubble Heritage Team (AURA / STScI) (cl); X-ray: NASA / CXC / SAO; Optical: NASA / STScI; Radio: NSF / NRAO / AUI / VLA (c)
126 Adam Block: Mount Lemmon SkyCenter / University of Arizona (adamblockphotos.com) (cl). **Philip Perkins:** (cb)
128 Adam Block: Mount Lemmon SkyCenter / University of Arizona (cr). **ESA:** Hubble & NASA (tr)
129 Jim Thommes www.jthommes.com: (br)
130 Adam Block: Fred Calvert / NOAO / AURA / NSF (clb); Mount Lemmon SkyCenter / University of Arizona (adamblockphotos.com) (bl). **NASA:** X-ray: NASA / CXC / RIKEN / D.Takei et al; Optical: NASA / STScI; Radio: NRAO / VLA (bc)
132 NASA: ESA, W. Keel (University of Alabama), and the Galaxy Zoo Team (bc)
133 Adam Block: Mount Lemmon SkyCenter / University of Arizona (adamblockphotos.com) (bc)
134 ESO: O. Maliy (cb). NASA: ESA and The Hubble Heritage Team (STScI / AURA) (bl)
136 NASA: The Hubble Heritage Team (AURA / STScI) (tl)
138 NASA: and The Hubble Heritage Team (STScI / AURA) (tc)
139 Daniel Verschatse - Observatorio Antilhue - Chile: (tl)
141 NASA: ESA, and the Hubble SM4 ERO Team (cl); The Hubble Heritage Team (AURA / STScI) (c)
142 NASA: and The Hubble Heritage Team (STScI / AURA) (bc, bl); J. English (U. Manitoba), S. Hunsberger, S. Zonak, J. Charlton, S. Gallagher (PSU), and L. Frattare (STScI) (tc)
144 ESA: Hubble & NASA (cb). **NASA:** and The Hubble Heritage Team (STScI / AURA) (ca)
145 ESO: Y. Beletsky (bl)
146 NASA: and The Hubble Heritage Team (STScI / AURA) (bc)
147 ESO: (cr)
148 ESO
149 ESA: Hubble & NASA (tc)
151 Adam Block: Mount Lemmon SkyCenter / University of Arizona (adamblockphotos.com) (tc)
152 NASA: Bruce Balick (University of Washington), Jason Alexander (University of Washington), Arsen Hajian (U.S. Naval Observatory), Yervant Terzian (Cornell University), Mario Perinotto (University of Florence, Italy), Patrizio Patriarchi (Arcetri Observatory, Italy) (cl)
153 NASA: NOAO, ESA, the Hubble Helix Nebula Team, M. Meixner (STScI), and T.A. Rector (NRAO) (br)
155 NASA: ESA, the Hubble Heritage (STScI / AURA)-ESA / Hubble Collaboration, and B. Whitmore (STScI) (t); ESA (ca)
157 ESO. NASA: ESA, the Hubble Heritage (STScI / AURA)-ESA / Hubble Collaboration, and A. Evans (University of Virginia, Charlottesville / NRAO / Stony Brook University) (c)
158 NASA: ESA, the Hubble Heritage (STScI / AURA)-ESA / Hubble Collaboration, and A. Evans (University of Virginia, Charlottesville / NRAO / Stony Brook University) (tl)
159 ESO
160 NASA: ESA, The Hubble Heritage Team, (STScI / AURA) and A. Riess (STScI) (cl); JPL-Caltech (c)
162 Roberto Colombari and Federico Pelliccia: (r). ESO: IDA / Danish 1.5 m / R.Gendler, J.-E. Ovaldsen, and A. Hornstrup (c)
164 ESO: J. Emerson / VISTA.
165 ESA: NASA / JPL-Caltech / N. Billot (IRAM) (tr); XMM-Newton and NASA's Spitzer Space Telescope / AAAS / Science (b). **NASA:** JPL-Caltech / T. Megeath (University of Toledo, Ohio) (tl)
166 NASA: Andrew Fruchter and the ERO Team [Sylvia Baggett (STScI), Richard Hook (ST-ECF),y (STScI) (c). **NOAO / AURA / NSF:** N.A.Sharp / NOAO / AURA / NSF (clb)
168 Adam Block: Mount Lemmon SkyCenter / University of Arizona (adamblockphotos.com) (clb)
169 ESO: Akira Fujii (clb, cr)
171 Adam Block: Mount Lemmon SkyCenter / University of Arizona (adamblockphotos.com) (t). **NASA:** ESA, Hans Van Winckel (Catholic University of Leuven, Belgium) and Martin Cohen (University of California, Berkeley) (c)
172 ESO
173 Corbis: (tc)
174 Adam Block: Mount Lemmon SkyCenter / University of Arizona (adamblockphotos.com) (crb). ESO: VLT (cra)
175 NASA: X-ray: NASA / CXC / Univ of Michigan / R.C.Reis et al; Optical: NASA / STScI (crb)
176 NASA: ESA, and the Hubble Heritage Team (STScI / AURA) (clb)
178 ESO: Y. Beletsky (cr)
179 ESO
180 NASA: and The Hubble Heritage Team (STScI / AURA) (cl); CXC / Middlebury College / F.Winkler (bl)
181 ESO. NASA: X-ray: NASA / CXC / UVa / M. Sun, et al; H-alpha / Optical: SOAR (UVa / NOAO / UNC / CNPq-Brazil) / M.Sun et al. (tc)
182 ESA: Hubble & NASA (tc, tr)
183 ESO: Sergey Stepanenko (c). **NASA:** CXC / J. Forbrich (Harvard-Smithsonian CfA), NASA / JPL-Caltech L.Allen (Harvard-Smithsonian CfA) and the IRAC GTO Team (cra)
184 ESA: and Garrelt Mellema (Leiden University, the Netherlands) (cl)
186 NASA: The Hubble Heritage Team (AURA / STScI) (crb)
187 NASA: ESA, and R. Sharples (University of Durham) (tr)
188 ESO
189 ESO
190 ESO
191 ESA: Hubble & NASA (tl). **ESO**
192 ESO
194 ESO: B. Bailleul (bl)
195 NASA: STScI (bl)
196 NASA: ESA, and K. Noll (STScI) (tr)
197 ESA: Hubble & NASA (c). **NASA:** X-ray: NASA / CXC / IAFE / G.Dubner et al & ESA / XMM-Newton (clb)
198 NASA: ESA and The Hubble Heritage Team (STScI / AURA) (br)
199 ESA: Hubble & NASA (tr)
200 ESO. NASA: The Hubble Heritage Team (AURA / STScI) (c, cb)
202 ESO. NASA: ESA, R. O'Connell (University of Virginia), F. Paresce (National Institute for Astrophysics, Bologna, Italy), E. Young (Universities Space Research Association / Ames Research Center), the WFC3 Science Oversight Committee, and the Hubble Heritage Team (STScI / AURA) (bc)
204–205 NASA: ESA, and the Hubble Heritage Project (STScI / AURA)
206 ESO: E. Helder & NASA / Chandra (bl). **NASA:** ESA and The Hubble Heritage Team (STScI / AURA) (cla)
207 ESA: Hubble & NASA (bl). **NASA:** ESA, the Hubble Heritage (STScI / AURA)-ESA / Hubble Collaboration, and A. Evans (University of Virginia, Charlottesville / NRAO / Stony Brook University) (cl)
208 ESA: Hubble & NASA (br). **NASA:** ESA, the Hubble Heritage (STScI / AURA)-ESA / Hubble Collaboration, and A. Evans (University of Virginia, Charlottesville / NRAO / Stony Brook University) (tc)
209 ESO
211 NASA: ESA, E. Sabbi (STScI) (tl); X-ray: NASA / CXC / U.Mich. / S.Oey, IR: NASA / JPL, Optical: ESO / WFI / 2.2-m (bl). **Eckhard Slawik (e.slawik@gmx.net).** : www.spacetelescope.org / images / heic0411d (br)
212 NASA: X-ray: NASA / CXC / Univ of Hertfordshire / M.Hardcastle et al., Radio: CSIRO / ATNF / ATCA (br)
214 NASA: ESA (cl)
215 NASA: ESA and A. Nota (STScI / ESA) (br)
217 NASA: ESA, and the Hubble Heritage (STScI / AURA)-ESA / Hubble Collaboration (cr)
218 NASA: ESA, and D. Maoz (Tel-Aviv University and Columbia University) (bl). **Daniel Verschatse - Observatorio Antilhue - Chile**
219 NASA: CXC / SAO (c)
222 Kevin Reardon: INAF / Arcetri; AURA / National Solar Observatory
225 ESA: Rosetta / NavCam - CC BY-SA IGO 3.0 (br). **NASA:** Johns Hopkins University Applied Physics Laboratory / Southwest Research Institute (bc); JPL / USGS (bl)
226 NASA: ESA, J. Clarke (Boston University, USA), and Z. Levay (STScI) (bc)
227 NASA: SDO (tr, cra, cr)
228 NASA: Johns Hopkins University Applied Physics Laboratory / Carnegie Institution of Washington (c); JPL (br)
229 NASA: Caltech / MSSS (bc); JPL-Caltech / University of Arizona (br)
230 NASA: JPL (cr, cb); JPL / DLR (br)
231 NASA: JPL / Space Science Institute (cla); JPL (cr)
232 NASA
233 NASA
239 ESA: Hubble, NASA, HST Frontier Fields (bl)
241 NASA: ESA

Endpapers: *Front and back* **NASA:** ESA, and J. Maíz Apellániz (Institute of Astrophysics of Andalusia, Spain)

All other images © Dorling Kindersley
For further information see:
www.dkimages.com